Excited States and Free Radicals in Biology and Medicine

Excited States and Free Radicals in Biology and Medicine

Contributions from Flash Photolysis and Pulse Radiolysis

R.V. BENSASSON
Laboratoire de Biophysique,
Muséum National d'Histoire Naturelle,
Paris

E.J. LAND
C.R.C. Department of Biophysical Chemistry,
Paterson Institute for Cancer Research,
Christie Hospital NHS Trust,
Manchester

and

T.G. TRUSCOTT
Department of Chemistry,
Keele University

OXFORD NEW YORK TOKYO
OXFORD UNIVERSITY PRESS
1993

Oxford University Press, Walton Street, Oxford OX2 6DP
Oxford New York Toronto
Delhi Bombay Calcutta Madras Karachi
Kuala Lumpur Singapore Hong Kong Tokyo
Nairobi Dar es Salaam Cape Town
Melbourne Auckland Madrid
and associated companies in
Berlin Ibadan

Oxford is a trade mark of Oxford University Press

Published in the United States
by Oxford University Press Inc., New York

A catalogue record for this book is available from the British Library

Library of Congress Cataloging in Publication Data
Bensasson, R. V.
Excited states and free radicals in biology and medicine: contributions from flash photolysis and pulse radiolysis/
R. V. Bensasson, E. J. Land, T. G. Truscott.
Includes bibliographical references and index.
1. Free radicals (Chemistry) 2. Excited state chemistry.
3. Flash photolysis. 4. Pulse radiolysis. 5. Pathology, Molecular.
6. Molecular biology. 7. Cancer–Photochemotherapy. I. Land, E. J.
II. Truscott, T. G. III. Title.
QP527.B46 1993 574.19'283–dc20 92–44261
ISBN 0–19–855560–1

Typeset by Colset Pte Ltd., Singapore
Printed in Great Britain by
Bookcraft (Bath) Ltd.
Midsomer Norton, Avon

Preface

'On fait de la science avec des faits, comme on fait une maison avec des pierres, mais une accumulation de faits n'est pas plus une science qu'un tas de pierres est une maison.'

'Science is built with facts as a house is with stones. But a collection of facts is no more a science than a heap of stones is a house.'

Henri Poincaré 1854–1912
***La science et l'hypothèse* (1902)**

This book has its origin in a previous volume written more than a decade ago, entitled *Flash photolysis and pulse radiolysis: Contributions to the chemistry of biology and medicine.** Our present broader goal is to introduce the reader to the expanding field of excited states and free radicals in biology and medicine, a discipline at the borderline of physics, chemistry, biology, pharmacology, and medicine. The major challenge of this field is to find solutions at the molecular level to biological problems which involve very short-lived species such as excited states and free radicals. One way to approach these short-lived chemical intermediates is the use of fast kinetic and spectroscopic techniques: flash photolysis and pulse radiolysis. With these pulsed techniques, improved during the 1980s by new sophisticated and computerized methods of detection, one is able

(1) to produce excited states and free radicals in high concentrations,
(2) to study their structure and physical properties by various spectroscopic methods, and
(3) to determine the kinetics of their reactions in timespans ranging from the subpicosecond to the second.

Many interesting advances have taken place during the last decade. Examples include:

1. Considerable improvement in understanding oxygen-related damage in biological systems. This progress have been characterized by a hundredfold increase in the number of articles on activated forms of oxygen, such as $OH^{\cdot}$, $O_2^{\cdot -}$, $RO^{\cdot}$, $RO_2^{\cdot}$, $NO^{\cdot}$, and $^1O_2$ (the latter unambiguously detected via its phosphorescence), involved in pathological processes. Another outcome

*Bensasson, R. V., Land E. J., and Truscott, T. G. (1983). *Flash photolysis and pulse radiolysis: Contributions to the chemistry of biology and medicine.* Pergamon Press, Oxford.

was the formation of the Society for Free Radical Research in Europe in 1982, and of the Oxygen Society in the USA in 1988, as well as the birth of two new journals *Free Radical Research Communications* and *Free Radical Biology and Medicine*.

2. Determination of the crystal structures of the reaction centres of two photosynthetic bacteria. This determination and the knowledge gained from flash photolysis studies provided a picture in space and time of the multi-step electron transfer in bacterial photosynthesis.

3. Confirmation by kinetic spectroscopy of surprising theoretical predictions on electron transfer. After almost three decades of controversy, the Marcus 'inverted' free energy dependence of electron transfer rates has been proved in certain examples. Reductions of the electron transfer rate with increasing driving force were observed in a number of cases: electron transfer between radiolytically generated radical anions and hydrocarbons, charge recombinations, and charge separations. Such behaviour was also reported for intramolecular electron transfer linked by rigid aliphatic skeletons.

4. Mimicry of biological processes by new synthetic molecules: (i) artificial endonucleases composed of oligonucleotide–photosensitizer conjugates able to cleave DNA at specific sites, and (ii) multicomponent molecular devices mimicking photosynthetic electron transfer. The latter have created an excellent starting point to the major challenge of creating artificial systems able to store solar energy and to synthesize organic compounds from CO_2 and water.

5. The use of pulsed radiation techniques to mimic enzymic reactions involving, for example, quinones. In particular, the use of pulse radiolysis to initiate the one-electron oxidation reactions associated with the molecular mechanisms of melanogenesis, and in the understanding of the role of antitumour drugs.

6. Development of protein and DNA conformation studies using pulsed techniques. For example, transitions between conformational states of myoglobin and haemoglobin photoinitiated by photodeligandation of CO and O_2 derivatives have been actively investigated by various fast spectroscopic techniques, on a very wide range of time scales starting with the subpicosecond range.

The writing of this book has led us into fields touching the limits of our expertise. These incursions may have led to some omissions, errors, or misquotations for which we apologize. We have concentrated on the viewpoints of physical chemists who study oversimplified systems with one, two, or three molecules at the same time. Nevertheless, we hope to have chosen convincing examples which illustrate the pertinence of research on excited states and free radicals *in vitro*, for a better understanding of reactions occurring in cells containing 10^4–10^5 different kinds of molecules. Our wish is to be useful to students, research scientists, teachers as well as physicians.

Acknowledgements

We wish to express our sincere thanks to colleagues with whom we have discussed various sections of the manuscript. These include R.J. Cogdell, G. Czapski, C. Ferradini, A.J. Hoff, P.J. Hore, N.-H. Jensen, J.B. Little, M.A.J. Rodgers, M. Rougée, B. Valeur, and J.M. Warman.

We also thank G.P.F. Chantry of the Medical Illustrations Department at the Christie Hospital NHS Trust for the figures, and A. Billington, S. Rakotoarinjara, and, in particular, E.M. Guscott for the typing.

We are grateful to the Cancer Research Campaign for support of many of our studies referred to in this book.

R.V.B.
E.J.L.
T.G.T.

Contents

1

Light and ionizing radiation: generation of excited states and radicals

'LASER, inter eximia naturae dona numeratum plurimis compositionibus inseritur'
Plinius, (49, first century AD), *Naturae historia*, XXII

1.1 Nature of light and light absorption

The sun and light were associated with life and medicine in the most ancient cultures. The Indian sacred book Atharva Veda and the Buddhist Bower manuscript described, 34 centuries ago, a phototreatment of leukoderma or vitiligo, a disease characterized by depigmented patches of skin (see Chapter 9). The deleterious effects of sunlight were also known thousands of years ago.

This century we have learned what light is, what are the excited states and the radicals formed by the interaction of light with molecules, and what are the other possible sources of excited states and radicals. Such other sources include reactions induced by enzymes, high energy radiation, electrolysis, and thermal homolysis. Moreover, during the last decade, the large gap between biochemical processes *in vitro* and *in vivo* has narrowed.

1.1.1 *What is light?*

Light corresponds to the visible portion of the electromagnetic spectrum (Fig. 1.1). Following a number of the early optical studies including those of the rainbow by Maurolico (1553) in Sicily and of the refraction of light by Descartes (1637), Isaac Newton (1672) described sunlight as a mixture of colours (wavelengths). Of course, it is the wave-like nature of radiation which explains diffraction, interference, refraction, and polarization. However, it was the work of the Edinburgh physicist, James Clerk Maxwell (1861a,b) which led to an understanding of the nature of the wave and its propagation. Maxwell's electromagnetic field theory describes light in terms of oscillating electric, **E**, and magnetic, **H**, vectors perpendicular to each other and to the direction of propagation. Such electromagnetic radiation is characterized by its wavelength, λ, or its frequency, ν, which are related by $\lambda\nu = c$ where c is the propagation velocity or speed of light, $3 \times 10^8\,\text{ms}^{-1}$ *in vacuo*. The oscillations of the electric and magnetic fields can be expressed by sine functions:

$$\mathbf{E}_y = A \sin 2\pi (x/\lambda - \nu t)$$

$$\mathbf{H}_z = A(\varepsilon/\mu)^{1/2} \sin 2\pi (x/\lambda - \nu t)$$

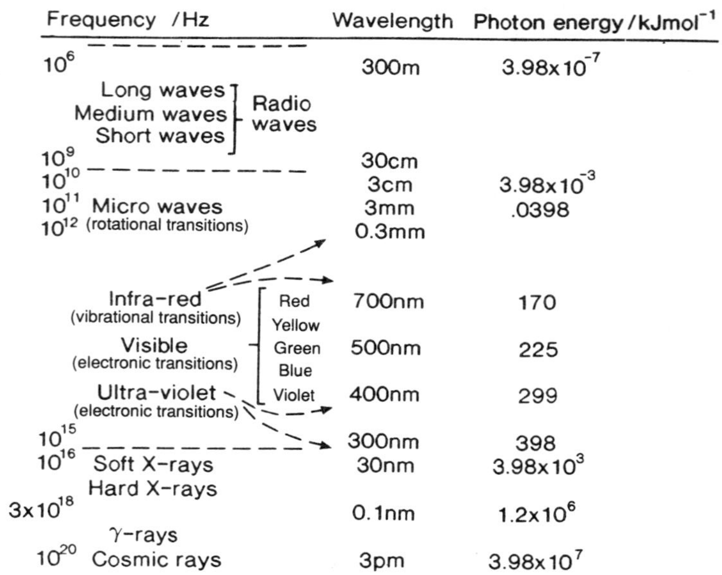

Fig. 1.1 The electromagnetic spectrum.

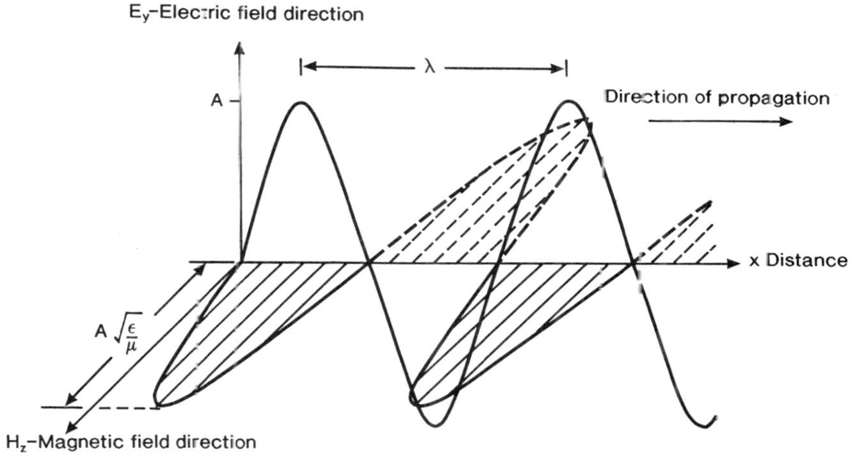

Fig. 1.2 Representation of electric and magnetic fields in a plane polarized electromagnetic wave.

using the coordinates of Fig. 1.2 where x is the distance of propagation at time t. In these equations $\mathbf{E}_y$ is the electric vector in the xy plane and $\mathbf{H}_z$ the magnetic vector in the xz plane. The intensity of radiation is proportional to the square of the electric vector amplitude (A). The amplitudes of the two fields are related by the term $(\varepsilon/\mu)^{1/2}$ with μ being the magnetic permeability and ε the relative permittivity of the medium traversed ($\mu = \varepsilon = 1$ *in vacuo*). It may be shown that the velocity of light in a given medium (c') is related to the velocity *in vacuo*, c, by $(c/\varepsilon\mu)^{1/2}$ [$c' = c/n$ (n is the refractive index)].

The absorption of light arises from the interaction of its electric vector with the electrons of the molecule. (The magnetic vector is of importance in nuclear magnetic resonance (NMR) and electron spin resonance (ESR) spectroscopies.) It should be noted that while, strictly, light is the visible part of the electromagnetic spectrum with wavelengths between 400 and 700 nm (Fig. 1.1), the term 'light' is often extended to include the ultraviolet regions of the spectrum (UVA, 320–400 nm; UVB, 280–320 nm; UVC, 200–280 nm). Hence the distinction between the 'light' and 'ionizing radiation' (Section 1.2) is arbitrary because UV, and even visible light in some cases, can lead to ionization. Nevertheless, the terms 'light' and 'ionizing radiation' are commonly used only in photochemical and radiolytic terminology, respectively.

Electromagnetic radiations, including light, are not fully described by electromagnetic waves. In order to understand the interaction of light with matter it has also been necessary to consider light as composed of packets of energy called 'light quanta' (from the latin quantus meaning 'how much') and later 'photons' (Lewis 1926) behaving as particles. The energy per photon is related to the wavelength and frequency by:

$$E = h\nu = hc/\lambda$$

where h is Planck's constant (6.626×10^{-34} Js). Of course, if the energy, E, of a given light source is known, this equation allows us to obtain the number of photons, n_p, emitted in a given time: $n_p = E(\mathrm{J})\lambda(\mathrm{m})/h(\mathrm{Js})c(\mathrm{ms}^{-1})$. For example, a 5 W (all lines) argon ion laser emitting 1.5 W at 514.5 nm and 1.5 W at 488 nm (see later) generates 3.88×10^{18} and 3.68×10^{18} photons in 1 s at 514.5 nm and 488 nm, respectively.

This equation, associated with the quantum theory of light, stems from a number of important experimental results obtained late in the last century and early in this century which the wave theory of classical physics cannot explain. These experiments concern, for example, black-body radiation, the photoelectric effect, the Compton effect, and typical atomic and molecular spectral observations. In general, they led to the conclusion that classical physics is not correct in considering radiation (or any system) to possess arbitrary amounts of energy.

A 'black-body' is a perfect absorber and a perfect emitter of radiation. The wavelength distribution of the energy emitted is a function of temperature and moves to shorter wavelength as the temperature rises. Thus a 'red hot' poker becomes 'white hot' on further heating. Classical physics failed to describe the complete wavelength spectrum of black-body radiation, with relationships developed by Wien only being satisfactory at short wavelengths and by Rayleigh and Jeans only being satisfactory at long wavelengths. At short wavelengths, the Rayleigh-Jeans formula predicted an increase of energy, implying that all objects should glow in the dark; this is known as the 'ultraviolet catastrophe' of classical physics. However, in the last week of the last century Max Planck (1900, 1901) suggested the revolutionary idea that the energy of the oscillators in the atoms of the 'black body' walls did not vary continuously but was 'quantized', i.e. only discrete values of energy were allowed. This theory introduced the concept of a unit of energy or 'quantum' emitted when the oscillator jumped to a lower allowed energy level, this energy unit being $E = h\nu$, as already described.

On the basis of this relation, Planck derived for the energy density, ρ, of the black-body radiator, the expression:

$$\rho = \frac{8\pi hc}{\lambda^5} \times \frac{1}{e^{hc/\lambda k_B T} - 1}$$

where k_B is Boltzmann's constant, 1.380×10^{-23} JK^{-1} and h is the Planck's constant calculated from this equation to fit the experimental curve. This expression was a satisfactory description of the energy distribution of black-body radiators over the entire range of temperatures (300–1700 K) and wavelengths (~500–5000 nm) tested. Planck's equation becomes Wien's equation at short wavelengths and Rayleigh's and Jeans' equation at long wavelengths.

Hertz (1987) observed that short wavelength light could liberate electrons from a metal surface. Some years later, this phenomenon, called the photoelectric effect, was studied by Lenard (1902) who established that the energy of the electrons emitted was independent of the intensity of the incident radiation but was proportional to its frequency. Einstein (1905, 1906) using Planck's then-new quantum theory proposed that the kinetic energy of the ejected electrons $m_e v^2/2$ was related proportionally to the incident radiation energy via the equation:

$$m_e v^2/2 = h\nu - \omega_o$$

ω_o being the minimum energy required to remove an electron from the metal (work function). This was tested on a number of metals by Ladenberg (1908), Richardson and Compton (1912), and Millikan (1916) who measured the proportionality constant and found it equal to Planck's constant.

Planck's result, associated with Einstein's (1905, 1906) developments of the photoelectric effect, resolved the dilemma of the two natures of light: light described as a wave motion and light described as a flow of particles or corpuscles regarded as quanta.

Before completing this section it is worth noting that the most convincing evidence of the quantum theory is the existence of spectral lines in that the radiation absorbed and emitted by atoms and molecules is at specific rather than continuous frequencies.

In fact, the classical concept of a localized particle must be abandoned not only for photons but for any particle when one considers the duality of particle and wave concepts. The first major expression of this was the relationship given by Louis de Broglie (1924) in his Paris thesis

$$\lambda = h/P$$

which shows that any particle with a momentum P has a wavelength λ. The location of a particle in quantum theory is described as its wavefunction (Ψ) and Ψ^2 at any point is the probability of finding the particle (e.g. electron) at that point. The wavefunction is the atomic orbital for an electron in an atom and is obtained from the well-known equation proposed in 1926 by Erwin Schrödinger:

$$H\Psi = E\Psi$$

where H is known as the Hamiltonian operator for the system, and E is the total energy of the system. This can be considered as the equivalent of $(KE + PE)\Psi = E\Psi$ where KE and PE are the kinetic energy and potential energy, respectively, and the term KE involves a second differential $d^2\Psi/dx^2$ (so that one cannot simply cancel Ψ from each side of the Schrödinger equation!). The apparent simplicity of this equation is illusory and it can only be solved rigorously for quite simple systems. For complex systems major assumptions are necessary—at the simplest level these are the valence bond and molecular orbital approaches.

1.1.2 *Light and molecules*

1.1.2.1 *Absorption and emission processes—lasers*

If we consider an atom having only two energy levels, an upper level (E_u) and a lower level (E_l), then normally we would expect the vast majority of atoms (N_l) to be in the E_l level (Sections 2.1.1 and 2.1.4). However, if radiation (light) of energy equal to $E_u - E_l$ impinges on the atoms or molecules, absorption can take place provided the frequency of the electromagnetic radiation, ν_{ul} equals $(E_u - E_l)$.

It is important to note that excitation from state l to u requires photons. However, for emission, two processes must be distinguished (Einstein 1917)—these are spontaneous emission and stimulated emission. The spontaneous emission is a random process with 'incoherent' light waves out of phase with each other in both space and time. The probability per unit time that a photon is emitted spontaneously is called the Einstein A_{ul} coefficient. If there are N_u atoms or molecules in the upper energy level, the spontaneous emission rate is $N_u A_{ul}$. It is this emission process which mainly arises in photochemistry, and processes such as fluorescence and phosphorescence will normally refer to spontaneous luminescence phenomena. However, as with the absorption process, the emission can also be stimulated to occur in the presence of a photon of frequency ν_{ul}. Photons emitted by stimulated emission are in phase with the stimulating photons, travel in the same direction, have the same polarization, and have the same energy. Stimulated emission is thus a source of 'coherent' light. The waves add in a constructive manner, and it is this which leads to the light amplification in the laser (**L**ight **a**mplification by **s**timulated **e**mission of **r**adiation). The coherence of such laser emission has many practical applications.

Both for absorption (which is, of course, stimulated) and for stimulated emission, the rate is proportional to the radiation density, ρ, with the proportionality constant being the Einstein 'B' coefficient (B_{lu} for absorption and B_{ul} for stimulated emission). In fact $B_{lu} = B_{ul}$ and the rates of these processes can be written:

$$\text{absorption rate} = N_l B;\ \text{stimulated emission rate} = N_u B$$

It can immediately be seen that under normal conditions absorption vastly predominates over stimulated emission (Boltzmann equation, see Section 2.1.1). However, if a so-called 'population inversion' in which ($N_u > N_l$) is achieved, stimulated emission will predominate and this leads to the laser.

It can be shown that the Einstein coefficients A and B are related by the equation:

$$N_l B = N_u B + N_u A_{ul}$$

and

$$A_{ul} = B \frac{8\pi h \nu^3}{c^3}$$

and these equations can be used to calculate the ratio of the spontaneous to stimulated emission rate.

In practice, most lasers are considered as three-level (e.g. ruby) or four-level (e.g. Nd–YAG) systems depending on the precise mechanism of laser action as illustrated in Fig. 1.3. In general, since the three-level laser emission terminates in the ground state, more than half the atoms must be excited to achieve a population inversion, and the four-level system, in which the lower laser level is an unpopulated state, is therefore more efficient.

Both continuous wave (CW) and pulsed lasers are important in photochemistry and photobiology. Often the ruby and Nd–YAG are operated in a pulsed mode with techniques used (Q switching and mode locking) to obtain pulse widths of nanoseconds, picoseconds or shorter.

Important CW lasers include the helium/neon gas mixture, the gold vapour laser, the argon ion laser, the titanium sapphire laser, the Alexandrite laser, and tunable dye lasers. Table 1.1 gives some of the most used laser transitions.

It is also important to note that shorter wavelengths can be obtained by harmonic generation techniques and frequency mixing methods using non-linear optical materials (crystals which have a non-linear response to an electric field). Longer wavelengths can be obtained by Raman shifting techniques using gas cells. Thus, for example, the Alexandrite laser may be continuously tuned from 200–2400 nm by using so-called 'frequency doubling' techniques. For

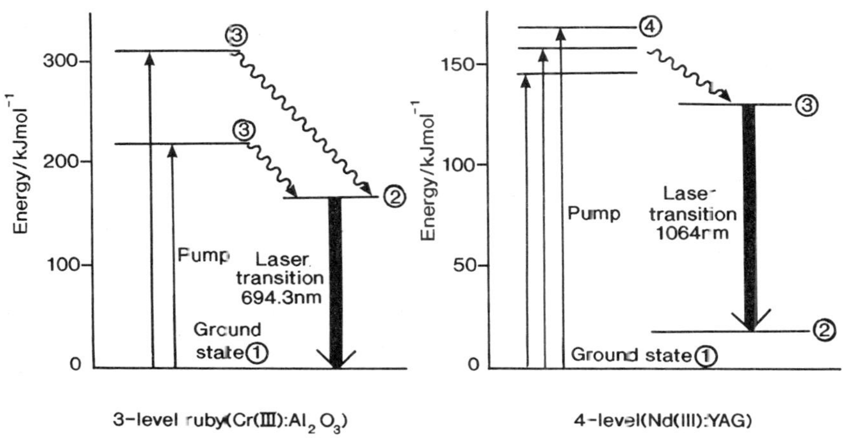

Fig. 1.3 Mechanism of laser action.

Table 1.1 Lasers and their transition wavelength

Laser	Wavelength (nm)
Solid	
Ruby	694.3
Neodymium: YAG	1064 (1314)
Alexandrite	720–800 (tunable)
Titanium: Sapphire	660–1070 (tunable)
Diodes, e.g. GaAlAs	670–690
GaAs	780–904
GaInPAs	1300–1550
Lead Salt	4–5.5 μm
Liquid	
Dye	311–1285 (tunable)
e.g. Rhodamine 6G	536–633
*Gas*1	
Excimer, e.g. ArF	193.0
KrF	248.4
XeCl	308.0
XeF	351.0
Nitrogen	337.1
Helium–Cadmium	441.6 and 325
Argon Ion	488, 514.5 (group of lines 333.6–528.7)
Copper	578
Gold	628
He/Ne	632.8 (group of lines 543.3 nm–3.39 μm)
CO_2	10.6 μm, 9.6 μm (lines 9.2 μm–10.8 μm)

Nd–YAG, 532, 355, and 266 nm may be obtained by frequency doubling, tripling, and quadrupling.

1.1.2.2 *The Beer–Lambert law*

The first quantitative approaches to the study of light absorption were described by the French mathematician Pierre Bouguer (1729, 1760) and the German physicist/astronomer Johann Lambert (1760), so that this law should, more correctly, be called the Beer–Lambert–Bouguer law. These workers showed that the fraction of incident, parallel light absorbed, dI, by a medium is proportional to the thickness of the medium traversed, dx, and is independent of the incident intensity of light, I:

$$-\mathrm{d}I = aI\mathrm{d}x$$

a being the absorption coefficient or extinction coefficient.

Beer (1852) observed that this coefficient, a, is proportional to the concentration, c, of the absorbing molecules:

$$-\mathrm{d}I = bcI\mathrm{d}x$$

This equation, where b is a constant, can be written *via* integration

$$\ln(I_o/I) = bcx$$

or in logarithms to the base 10

$$\log_{10}(I_o/I) = \varepsilon cl = \text{absorbance } (A)$$

where c is in moles per litre; l, the optical path, is in cm; and ε, the molar decadic absorption coefficient, is in $dm^3\,mol^{-1}\,cm^{-1}$, and was called the 'extinction coefficient' in the past.

As will be shown later, the value of ε for a given absorption band may be a useful indication of the type of transition (e.g. $\pi\pi^*$ or $n\pi^*$, see, for example, Section 2.1.4), that is, whether the spectroscopic transition is 'allowed' or 'forbidden' by the selection rules.

Photochemistry is, of course, concerned with the photon intensity absorbed (I_a). This is related to the incident (I_o) and transmitted (I) intensities as

$$I_a = I_o - I = I_o(1 - 10^{-\varepsilon cl})$$

For dilute solutions ($\varepsilon cl \leqslant 0.01$), $1 - 10^{-\varepsilon cl} = 2.3\,\varepsilon cl$ since $(1 - a^x) \approx -x\log_e a$ for small x, and thus I_a and I_o are linearly related as $I_a \approx 2.3\,I_o\varepsilon cl$.

The effects of higher concentration on the divergence from linearity is important to note and is illustrated by Table 1.2 which assumes $l = 1$ cm and $\varepsilon = 10\,000\ dm^3\,mol^{-1}\,cm^{-1}$. This table shows that with concentrations up to $10^{-6}\ mol\,dm^{-3}$, corresponding to $\varepsilon cl \leqslant 0.01$, the intensity absorbed is linearly proportional to the incident light intensity.

The absorbance, A, called also, less desirably, the optical density, is

$$A = \log_{10}(I_o/I) = -\log_{10}T = OD = \varepsilon cl$$

T being the fraction of incident radiation transmitted, called transmittance. This equation is the Beer–Lambert–Bouguer law and holds only in certain con-

Table 1.2 Divergence from linearity of photon intensity absorbed for high concentrations

$(mol\,dm^{-3})$ c	εcl	$1\text{–}10^{-\varepsilon cl}$	$2.303\ \varepsilon cl$
10^{-6}	0.01	0.0227	0.0230
10^{-5}	0.1	0.2056	0.230
10^{-4}	1	0.9	2.3

ditions. Absorbance, A, is equal to OD when the absorbing molecules do not aggregate, when the incident light is collimated, when the medium is homogeneous, and when reflection or scattering do not contribute to the loss of transmitted light. In a mixture of absorbing species, a, b. . . ., the total absorbance is given by

$$A_t = A_a + A_b + \ldots = (\varepsilon_a c_a + \varepsilon_b c_b + \ldots)l$$

and the intensity absorbed by component a is

$$I_a = I_o \frac{A_a}{A_t}(1 - 10^{-A_t})$$

Furthermore, the Beer–Lambert–Bouguer law is only obeyed for monophotonic absorption. For high light intensities including laser light, multiphotonic absorptions can produce 'non-linear' effects leading to photothermal, electromechanical or photoablative processes which are used in medical and surgical applications (Boulnois 1986, 1989).

1.1.2.3 *The two laws of photochemistry*

Grotthus (1815) and Draper (1841) stated the first law of photochemistry: 'Only light absorbed by a molecule can produce photochemical change in that molecule'. Later, Stark (1908) and Einstein (1912) stated what is now the second law of photochemistry: 'One quantum of light is absorbed per molecule which emits luminescence or is chemically altered after light absorption'. To express the efficiency of a photochemical reaction, Einstein introduced the concept of quantum yield, ϕ

$$\phi_{\text{Product}} = \frac{\text{number of molecules of product formed}}{\text{number of quanta absorbed}}$$

$$= \frac{\text{rate of formation of product}}{\text{intensity of absorbed radiation}}$$

As discussed later, numerous processes can arise following the absorption of a quantum of light and the concept of quantum yield is applied to any such physical or chemical process following light absorption. It is possible to deduce a principle from the second law of photochemistry: 'The sum of the quantum yields for each individual photophysical and photochemical process must be unity' This principle is only valid when the primary photochemical act is monophotonic. With high intensity light sources, biphotonic and multiphotonic effects may occur and the concept of quantum yield for each photophysical and photochemical process has no rigorous meaning.

1.2 Nature of ionizing radiation and its absorption

As stated above there is no clear-cut classification of ionizing radiation since, under suitable conditions, ultraviolet or even visible light can lead to ionization of a molecule. Nevertheless, such near-visible radiation is not normally considered as 'ionizing radiation', instead this term is used for high energy photons (X-rays and γ-rays) and energetic particles such as electrons, protons, neutrons, α particles, and heavy ions. Often the classification reflects the mode of generation of the radiation, such as X-rays and synchrotron radiation, rather than the energy.

We have already noted the major development in the physical sciences around the beginning of the twentieth century and closely linked to this was the beginning of the modern study of ionizing radiation. It was in late 1895 in Germany that Roentgen unexpectedly discovered X-rays, noted their penetrating power and even demonstrating their use in imaging the bone structure in his wife's hand! The first attempt to use X-rays to treat cancer followed Roentgen's discovery almost immediately, and in 1896 Grubbé treated a cancer patient using radiotherapy (Grubbé 1933). Rapid developments in radiation chemistry followed Roentgen's observations and noteworthy is the discovery of natural radioactivity in Paris by Henri Bequerel in 1896, the isolation of radium by the Curies, also in Paris, the distinction of α and β radiation by Rutherford at Manchester (1899, 1911), the decomposition of water with radium salts (Curie and Debierne 1901), and the formation of ozone from oxygen irradiated with α particles (Lind 1921).

In the 1920–30 period there were some useful studies of the radiolysis of water, and studies of the radiation-induced oxidation of Fe^{II} sulphate in aqueous solution led to the development of the Fricke dosimeter which is still used today to determine the dose delivered by ionizing radiation sources.

However, it was not until the mid-1940s that reactive intermediates such as the hydroxyl radical [OH·] and the hydrogen atom [H·] were accepted to arise from water radiolysis, although André Debierne suggested that water radiolysis leads to hydrogen atoms and hydroxyl radicals about 30 years earlier. The next major advance was the detection of the solvated electron, based on the ESR signal in an alkaline glass at 77 K, and by the pulse radiolysis of liquid water in London and Manchester (Hart and Boag, 1963; Keene 1963). Many other studies have shown, for example, that N_2O converts the solvated electron into OH· radicals

$$e^-_{aq} + N_2O + H_2O \rightarrow OH^{\cdot} + OH^- + N_2$$

and such studies have led to the mechanism of water radiolysis becoming firmly established. It is the good understanding of this system which has allowed major progress in the radiation chemistry of biological systems.

While the use of high energy radiation to generate and understand the reactivity of (short-lived) radicals has developed over the last 30 years, radicals in chemistry have a long and polemical history extending over the last century (see Section 2.3).

1.2.1 *Ionizing radiation sources*

Sources of ionizing radiation may be continuous or pulsed. Isotopes produce continuous radiation and some important radioisotope sources are ^{60}Co, ^{137}Cs, ^{90}Sr, ^{32}P, ^{241}Am, and ^{226}Ra.

A wide range of machine sources exists including the X-ray tube, cyclotron, Van de Graaff accelerator, and, most important, the linear electron accelerator (LINEAC). Machine sources can provide continuous (e.g. X-ray tube) or pulsed (LINEAC) radiation, with the Van de Graaff generator able to produce either continuous or pulsed radiation. With the LINEAC, pulses of electrons are injected into a linear tube in which they are accelerated by radio frequency waves which are generated by a klystron. Such LINEAC machines are readily operated in the nanosecond and picosecond time domains.

1.2.2 *Interaction of high energy photons with matter*

X-Rays and γ-rays (Fig. 1.1) are absorbed by matter according to an exponential law rather similar to that devised for light by the Beer–Lambert (and Bouguer) Law (see Section 1.1.2.2)

$$I = I_o e^{-\mu x}$$

where I_o and I are the intensities of the incident and transmitted radiation, x is the thickness of the absorber, and μ is the linear absorption coefficient. For 1 MeV photons, the thickness corresponding to an attenuation of the dose by 50 per cent is 11 cm for water and biological tissue, and 1.6 cm for lead.

The total absorption coefficient is made up of contributions by which energy can be transferred from the electromagnetic radiations to the absorbing medium. These are the photoelectric effects (see Section 1.1.1), the Compton effect, and electron–positron pair production.

1.2.2.1 *The photoelectric effect*

In the photoelectric effect the entire energy of the X- or γ-photon is transferred to an electron with the subsequent ejection of the electron (see Fig. 1.4(a)). The photoelectrons are, in this case, ejected by the same process as that occurring with UV and visible light. As expressed by Einstein's photoelectric equation (see Section 1.1.1) the kinetic energy (E_{kin}) of the ejected electron depends on the energy of excitation ($h\nu$) and the binding energy of the electron (W_o) so that

$$E_{kin} = 1/2 m_e v^2 = h\nu - W_o$$

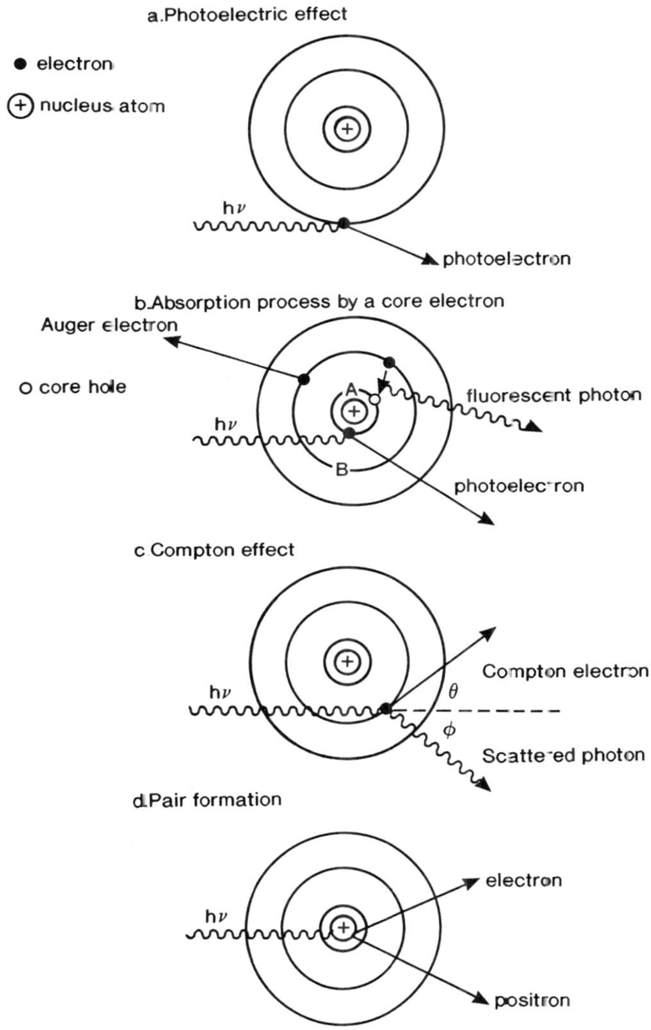

Fig. 1.4 Schematic representation of different interactions of photons with atoms. (a) Photoelectric effect; (b) absorption process by a core electron (this is a special case of the photoelectric effect); (c) Compton effect; (d) pair formation.

Since various electrons in an atom have different values of W_0, the kinetic energies of the ejected electrons (photoelectrons) vary. The maximum probability of ionization arises when the energy of the photon is the same as W_0 and this leads to sharp 'edges', such as the K-edge and M-edge in spectra in which the photoelectric absorption coefficient is plotted against the photon energy. This effect is most pronounced for heavy elements such as lead.

Moreover in the case of the ionization of an inner shell (A), the core electron leaves a hole and a less strongly bound electron can fall from a lower binding energy level (B) into shell (A). The corresponding release of energy may then result in X-ray fluorescence and/or ejection of an extra (Auger) electron (Fig. 1.4b). For K-edges (1*s* initial state), the radiative X-ray fluorescence process is lowest for low Z atoms and increases at the expense of the non-radiative Auger effect with increasing Z (Auger 1925, 1926).

1.2.2.2 *The Compton effect*

In the Compton effect the high energy X or γ photon gives up only part of its energy to an electron which is ejected forward at an angle θ to the primary ray and scattered radiation emerges at the opposite side with an angle ϕ to the primary ray (Fig. 1.4(c)). The kinetic energy of the recoiling electron is taken from the incident radiation and the scattered photons thus have lower energy than the incident ones. The corresponding increase in wavelength is:

$$\Delta\lambda = h/m_e c(1 - \cos\phi)$$

the kinetic energy of the recoiling electron being a function of θ (Compton 1923). The angular distribution of the Compton-scattered photons is a function of the energy of the incident photons.

1.2.2.3 *Pair formation*

At higher incident photon energies pair formation arises from the interaction of the photon with the nucleus of an atom in which an electron plus a positron (e^+) are formed, and any excess energy of the incident photon appears as kinetic energy of the electron and positron (see Fig. 1.4(d)). This is an example of a physical process in which energy is converted into mass and it arises from the complete disappearance of the incident photon in producing the pair. Since the rest mass energy of the electron and positron is 0.51 MeV, the minimum energy for pair production is 1.02 MeV but, in practice, pair production only becomes significant compared to the Compton effect at much higher incident energies (~3 MeV for water).

1.2.2.4 *Contribution of the different processes as a function of photon energy*

The probability of the different processes, photoelectric effect, Compton effect, pair formation, depends on the energy of the photon and the atomic number Z of the material involved. In general, the photoelectric effect is most important at lower energies, the Compton effect at intermediate ones, and pair production at higher energies. This is shown for the case of carbon in Fig. 1.5. In general, whatever the mechanism of the energy loss of the high energy photon, secondary electrons are produced and this is summarized in Fig. 1.4(a)–(d). Absorption of these photons will then be rather similar to

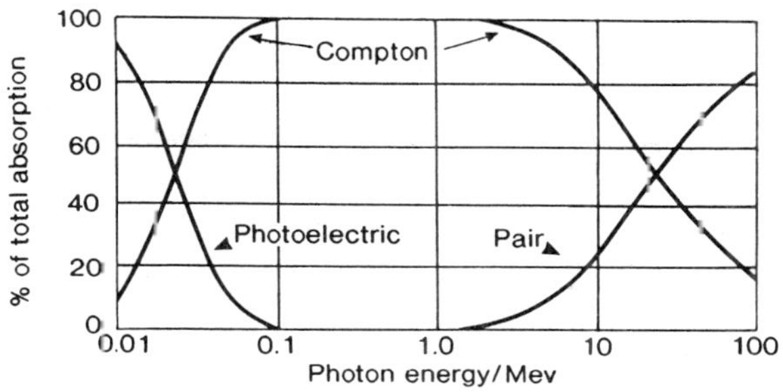

Fig. 1.5 Contributions of the different effects of X and γ photons as a function of photon energy for carbon (Fano 1954).

that which arises when a system is irradiated directly with electrons, which is discussed briefly in the next section.

The induction of chemical changes occurs more frequently by exciting or ejecting the external electrons involved in molecular bonds. Thus, as bond energies range from 1 to about 10 eV, the actual chemical changes are produced when the energy of the X- and γ-ray photons are 'degraded' to 1–10 eV.

1.2.3 *Interaction of charged particles—energy loss of electrons*

In radiation chemistry, most studies are concerned with energetic electrons, rather than heavier particles, as the source of radiation and we shall restrict our discussion mainly to the interaction of electrons with matter. Such interaction arises in a number of ways, particularly the emission of radiation and inelastic and elastic collisions.

At relatively low energies typically below a few hundred eV, the electrons lose their energy by inelastic collisions with the electrons of the stopping material causing ionization and electronic excitation. The rate of energy loss of the electron, or other particle, $(-\mathrm{d}E/\mathrm{d}x)$ by this mechanism, known as Linear Energy Transfer (LET), is described by the Bethe equation, which in a simplified form can be written as

$$-\frac{\mathrm{d}E}{\mathrm{d}x} = \frac{Az^2MNB}{E}$$

where A is a constant, z is the charge on the electron, M is the mass of the electron, N is the number of electrons per unit volume, B is a logarithmic term known as the stopping factor, and E is the energy of the electron. The main feature of this equation is that the LET decreases with increasing energy and decreasing mass and charge of the particle. Some typical LET values for water are 0.2 eV nm^{-1} for 2 MeV electrons and for electrons produced by 1.25 MeV γ-rays, 17 eV nm^{-1} for 2 MeV protons, and 190 eV nm^{-1} for 1 MeV

α particles. Note the high values for α-particles which have relatively large mass (four times that of a proton) and a charge of two, with this latter factor appearing as z^2 in the Bethe equation. In the case of Ni ions, the LET can reach 5×10^3 eV nm^{-1}.

For low-energy electrons in materials of high atomic number the phenomenon of 'elastic collisions' arises in which there is no overall energy loss but the deflection of the electron leads to a non-linear passage of the electron through the material.

At higher energies (10–100 MeV) the main loss of energy for charged particles is an emission of radiation known as 'bremsstrahlung' which arises due to the deceleration of the particle as it passes close to the nucleus of an atom. An additional radiative emission effect was discovered in 1934 by Čerenkov and Vavilov in Leningrad. These workers showed that when the velocity of an electron exceeds the speed of light in a given medium (speed of light = c/n where c is the speed *in vacuo* and n is the refractive index [=1.33 for water]) a bluish light is emitted which is known as Čerenkov emission. There is some work which has shown the contribution of Čerenkov radiation to both DNA damage and DNA repair *via* a photoreactivating enzyme (as discussed by von Sonntag 1987a).

In the technique of pulse radiolysis, and particularly when this technique is used to study light emission processes such as the near infra-red emission from singlet oxygen (see Chapter 4), the Čerenkov radiation is an experimental difficulty. On the other hand, Hunt and co-workers (Bronskill *et al.* 1970) used Čerenkov radiation as a ps monitoring source in a stroboscopic method of pulse radiolysis.

Thus, to summarize, as electrons pass through a typical solvent such as water, energy is lost either by radiation, or at somewhat lower energies ($\approx$2 MeV) by inelastic scattering. Such inelastic scattering produces secondary electrons along the primary 'track' of the inelastic scattering, and any which have relatively low energies (say $\approx$100 eV) will be dissipated rapidly as ionization and excitation in a short range (known as a spur), typically over a few nm. The energy necessary to create an ion pair is $\leqslant$30 eV so that more than three such pairs tend to be produced in each spur. More energetic secondary electrons will have low LET and will form 'tracks' of their own.

Three different track entities—spurs, blobs, short tracks—are distinguished by deposited energy criteria (Mozumder and Magee 1966; Chatterjee 1987):

spur:	6–100 eV (low LET)
blob:	100–500 eV
short track:	500–5000 eV

Track structure of charged particles can be visualized and recorded in gases (cloud chambers), in liquids (bubble chambers), and in solids (nuclear emulsions and other solid state nuclear detectors). Typically the distance, D, in cm reached by high energy electrons between 1 and 50 MeV is estimated as $D = E_{\mathrm{MeV}}/2$ in water.

1.2.4 *Generation of excited states, ionic intermediates, and radicals*

The species created in the track during the physical stage of radiation action are excited states, ions, electrons, and radicals. These start the chemical stage of the radiation action and are responsible for its ultimate biological effects.

1.2.4.1 *The spatial and temporal distribution of primary species*

The spatial and temporal distribution of primary species is a difficult but essential question for condensed matter and we shall describe what happens in liquid systems. The inhomogeneous distribution of the primary species is related to the Linear Energy Transfer of the radiation and to the physical properties of the liquid. Dilute solutes do not intervene in the early stages of

Table 1.3 Time-scale of radiation events in water at room temperature

Processes	Species	Time (s)
	H_2O	
Excitation an ionization	H_2O^* ; $H_2O^{\cdot +} + e^-_{dry}$	10^{-17}–10^{-16}
	H_2O	
Protonation	$H_3O^+ + OH^{\cdot}$	10^{-14}
Dissociation	$H^{\cdot} + OH^{\cdot}$	10^{-13}
Solvation	e^-_{aq}	0.3×10^{-12}
Dielectric relaxation time of water		10^{-11}
Formation of molecular products (H_2, H_2O_2) in the spurs and 'spur expansion' (diffusion of radicals)		10^{-7}
Radical reactions with scavenger at millimolar concentration		$>10^{-7}$

the primary processes. In a solvent such as water, the energy of the electrons is lost by ionization and excitation of water molecules. The time scale of events is shown in Table 1.3. Some water molecules are excited to upper electronic states from which they can autoionize, dissociate or fall back to the ground state. The ionization events occur in the same time scale, 10^{-16} s.

The formation of molecular products in the spurs takes place in a few nanoseconds and occurs via the reactions:

		$k\ (10^{10}\ \mathrm{dm^3\,mol^{-1}\,s^{-1}})$
$e^-_{aq} + e^-_{aq} \rightarrow$	$H_2 + 2OH^-$	0.54
$e^-_{aq} + OH^{\cdot} \rightarrow$	OH^-	3.0
$e^-_{aq} + H_3O^+ \rightarrow$	$H^{\cdot} + H_2O$	2.3
$e^-_{aq} + H^{\cdot} \rightarrow$	$H_2 + OH^-$	2.5
$H^{\cdot} + H^{\cdot} \rightarrow$	H_2	1.3
$H^{\cdot} + OH^{\cdot} \rightarrow$	H_2O	3.2
$OH^{\cdot} + OH^{\cdot} \rightarrow$	H_2O_2	0.53
$OH^- + H_3O^+ \rightarrow$	$2H_2O$	14.3

As time goes by, these reactions are less and less probable because the initial species diffuse and their concentration in the spur decreases to reach a homogeneous distribution. Subpicosecond laser techniques have shown that the orientation of the water molecules around the thermalized electron (so-called dry electron) occurs in less than 10^{-12} s (Wiesenfeld and Ippen 1980; Migus *et al.* 1987; Long *et al.* 1990) while the dielectric relaxation time of water is 10^{-11} s. This difference is probably due to the electron finding a preformed site in water satisfying the solvation requirements. Moreover, the role of the intense electron field of the excess electrons in shortening the dielectric relaxation time should be noted (Mozumder 1969). Despite a large number of theoretical and experimental investigations in about one hundred pure polar solvents, the mechanism of electron solvation is still not completely understood (Jay-Gerin and Ferradini 1991).

Regions of high ion and radical concentrations have been directly observed during the decay of solvated electrons produced in the radiolysis of water by 3 ns pulses of 3 MeV by J.K. Thomas and Bensasson (1967). The rapid disappearance of the e^-_{aq} in this work during the first $\approx$70 ns following the pulse has been since considered as the duration of the spur reactions in water in the 'spur diffusion model' (Schwarz, 1969). In the frame of this spur diffusion model, a scavenger S will compete with the spur reactions of e^-_{aq}, $H^{\cdot}$, $OH^{\cdot}$ if the concentration of S, multiplied by the rate constant k with e^-_{aq}, $H^{\cdot}$, $OH^{\cdot}$, $k[S]$ is $>10^7\ s^{-1}$.

For low LET radiations, commonly used in radiation chemistry, the values of the primary yields (G = number of species per 100 eV absorbed) in the pH range 3 to 11 for the radicals and the molecular products are: $G(e_{aq}^-) = G(OH^\cdot) = G(H_3O^+) = 2.7$, $G(H^\cdot) = 0.6$, $G(H_2) = 0.40$, and $G(H_2O_2) = 0.7$.

1.2.4.2 *Production of excited states*

The concept of energy necessary to produce an ion pair in a gas was introduced by L. H. Gray (1929). This energy, (W), is much greater than the lowest ionization potential of the molecules concerned, the additional production of excited states explaining this important difference.

The absorption of energy either from a γ-ray photon or from the electric field of a fast moving electron excites molecules (of low atomic number, $Z < 10$), via transitions in which the total spin of the system does not change, to the excited singlet state (Lipsky 1981). Energetic (above 50–100 eV) electrons also produce singlet excited states. With lower energies but still above the ionization potential, ionization can take place immediately or with a delay of 10^{-14} to 10^{-12} s, called autoionization. The molecules above the ionization potential and which do not autoionize are in a superexcited state. If the delay of autoionization is long, the molecule has enough energy to break a bond, and may dissociate into neutral molecular radical fragments. This is the case for all repulsive states which have no stable configuration. If the excited states are not dissociative, they deactivate by internal conversion to the lowest singlet excited state, within the picosecond range, and they then follow the familiar photochemical processes: fluorescence, intersystem crossing, internal conversion, and phosphorescence (see Section 2.1.6). The decelerated electrons, at energies below 100 eV, may start to produce triplet excited states via optically forbidden processes. However, more importantly, excited states are produced by anion and cation recombinations. The G values of excited states (number per 100 eV absorbed) have been determined for different solvents by fluorescence

Table 1.4 Excited state G values[a] for typical solvents measured by pulse radiolysis

	G (singlet)	G (triplet)
Benzene	1.6	4.2
Toluene	1.35	2.8
Benzonitrile	0.7	1.4
Benzyl alcohol	0.7	1.1
Dioxan	1.4	0
Cyclohexane	1.45	—
Acetone	0.34	1
Tetrahydrofuran	0.1	<0.1
Ethanol	<0.1	<0.1
Acetonitrile	—	<0.1

[a] For cyclohexane from Choi *et al.* (1982); for other solvents from a review by Rodgers (1983*a*).

and triplet–triplet absorption studies and are given in Table 1.4.

For polar solvents, the ions are solvated and thus the formation of excited states is very low. For solutes, the excited states can be formed by Čerenkov radiation, subexcitation electrons, and mainly via electronic energy transfer in aromatic hydrocarbon solvents and ion recombinations in aliphatic hydrocarbons. For solid organic films, the energy transfer from low-energy electrons (0–30 eV) has been measured by Sanche *et al.* (1982), with energy transfer correlating with both the singlet and triplet gas-phase energy levels.

1.2.4.3 *Production of ionic species*

Free ion yields have been measured by conductivity experiments, pulse radiolysis or product analysis (Allen 1976). A free electron is an electron which has escaped from the interaction of its counterion. The probability of escape P_{esc}, from the parent ion, is related to the relative permittivity of the liquid and the distance between the electron and its positive counterion (see Section 2.5.3).

The solvated electrons can be observed via their optical properties. Their absorption spectra are found in the visible and infra red regions of the spectrum. The absorption spectra of the solvated electron in various pure solvents at room temperature are shown in Fig. 1.6 (Dorfman and Jou 1973). Using

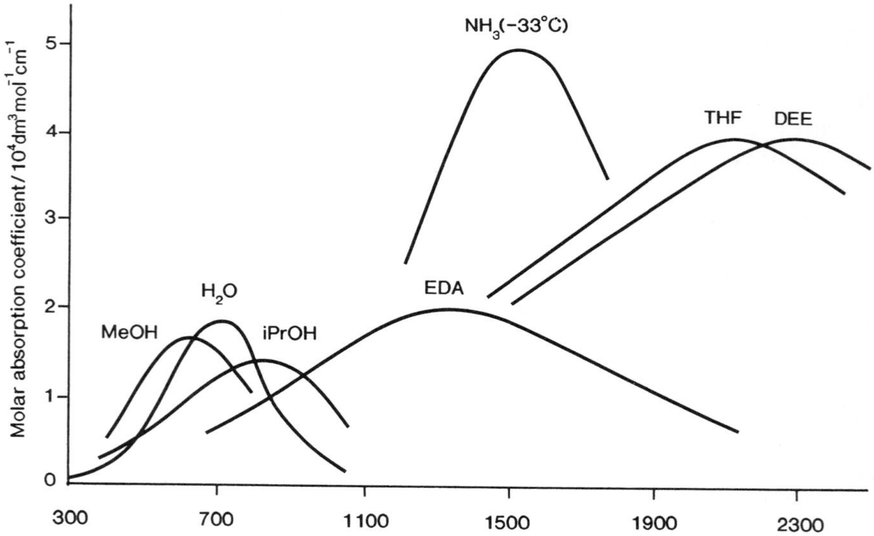

Fig. 1.6 Absorption spectra of the solvated electron in various pure solvents at room temperature: methanol (MeOH), water (H_2O), isopropanol (iPrOH), ethylenediamine (EDA), ammonia (NH_3), tetrahydrofuran (THF), and diethylether (DEE) (from Dorfman and Jou 1973).

these absorption spectra it has been possible to quantitate the reactions of solvated electrons with a large number of molecules of biological interest.

1.2.4.4 *Production of neutral radicals*

Neutral free radicals, molecular fragments with one unpaired electron (see Section 2.3), are formed from the unimolecular dissociation of excited states and from protonation or deprotonation of radical ions.

The yields reported below deal with radicals which have escaped from the spur reactions and are homogeneously distributed. They are usually detected by physical methods including ESR techniques and optical absorption spectroscopy in pulse radiolysis experiments as described in Chapter 3. They have also been identified by chemical methods using scavengers such as iodine and hydrogen iodide, via the reactions:

$$R^{\cdot} + I_2 \rightarrow RI + I^{\cdot}$$

$$R^{\cdot} + HI \rightarrow RH + I^{\cdot}$$

or by radioactively labelled radicals such as $^{14}CH_3I$, generated by dissociative electron attachment (see Section 2.3.3.1) of $^{14}CH_3I$, via the reactions:

$$^{14}CH_3I + e^-_{aq} \rightarrow {}^{14}CH_3^{\cdot} + I^-$$

$$^{14}CH_3^{\cdot} + R^{\cdot} \rightarrow {}^{14}CH_3R$$

We have described above the radicals observed in the radiolysis of water. It is interesting to describe what is known for alkanes, alkenes, and aromatic simple hydrocarbons as they can serve as solvents for *in vitro* studies and as models of biological molecules.

In the case of alkanes, the total radical yield per 100 eV absorbed, *G*, is about 5. Yields of individual alkyl radicals can be expressed as varying functions of the number of carbon atoms in the molecule (Schuler and Kuntz, 1963; Holroyd, 1966). Neutral excited molecules are probably the main precursors of these radicals as it has been observed that the photolysis of an alkane gives the same ratio of yields of fragment radicals as that produced by radiolysis (Holroyd 1966).

In alkenes, the yields of total radical are 4–5, the most important radical being the allylic species $CH_2 = CHCH_2^{\cdot}$, the second most important being the H-atom adduct, thus butyl is formed in butenes, propyl in propenes, etc. (Holroyd and Klein 1965).

In benzene, the total radical yield is 0.7–1.0, that is, lower than that for alkanes or alkenes. The radicals detected are the cyclohexadienyl, $C_6H_7^{\cdot}$, resulting from the addition of an atom of H to benzene, and the phenylcyclohexadienyl radical resulting from the addition of the phenyl radical to benzene (Cherniak *et al.* 1964).

1.3 Other sources of excited states and radicals

We have noted several times that both UV/visible light and high energy radiations can lead to the production of excited states and/or radical species. However, several other routes to excited states and radicals exist. These can be physical or chemical.

A typical example of a physical route is frictional forces (triboluminescence) such as the crushing of uranyl nitrate crystals. The most well known example of chemiluminescence is the oxidation of alkaline solutions of luminol leading to a bluish green emission from the excited state of an aminophthalate ion. Another example, which is discussed later (Chapter 4) is the 'dimol' emission of two singlet excited oxygen molecules leading to 634 nm light in many oxidation reactions involving organic peroxides or hydroperoxides. The most efficient chemiluminescent systems reported to date are based on the generation of an energy rich dioxetane-type intermediate, (Diox):

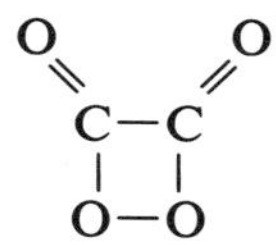

which reacts with substrates (S) as

$$S + \text{Diox} \rightarrow S^* + 2CO_2$$

$$S^* \rightarrow \text{Fluorescence}$$

The commercial (pop-art) light sticks are based on such systems and there are also a few useful related analytical techniques.

In the gas phase there have been many studies of chemiluminescence based on atom recombination reactions and these are often detectable in flames. Bioluminescence is also well known, with the best examples being the firefly (luciferin/luciferase) system.

Other sources of radicals, of chemical and biological interest, are also known: thermal homolysis, sonolysis, redox reactions (Fenton, Haber–Weiss), and enzymatic processes.

Thermal homolysis: Peroxides ROOR can decompose easily as the peroxidic bond O—O has one of the lowest average bond dissociation energies D of 143 kJ mol^{-1} (1.48 eV) to be compared with that of the C—C and C—H, 344 and 415 kJ mol^{-1} (3.5 eV and 4.3 eV), respectively (D values from Pauling and Pauling 1975), the unimolecular homolytic scission giving two alkoxy radicals RO·.

Sonolysis: Radicals can also be generated by ultrasound in liquids (Riesz *et al.* 1985) via transient cavitation in which a bubble violently collapses producing high temperatures and pressures. In aqueous solutions cavitation produces

H· and OH· radicals. Using spin trapping and electron spin resonance techniques, Christman *et al.* (1987) have shown that diagnostic imaging ultrasound leads, in aqueous solutions, to the formation of these radicals.

Redox reactions: At the end of last century, Fenton (1894, 1899) described the powerful oxidizing properties of a mixture of hydrogen peroxide and ferrous salts. Forty years later, Haber and Willstäter (1931) and Haber and Weiss (1934) proposed a decomposition of hydrogen peroxide catalysed by iron salts *via* four reactions in acidic or neutral solutions:

$$Fe^{2+} + H_2O_2 \rightarrow Fe^{3+} + OH^{\cdot} + OH^-$$

$$OH^{\cdot} + H_2O_2 \rightarrow H_2O + HO_2^{\cdot}$$

$$HO_2^{\cdot} + H_2O_2 \rightarrow O_2 + H_2O + OH^{\cdot}$$

$$OH^{\cdot} + Fe^{2+} \rightarrow OH^- + Fe^{3+}$$

Since, the third reaction (but with $HO_2^{\cdot}$ deprotonated to $O_2^{\cdot -}$) has become known as the 'Haber–Weiss' reaction. This reaction has been postulated to explain the toxicity of $O_2^{\cdot -}$ *in vivo* (see Sections 4.2.3.1 and 4.4.3.5). In this hypothesis, the superoxide anion produced biologically is able to produce hydrogen peroxide by dismutation and yield a highly oxidizing species, $OH^{\cdot}$. It must be emphasized that such reactions only occur at measurable rates in the presence of metal ions, the key processes being (Halliwell and Gutteridge 1989):

$$Fe^{3+} + O_2^{\cdot -} \rightarrow Fe^{2+} + O_2$$

$$\underline{Fe^{2+} + H_2O_2 \rightarrow Fe^{3+} + OH^{\cdot} + OH^-}$$

$$\text{net } O_2^{\cdot -} + H_2O_2 \rightarrow OH^{\cdot} + OH^- + O_2$$

The spontaneous oxidation of ferrous ion can compete with the oxidation of the organic substrate investigated. In order to avoid this rapid loss of ferrous ions, a chelate is used: Ferrous-DTPA (DTPA = diethylenetriamine pentaacetate). This is the chelate of choice used by biochemists to conduct a Fenton reaction (Cohen 1985) and it is relatively stable to autoxidation. However, it should be emphasized that some new data have established that, with such a chelate, the oxidizing species is not the hydroxyl radical but an iron–oxo species such as the ferryl ion (FeO^{2+} –chelate) (Koppenol and Liebman 1984; Rahhal and Richter 1988).

Enzymatic reactions: In vivo, certain cytochrome oxidases and xanthine oxidase can produce the superoxide anion $O_2^{\cdot -}$ (Cadenas 1989), leading to the classical radical components of oxidative stress $OH^{\cdot}$, $RO^{\cdot}$, $ROO^{\cdot}$, as described in Chapter 4. Another enzymatic source of free radical intermediates *in vivo* is linked to the metabolism of toxic chemicals (R. P. Mason, 1982).

2

Nature and reactivity of excited states and radicals

'Forsan et haec olim meminisse invabit'

(*Ovid*, *Aeneid*, **1**, 203)

2.1 Nature and reactivity of excited states

2.1.1 *Translational, rotational, vibrational, and electronic energies*

When light (or any electromagnetic radiation) falls on a sample it may be transmitted, scattered or absorbed. If an atom or molecule is in the path of electromagnetic radiation of a range of frequencies certain energies will be absorbed, corresponding to the energy difference between two energy levels, i.e. at the resonant frequency—this applies whether the two levels are rotational, vibrational or electronic.

Electronic transitions involve energy differences which correspond to the visible and ultraviolet regions of the spectrum. Typically, 150 kJ mol^{-1} to 600 kJ mol^{-1} correspond to 800 nm to 200 nm, respectively. So that, for example, the longest wavelength absorption band of haematoporphyrin is at 620 nm (see Chapter 12) corresponding to 190 kJ mol^{-1}. Vibrational transitions involve energy differences corresponding to the infrared (120 to 10 kJ mol^{-1}) region of the spectrum.

The Boltzmann distribution law relates the population of a given energy level to the energy of that level, and can be written as:

$$N_u/N_l = e^{-\Delta E/RT}$$

where N_u is the number in the upper level U and N_l is the number in the lower level l, ΔE is the energy difference between levels in J mol^{-1}, T is the temperature (Kelvin), R the gas constant = $k_B N_A$ = 8.314 J K^{-1} mol^{-1} where k_B is the Boltzmann constant = 1.381×10^{-23} J K^{-1} and N_A is the Avogadro constant = 6.022×10^{23} mol^{-1}.

This equation allows us to calculate the population of the ground and excited states for all types of energy spacing, i.e. that associated with translational, rotational, vibrational, and electronic energy. At room temperature (298 K) the value of RT is 2.48 kJ mol^{-1}. Thus it can be seen that excited translational levels (typically $\Delta E = 10^{-21}$ kJ mol^{-1}) and excited rotational levels (typically $\Delta E = 0.05$ kJ mol^{-1}) are populated, whereas the vibrational (typically $\Delta E \geqslant 10$ kJ mol^{-1}) and electronic (typically $\Delta E > 150$ kJ mol^{-1}) excited states are effectively not populated (~1.77% for 10 kJ mol^{-1} vibrational energy).

As an example, use of the Boltzmann distribution equation readily allows the calculation of the ratio of molecules, at 298 K, in the first electronic excited state of haematoporphyrin, located at about 190 kJ mol^{-1}. Thus, $N_u/N_l = e^{-190/RT} = e^{-77} = 10^{-33}$, so that effectively every molecule in a given sample is in the ground electronic state at room temperature.

2.1.2 *Potential energy diagram*

A diatomic molecule can be considered as a harmonic oscillator with the atoms attached by a spring and vibrating along the bond axis, the restoring force being proportional to the interatomic distance r. The potential energy of this system can then be approximated by a parabola according to Hooke's law

$$PE = \int_0^x kx\,dx = 1/2\,kx^2 = 1/2\,k(r - r_e)^2$$

where k is the force constant and r_e is the internuclear displacement at equilibrium. The vibrational energy is quantized with values $E = (v + ½)h\nu$ where v is the vibrational quantum number with values 0, 1, 2–n. In the parabolic approximation of the potential energy curve (Fig. 2.1), the lower vibrational energy levels are a series of equally spaced lines. At high values of $(r - r_e)$, the potential energy curve is assymmetric as also shown in Fig. 2.1 and will rather be represented by a more complex empirical curve. Associated with this anharmonic curve is a reduction of the spacing of the vibrational levels, especially at higher energies.

For polyatomic molecules, polydimensional surfaces (not easily visualized) would be needed to describe the positions and energies of the nuclei. Usually the features of the two dimensional energy curve, dealing with the structure, energetics, and dynamics of a diatomic molecule, are extrapolated and applied to the energy surfaces of polyatomic molecules.

Associated with vibrational transitions are rotational transitions, so that under high resolution (in the gas phase), each vibrational line contains rotational fine-structure (the so-called P, Q, and R branches—usually Q is not observed). Thus it is usual to consider such vibration–rotation spectra to consist of bands rather than discrete lines.

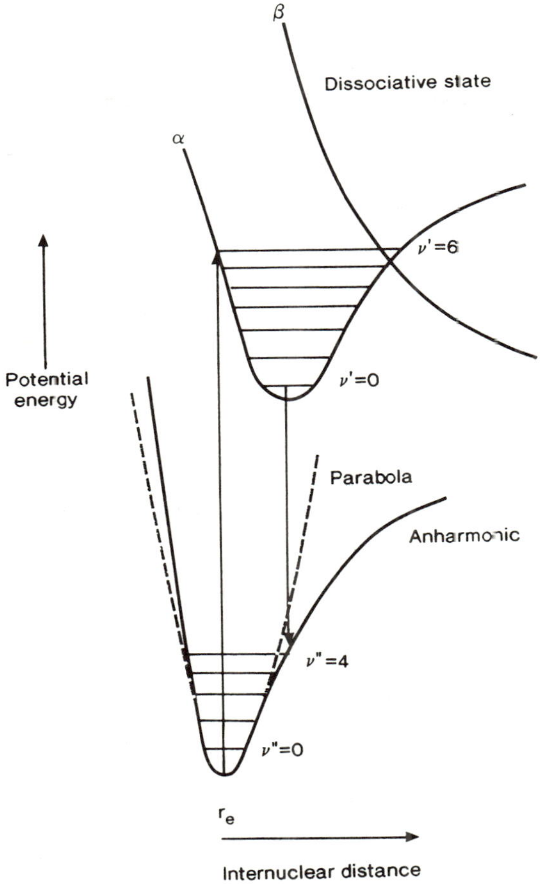

Fig. 2.1 Potential energy curves versus internuclear distance for the ground state and two excited states (α and β) of a diatomic molecule, showing absorption and emission transitions.

In the condensed phase the rotational fine structure is lost and the vibrational frequencies are characteristic of different groups in a molecule, so that infra-red spectroscopy is an important analytical technique. Typically the following species can be identified in a complex molecule, the carbonyl group at $\sim 1700\,cm^{-1}$, amines at $\sim 1600\,cm^{-1}$, C—H groups at $\sim 3000\,cm^{-1}$ and $\sim 1400\,cm^{-1}$, and O—H groups at $\sim 3600\,cm^{-1}$. The region 900–1400 cm^{-1} is often called the fingerprint region and to identify an unknown compound one has only to compare its infra-red spectrum in this region with a set of standard spectra. Nowadays this is often done via computer library searching.

2.1.3 *Spin multiplicity of electronic transitions*

One of the most important aspects of electronic transitions concerns the spin multiplicities of the two states involved in the transition. Each orbital can be occupied by one or two electrons and, according to Pauli's exclusion principle, these must have opposite spin orientation. Thus, when all occupied orbitals contain two electrons, there are equal numbers of electrons of opposite spin and a net zero electron spin. The concept of electron spin was originally based on empirical considerations, but subsequently Paul Dirac showed that the concept arises naturally from the theory of wave mechanics. In fact, electron spin and nuclear spin are phenomena which only arise in quantum mechanics, and there are no strict analogues in classical mechanics. Thus we cannot truly think of electron spin as the actual spinning of a particle when we think of the electron in its orbit not as a particle but as a standing wave. Nevertheless, from a simplistic point of view, the concept of an electron with two spin quantum numbers $+1/2$ and $-1/2$ (sometimes called an α-electron and a β-electron) is sufficient.

In a ground state, the highest occupied molecular orbital (HOMO) is occupied by two electrons of opposite spin. During a transition one of the electrons will be promoted to an upper excited state. The Wigner rule of spin conservation indicates that the spin of the promoted electron is conserved. The excited state and the ground state are both singlet states, as their total spin, obtained by adding the individual spin quantum number, $(S_1 + S_2)$, is $S = 1/2 + (-1/2) = 0$. The spin multiplicity M of any state is given by the

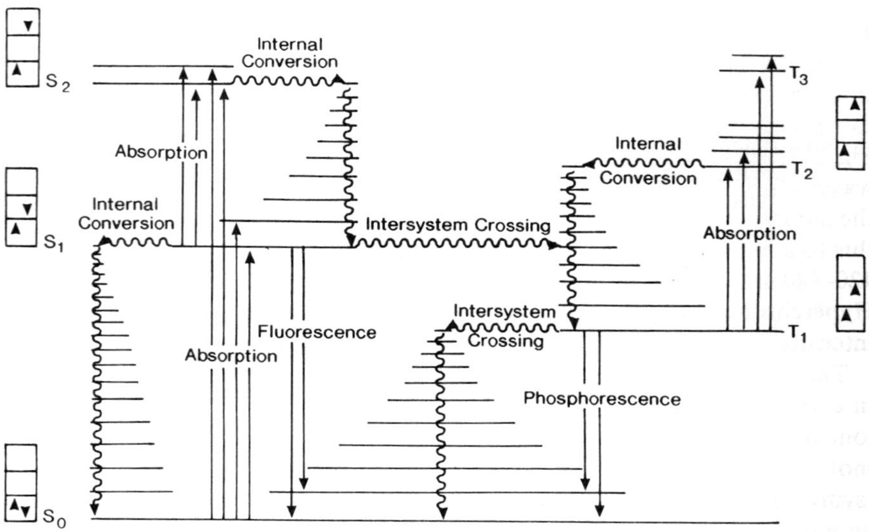

Fig. 2.2 Jablonski diagram.

relation $M = (2S + 1)$ where $S = 1/2$ for each unpaired electron, and thus the spin multiplicity of the singlet state is, of course, one. However, the Wigner rule can be violated by the spin of the electron being influenced by its orbit about the nucleus. This spin–orbit coupling can lead to the spin of the promoted electron being reversed from the direction it had in the ground state. This reversal occurs from the interaction of the magnetic moment associated with the electron spin and the magnetic moment due to the orbital motion of the electron. In this case the total spin is now $S = 1/2 + 1/2 = 1$, the spin multiplicity is $2S + 1 = 3$, and the excited state is called a triplet state (see boxed arrows in Jablonski diagram in Fig. 2.2).

2.1.4 *Nomenclature of electronic excited states*

Five types of molecular orbitals are found in organic molecules: the sigma-bonding (σ) and the sigma-antibonding (σ^*) orbitals involved in single bonds, the pi-bonding (π) and the pi-antibonding (π^*) orbitals involved in π or multiple bonds, the non-bonding (n) orbitals localized in heteroatoms (e.g. oxygen, nitrogen, and sulphur) without anti-bonding counterpart. To these five orbitals, correspond five energy levels and several different transitions $\sigma \rightarrow \sigma^*$, $\pi \rightarrow \pi^*$, $n \rightarrow \sigma^*$, $\pi \rightarrow \sigma^*$, and $n \rightarrow \pi^*$ (Kasha nomenclature, Kasha 1950) due to absorption of UV and visible radiation. For the photobiologist the $n \rightarrow \pi^*$ and $\pi \rightarrow \pi^*$ are the most important, and as a shorthand, these transitions are denoted $n\pi^*$ and $\pi\pi^*$.

For an unconjugated double bond the energy of a $\pi\pi^*$ transition is about 680 kJ mol^{-1} ($\sim$7 eV) and results in a UV absorption around 180 nm as in the case of ethene (Fig. 2.3). Conjugation shifts the $\pi\pi^*$ transition to the visible, as the energy gap between the highest occupied molecular orbital (HOMO) and the lowest unoccupied molecular orbital (LUMO) becomes progressively smaller. The interaction of two ethenic bonds leads to the π orbitals combining to form two new orbitals (π_1 and π_2) and two anti-bonding (π_3^* and π_4^*), see Fig. 2.3. In butadiene this leads to a so-called bathochromic shift to longer wavelength (220 nm $\equiv$ 560 kJ mol^{-1}). This bathochromic shift increases with the number of double bonds for polyenes, and the carotenoids (Chapter 7) show this to a marked effect. Thus, for example, β-carotene has a $\pi\pi^*$ transition at 420–480 nm. (It is worth noting some other of the older spectroscopic terms: Hyperchromic $\equiv$ increased intensity of absorption, Hypochromic $\equiv$ reduced intensity of absorption.)

The electronic transition responsible for the longest wavelength absorption in carbonyls is associated with a non-bonding electron initially in the oxygen lone pair. Excitation leads to an $n \rightarrow \pi^*$ promotion with typical energies 340 kJ mol^{-1} ($\sim$280 nm), Fig. 2.3. It is important to note that $n\pi^*$ transitions are 'symmetry forbidden' so that the corresponding molar absorptivities are very much smaller ($<$100 dm^3 mol^{-1} cm^{-1}) than for the 'allowed' $\pi\pi^*$ transitions (usually $>$ 1000 dm^3 mol^{-1} cm^{-1}). An example of an $n\pi^*$ transition is

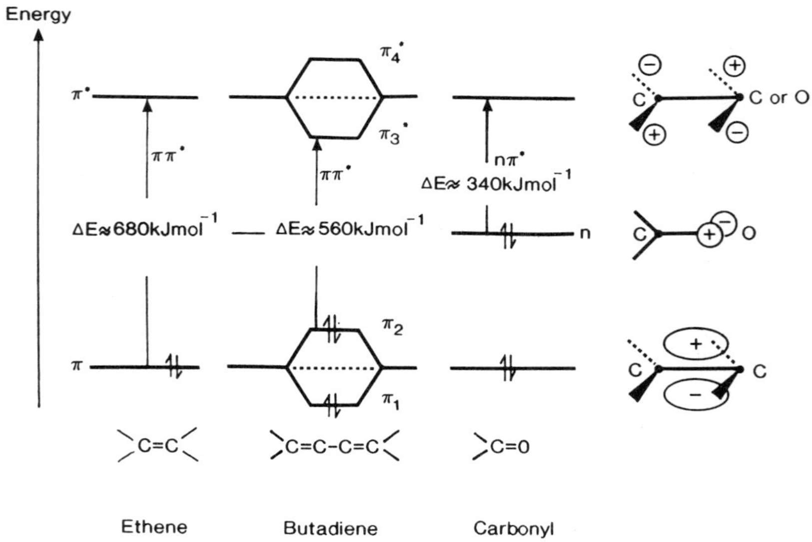

Fig. 2.3 $\pi\pi^*$ transitions for ethene and butadiene and $n\pi^*$ transitions for carbonyl-containing compounds.

exhibited by the absorption spectrum of retinal, the chromophore of the visual pigment (Chapter 7). Such $n\pi^*$ states can lead to hydrogen abstraction reactions. Apart from the substantial differences in absorptivities for $\pi\pi^*$ and $n\pi^*$ transitions as noted above, a useful way of distinguishing such processes is the effect of solvent. Thus the energy of the $\pi\pi^*$ transition decreases while that of $n\pi^*$ transition increases as the solvent polarity increases, corresponding to a red shift and blue shift, respectively (see Section 2.1.7). For a comparison of the differing fluorescence properties see Section 7.2.2.2.

Group theory nomenclature for electronic transitions, other than $n\pi^*$ and $\pi\pi^*$, is also frequently used, for simple systems (atoms and diatomic molecules like oxygen), and for polyatomic molecules such as the linear polyenes (carotenoids), these we refer to later in this text. Furthermore, transition metal complexes often exhibit colour (visible absorption) due to so-called '*d–d*' transitions in which the ligands bound to the central metal split the *d* orbitals into at least two sets at different energies with the precise splitting depending on the geometric ligand arrangement. For example, for the octahedral $[Ti(H_2O)_6]^{3+}$ the *d–d* splitting leads to an absorption band at ~500 nm. Transition metal complexes may also absorb light due to 'charge transfer' processes in which electron density is transferred from ligand to metal or vice versa. Such charge transfer transitions (e.g. the intense purple colour of permanganate) occur with very much higher absorptivities than the rather weak *d–d* transitions.

Charge transfer (CT) transitions are often observed in solutions of mixtures of electron donors and electron acceptors. Electron donors possess a low ionization potential, usually with electron donating substituents (such as OH, NH_2, OCH_3, CH_3), and electron acceptors a high electron affinity with electron withdrawing substituents (such as NO_2, SO_3H, CHO, Cl). Charge transfer complexes are common for biological molecules in solution (Slifkin 1971). They are often characterized by broad charge transfer absorption bands located at the long wavelength side of the free component absorption. The transition from a ground state oxygen–organic molecule complex to an oxygen–organic molecule charge-transfer state is of great importance in the photo-oxidation damage of the organic molecule (Ogilby *et al.* 1990; Kristiansen *et al.* 1991).

2.1.5 *The Franck–Condon principle*

The absorption of radiation is not an instantaneous process, it takes a period of time (t) given by:

$$t = \nu^{-1}$$

and is thus related to the energy separation of the two states involved and to the wavelength of the corresponding transition. Thus

$$t = \nu^{-1} = h/E = \lambda c^{-1}$$

and for the longest wavelength transition of haematoporphyrin at 620 nm the value of t is 2.07×10^{-15} s. Thus, absorption is time-dependent and the quantum mechanical description involves the use of the time-dependent Schrödinger equation.

Compared to this very short absorption time, the time for vibrational motion of the nuclei is longer (10^{-13} s). This leads to the statement of the Franck–Condon principle: during an electronic transition the nuclei keep the same relative position. This was first described in 1925 (the year before the development of the Schrödinger equation) by James Franck and later (1929) in quantum mechanical terms by Condon and Morse. It is this principle that allows us to understand the vibrational structure of an absorption or emission band. This is simply based on the fact that the line illustrating a transition between two electronic states on a potential energy diagram must be vertical, corresponding to a constant internuclear distance. Figure 2.1 shows this for the usual case of the excited state corresponding to longer equilibrium bond lengths (more anti-bonding character) compared to the ground state. For the situation, as illustrated in Fig. 2.1, the strongest absorption band is from $\nu'' = 0$ to $\nu' = 6$, and the strongest emission is from $\nu' = 0$ to $\nu'' = 4$ (Stokes shift). For the case where there is no difference in the internuclear distance, the so-called (0, 0) transitions corresponding to $\nu' = 0 \leftrightarrow \nu'' = 0$ will show the highest intensity.

If the transition is to a dissociative state (Fig. 2.1) the vibrational spectra are lost and a diffuse region results. Clearly, the vibrational structure of the absorption or emission spectra depends on the relative displacement of the two potential energy curves, and the more the displacement the greater the vibrational structure.

2.1.6 *Physical deactivation of electronically excited states*

Perrin (1929), Jablonski (1933, 1935), and Lewis and Kasha (1944, 1945) laid the foundations of a general sequential scheme for the physical activation and deactivation processes of electronically excited states (Fig. 2.2) produced by monophotonic excitation. The two states that have the longest lifetime, and which can be most readily observed usually via absorption or emission, are the first excited singlet state, S_1, and the lowest triplet, T_1. The lifetime, or reciprocal first-order rate constant, in solution of the processes of Fig. 2.2 are often as given below:

S_0	$\longrightarrow$	S_n, absorption	10^{-15} s
S_n	$\longrightarrow$	S_1, internal conversion	10^{-11} to 10^{-14} s
S_1	$\xrightarrow{k_f^0}$	$S_0 + h\nu$, fluorescence	10^{-7} to 10^{-11} s
S_1	$\xrightarrow{k_{isc}}$	T_n, intersystem crossing	10^{-8} to 10^{-11} s
S_1	$\xrightarrow{k_{ic}}$	S_0, internal conversion	10^{-7} to 10^{-11} s
T_n	$\longrightarrow$	T_1, internal conversion	10^{-11} to 10^{-14} s
T_1	$\longrightarrow$	$S_0 + h\nu$, phosphorescence	10^{1} to 10^{-5} s
T_1	$\longrightarrow$	S_0, intersystem crossing	10^{1} to 10^{-5} s

where k_f^o, k_{isc}, k_{ic} are the rate constants of S_1 deactivation processes. An electronically excited species will decay radiatively to the ground state with a natural radiative lifetime τ_0 and a rate constant k_f^o if no decay other than radiative exists. Actually, the fluorescence of the S_1 state is in competition with internal conversion, intersystem crossing, and possibly with a chemical or physical reaction with pseudo first-order rate constant, k_r. The actual lifetime of S_1 (as measured by the decay of fluorescence) will be $\tau_f = 1/\Sigma\ k_i$ with k_i being the sum of all radiative and non-radiative decays $= k_f^o + k_{isc} + k_{ic} + k_r$.

The radiative lifetime, τ_0, which can be calculated theoretically from the Einstein coefficient for spontaneous emission, is related to the actual measured lifetime τ_f via the quantum yield of fluorescence, ϕ_f, by the expression $\phi_f = \tau_f/\tau_0$. The related process of delayed fluorescence can occur via a number of different mechanisms including E-type delayed fluorescence produced by a thermally activated reverse intersystem crossing from vibrationally excited levels of the T_1 state, and P-type delayed fluorescence produced by the formation of an S_1 state by triplet–triplet annihilation (C. A. Parker 1964; Leach and Migirdicyan 1966).

Two important rules are related to the Perrin–Jablonski–Lewis–Kasha diagram (usually called the Jablonski diagram) of Fig. 2.2:

1. *Kasha's rule* (1950) which states that the fluorescence and the phosphorescence of an aromatic hydrocarbon occurs from the lowest excited π-electronic singlet state, S_1, and triplet state, T_1, respectively, and is independent of the exciting wavelength. Often, the vibrational structure of the $S_1 \rightarrow S_0$ emission spectrum is approximately the mirror image of the vibrational structure of the absorption band $S_0 \rightarrow S_1$, because the vibrational level spacings are rather similar for S_1 and S_0 of large organic molecules. A good example exhibiting this phenomenon is anthracene.
2. *Vavilov's rule* (1922) which states that the fluorescence quantum efficiency is independent of the excitation wavelength.

These rules are due to the fast vibrational relaxation among higher excited states ($S_n \rightarrow S_1$ and $T_n \rightarrow T_1$). As pointed out by Longworth (1982), the invariance of luminescence emission spectrum with the excitation energy had been observed more than three centuries earlier by Zucchi (1652) and later by Dessaignes (1810) and Nichols and Meritt (1910). Of course, the usefulness of fluorescence (and phosphorescence) as an analytical technique is related to this invariance. Improvements in fluorescence detection techniques have allowed the detection of the very weak radiative transitions from higher excited states (Birks 1975).

2.1.7 *Solvent effects*

2.1.7.1 *UV and visible absorption spectra*

The position, intensity, and shape of a UV or visible absorption band of a solute is usually modified by solvents of different polarity. This effect is called solvatochromism. A hypsochromic or blue shift with increasing solvent polarity is a negative solvatochromism, while a bathochromic or red shift is a positive solvatochromism. The effect of solvent is related to the interaction forces of solvent and solute. The solvatochromic shift represents the difference in the solvation energies of the excited state and ground state of a transition, S_E and S_G respectively, as shown in Fig. 2.4. Apart from some symmetrical compounds (such as tetrachloromethane) and a few hydrocarbons (*n*-hexane, cyclohexane, benzene), common organic solvents and organic solutes usually have unsymmetrical charge distribution and possess a permanent molecular dipole moment. If the magnitude of the two equal and opposite charges of this molecular dipole is called q and the distance of separation r, the dipole moment μ is qr, with the vector directed towards the positive charge and measured in Coulomb metre (C m) or in Debye units (D) ($1\,D = 3.336 \times 10^{-30}$ C m).

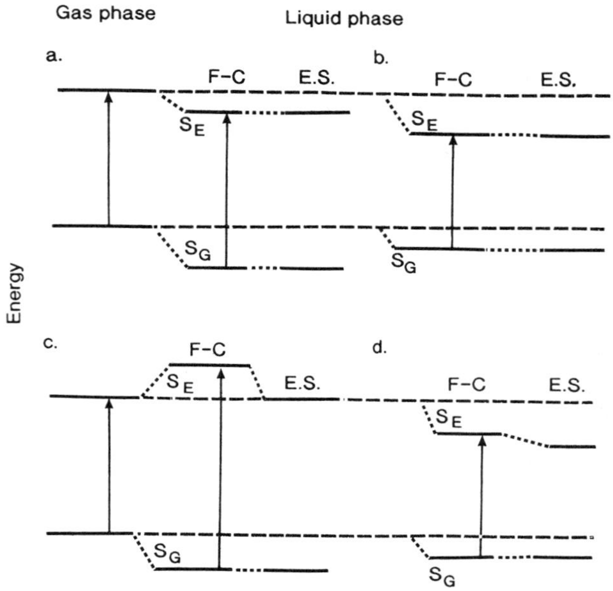

Fig. 2.4 Energy diagram of a solvatochromic shift for (a) polar solute and non-polar solvent with $\mu_g > \mu_e$, (b) with $\mu_g < \mu_e$, (c) polar solute and polar solvent with $\mu_g > \mu_e$, and (d) with $\mu_g < \mu_e$. F–C is the Franck–Condon state, E.S. is the equilibrium relaxed state, and S_E and S_G are the solvation energies in the excited and ground states, respectively.

Molecules with permanent electric dipole moments are known as polar molecules and should be called 'dipolar' molecules. Common polar organic solvents possess a permanent dipole moment ranging between 0 and 18×10^{-30} C m. According to Bayliss and McRae (1954), there are four cases describing the solvent–solute interaction:

(1) Non-polar solute in a non-polar solvent;

(2) non-polar solute in a polar solvent;

(3) polar solute in a non-polar solvent;

(4) polar solute in a polar solvent.

In both cases (1) and (2), there is no permanent dipole in the solute. Thus, dispersion forces, which are instantaneous dipole-induced dipole interactions, often called London–van der Waals forces, are the only attractive forces. These forces are due to the fact that even in atoms and molecules which do not possess a permanent dipole moment, the continuous electronic movement creates an array of disymmetrical distributions (i.e. a series of oscillating instantaneous dipoles) which on average cancel each other. Thus, at any instant, a non-polar molecule possesses an instantaneous dipole which can fluctuate and polarize the electron system of a neighbouring molecule. In both cases (1) and (2), dispersion forces contribute to the solvation of the solute and cause a small bathochromic shift. This red shift, relative to that in the gas phase, is a function of the refractive index (n) of the solvent.

In case (3), in the vicinity of a molecule with a permanent dipole moment, a non-polar molecule becomes polarized and acquires an induced dipole moment. In this case the solvation forces are dispersion and dipole-induced dipole. The dipole-induced dipole forces differ between the ground and the excited state in two possible ways: either the dipole moment decreases ($\mu_g > \mu_e$) or increases ($\mu_g < \mu_e$) during the electronic excitation. When the solute dipole moment decreases during the excitation, the Franck-Condon excited state is less solvated and a blue shift of the spectrum is expected relative to that in the gas phase. The relative magnitude of this blue shift and of the red shift due to the dispersion forces determines the ultimate shift (an ultimate blue shift is shown as an example in Fig. 2.4(a)).

When the solute dipole moment increases, the Franck–Condon excited state is more solvated. This results in a greater lowering of the energy of the excited state as compared to the ground state and a consequent red shift is observed. The magnitude of these shifts depends on n, the refractive index of the solvent, and ε, the bulk state permittivity (or dielectric constant) of the solvent. An additional factor is the red shift due to the dispersion forces (Fig. 2.4(b)).

In case (4) the ground state solvation is mainly due to dipole–dipole forces. The orientation of the polar solvent molecules around the polar solute molecules causes an orientation strain for the excited molecules which have longer bond lengths than ground states. If the solute dipole moment decreases ($\mu_g > \mu_e$), the Franck–Condon excited state is not efficiently stabilized by the oriented dipoles of the solvent molecules, in the so-called 'strained' solvent cage. The energy required to reorientate the solvent molecules around the less polar solute excited molecule provides a negative energy term (Pauling 1929) to the solvation energy and thus an increase of the excited state energy (Fig. 2.4(c)) and hence a blue shift of the spectrum is expected. As the superimposed red shift of the polarization is smaller relatively, this causes a net blue shift (Fig. 2.4(c)). If the solute dipole increases ($\mu_g < \mu_e$) during the electronic transition, a better stabilization of the excited state relative to the ground state arises. This increases with increasing solvent polarity and will give a bathochromic or red shift (Fig. 2.4(d)).

The rules for solvent effects given above for $\pi \rightarrow \pi^*$ absorption bands are also valid for $n \rightarrow \pi^*$ and $n \rightarrow \sigma^*$ absorption bands. For example, the lowest singlet $n \rightarrow \pi^*$ transition of benzophenone leads to an excited state which is less polar ($\mu_e = 5 \times 10^{-30}$ C m or 1.5 D) than the ground state ($\mu_e = 10 \times 10^{-30}$ C m or 3 D) (Liptay 1974). This is because a non-bonding n-electron on the oxygen atom has been promoted to a π^* antibonding orbital delocalized over the whole molecule. This decrease in dipole moment results in a hypsochromic shift, or

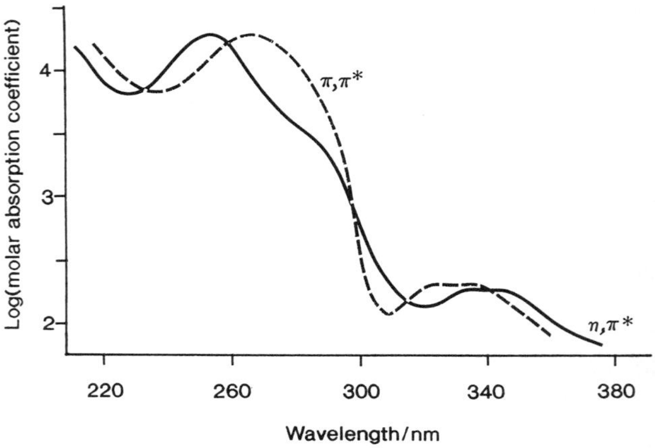

Fig. 2.5 Absorption spectra of benzophenone in ethanol (-----) and cyclohexane (——) at room temperature.

negative solvatochromism, with increasing polarity of the solvent. Furthermore, this blue shift for n → π* absorption bands of carbonyl compounds with solvent polarity can be increased by hydrogen bonding. On the other hand, the dipole of the π → π* absorption of benzophenone is greater in the excited state ($\mu_e = 16 \times 10^{-30}$ C m) than in the ground state ($\mu_g = 10 \times 10^{-30}$ C m) (Exner, 1975). Thus the π → π* absorption undergoes a red shift while the n → π* absorption undergoes a blue shift when the polarity of the solvent increases (Fig. 2.5).

2.1.7.2 *Fluorescence spectra*

The emission from fluorophores occurs usually at longer wavelengths than those of the absorption. This shift in energy between absorption and emission, called Stokes' shift, is due to a number of different processes, including energy losses due to vibrational energy: rapid decay to the lowest vibrational levels of the first excited state S_E and fluorescence to excited vibrational levels of the ground state S_G and interactions between the excited fluorophore and the solvent molecules.

The interactions between the excited state of the solute molecules with the surrounding solvent molecules may modify the quantum yield and the lifetime of the fluorescence emission as well as the shape and spectral position of the emission bands.

On excitation, within 10^{-15} s, a solute molecule reaches a 'Franck–Condon excited state' with a new dipole moment, usually with different magnitude and orientation compared to the ground state. The position of the emission bands is related to the dielectric relaxation time τ_R for the rearrangement of the solvent molecules surrounding the solute in its Franck–Condon excited state, and to the lifetime τ_f of the solute in its singlet excited state. If $\tau_R \gg \tau_f$, the emission will occur from the Franck–Condon excited state before any reorientation of the solvent molecules surrounding the solute. When $\tau_R \ll \tau_f$, the 'equilibrium' or 'relaxed' rearrangement of the solvent molecules arises before the emission and the fluorescence occurs from this equilibrium excited state. In the case of a polar solute in a non-polar medium, there is little or no difference between the energy of the Franck–Condon state and the equilibrium state (Fig. 2.4(a) and (b)), in contrast to the case of a polar solute in a polar solvent (Fig. 2.4(c) and (d)). Moreover, in an emission, the final state is a Franck–Condon ground state located above the ground state at equilibrium. Time-resolved fluorescence spectra can show the dipolar solvent relaxation; usually two observations can be made:

(1) the emission maxima shift with time to longer wavelengths as the emitting state is relaxing towards an equilibrium state of lower energy, and

(2) the relative fluorescence intensity decreases at the red edge of the spectrum as the population in the excited state decreases with time (Sommer *et al.* 1990).

2.1.7.3 *Effect of solvent upon solute absorption and emission band maxima*

The McRae–Bayliss model (McRae 1957) developed according to the theory of dielectrics (Onsager 1936) is considered one of the most successful for correlating solvatochromic effects with solvent parameters and remains a benchmark (Brady and Carr 1985). After completion of the solvent relaxation, equations for the dipole–dipole interaction solvatochromic shifts can be expressed within the simple model of spherical-centred dipoles and within the assumption of equal dipole moments in Franck–Condon (F–C) and relaxed equilibrium states (ES). These expressions can be given as functions of the refractive index of the solvent (n), the static permittivity or static dielectric constant (ε), and the dipole moments of the excited state and of the ground state expressed as vectors, $\vec{\mu}_e$ and $\vec{\mu}_g$. The short range interactions such as H-bonding being neglected, the solvatochromic shift for absorption and emission are expressed by:

$$\tilde{\nu}_a = -(2/4\pi\varepsilon_o hc)\vec{\mu}_g(\vec{\mu}_e - \vec{\mu}_g)a^{-3}\Delta f + \text{const}$$

$$\tilde{\nu}_f = -(2/4\pi\varepsilon_o hc)\vec{\mu}_e(\vec{\mu}_g - \vec{\mu}_e)a^{-3}\Delta f + \text{const}$$

where the parameters are in SI units: ε_o is the vacuum permittivity $= 8.854 \times 10^{-12}\,\text{J}^{-1}\,\text{C}^2\,\text{m}^{-1}$, $\vec{\mu}_e$ and $\vec{\mu}_g$ are the dipole moments in Coulomb metre (C m), h the Planck constant (J s), c the velocity of light (m s^{-1}), a is the radius (m) of the Onsager cavity in which the solute resides, and Δf is the orientation polarizability expressed as:

$$\Delta f = f(\varepsilon) - f(n^2) = [(\varepsilon - 1)/(2\varepsilon + 1)] - [(n^2 - 1)/(2n^2 + 1)]$$

The Stokes' shift is given by the subtraction leading to the equation of Lippert (1957):

$$\Delta\tilde{\nu} = \tilde{\nu}_a - \tilde{\nu}_f = (2/4\pi\varepsilon_o hc)\,(\vec{\mu}_e - \vec{\mu}_g)^2 a^{-3}\,\Delta f + \text{const}$$

The inadequacies of different expressions of solvatochromism, including the model of McRae–Bayliss as well as the formalism of Kamel, Taft, and Abboud who use empirical solvent property scales, have been analysed by Brady and Carr (1985). The most recent general theory has been presented by Brunschwig *et al.* (1987). This new expression of the solvatochromic shift is shown to reduce to previous equations including that of McRae and Bayliss, when point-dipole approximations are made and values of the solute polarizability are assigned.

Discussions on the intermolecular forces between solute and medium can be found in reviews by R.S. Becker (1969), Reichardt (1990), and Suppan (1990).

2.1.7.4 *Biological applications*

In general, changes in fluorescence are more readily detected than small changes in absorption. Furthermore, fluorescence is much more sensitive to the environment than absorption, in part because it can last 10^{-9} and even 10^{-8} s in contrast with the 10^{-15} s of the absorption process. A number of fluorescent molecules have been used as probes of physical and structural properties of the environment, including polarity, fluidity, order, molecular mobility, pH, electric potential, redox state, site accessibility, deformation due to intercalating agents, distances, visualization of membranes, proteins, and DNA in living cells (Valeur 1993). The applications of fluorescent probes have been reported many times, for example in the study of biological membranes by Hilderson (1988), of proteins by Szabo (1989), of nucleic acids by Le Pecq (1975) and Steiner and Kubota (1983), and of living cells by Wang and Taylor (1989).

2.1.8 *Acidities and redox potentials of excited states*

2.1.8.1 *Acid–base properties of excited states*

We noted in the previous section how the dipole moment may be changed on excitation. Other modified parameters include the shape of the molecule and its acid–base properties. Electronic transitions alter the charge distribution in a molecule and thus can lead to a modification of the acid–base equilibrium constant, K_a. In the case of phenol, the pK_a is 10.0 and 5.7 in the ground and singlet excited state, respectively.

The acid–base equilibrium between the excited phenol $(\Phi OH)^*$ and its conjugate base $(\Phi O^-)^*$ can be written as:

$$\begin{array}{ccccc} \Phi OH \xrightarrow{h\nu} & \Phi OH^* & \rightleftharpoons & \Phi O^{-*} & + H^+ \\ & \swarrow & & \downarrow & \\ \Phi OH + h\nu' & & & \Phi O^- + h\nu'' & \end{array}$$

The pK_a values of singlet excited states may be measured by spectrofluorimetry, when both ΦOH^* and ΦO^{-*} are fluorescent with distinct emission spectra.

Additionally, a thermodynamic calculation of pK_a may be used, known as the Förster–Weller cycle. Overall the increased acidity of singlet excited phenols is due to electron transfer from the oxygen to the aromatic ring, so that the oxygen has a more positive charge in the excited state. In the same way pK_a^* values for triplet excited states have been determined from phosphorescence studies using flash photolysis techniques. Often the pK_a values of triplet states are similar to the ground state.

Classical examples of acid–base properties of excited states have been reviewed by Vander Donckt (1970).

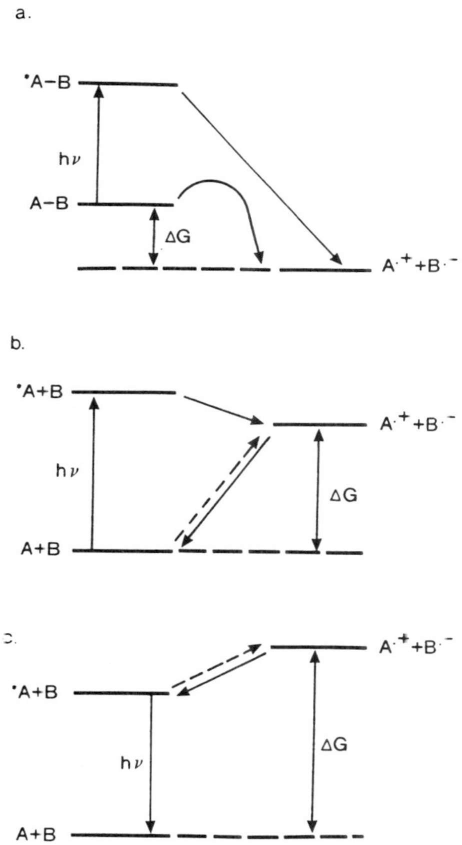

Fig. 2.6 Energy levels and electron transfer processes between molecules A* and B: (a) photocatalysis; (b) conversion of light into chemical energy, and between ions $B^{\cdot -}$ and $A^{\cdot +}$; (c) conversion of chemical energy into light (taken from Balzani and Scandola 1981).

2.1.8.2 *Redox potentials of excited states*

The excited states have an extra free-energy content which is approximately equal to the energy difference between the excited A* and ground state A molecules lying at their zero vibrational level, the 0–0 spectroscopic energy, E^{00} (A*, A). Thus, the redox potentials of A* are:

$$E^0(A^{\cdot+}/A^*) = E^0(A^{\cdot+}/A) - E^{00}(A^*, A)$$

and

$$E^0(A^*/A^{\cdot-}) = E^0(A/A^{\cdot-}) + E^{00}(A^*, A)$$

Excited states are more easily oxidized and more easily reduced than ground state molecules. They are important redox reactants as shown in Section 2.2.4 and Fig. 2.6.

2.1.9 *Electronic energy transfer*

The exchange of electronic energy between an excited donor and an acceptor was first demonstrated by Cario and Franck (1923) using mixtures of mercury and thallium vapours. In electronic energy transfer, an excited donor molecule D* is deactivated to its ground state D, while transferring its electronic excitation energy to an acceptor molecule A via:

$$D^* + A \rightarrow D + A^*$$

Two major types of mechanism can be distinguished:

1. The radiative 'trivial' electronic energy transfer which involves an emission of the donor (D) and the subsequent absorption of the emitted photon by the acceptor (A):

$$D^* \rightarrow D + h\nu$$
$$A + h\nu \rightarrow A^*$$

   It follows a dependence on donor–acceptor distance r proportional to r^{-2}. The most important example of this type of transfer is of course light absorption on Earth of the emission from the Sun.

2. The non-radiative energy transfer which can be either a long range transfer (several nanometres, up to ≈ 10 times the collision diameter) by dipole–dipole coupling interactions described by Förster (1948, 1959), or a short range transfer (collision transfer by electron exchange), described by Dexter (1953).

2.1.9.1 *Selection rules*

In the case of the dipole–dipole coupling, also known as 'coulombic interaction', the spin selection rules state that no changes of spin in either D or A are allowed, and that the multiplicities of D* and A* must be equal to those of D and A, e.g. the transfer:

$$^1D^* + {}^1A \rightarrow {}^1D + {}^1A^*$$

is fully allowed.

For an exchange mechanism (Dexter), the D, A system follows Wigner's spin correlation rule which states that if S_1 and S_2 are the initial total spins of the electrons of the colliding species, the resultant total spins of the two systems taken together must have one of the values $|S_1 + S_2|$, $|S_1 + S_2 - 1|$, $|S_1 + S_2 - 2| \ldots |S_1 - S_2|$. It follows therefore that the total spins of the resulting species can only have values S_3 and S_4 if at least one of the numbers $|S_3 + S_4|$, $|S_3 + S_4 - 1| \ldots |S_3 - S_4|$ is common to the series above. This is easier to understand by considering some simple examples.

In the electron exchange mechanism the main spin allowed transitions are: singlet–singlet energy transfer

$$^1D^*|S_1 = 0| + {}^1A|S_2 = 0| \rightarrow {}^1D|S_3 = 0| + {}^1A^*|S_4 = 0|$$

(total spin of colliding species and of resulting species have both a value of 0) and triplet–triplet energy transfer

$$^3D^*|S_1 = 1 + {}^1A|S_2 = 0| \rightarrow {}^1D|S_3 = 0| + {}^3A^*|S_4 = 1|$$

(total spin of colliding species and of resulting species is 1), whereas the process

$$^3D^*|S_1 = 1| + {}^1A|S_2 = 0| \rightarrow {}^1D|S_3 = 0| + {}^1A^*|S_4 = 0|$$

(total spin of colliding species is 1 and that of resulting species is 0) is not allowed.

Energy pooling can occur when the acceptor is also an excited state:

$$^3D^*|S_1 = 1| + {}^3D^*|S_2 = 1| \rightarrow {}^1D|S_3 = 0| + {}^1D^*|S_4 = 0|$$

(total spin of colliding species can be 2, 1, and 0 and that of resulting species 0). This last reaction is known as triplet–triplet annihilation, and as noted above, fluorescence from $^1D^*$ is known as 'P-type' delayed fluorescence (C. A. Parker, 1964) whose emission intensity is proportional to the square of the incident intensity. This distinguishes it from E-type delayed fluorescence associated with thermal repopulation of S_1 state from T_1 state, which is directly proportional to the incident intensity. Allowed excited state interactions of large polyatomic molecules with ground state molecular oxygen, which is a triplet state O_2 ($^3\Sigma_g^-$), are often associated with the formation of a singlet excited state of oxygen O_2 ($^1\Delta_g$) (often called 'singlet oxygen'). Provided the energy difference between S_1 and T_1 of the donor exceeds 7786 cm^{-1}, which is the energy difference between $O_2(^3\Sigma_g^-)$ and $O_2(^1\Sigma_g)$, an allowed energy transfer process leads to the formation of singlet oxygen (Stevens and Algar 1968):

$$^3D^* + O_2(^3\Sigma_g^-) \rightarrow {}^1D + O_2(^1\Delta_g)$$

Interaction of a singlet excited state, $^1D^*$, with ground state oxygen leads to a process known as oxygen-induced or oxygen-catalysed intersystem crossing (Algar and Stevens 1970; Potashnik *et al.* 1971):

$$^1D^* + O_2(^3\Sigma_g^-) \rightarrow {}^3D^* + O_2(^3\Sigma_g^-)$$

These reactions involving oxygen are more extensively discussed in Chapter 4. We now consider the two non-radiative modes of energy transfer.

2.1.9.2 *Coulombic interaction (Förster mechanism)*

As noted above, this process can take place when the donor and acceptor are separated by more than their collisional diameters. Of course, all energy transfer processes must compete against the spontaneous decay (emission) of the donor excited state. For polycyclic aromatics such as anthracene, in solvents such as benzene, equal rates arise at solute concentrations corresponding to a distance (r) of ~ 5 nm between donor and acceptor.

Förster has shown that the rate constant of energy transfer by this mechanism is related to r^6 and the donor lifetime, τ_D, by the relation:

$$k_{ET} = \frac{1}{\tau_D} \times \left(\frac{r_o}{r}\right)^6$$

where r_o is known as the critical transfer distance at which the transfer efficiency equals that for spontaneous decay.

Except for low concentrations ($\leqslant 10^{-4}$ mol dm^{-3}) requiring the diffusion of the molecules to reach the critical distance r_o, k_{ET} is not diffusion controlled and values > 10 times that of the diffusion controlled rate constant can arise. An interesting example of the Förster mechanism is the energy transfer between chlorophyll molecules in the photosynthetic antenna complexes (see Chapter 7).

Also, based on the Förster mechanism, there have been many energy transfer studies using fluorescence techniques. These have allowed distances between moieties (1–8 nm) within biological macromolecules such as immunoglobins, oligopeptides, and rhodopsin to be measured.

2.1.9.3 *Electron exchange interaction (Dexter mechanism)*

Exchange interaction occurs by collision producing a spatial overlap of the donor and acceptor wave functions. For a vibrationally relaxed system, Dexter calculated k_{ET} for an exchange interaction:

$$k_{ET} = KJ \exp(-2r/L)$$

where K is related to the specific orbital interactions, J is a spectral overlap integral normalized for the absorption coefficient of the acceptor, r is the distance between D* and A, and L is a constant. The energy transfer occurs in a collision complex where electrons are exchanged, provided the Wigner spin correlation rule is obeyed as described above.

It should be noted that the process:

$$^3D^* + A \rightarrow {}^1D + {}^3A^*$$

is doubly forbidden by spin selection rules of the Förster mechanism, the multiplicities (M) being such that $M_D^* \neq M_D$ and $M_A \neq M_A^*$, but this process is allowed by Wigner's rule and thus can only occur by an exchange interaction. The rate constant will be near the diffusion controlled limit ($\sim 10^{10}$ dm^3 mol^{-1} s^{-1}) for most common solvents at room temperature if the triplet energy levels of the donor and the acceptor are separated by more than 12 kJ mol^{-1} and no back energy transfer is possible. In fact, by measuring this rate constant for a range of donor molecules of known triplet energy level, it is possible to estimate the triplet energy level of the acceptor (corresponding to the transfer rate constant falling well below the diffusion limit (Bäckstrom and Sandros 1958)).

2.2 Elementary chemical reactions of excited states

Some typical reactions are described below.

2.2.1 *Dissociative excited states*

A molecule AB can be dissociated by transitions which produce an excited state with a vibrational energy above the energy required to dissociate the molecule into A + B* (one of the fragments being in an excited state). After such a transition the molecule dissociates within the period of a single vibration.

If a stable vibrational level of an excited state α overlaps a dissociative excited state β, the molecule dissociates although the vibrational level of the state α lies below the dissociation limit of that state (see Fig. 2.1). This phenomenon, called predissociation, does not take place immediately, i.e. it does not take place in a time of a vibrational period as for normal photodissociation. It requires 10^2 or 10^3 or even more such periods and is thus a delayed dissociation. Such a process reduces the fluorescence yield and the lifetime of an excited molecule.

The natures of the photofragments A and B are related to the possible distributions created by a σ bond cleavage, and is discussed in detail by, for example, Michl (1972 *a,b,c*). One point that should be emphasized is that if diradicals or radical pairs of triplet character are formed, which should be able to recombine with a very low activation energy, they do not in fact recombine, because the formation of a σ bond requires the spins to be antiparallel. As the time for a triplet to singlet interconversion is long, these radicals diffuse out from their geminate cages. The probability of photodissociation of a molecule, in a condensed phase, is much less than expected if one compares bond energies and photochemical excitation energies. This is due to the barrier of solvent molecules surrounding the fragments, sometimes called the

Franck–Rabinowitsch cage (1934). The chance to escape the 'cage effect' is greater for small fragments such as H· atoms.

2.2.2 *Photoionization*

Molecules are photoionized in the gas phase by photons with an energy larger than ~7 eV. However, in polar solvents, photoionization occurs at energies lowered by the solvation energy and is thus below 5–6 eV. Two kinds of photoionization in solution can be distinguished (Mialocq 1988). The first corresponds to an intermolecular electronic transition such as that observed in a donor–acceptor complex where the acceptor can be the solvent. The second corresponds to an intramolecular electronic transition, and includes the case of the important biological amino acid tryptophan. The photoionization of tryptophan, tyrosine, and their simplest models, indole and phenol, respectively, are among the most studied, since these amino acids are endogenous chromophores absorbing in the near-UV region. The mechanisms of aromatic amino acid photoionization are still controversial and are discussed in Chapter 6.

There are many mechanisms of photoionization of organic molecules in solution which have been observed or proposed. Often they are contradictory, probably on account of the very different techniques used: steady state irradiation using N_2O to scavenge the solvated electron, pulsed lasers of different energy, power, and wavelength. The main mechanisms proposed are:

1. Monophotonic electron ejection occurring from the lowest relaxed excited singlet state or from a Franck–Condon singlet excited state, before its relaxation and before the solvent reorganization.
2. A biphotonic process where the first photon is absorbed either by the relaxed or non-relaxed singlet or by the triplet.

Photoionization studies performed on nucleic acid bases and aromatic amino acids are discussed in Chapters 5 and 6. In Section 2.5 the difficulty of observing photoionization in non-polar solvents is emphasized.

2.2.3 *Excimers and exciplexes*

The simple molecular deactivation described by the Jablonski diagram (Fig. 2.2), becomes more complex in the case of chromophores close enough to interact electronically. In such cases, an excited dimer $^1D^*$ (so-called excimer), can be formed by interaction between a singlet excited molecule $^1M^*$ and an unexcited one: $^1M^* + {}^1M \rightleftarrows {}^1D^*$. This excimer has a dissociative ground state, its fluorescence being structureless and at lower energies than the parent monomer. When the excited state interaction occurs between two different molecules, a transient complex, called exciplex, can be produced, with a red-shifted emission. Most of these excited complexes are stabilized by charge transfer interactions (Förster 1975).

Exciplex and excimer emissions have been observed in many di- and polynucleotides, with nanosecond lifetimes in contrast with the picosecond lifetimes of the monomers. These longer-lived excited states are considered as the precursors of the photoadducts produced within polynucleotides (Cadet and Vigny 1990).

2.2.4 *Redox reactions*

Three fundamental redox reactions can be distinguished (Fig. 2.6):

(a) photocatalytic reactions where excited states are used in order to increase the efficiency or the selectivity of thermodynamically feasible reactions;

(b) conversion of light into chemical energy in the form of charge separation as in photosynthesis; and

(c) conversion of chemical energy into light, which in bioluminescence is a very efficient process (McCapra 1990). For example, bioluminescence by the firefly can reach a 90 per cent quantum yield (Seliger and McElroy 1960).

2.2.5 *Photo-oxidation*

Photo-oxidative reactions of biomolecules such as membrane lipids (Chapter 4), nucleic acids (Chapter 5), and proteins (Chapter 6) contribute to the molecular basis of the damage produced by UV and visible light on skin (Chapter 9). The definition of the Type I and Type II photosensitized oxidations has been controversial (Foote 1991). These processes are produced via a photosensitizer in its singlet excited state or, more generally, in its triplet excited state (Sens*):

Type I

Sens* + substrate or solvent → neutral radicals or radical ions which can react with oxygen to produce oxygenated products

Type II

$$\text{Sens}^* + O_2 \rightarrow \text{Sens} + {}^1O_2$$

followed by reactions of this activated oxygen species with substrate leading to oxygenated products (Foote 1991). The latter reference also proposes that the reaction:

$$\text{Sens}^* + O_2 \rightarrow \text{Sens}^{\cdot +} + O_2^{\cdot -}$$

is also a Type II mechanism. This proposition is not appropriate since radical anions formed in Type I processes can often react with oxygen to form $O_2^{\cdot -}$.

The term Type II should be exclusively reserved for energy transfer to produce singlet oxygen, thus avoiding this confusion.

2.2.6 *Photocycloaddition*

A cycloaddition produces a ring by the addition of m atoms of one molecule with the n atoms of a second molecule, and is called a $m + n$ cycloaddition. Most photocycloaddition reactions involve intermediates such as diradicals, zwitterions or exciplexes.

Some photocycloadditions are concerted reactions. They obey selection rules and are stereospecific, i.e. producing one configurational isomer of the cycloadduct (Woodward and Hoffman 1970). It is possible to distinguish suprafacial additions corresponding to both new bonds being formed on the same face of the original bond and antarafacial addition corresponding to new bonds being formed on the opposite faces of the bond. For π bonds, these terms are *syn* (suprafacial) addition and *anti* (antarafacial) addition. Photodimerization of thymine, and photoaddition of a thymine with furocoumarin, are examples of formation of cyclobutane rings with different isomeric forms (see Chapters 5 and 9).

Formation of endoperoxides by a Diels–Alder reaction of dienes with $^1O_2$ is an example of a supra–supra cycloaddition (Chapter 4).

2.2.7 *Photoisomerization*

Photochemical *cis–trans* isomerization is one of the most important rearrangements. Its mechanism can be either direct or sensitized, from the lowest excited singlet or triplet states, as shown for example in Fig. 2.7. In the case of triplet photosensitized *cis–trans* isomerization of stilbenes, Hammond and Saltiel (1963), Hammond *et al.* (1964), Herkstroeter and Hammond (1966), and Saltiel *et al.* (1984) have observed that rate constants for triplet energy transfer from a series of triplet donors remained abnormally high for triplet donors with energy levels below that of the *cis* acceptor by about 40 kJ mol^{-1}. This result was explained as due to a 'non-vertical' transition to a twisted form of the *cis* triplet state of stilbene which was called a 'phantom' triplet. An interesting explanation of the 'non-vertical' energy transfer has been the 'hot band' model of Ramamurthy and Liu (1976). These authors have determined the photostationary state composition of β-ionol as a function of sensitizer excitation energy as shown in Fig. 2.8. For this polyene, as with the stilbenes, and other olefinic systems, decay to the *cis* is slightly favoured, resulting in the photostationary state containing > 50 per cent of the *cis* isomer (63 per cent in this case). The shape of the curve (Fig. 2.8) implies that the triplet energy of *cis*-ionol is 309 kJ mol^{-1}. Sensitizers with triplet energy levels (E_T) below this value should transfer to *cis* in an inefficient manner. As a consequence, a 'classical' energy transfer should have given a photostationary state with an exclusive presence

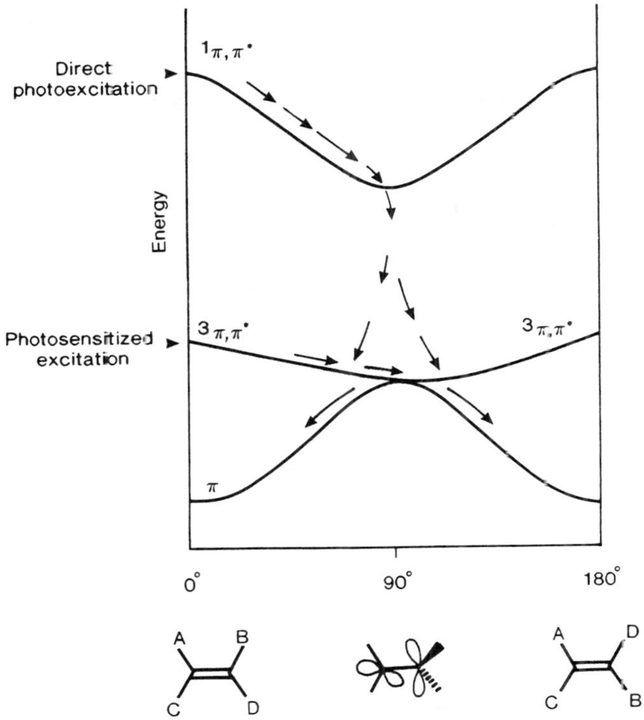

Fig. 2.7 Potential energy diagram for twisting around a C=C double bond.

of *cis* isomer. Fig. 2.8 shows a 'non-classical' behaviour of endothermic efficient energy transfer to *cis*-ionol between 309 and 267 kJ mol^{-1}. The model of Ramamurthy and Liu (1976), which explains such a 'non-classical behaviour' is illustrated by Fig. 2.9 which shows the ground and excited triplet torsional potential curves around the 7,8 double bond of ionol. The so-called 'non-vertical' excitation transfer is the result of an excitation transfer to *cis*-ionol acceptor molecules thermally excited along the relevant distortion coordinate as expected from the Boltzmann distribution law. Figure 2.9 shows how endothermic energy transfer to *cis* involves vertical excitation of vibrationally excited ground state molecules. This vertical excitation from a thermally activated structural deformation (torsional motion) in the ground state G prior to vertical energy transfer lowers the polyene triplet excitation energy by an amount ΔE_{G}^{trans} and simultaneously the excited triplet potential curve by an amount ΔE_{T}^{trans}. Thus, this model describing endothermic energy transfer to *cis* via vertical excitation, obeys the Franck–Condon principle. Similar behaviour has also been found for the photosensitized isomerization of another flexible polyene, retinal (Jensen *et al.* 1989).

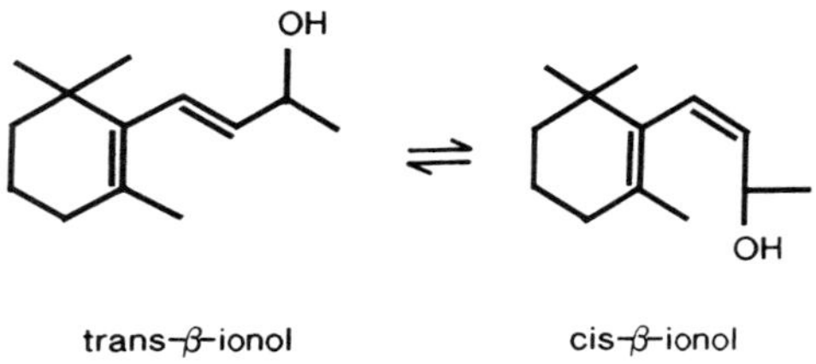

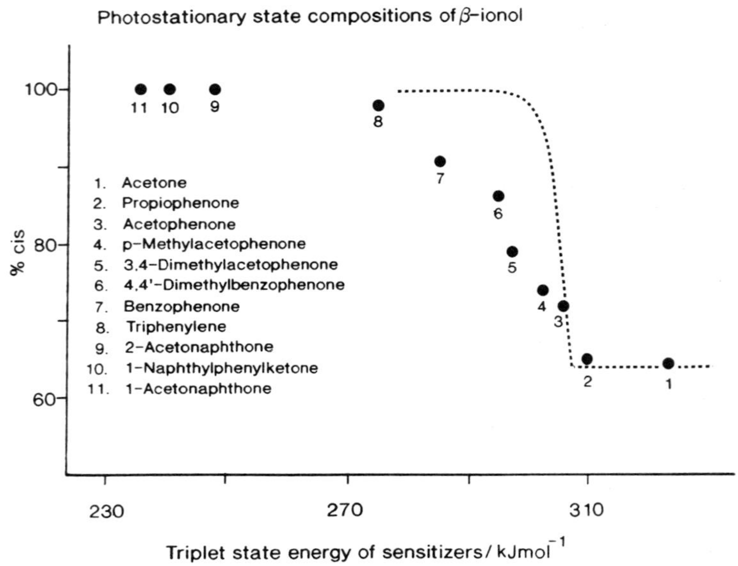

Fig. 2.8 Photostationary state composition of β-ionol as a function of sensitizer triplet state energy; ●, experimental results. The dashed line corresponds to values predicted by a classical endothermic energy transfer process. (adapted from Ramamurthy and Liu 1976).

The most studied photoisomerizations occurring in biological molecules are the 11-*cis* → all-*trans* torsional isomerization of the retinal prosthetic group of rhodopsin during the primary step in vision (Schoenlein *et al.* 1991), and the bilirubin system (see Section 9.8.1).

2.2.8 *Photorearrangement*

Vitamin D_3 in mammalian skin is photoproduced via an electrocyclic (Woodward and Hoffman 1970) opening of 7-dehydrocholesterol upon UV irradiation to yield a 6,7-*cis*-hexatriene derivative or previtamin D_3. This compound gives vitamin D_3 via thermal isomerization and can also under irradia-

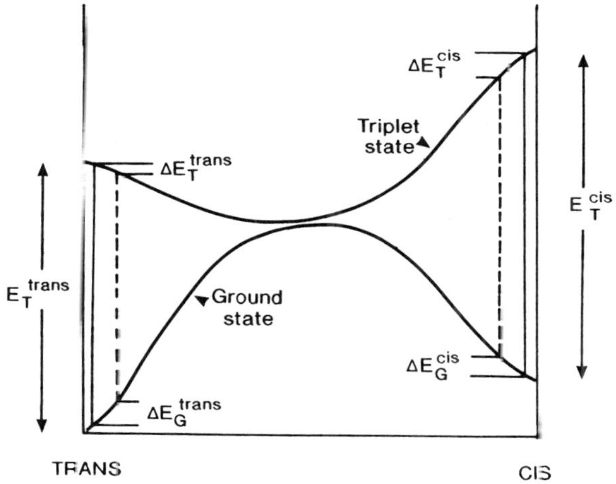

Fig. 2.9 Idealized torsional potential curves of the ground and the triplet states of *trans*- and *cis*-β-ionol. Excitations involving exothermic (solid lines) and endothermic (dashed lines) energy transfers are depicted (adapted from Ramamurthy and Liu 1976).

tion reform the parent compound via ring closure or its stereoisomer, lumisterol, or via isomerization, form tachysterol. A quasi-photoequilibrium is reached (Holick *et al.* 1982).

2.3 Nature and reactivity of radicals

2.3.1 *Historical introduction*

At the end of the eighteenth century Antoine-Laurent Lavoisier and many other chemists described radicals as part of molecules, as groups of atoms with an independent existence. Thus Joseph-Louis Gay-Lussac (1815), at the Muséum National d'Histoire Naturelle in Paris, observed that the cyano radical CN constituted an indivisible entity. He obtained cyanogen, the dimer of the cyano radical, by heating mercuric cyanide. However, in search of the molecular weight (molar mass, *M*) of radicals by the vapour density method, chemists found the molecular weight of the dimers of the radical; for example, ethane ($M = 30$) instead of that of the methyl radical. In 1896, Ostwald stated 'The very nature of organic radicals is such as to preclude the possibility of isolating them' (quoted by Pryor 1976). Two experiments by Gomberg (1900) and Paneth and Hofeditz (1929) changed these ideas. Gomberg, attempting to prepare hexaphenylethane from solutions of triphenylmethylchloride treated with silver or zinc powders, obtained a yellow solution:

$$(C_6H_5)_3C{-}Cl + Ag \rightarrow (C_6H_5)_3C^{\cdot} + AgCl$$

This yellow solution was rapidly decolourized by a number of materials including air (oxygen) and iodine:

$$(C_6H_5)_3C^{\cdot} + I_2 \rightarrow (C_6H_5)_3CI + I^{\cdot}$$

Yellow Colourless

Gomberg correctly concluded that the decolouration was due to the reaction of a coloured sterically hindered triphenylmethyl radical.

The next step in the observations of radicals as transient species with a measurable lifetime was elegantly performed by Paneth and Hofeditz (1929). Tetramethyl lead, $(CH_3)_4Pb$, in a current of nitrogen, was heated by a movable furnace and a lead mirror deposited. Afterwards the tube was heated at a second position. A new mirror was deposited while the original mirror was removed by the methyl radicals $CH_3^{\cdot}$ whose half-lives were calculated as being approximately 10^{-3} s. These reactions were the starting point of a large number of studies on 'free' radical reactions.

After the demonstration of the existence of free radicals in organic chemistry, free radicals in biology had an initially very controversial history, because of difficulties in their experimental detection *in vivo*. Since the first ESR studies of Commoner *et al.* (1954), it has been unambiguously demonstrated that free radicals are present in a number of biological processes (Halliwell and Gutteridge 1989).

The evidence, direct and indirect, of free radicals in chemistry and biology, as well as a large number of their spectroscopic and chemical properties, have been mainly obtained from various techniques, e.g. electron paramagnetic resonance, flash photolysis, pulse radiolysis, and various trapping methods. These techniques, and particularly flash photolysis and pulse radiolysis, are discussed in Chapter 3.

2.3.2 *Paramagnetic properties*

When a substance is placed in a magnetic field of strength B (more correctly called induction), the intensity of magnetization per unit volume M is given by $M = \chi B$ where χ is the magnetic susceptibility of magnetization per unit volume. This susceptibility can be either positive and the substance is called paramagnetic, or negative and the substance is called diamagnetic. Some materials, such as iron and nickel, have extremely high positive susceptibilities; they are called ferromagnetic. Most molecules are diamagnetic. However, for a species with an unpaired electron, the magnetic susceptibility is positive and much higher than the diamagnetic term for typical organic molecules, and thus it appears that paramagnetism is a unique property of free radicals. Zavoisky (1945) measured the paramagnetism due to an unpaired electron by Electron Spin Resonance (ESR).

An electron has a spin, with an associated magnetic moment, having two possible orientations corresponding to two values of the spin angular momentum $m_s = \pm 1/2$. In a magnetic field of strength B, the single energy level of the electron is split into two different levels separated by $\Delta E = g\mu_B B$, with μ_B being a conversion factor called the Bohr magneton ($\mu_B = 9.274 \times 10^{-24}$ JT^{-1}), and g a proportionality constant called the gyromagnetic ratio. For a free electron, g has the value of 2.0023. In organic radicals, g varies from 2.002 to 2.006.

Transitions between the two levels occur with radiation of frequency $\nu = g\mu_B B/h$, where h is Planck's constant, B is measured in Tesla, and ν is expressed in Hertz (Hz). The technique of ESR is the study of unpaired electrons of radicals by the determination of the magnetic fields for which a resonance is observed. For technical reasons, fields of about 0.3 Tesla (3000 Gauss) are used with an electromagnetic field of frequency 10 GHz (10^{10} Hz), and wavelength 3 cm which is in the microwave region (see Fig. 1.1). The ESR spectrum is observed by monitoring the microwave absorption as the field is modified. The display of the spectrum is, in fact, usually given as its first derivative. Most of the free radicals give spectra with more than one line. A hyperfine structure of the spectrum results from the interaction of the unpaired electron with the magnetic moments of the neighbouring atoms. A nucleus splits the spectrum into $(2I + 1)$ lines, where I is the nuclear spin. Thus, a hydrogen atom has two lines and a methyl radical has four lines as $I = 1/2$ for the proton. With n equivalent protons there are $n + 1$ hyperfine lines with a binomial intensity, given by Pascal's triangle:

1:2:1	for 2 protons
1:3:3:1	for 3 protons
1:4:6:4:1	for 4 protons, and so on. . . .

ESR is a uniquely powerful method for the determination of radical structures as discussed in Chapter 3.

2.3.3 *Examples of elementary radical reactions*

Pulse radiolysis is able to generate specific radicals (see Chapter 3) and has thus facilitated the determination of rate constants of reactions of radicals and of their mechanisms. Representative rate constants for radical reactions of biologically important molecules have been reviewed by Neta and Ross (1988). Some examples of radical reactions are given below.

2.3.3.1 *Solvated electron reactions*

The solvated electron (e^-_{aq}) is the simplest radical. Its redox potential in water has been estimated to be $E^\circ = -2.9$ V versus NHE. The rate constants for its reaction with thousands of inorganic and organic compounds in aqueous solution have been compiled and critically reviewed by Buxton *et al.* (1988). Typical reactions are:

(1) addition of a single electron: for example, the valence of a metal is thus reduced by one;

(2) removal of a proton

$$RCOOH + e^-_{aq} \rightarrow RCOO^- + H^\cdot$$

(3) addition followed by protonation

$$>C=C< + e^-_{aq} \rightarrow >C^- - C^\cdot<$$

$$>C^- - C^\cdot< + H^+ \rightarrow >CH - C^\cdot<$$

(4) dissociative electron attachment

$$RI + e^-_{aq} \rightarrow R^\cdot + I^-$$

2.3.3.2 *Abstraction of hydrogen*

The enthalpy of reaction for hydrogen abstraction ΔH, is equal to the difference in the dissociation energy of the bond being broken, D_B, and the dissociation energy of the bond being formed, D_F: $\Delta H = D_B - D_F$. The energy necessary to break a C—H bond is very high (approximately 415 kJ mol^{-1}). By comparison, it is interesting to note that the S—H bond is more labile than a C—H bond, with a dissociation energy of approximately 259 kJ mol^{-1}, and thus a thiol R′SH should be able to easily repair a carbon centred free radical R$^\cdot$:

$$R'SH + R^\cdot \underset{k_{21}}{\overset{k_{12}}{\rightleftharpoons}} R'S^\cdot + RH$$

However, thiyl radicals (RS$^\cdot$), considered in the past as biologically harmless, are oxidants with redox potential in the order of + 0.6–0.8 V versus NHE (Surdhar and Armstrong 1986). They may diffuse *in vivo* and cause pathological reactions at other sites. Indeed, the relative values of k_{21} have been measured for a series of hydrogen donors including alcohols, ethers, amines, amides, sugars, and other components of biologically relevant substrates (Schöneich *et al.* 1991). Thiols can either enhance or protect against radical damage. For instance, at low cysteine concentration, the equilibrium below shifts to the right, causing inactivation of the enzyme papain; at high thiol concentration it will shift to the left, resulting in protection:

$$\text{papain-SH} + CyS^\cdot \rightleftharpoons \text{papain-S}^\cdot + CySH$$

This will be discussed further in Chapter 10.

2.4 Electron transfer—Marcus theory

Electron transfer reactions (ET, either intermolecular between separated molecules, or intramolecular from one site to another of covalently linked donor and acceptor groups) are processes of great chemical and biochemical interest. Biological conversion and storage of energy are dependent on electron transfer reactions. Major examples are:

1. In photosynthesis, where electrons, produced from electronically excited bacteriochlorophylls or chlorophylls, are transferred by fast and efficient steps to a series of electron transport molecules and produce high energy redox products storing the light energy.
2. In the dioxygen metabolism where electrons are transferred via a multienzyme system called the 'respiratory chain'.

Theoretical descriptions of ET have been developed along a classical approach of the transition state theory (Marcus 1964, 1965, 1982; Marcus and Sutin 1985; Nadler and Marcus 1987), and a quantum mechanical approach (Levich 1965). Extensive work has been performed in organic and bioinorganic chemistry (Taube and Gould 1969; Taube 1970, 1984; Hush 1982) and has been extended to photoexcited organic molecules (Rehm and Weller 1970), photoexcited metal complexes (Sutin 1973; Balzani *et al.* 1978, 1979), covalently linked porphyrin–quinone systems (review of over a hundred porphyrin–quinone systems by Connolly and Bolton 1988) and to proteins (Gray 1986; Mayo, S. L. *et al.* 1986; McLendon *et al.* 1986; McLendon 1988; Palmer 1991; J. S. Zhou and Rodgers 1991).

Marcus theory, as well as related theories of electron transfer, is based upon a relationship between the standard free energy of the reaction, $\Delta G^{\ominus}$, and the energy of activation of the reaction $\Delta G^{\ominus\neq}$. It can also be expressed as a relationship of the rate constant of electron transfer, k_{ET}, to both $\Delta G^{\ominus}$ and $\Delta G^{\ominus\neq}$.

The aim of the present section is to describe in a simple way:

(1) the Marcus model of ET which predicts that the rate constant of ET should rise with increasing driving force, $-\Delta G^{\ominus}$, in the endergonic, so called 'normal' free energy region, but should decrease as $\Delta G^{\ominus}$ becomes more negative in the so-called exergonic, 'inverted region';

(2) the Rehm and Weller (1970) empirical relationship, dominated by diffusional processes, and at variance with the Marcus model; and

(3) the Miller and coworkers study which confirmed, after almost three decades, the Marcus prediction of an 'inverted region' for covalently linked donor–acceptor systems.

2.4.1 *The kinetic expression of an ET reaction*

As with most theories of absolute rate, the Marcus theory is based on Eyring's transition state theory where the rate constant, k, for a reaction is given by:

$$k = \kappa Z \exp - (\Delta G^{\ominus\neq}/RT)$$

where κ is a transmission coefficient related to the probability that the reaction will occur, Z is a universal frequency factor ($Z = k_B T/h$, k_B and h being the

Boltzmann and Planck constants, respectively) equal to $6.215 \times 10^{12}\,s^{-1}$ at 298 K; and $\Delta G^{\ominus\neq}$ is the standard free energy difference between the initial state and the transition state. In the original hypothesis of Marcus, κ is very near to unity, which means that the ET should be adiabatic. In an adiabatic reaction, there is a weak interaction between the reactants in the transition state with an upper limit of $\sim 4\,kJ\,mol^{-1}$. In the adiabatic limit the frequency factor is proportional to the inverse longitudinal dielectric relaxation time, i.e. $Z \alpha \tau_L^{-1}$, where $\tau_L = (\varepsilon_\infty/\varepsilon_s)\,\tau_D$, with ε_∞ and ε_s being the optical and static relative permittivities, respectively, while τ_D is the Debye dielectric relaxation time of the solvent. This indicates the importance of solvation dynamics for the rate of electron transfer reactions. New theoretical models using a distribution of relaxation times, τ, have been reviewed by Barbara and Jarzeba (1990). Outside the weakly adiabatic condition of Marcus theory, quantum mechanical methods are necessary to calculate ET rate constants (Larson 1981; Siders and Marcus 1981).

The scheme of the electron transfer between a donor molecule, D, and an acceptor molecule, A, can be represented in the simplest case by the 'collide and react' model or by a more complex model proposed initially by Rehm and Weller (1970) for the study of fluorescence quenching by electron transfer.

The collide and react model: The two reactants, D and A, diffuse together with a rate constant, k_{12}, to form an encounter complex, [D..A], which is in equilibrium with the reactants and which undergoes the ET reaction with a rate constant, k_{23}.

$$D + A \underset{k_{21}}{\overset{k_{12}}{\rightleftharpoons}} [D...A] \xrightarrow{k_{23}} P$$

Using the steady state approximation applied to the encounter complex [D...A], the observed rate constant k_q for the appearance of P or the disappearance of one of the reactants is:

$$k_q = \frac{k_{12}}{1 + k_{21}/k_{23}}$$

The kinetic scheme of Rehm and Weller: Interest in electron transfer was stimulated among photochemists by the work of Leonhardt and Weller (1963), who reported the formation of charge transfer complexes between the singlet excited state of perylene and the ground state of dimethylaniline, which were later called exciplexes (Förster 1975). Rehm and Weller (1970) showed a correlation of the fluorescence quenching rate constant of scores of typical electron donor–acceptor systems with the free energy change $\Delta G^{\ominus}$ involved in the actual electron transfer process.

In their scheme, the acceptor A is excited to its lowest excited singlet state $^1A^*$ in a polar medium. The fluorescence is quenched by an electron donor D to give, via a reversible reaction, an encounter complex $[^1A^*....D]$ in equilibrium with an ion pair $[^2A^{\cdot -}....{}^2D^{\cdot +}]$. This ion pair can back transfer or give products P which include either free ions or triplet and/or ground states:

$$A + D \underset{k_{21}}{\overset{k_{12}}{\rightleftharpoons}} [^1A^* \ldots . D] \underset{k_{32}}{\overset{k_{23}}{\rightleftharpoons}} [^2A^{\cdot -} \ldots . {}^2D^{\cdot +}] \xrightarrow{k_{30}} P$$

According to this scheme, and with the steady state approximation applied to the encounter complex $[^1A^* \ldots . D]$ and the ion pair $[^2A^{\cdot -} \ldots . {}^2D^{\cdot +}]$, the rate constant of quenching observed, k_q, is given by:

$$k_q = \frac{k_{12}}{1 + k_{21}/k_{23} + k_{21}/k_{30} \times 1/K_{23}}$$

The classical approach gives:

$$K_{23} = k_{23}/k_{32} = \exp(-\Delta G^{\ominus}_{23}/RT)$$

and

$$k_{23} = k^{\circ}_{23} \exp(-\Delta G^{\ominus\neq}_{23}/RT)$$

where k°_{23}, $\Delta G^{\ominus}_{23}$, and $\Delta G^{\ominus\neq}_{23}$ are the frequency factor, the standard free energy change, and the standard free activation energy of the electron transfer step. Using the above expressions for K_{23} and k_{23}, the equation for k_q becomes:

$$k_q = \frac{k_{12}}{1 + (k_{21}/k^0_{23}) \exp(\Delta G^{\ominus\neq}_{23}/RT) + (k_{21}/k_{30}) \exp(\Delta G^{\ominus}_{23}/RT)}$$

The values of the rate constants k_{12}, k_{21}, k_{30} and the frequency factor k^0_{23} can be considered as unchanged for the reaction of a given acceptor with a series of related donors or for a given donor with a series of related acceptors. Thus k_q depends only on $\Delta G^{\ominus}_{23}$ and $\Delta G^{\ominus\neq}_{23}$.

The kinetic expression is equally applicable for D being the excited state (singlet or triplet) instead of A or for a thermal electron transfer.

2.4.2 *The free energy change $\Delta G^{\ominus}$ of an ET reaction*

The free energy change $\Delta G^{\ominus}_{23}$ involved in the electron transfer process is given by

$$\Delta G^{\ominus}_{23} = F[E(D^{\cdot +}/D) - E(A/A^{\cdot -})] - e^2/\varepsilon_r a - \Delta E^{00}$$

where F (Faraday) is 96.5 kJ mol^{-1}, $E(D^{\cdot +}/D)$ and $E(A/A^{\cdot -})$ are the oxidation potential of the donor and the reduction potential of the acceptor, and $e^2/\varepsilon_r a$ is the free energy gained by bringing together the two radical ions at the encounter distance a (0.7 nm) in a solvent of relative permittivity ε_r (also called the dielectric constant). In highly polar solvents, this coulombic attraction term becomes negligible: in water $e^2/\varepsilon_r a$ is 0.023 eV or 2.2 kJ mol^{-1} and in acetonitrile 0.06 eV or 5.8 kJ mol^{-1}. ΔE^{00} is the electronic excitation energy.

The electron transfer can be represented in the Marcus model as in Fig. 2.10. It occurs between the donor–acceptor precursor complex described by the D . . .^{1}A* parabolic potential energy and the ion pair, $D^{\cdot +}$ . . . $A^{\cdot -}$ successor complex described by a second isostructural parabolic potential. As noted above, the parabola of the precursor complex can represent a ^{1}D* . . . A or a D . . . A complex. These parabolic curves express a quadratic relation between the potential energy of the reaction versus the nuclear topology or, in other terms, the nuclear geometry related to bond lengths, bond angles, molecular orientation, interaction with solvent dipoles, etc. The $\Delta G^{\ominus}$ of the reaction is represented by the difference in energy between the minima of the D . . . ^{1}A* parabola and the $D^{\cdot +}$ . . . $A^{\cdot -}$ parabola. The free energy of activation $\Delta G^{\ominus\neq}$ is related to the intersection of the two parabola where the non-equilibrium nuclear arrangement is that of the transition state.

2.4.3 *Relationships between the free energy of activation $\Delta G^{\ominus\neq}$ and the free energy change $\Delta G^{\ominus}$*

To describe the ET, Marcus has derived the equation

$$\Delta G^{\ominus\neq} = (\Delta G^{\ominus} + \lambda)^2/4\lambda$$

λ being a reorganization energy (with the $\Delta G^{\ominus}$ and the $\Delta G^{\ominus\neq}$ being equivalent to the $\Delta G^{\ominus}_{23}$ and $\Delta G^{\ominus\neq}_{23}$ used above). Such a relation has also been successfully applied to other transfers (proton, hydride, methyl, sulphuryl, phosphoryl; see Marcus and Sutin 1985).

Figure 2.10(a) shows the reorganization energy λ required to move from the D . . . A* parabola to the $D^{\cdot +}$ . . . $A^{\cdot -}$ parabola for a vertical transition at the same nuclear configuration (solvent and intramolecular). For isoenergetic parabola, as in Fig. 2.10(a), geometric arguments show that $\Delta G^{\ominus\neq}(0) = \lambda/4$ with $\Delta G^{\ominus} = 0$. $\Delta G^{\ominus\neq}(0)$ corresponds to an intrinsic barrier related to the internal and solvent nuclear rearrangements of the electron transfer.

It can be seen from Fig. 2.10(b) and (c) that as the driving energy $\Delta G^{\ominus}$ becomes more and more negative, $\Delta G^{\ominus\neq}$ decreases and the rate increases to a maximum at $\Delta G^{\ominus\neq} = 0$ when $-\Delta G^{\ominus} = \lambda$. Then the rate decreases when $\Delta G^{\ominus}$ becomes even more negative (Fig. 2.10(d)). For almost 30 years no experimental evidence was found to follow the Marcus prediction illustrated by Fig. 2.10(d). However this rather surprising prediction, the decrease in electron transfer rate with increasing reaction exergonicity in the so-called Marcus 'inverted region', has recently been observed (see later).

If we replace λ by $4\Delta G^{\ominus\neq}(0)$, the Marcus equation becomes:

$$\Delta G^{\ominus\neq} = \Delta G^{\ominus\neq}(0)\{\Delta G^{\ominus}/4\Delta G^{\ominus\neq}(0) + 1\}^2$$

From a series of experimental results, dealing with the fluorescence quenching observed for a large number of excited acceptor–donor systems in acetonitrile, with $\Delta G^{\ominus}_{23}$ varying between -104 and $+25$ kJ mol^{-1}, Rehm and Weller (1970)

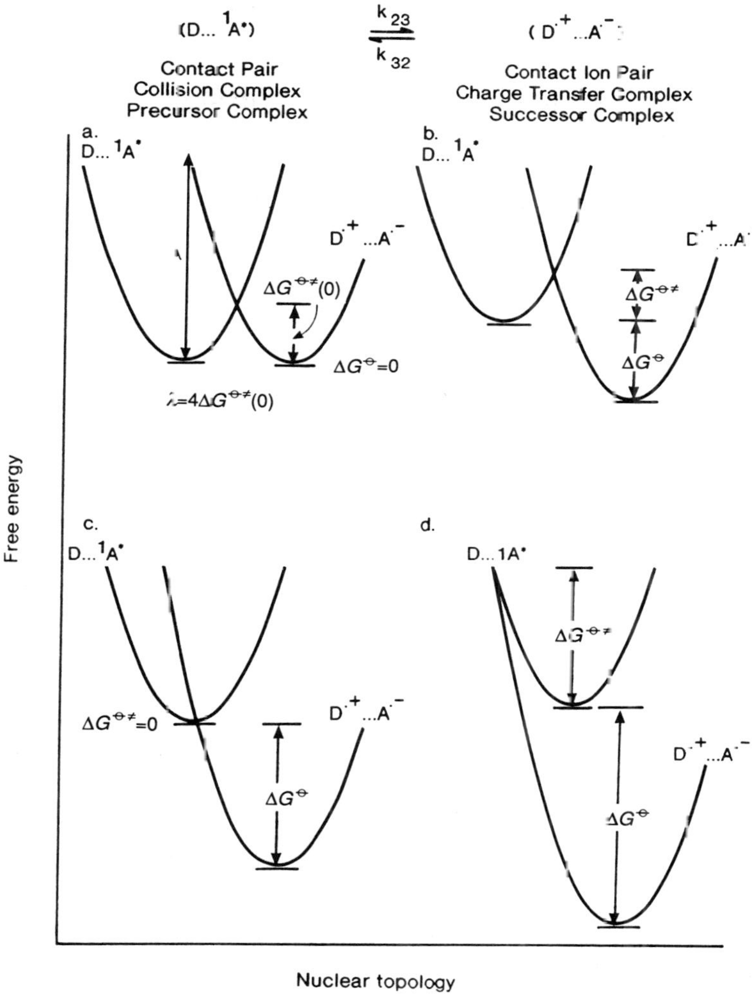

Fig. 2.10 Potential energy diagrams depicting a photosensitized electron transfer between a collision complex (D . . . $^1A^*$) and a charge transfer complex ($D^{\cdot +}$ . . . $A^{\cdot -}$) (a) for $\Delta G^{\ominus} = 0$, $\Delta G^{\ominus\neq}(0) = \lambda/4$, (b) for $\Delta G^{\ominus} > -\lambda$, $\Delta G^{\ominus\neq} > 0$, (c) $\Delta G^{\ominus} = -\lambda$, $\Delta G^{\ominus\neq} = 0$, (d) for $\Delta G^{\ominus} < -\lambda$, $\Delta G^{\ominus\neq} > 0$ in the 'inverted' Marcus region. (Adapted with permission from Kavarnos and Turro, *Chem. Rev*, **86**, 401–49. Copyright 1986 American Chemical Society.)

proposed an empirical relationship relating $\Delta G^{\ominus\neq}$ and $\Delta G^{\ominus}$ for the step 23 of the scheme given above:

$$\Delta G^{\ominus\neq} = \Delta G^{\ominus}/2 + \{[\Delta G^{\ominus}/2]^2 + [\Delta G^{\ominus\neq}(0)]^2\}^{1/2}$$

With the empirical value of $\Delta G^{\ominus\neq}(0) = 10\,\text{kJ mol}^{-1}$, this empirical relationship fitted well with the Marcus prediction for the endergonic region but not

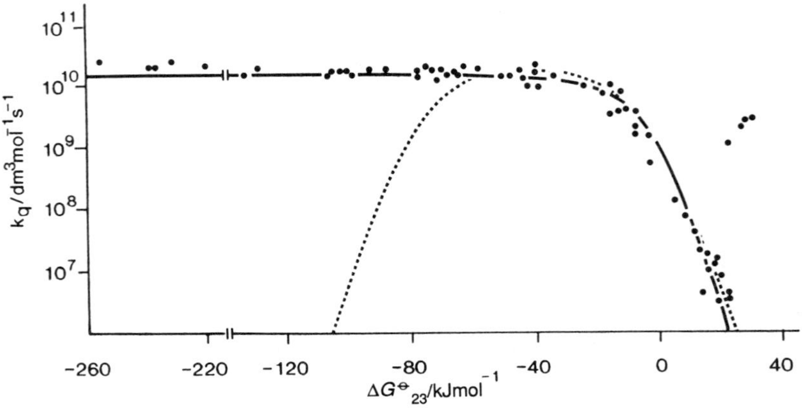

Fig. 2.11 Log of the fluorescence quenching rate constant k_q as a function of the $\Delta G^{\ominus}_{23}$ of the electron transfer process. The experimental points are from Rehm and Weller (1970), the experimental curve —— is calculated from the Rehm and Weller empirical expression of $\Delta G^{\ominus\neq}_{23}$, the curve ----- is calculated from the Marcus theoretical expression of $\Delta G^{\ominus\neq}_{23}$, both for an intrinsic energy barrier $\Delta G^{\ominus\neq}(0) \approx 10\,\text{kJ mol}^{-1}$ (adapted from Rehm and Weller 1970).

the inverted region as shown in Fig. 2.11. The plot of Rehm and Weller is characterized by a linear part for $\Delta G^{\ominus}_{23} > 0$ with a slope equal to $1/2.3\,RT$ and a horizontal part for very negative $\Delta G^{\ominus}_{23}$ values corresponding to the diffusion limit of the rate constant. Scandola *et al.* (1981) have proposed an expression using a relation established for atom transfer by Agmon and Levine (1977):

$$\Delta G^{\ominus\neq} = \Delta G^{\ominus} + \lambda/4\ln 2 \times \ln[1 + \exp(-4\,\Delta G^{\ominus}\ln 2/\lambda)]$$

which gives results similar to the Rehm and Weller empirical formula and has the advantage of better theoretical support.

2.4.4 *The problem of the 'inverted' region*

After the work of Rehm and Weller, other experimental investigations have correlated the rate of light-induced electron transfer with the free energy change, $\Delta G^{\ominus}$, in reactions of freely diffusing species. These results were still at variance with the inverted region of Marcus theory (see references in Rau *et al.* 1986; Legros *et al.* 1991).

However, a striking confirmation of the 'inverted region' was found by Miller *et al.* (1984) and Closs and Miller (1988) in the case of intramolecular electron transfer between a donor $D^{\cdot-}$ anion, generated by pulse radiolysis, and an acceptor A separated by a hydrocarbon spacer. Moreover, Wasielewski *et al.* (1985) reported that the rate constants for ion pair recombination in a series of restricted-distance porphyrin–quinone donor–acceptor molecules, displayed an inverted Marcus region. Other observations of the Marcus prediction for charge recombinations (CR) also include studies by Gould *et al.* (1987, 1988).

Kakitani and Mataga (1987) have described a model predicting a pronounced inverted region for CR reactions, no inverted region for charge separation (CS) reactions in which two neutral species yield two ionic species, and intermediate behaviour for charge shift (CSH) in which no net charge change takes place. However, Gould *et al.* (1989) observed a decrease in rate with increasing exothermicity for CR reactions between cyanoanthracene radical anions and alkylbenzene radical cations as well as for CSH reactions between *N*-methylacridinium radicals and alkylbenzene radical cations. Another example of a CS reaction showing the full range of behaviour predicted by Marcus theory is that of the radical pair generated in benzylic ester photochemistry (DeCosta and Pincock 1989). (For others references to Marcus theory see the special issue dedicated to Marcus of *J. Phys. Chem.*, 1986; **90**, 3454–3862; Kavarnos and Turro 1986; Eberson 1987; Pouliquen and Wintgens 1988; Fox 1990; Mattay 1991.)

Applications of the Marcus theory and the Rehm–Weller empirical relationship will be found in the analysis of singlet oxygen quenching by electron donors (Chapter 4), of repair of thymine dimers in DNA (Chapter 5), of long-distance electron transfer in proteins and model systems (Chapter 6), and photoinduced electron transfer in primary photosynthetic reactions and in model systems (Chapter 7).

2.5 Comparison of radiation chemistry and photochemistry

It is useful at this point to compare high energy radiation excitation with photochemical excitation.

2.5.1 *Absorption and spatial distribution of primary species*

The interaction of high energy radiation depends on the atomic composition and the number of electrons per unit mass and generally not the molecular structure. In radiation chemistry, the absorption is non-specific and in solution each component will absorb in proportion to its electron fraction. In a dilute solution, this means that virtually all of the energy is initially absorbed by the solvent rather than the solute. Thus, the direct effect from ionizing radiation is negligible compared to the indirect effect due to the interaction of the solute with the excited species, ions, and radicals formed in the solvent. In the absorption process, the high energy radiation successively undergoes energy loss processes so that one initial high energy photon or particle causes the excitation and ionization of many molecules. Furthermore, with high energy radiation, an inner electron can be ejected, followed by an Auger process (see Fig. 1.4) which produces a multiply charged ion. In contrast, visible or ultraviolet radiation can only eject valence electrons.

In photochemistry, UV and visible radiation are absorbed by the solute in a specific manner and normally as a single photon event. This leads to the initial production of only one solute excited state with the absorption process following the selection rules (Selection 2.1.3), the solvent being transparent. The units used for absorbed (dose) of energy are kJ mol^{-1} in photochemistry and J kg^{-1} (Gray, Gy) in radiation chemistry.

The distribution of primary species produced by high energy electrons along tracks, blobs, and spurs varies markedly both laterally and longitudinally and varies with the LET of the radiation (see Section 1.2.3). However, for radiations with energies of the order of 1 MeV, it is possible to consider that the distribution of the 'primary' species from water is homogeneous after 10^{-7} s. There is no intervention of a dilute solute with a 'primary' species formed in the tracks, blobs, and spurs before the diffusion of the 'primary' species to a homogeneous distribution. In the spur diffusion model for aqueous solutions, a scavenger S will compete with the spur reactions of e^-_{aq}, $H^{\cdot}$, and $OH^{\cdot}$ if the concentration of S multiplied by the second-order rate constant k for reaction with these radicals is greater than $10^7\,s^{-1}$.

On the other hand, the spatial distribution of excited species generated in a solution by a beam of exciting light decreases exponentially with the concentration of the absorbing solute, its molar absorption coefficient, and the thickness of the solution.

2.5.2 *The species produced*

The nature of the species, excited singlet or triplet states, neutral, cationic or anionic radicals, produced from solutes in solution are usually similar, although they result from an indirect effect for ionizing radiation and from a direct effect for UV and visible radiation. This similarity will be discussed in Chapter 3 where the two complementary techniques of pulse radiolysis and flash photolysis are described.

Because excitation of molecules via high energy radiation is not subject to selection rules, direct production of triplet states can arise when direct photoproduction of triplet states is spin forbidden (Section 2.1.3). An instructive example of the value of this is the pulse radiolysis study of carotenoid triplets (Chapter 7), for which the $S_1 \rightarrow T_1$ intersystem crossing yield is near zero.

Overall there is a much greater complexity of high energy radiation processes compared to photochemical processes, but the great strides in elucidating the primary products of radiation chemistry made since the early 1970s have allowed the results obtained from radiation chemistry to be of value to photochemistry and photobiology, as well as several other branches of chemistry. The next chapter will show, for instance, that pulse radiolysis has been able to determine molar absorption coefficients of triplet excited states more easily than flash photolysis and one-electron redox potentials often more precisely than by electrochemical techniques.

2.5.3 *Free ions and solvated electrons in polar and non-polar solvents*

The escape probability of the electron at a distance r_0 from its counterion is given by Onsager theory (1938):

$$P_{esc} = \exp(-r_c/r_0)$$

where r_c, the distance at which the potential energy of an ion pair is equal to the thermal energy $k_B T$, is given by the Bjerrum's length

$$r_c = e^2/4\pi\varepsilon_0\varepsilon_r k_B T$$

r_c is in metres, e is the charge on the electron (1.602×10^{-19} C), ε_0 is the permittivity of free space (8.854×10^{-12} C^2 J^{-1} m^{-1}), ε_r is the dielectric constant of the medium (a pure number), k_B is the Boltzmann constant (1.381×10^{-23} J K^{-1}), and T is the absolute temperature (K). On thermalization, the distance of the electron from its parent positive ion is about 2 nm for polar liquids and about 5–20 nm for hydrocarbons. In non-polar solvents such as *n*-hexane, where r_c is 30 nm at room temperature, P_{esc} is negligible ($<5.5 \times 10^{-4}$) at a distance < 4 nm and is 0.22 at 20 nm. In water, where r_c is 0.75 nm, P_{esc} is 0.83 even at 4 nm.

Pulse radiolysis produces much higher yields of free ions in non-polar solvents than flash photolysis because the solvent and not the solute molecules are directly ionized. For this reason, the investigation of the spectroscopic and kinetic properties of free ions and solvated electrons in non-polar solvents has mainly been possible via pulse radiolysis.

Electron transfer studies by time-resolved techniques have introduced different types of ion pairs. Depending on solvent polarity, solvation, and degree of dissociation, the ions can be either the free solvated radical ions (FRI) or two kinds of ion pairs, the contact ion pair (CIP) and the solvent separated ion pair (SSIP) also called loose ion pair. Marcus uses the terms inner-sphere ion pair and outer-sphere ion pair for CIP and SSIP, respectively. Weller uses the word exciplex for CIP and radical ion-pair for SSIP. These species may exist in equilibrium (Reichardt 1988; Mattay and Vondenhof 1991).

$$(A^{\cdot -}D^{\cdot +})_s \equiv \text{CIP} \rightleftharpoons (A_S^{\cdot -}D_S^{\cdot +}) \equiv \text{SSIP} \rightleftharpoons A_S^{\cdot -} + D_S^{\cdot +} \equiv \text{FRI}$$

2.5.4 *Multiphoton photochemistry*

The advent of high intensity lasers has led to new studies of biphotonic and multiphotonic non-linear effects which do not obey the laws of monophotonic photochemistry. There are two kinds of biphotonic effects:

(1) the sequential absorption of two photons, the first photon by the ground state and the second by an excited state, either singlet (S_1) or triplet states (T_1),

(2) the simultaneous absorption by a molecule of two (or more) photons at wavelengths for which there is no monophotonic transition.

In the first process, the rate of the excitation must be higher than the rate of deactivation of the intermediate level S_1 or T_1. This is achieved by using intense (MW cm^{-2}–GW cm^{-2}) laser pulses or by increasing the lifetime of T_1 using low temperature matrices. The kinetics and the intensity dependence of a two-step biphotonic process has been analysed by Lachish *et al.* (1976).

In the second process, the laser frequency does not match a specific excitation band of the chromophore. The interaction mode with matter is peculiar with new physical processes appearing at high power, such as photoablative and electromechanical interactions (Boulnois 1986, 1989). At intensities around one TW cm^{-2} for ps pulses and around ten GW cm^{-2} for ns pulses, lasers generate locally high electric fields (10^6–10^7 V cm^{-1}) and produce optical dielectric breakdown associated with physical processes such as the formation of a microplasma with local temperatures above 10 000 K, the generation of an acoustic shock wave and of cavitation phenomena, and the formation of large concentrations of free electrons up to 10^{21} cm^{-3} (Boulnois 1986, 1989). The latter processes are essential in the use of lasers for tissue-incision (Zysset *et al.* 1989).

In a solvent transparent to the wavelength of the excited photons, high energy lasers can produce species very similar to those produced by high energy radiation. In water, $H^{\cdot}$, $OH^{\cdot}$, and hydrated electrons have been produced with powerful picosecond pulsed lasers (10^9–10^{10} W cm^{-2}, 30 ps) emitting at 266 nm (Nikogosyan and Angelov 1981; Oraevsky and Nikogosyan 1985). Similar experiments resulted in the photoionization of water by high intensity 310 nm laser pulses and allowed Gauduel *et al.* (1986) to determine a relaxation time of 240 ps for electron solvation in water.

Another feature of multiphoton spectroscopy is that transitions forbidden by conventional one-photon spectroscopy become observable. Once again, the carotenoids are an example of the use of this effect (see Section 7.2.1).

2.5.5 *Quantitative effects*

The quantitative effects of light are measured by the quantum yield ϕ (Section 1.1.2.3), while the quantitative effects of ionizing radiation are expressed as *G* values. The term *G* is defined as the number of radicals, excited states or molecular products produced or transformed per 100 eV of absorbed energy (see Section 1.2.4.1). The *G* values for the species produced in the primary interaction with ionizing radiation are usually up to 3, much higher values indicating the presence of a chain reaction. The absorbed dose was originally expressed in rads (1 rad = 100 erg g^{-1} = 6.242 4 10^{13} eV g^{-1}). Subsequently a new unit of absorbed dose, the Gray (Gy) was adopted in 1975. The Gy is the absorption of 1 J kg^{-1} so that 1 Gy = 100 rads.

2.6 Microheterogeneous media—supramolecular chemistry

The confinement of the reactants in intracellular compartments modifies the deactivation of excited states and the reaction pathways of radicals. For this reason, many investigations are conducted using various membrane-mimetic systems such as micelles, microemulsions, monolayers, bilayers, vesicles, host–guest systems, polyions, and liquid crystals. Many reviews have been written on this theme (Fendler 1982; Thomas, J.K. 1984; Turro *et al.* 1985; Kalyanasundaram 1987; Almgren 1991).

Moreover, biological reactions occur in organized assemblies, very sensitive to their anisotropic order. Thus, distance and orientation of biomolecules are essential parameters ruling biochemical reactions *in vivo*. Synthetic models in which covalent bonds link different molecules have allowed the mimicry of orientation and proximity of biomolecules in their biological environment. This strategy has been used to mimic and elucidate certain aspects of photosynthetic electron transfer and singlet or triplet energy transfer (see review by Gust and Moore 1989). This covalent linking of molecular components constitutes a part of a new research area: supramolecular photochemistry (Balzani and Scandola 1991).

3

Pulsed radiation techniques: flash photolysis and pulse radiolysis

'Axioms in philosophy are not axioms until they are proved upon our pulses'

John Keats

3.1 Monitoring via conventional spectrophotometry in solution

The technique of flash photolysis was introduced in 1949 (Norrish and Porter 1949; Porter 1950). It is a powerful method for studying, as a function of time, transient species (mainly excited states and radicals) produced via electronic excitation of molecules by short pulses of UV or visible radiation. About a decade later pulse radiolysis, using the same principle, was applied to the study of similar species produced by pulses of high energy radiation (Boag and Steel 1960; Keene 1960; M.S. Matheson and Dorfman 1960; McCarthy and MacLachlan 1960).

Conventional absorption spectroscopy remains the most widely used technique for monitoring spectra and kinetics of transient intermediates, with the classical configuration shown in Fig. 3.1. The absorbance change induced in the sample solution by the UV or visible exciting flash, or the pulse of electrons, is monitored by an analysing light beam passing through the sample and reaching a detector (photomultiplier or diode) at selected wavelengths via a monochromator. The detector converts changes of the analysing light intensity into electrical signals which are digitized and displayed, then stored and treated by microcomputer. The transient absorbances are calculated as a function of time and wavelength. Corresponding spectra are displayed on the microcomputer either in two dimensions, absorbance-against-wavelength, at preselected times after excitation, or in three dimensions including time. Full spectra can be obtained after a single excitation over a wide range of wavelengths using a spectrograph and photographic detection as in the past, or more recently using an optical multichannel analyser. In the spectral region where the transient species, t, and the original molecule, G, absorbances overlap, a difference spectrum is obtained, the change in absorbance being:

$$\Delta A = (\varepsilon_t - \varepsilon_G)cl$$

where ε_G and ε_t are the molar decadic absorption coefficients (previously

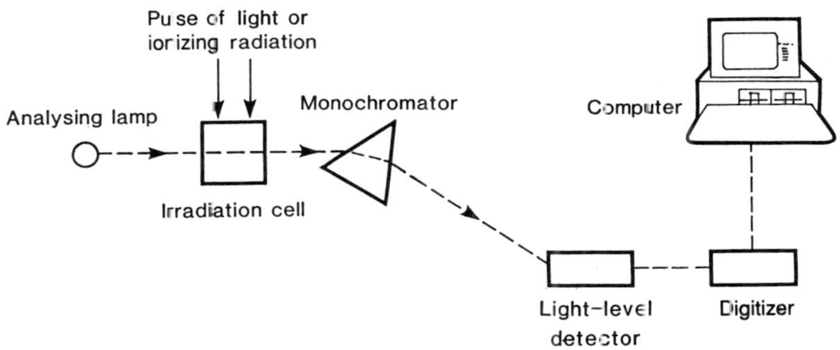

Fig. 3.1 Principle of flash photolysis and pulse radiolysis.

called extinction coefficients) of the molecule in the ground state and the transient species, respectively, c is the concentration of ground state converted to transient, and l is the optical path length.

In some cases emission from the transient species is monitored. The analysing light is then not used, the optics collecting the photons emitted by the sample and the photodetector being modified accordingly. An example of such a time resolved emission set-up is the arrangement used for singlet oxygen detection following laser excitation (Ogilby and Foote 1982; J.G. Parker and Stanbro 1982; Rodgers and Snowden 1982) or following pulse radiolysis (Gorman *et al.* 1989).

Since the introduction of flash photolysis and pulse radiolysis, the pulsed sources of radiation and detection systems employed have continually been improved. The first type of sources used in flash photolysis were gas-filled discharge lamps having durations of the order of 2–3 ms, subsequently reduced to the microsecond range. The short flash durations were obtained with low

Table 3.1 Pulsed lasers for nanosecond flash photolysis (adapted from Kobayashi 1982)

Laser	λ (nm)	Typical repetition rate (Hz)	Typical single-pulse energy (J)	Pulse width (ns)
Gas				
N_2	337.1	100–1	$(0.3–10) \times 10^{-3}$	3–10
H_2	156.7–161.3	<10	$(10–50) \times 10^{-6}$	1–5
Solid-state				
Ruby	694.3	Single shot	5	20–30
Frequency doubled	347.2	Single shot	1	20–30
Nd/glass	~1060	Single shot	1–3	15
Frequency doubled	~530	Single shot	0.2–0.6	15
Nd/YAG	1064	10	0.1–1	10–20
Frequency doubled	532	10	0.03–0.3	10–20
Frequency tripled	355	10	0.1	10–20
Frequency quadrupled	266	10	0.1	10–20
Excimer				
ArF	193	10–20	0.02	20
KrF	248	10–20	0.1	20
XeCl	308	10–20	0.05	20
XeF	351	10–20	0.04	20–30

capacity, *C*, and low self induction, *L*, of the discharge circuit, the duration of the flash being proportional to $(LC)^{1/2}$ (see, for example, Claesson 1976). When pulsed lasers became available in the 1965–70 period, nanosecond laser flash photolysis was developed in many laboratories (Kosonocky *et al.* 1965; Linqvist 1966; Danziger *et al*, 1967; Novak and Windsor 1967; Porter 1967; Bonneau *et al.* 1968, 1969; Porter and Topp 1968, 1970; Bensasson *et al.* 1970; Goldschmidt and Ottolenghi 1970; Kliger and Albrecht 1970). Various pulsed lasers used for nanosecond flash spectroscopy are given in Table 3.1. Such equipment employs similar photomultipliers or photodiodes as used previously for conventional flash photolysis with gas discharge lamps. The advent of the mode-locked solid state laser together with new detection methods, including the optical multichannel analyser and streak camera, extended the time resolution to the picosecond domain (Table 3.2) (for reviews, see Bradley 1982; Fleming and Siegman 1986). In the 1980s flash photolysis entered the subpicosecond and femtosecond (10^{-15} s) time domain, leading to observation of the actual transition state and the process of bond breaking (Martin *et al.* 1983; Shank 1983; Zewail 1988; Zewail and Bernstein 1988; Potter *et al.* 1992).

Table 3.2 Classification of flash photolysis in the nanosecond to subpicosecond time region (adapted from Kobayashi 1982)

Time domains	Typical light source	Monitoring light source	Typical detector systems
Nanosecond (1000–1 ns)	Q-switched ruby Nd/YAG, N_2, H_2 excimer and dye lasers N_2, H_2 and D_2 flash lamps	Xe flash lamp; air breakdown	Photomultiplier; wide-band oscilloscope; box car integrator; sampling oscilloscope
Subnanosecond (1–0.1 ns)	Mode-locked gas (He–Ne, Ar ion) solid-state lasers; N_2 and dye lasers	Xe flash lamp	Sampling oscilloscope, fast diode; very wide band oscilloscope; streak camera
Picosecond (1000–1 ps)	Mode-locked solid state and dye lasers	Picosecond continuum; nanosecond laser; nanosecond flash lamp	Echelon and CMA, streak camera; Kerr gate, variable delay, and photomultiplier
Subpicosecond	Passively mode-locked dye laser	Subpicosecond continuum	Kerr gate, variable delay, and photomultiplier

Table 3.3 Operating parameters of some typical electron accelerators (adapted from L.K. Patterson 1987)

	Lineacs			Febetrons		Van de Graaffs	
Pulse length (ns)	0.03	5	10	4	50	1	100
Energy (MeV)	20	12	10	0.5	1.8	3	3
Dose (Gy)	20	500	100	8 000	20 000	10	1000

The operating parameters for some representative pulsed sources of high energy electrons are reported in Table 3.3. The linear accelerators operate with the shortest pulses in the 20 ps range. More than 20 years ago, Hunt and co-workers (Bronskill and Hunt 1968; Aldrich *et al.* 1972) developed a stroboscopic pulse radiolysis technique employing the picosecond fine structure of a linear accelerator as an exciting pulse, and the Čerenkov radiation as the analysing light source. The transient species were probed at prechosen times within 350 ps, the time interval between the fine structure pulses of the University of Toronto linear accelerator. This novel technique avoided the need for expensive fast electronic detection systems. Up to 1991, the few other laboratories that have developed picosecond pulse radiolysis facilities are: Argonne National Laboratory in the USA (Jonah 1975; Sauer *et al.* 1987), the University of Tokyo in Japan (Kobayashi *et al.* 1981; Tabata *et al.* 1985) and the Institute of Chemical Kinetics and Combustion, Novosibirsk, in the USSR (Grigoryants *et al.* 1989)

For pulse radiolysis and flash photolysis using, as monitoring techniques, either absorption spectrophotometry, or any alternative method as described below, recent data acquisition methods have greatly facilitated the experimental procedure. The signals are now usually controlled, digitized, stored, and analysed by microcomputers (see, for instance, Foyt 1982).

Another source of pulsed electromagnetic radiation is the synchrotron. High energy electrons, circulating in 'bunches' around a synchrotron ring, are bent and accelerated each time they pass through a magnet, and as a result emit flashes of electromagnetic radiation covering a very broad range from X-rays to infra-red. The duration of the pulses is 1 ns or less, and the pulse-to-pulse separation can vary from 10 ns to a few microseconds. One of the major advantages of synchrotron radiation is that the pulses have virtually no jitter. Although the number of photons emitted per pulse is not sufficient for absorption spectroscopy, other detection techniques have been successfully applied to the study of biological molecules. These methods include fluorescence, photoacoustic, circular dichroism, and extended X-ray absorption fine structure (EXAFS) spectroscopies (see for example, Castellani and Quercia 1979).

3.1.1 *Assignment of transient absorptions*

This section is intended to provide a practical guide to the identification and characterization of transient species.

3.1.1.1 *Kinetics*

Transient species observed on flash photolysis or pulse radiolysis are usually due either to excited states or to free radicals. The latter often tend to be longer-lived. If the transient absorption decays by predominantly first-order kinetics, an excited state is often indicated, whereas second-order decay kinetics often means a radical is involved. There are many exceptions, however, to this rule. For example, when there is a component of triplet decay due to triplet–triplet annihilation, second-order kinetics arise. Also, when a radical decays intramolecularly, for example to another radical with a different absorption spectrum, as in water elimination from the $OH^{\cdot}$ adduct of phenol to give the phenoxyl radical (G.E. Adams *et al.* 1966), first-order kinetics are observed. Moreover, quite often using flash photolysis or pulse radiolysis, both excited state *and* radical intermediates are formed. A simple way of testing whether a transient absorption decays by mainly first or second-order kinetics is to change the photolysing light intensity (flash photolysis) or ionizing radiation dose (pulse radiolysis). Such a change does not affect the half-life of a first-order decay but alters the first half-life of a second-order decay.

In the absence of added quenchers, triplet excited state lifetimes as measured by flash photolysis or pulse radiolysis are primarily determined by properties of the solvent and the presence of trace impurities. They are usually in the micro-millisecond time domain, much shorter than found in rigid media at low temperatures. Occasionally, e.g. for molecules containing paramagnetic metal ions, submicrosecond triplet lifetimes are found. All such lifetimes may be further shortened if very high photolysing flash energies or ionizing radiation doses are used, leading to the introduction of triplet–triplet annihilation processes. In the case of pulse radiolysis, quenching by relatively long-lived radicals formed at the same time as the solute triplet state can lead to a further reduction in triplet state lifetime. The lower the pulse dose used, the smaller this effect (see, for example, Land 1969). It is also often found that triplet lifetimes measured after laser flash photolysis tend to be shorter and more correct than those found using the same systems after flash photolysis excitation with conventional gas-filled lamps, due to the long emission tail with such lamps. The lifetime of certain solute triplet states becomes shorter as the solute concentration is increased, this shortening being due to reaction of the triplet with its own ground state.

If an absorption which is transient at room temperature can be shown to be stable indefinitely when the temperature is lowered to that of boiling liquid nitrogen (77 K), then the species causing this absorption cannot be due to an excited state. Lowering the temperature has little effect upon singlet excited state lifetimes, and can only lengthen triplet state lifetimes, in most cases, to tens of seconds.

There are ways of stabilizing free radicals other than by lowering the temperature. Steric hindrance can sometimes prevent radicals from decaying. For example, the anilino radical $C_6H_5\dot{N}H$ is very unstable, lasting only milliseconds at room temperature. However, if *tert*-butyl groups are attached to the 2, 4, and 6 positions of the aromatic ring, the corresponding substituted anilino radical is stable for hours at room temperature (Land and Porter 1961). Such stabilities enable the technique of ESR to be readily applied, the corresponding spectral hyperfine structures often leading to positive identification of radicals. Also, spin trapping agents, e.g. such as 5,5′-dimethylpyrroline-*N*-oxide (DMPO), are often used in conjunction with ESR studies to prove the involvement of radicals (see Section 3.2.8 and Chapter 4).

When mixtures of solute radical cations and anions are being prepared by pulse radiolysis of dilute solutions in certain non-polar solvents, the kinetics of formation of the radical cations and anions are different. In hexane and cyclohexane, solute radical anion formation is extremely fast ($k \sim 10^{12}\,dm^3\,mol^{-1}\,s^{-1}$) due to the exceptional mobility of the electron in such media. On the other hand, positive charge capture is slower, being diffusion controlled ($k \sim 10^{10} dm^3\,mol^{-1}\,s^{-1}$) in hexane although somewhat faster in cyclohexane (see for example, Dawe and Land 1975). Simultaneous conductivity experiments (see Section 3.2.6) can also assist in the identification of charged radicals.

3.1.1.2 *Solvent polarity*

In the case of a pulse radiolysis-induced transient, the polarity of the solvent can often help in its assignment. Polar solvents, e.g. water and methanol, support only very low yields of solute excited states and relatively high yields

Table 3.4 Major unstable solute species produced on pulse radiolysis in various solvents at room temperature

Solvent	Dielectric constant	Unstable solute species
Hexane	1.9	Mainly excited states and small yields of radicals
Cyclohexane	2.0	
Dioxane	2.2	
Benzene	2.3	
Acetone	20.7	Excited states and radicals in similar yields
Methanol	32.7	Reduced radicals only
Acetonitrile	36.5	Radicals only
Dimethylsulphoxide	46.6	Radicals only
Water	80.4	Oxidized and reduced radicals only

of radical ions, whereas non-polar solvents, e.g. benzene and cyclohexane, support high yields of solute excited states and relatively low yields of radical ions. Intermediate solvents, e.g. acetone, support similar yields of both excited states and radical ions (see Table 3.4).

In water, three different primary radicals are produced, e^-_{aq} and H· being reducing and OH· oxidizing. One method of generating exclusively reducing radical species in water is to add a high concentration of formate ion. This reacts with oxidizing OH· to form $CO_2^{\cdot -}$ which is a reducing species, although less powerful than e^-_{aq}:

$$OH^{\cdot} + HCO_2^- \rightarrow H_2O + CO_2^{\cdot -}$$

Alcohols, e.g. methanol or isopropanol, also convert oxidizing OH· into reducing $\dot{C}H_2OH$ or $(CH_3)_2\dot{C}OH$, respectively. *tert*-Butanol is also sometimes used to remove OH· since this alcohol reacts with OH· to form $\dot{C}H_2(CH_3)_2COH$ radicals which are normally very unreactive.

One method of generating almost exclusively oxidizing radical species is to irradiate the aqueous solution saturated with nitrous oxide which converts e^-_{aq} into extra OH·:

$$e^-_{aq} + N_2O + H_2O \rightarrow OH^{\cdot} + N_2 + OH^-$$

In the presence of N_2O, over 90 per cent of the water radicals formed under such conditions are oxidizing, the remainder being reducing H·.

The radical OH· is a vigorous oxidizing agent which also tends to add to solutes as well as abstracting electrons or hydrogen atoms. Milder oxidizing agents are the radicals $Br_2^{\cdot -}$, $(SCN)_2^{\cdot -}$, and $N_3^{\cdot}$, which can be produced from OH· by adding a high concentration of the corresponding halide or pseudohalide ion:

$$OH^{\cdot} + Br^- \rightarrow OH^- + Br^{\cdot}$$

$$Br^{\cdot} + Br^- \rightarrow Br_2^{\cdot -}$$

$$OH^{\cdot} + SCN^- \rightarrow OH^- + SCN^{\cdot}$$

$$SCN^{\cdot} + SCN^- \rightarrow (SCN)_2^{\cdot -}$$

Since the above oxidizing and reducing radicals are encountered so frequently, the spectra of e^-_{aq}, H·, OH·, $O^{\cdot -}$, $CO_2^{\cdot -}$, $(SCN)_2^{\cdot -}$, $Br_2^{\cdot -}$, and $N_3^{\cdot}$ are collected together in Fig. 3.2. Also included in this figure are the spectra of $O_2^{\cdot -}$ and its conjugate acid $HO_2^{\cdot}$. The superoxide anion is extremely important biologically, and is often prepared radiolytically using oxygen-saturated aqueous formate (see Section 4.2.2.1).

Methanol can be a very useful polar solvent for solutes insoluble in water. A mixture of radicals, including e^-_{MeOH}, $\dot{C}H_2OH$, $CH_3O^{\cdot}$, H·, OH·, and $\dot{C}H_3$, is formed initially on radiolysis of methanol. Apart from e^-_{MeOH} and $\dot{C}H_2OH$,

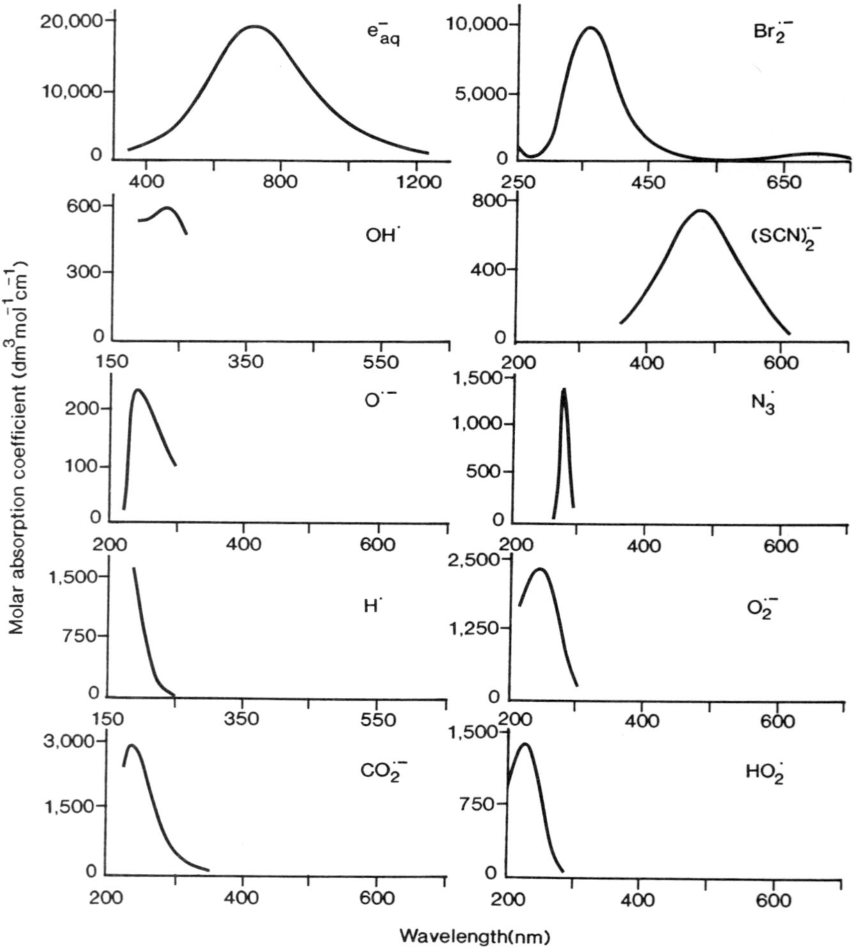

Fig. 3.2 Absorption spectra of various oxidizing and reducing radicals in water (taken from Hug 1981).

these react very rapidly with methanol to give more $\dot{C}H_2OH$, so, for studies of dilute solutes, the 'primary act' reduces to:

$$CH_3OH \rightsquigarrow e^-_{MeOH} + \dot{C}H_2OH$$

Both e^-_{MeOH} and $\dot{C}H_2OH$ are reducing radicals.

In the pulse radiolysis of non-polar solvents, triplet and singlet excited states, in particular of aromatic solutes, are the dominating feature. In cyclohexane these excited states are formed predominantly via fast geminate ion recombinations involving the solute directly. In benzene, which itself has easily accessible and reasonably long-lived singlet and triplet excited states, a main route for formation of solute excited states is energy transfer from excited solvent. In general, higher yields of triplet rather than singlet states are formed, most of the triplet states being formed independently of the excited singlet, i.e. not *via* intersystem crossing. As well as triplet and singlet excited states, low yields of solute radical ions can arise. These can only be effectively studied by pulse radiolysis when their molar absorption coefficients are high, for example, polyene radical ions (see Chapter 7, Table 7.3).

Flash photolysis of a solute in either a polar or non-polar solvent, on the other hand, results in the initial direct formation of a solute excited state. The excited singlet state can, for instance, cross over to the triplet state, fluoresce, vibrationally deactivate or dissociate, leading, for example, to the formation of two radicals. Triplet states may be relatively long-lived or may react with the solvent, or another solute present, by electron or H atom transfer to give radicals, or, as discussed below, undergo energy transfer

Variation in S → T intersystem crossing efficiencies with solvent polarity can often give information regarding excited state orderings ($\pi\pi^*$ or $n\pi^*$) as detailed in the chapters dealing with retinoids and pyrimidines.

3.1.1.3 *Energy transfer*

A useful way of defining whether or not a particular transient absorption is due to a triplet excited state employs the phenomenon of triplet–triplet energy transfer. If a transient species can be shown to sensitize a triplet state of another molecule, then the sensitizing species almost certainly must be a triplet state itself. The identification of duroquinone triplet by such means is illustrated in Fig. 3.3.

The triplet state of β-carotene is particularly useful for establishing the presence of other triplet states. All-*trans* β-carotene has a very low triplet energy level (< 95 kJ mol^{-1}), so most other triplet states can sensitize its triplet, and it has a very intense triplet–triplet absorption coefficient at 520 nm (ε = 240 000 dm^3 mol^{-1} cm^{-1}; Bensasson *et al.* 1977). Furthermore, since β-carotene has an extremely low intersystem crossing quantum yield, it cannot be directly excited itself photolytically, but only via sensitization by another triplet (Chessin *et al.* 1966). In the pulse radiolysis of dilute solutions of β-carotene, some β-carotene triplet is produced directly (Land *et al*, 1970), but the yield can be substantially increased by sensitization by another triplet whose ground state is present in much higher concentrations than the β-carotene.

3.1.1.4 *Light emission*

Both singlet and triplet excited states can emit light. In solution at room temperature many substances fluoresce but, due to competitive processes, the phosphorescence cannot normally be detected. When an absorption of a transient species is observed having a decay half-life of 1 ns to 1 μs, which matches

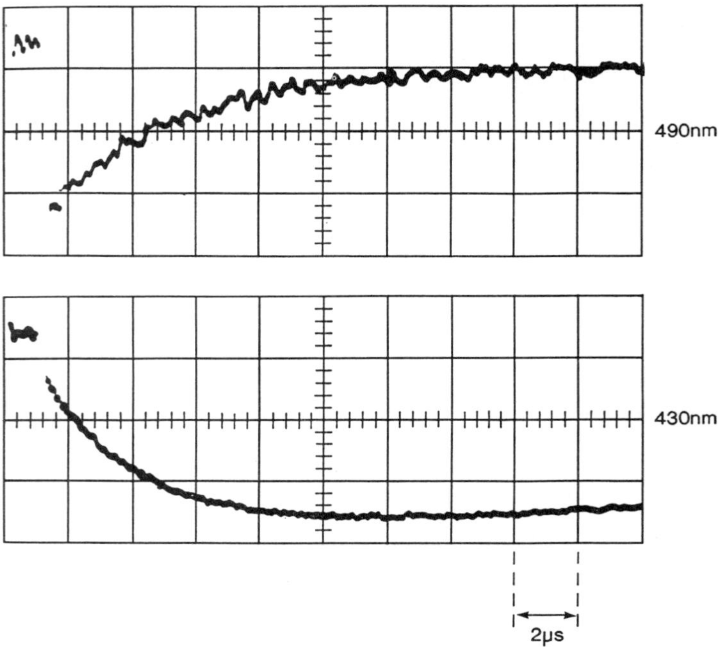

Fig. 3.3 Oscilloscope traces showing triplet energy transfer from duroquinone to anthracene. The solution contained 10^{-2} mol dm^{-3} duroquinone + 3 × 10^{-5} mol dm^{-3} anthracene in benzene. *Top*: decay of duroquinone transient at 490 nm. *Bottom*: corresponding formation of anthracene triplet absorption at 430 nm. Pulse radiolytic excitation, dose ~ 4 Gy, pulse length 0.1 μs. Horizontal displacements, time (1 large division = 2 μs). Vertically upward displacements, increasing transmission (1 large division = 2.5 per cent change in transmission (*top*) = 4 per cent change in transmission (*bottom*)) (taken from Land 1969).

the decay of the fluorescence, then this transient species is probably the corresponding singlet excited state. Similar arguments may be made for triplet states when an absorption decays at the same rate as the phosphorescence of the parent substance (see, for example, Fig. 3.4 for biacetyl in benzene). Since phosphorescence is rare at room temperature, biacetyl being one of the few examples, such measurements usually have to be made at low temperatures. Another important phosphorescence that can be studied at room temperature by flash photolysis or pulse radiolysis is that arising from $^1O_2$ at 1270 nm (see Chapter 4).

A weak chemiluminescence due to dye fluorescence resulting from the reaction in aqueous solution of e_{aq}^- with semioxidized dyes has also been observed after pulse radiolysis. This can be used to assign a transient absorption to a semioxidized radical cation (Prutz *et al.* 1966).

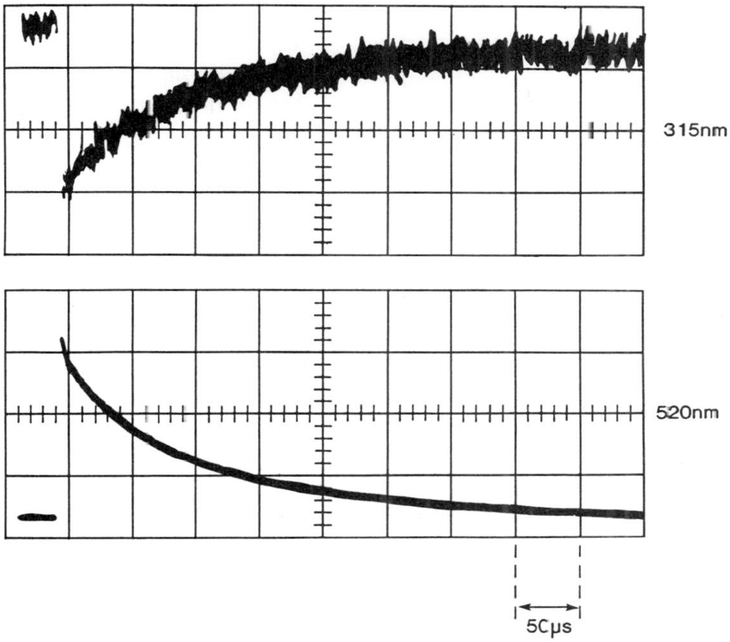

Fig. 3.4 Oscilloscope traces showing matching decay of biacetyl triplet absorption (315 nm, *top*) and biacetyl phosphorescence (520 nm, *bottom*). The solution contained 10^{-2} mol dm^{-3} biacetyl in benzene. Pulse radiolytic excitation; dose ~ 5 Gy; pulse length 2 μs; horizontal displacements, time; upward displacements, increasing transmission.

It is important to remember that in flash photolysis or pulse radiolysis experiments, emissions generated by the radiation pulse are superimposed on changes in transmission of the analysing light source. In addition, an emission may also arise in the case of pulse radiolysis from Čerenkov radiation (see Section 1.2.3). Since emissions result in signals (e.g. oscilloscope deflections) in the opposite direction to absorptions, the two must be subtracted. This is often performed by carrying out the experiment with and without the analysing light source. Sometimes an apparent growth of absorption immediately after a flash or pulse is really due to the decay of a superimposed emission. Failure to take their overlap into account has led to errors in the literature. Raising the intensity of the analysing light source tends to minimize such problems.

3.1.1.5 *Effects of oxygen and nitrous oxide*

Oxygen reacts with most excited states and radicals, so steps are normally taken to remove air from solutions. This is carried out either by repeatedly freezing the solution, evacuating the space above the frozen solid, and melting, i.e. the 'freeze–pump–thaw' technique, or by the more rapid method of bubbling the solution with pure nitrogen, argon or helium. The rate of reaction of a transient absorption with oxygen is also sometimes useful in helping to decide its assignment. Oxygen tends to quench most triplet excited states very rapidly by energy transfer (rate constant $\sim 10^9$ dm^3 mol^{-1} s^{-1}). Lowest singlet excited states are not as readily quenched by oxygen despite a rate constant for the quenching reaction of $\sim 10^{10}$ dm^3 mol^{-1} s^{-1}, simply because the S_1 lifetimes are too short.

In addition, oxygen reacts very readily with the hydrated electron and thus the presence of oxygen precludes the formation of solute radical anions. Thus, if a given experiment were to lead to the production of a mixture of the triplet state, radical anion, and radical cation of the solute, the use of oxygen to remove both the triplet and radical anion, and (separately) nitrous oxide to remove only the radical anion, could be useful in assigning the transients due to each species. The radical anions and cations of several polyenes (Dawe and Land 1975; Lafferty *et al.* 1977) and porphyrins (Bonnett *et al.* 1982) have been distinguished by such means.

3.1.2 *Physico-chemical parameters of excited states*

A knowledge of the molar absorption coefficients of singlet or triplet excited states is essential when quantitative data concerning their yields are required. For example, the comparative method of determining singlet-triplet intersystem crossing quantum yields (see later in this section) presupposes a knowledge of the corresponding triplet molar absorption coefficients.

3.1.2.1 *Molar absorption coefficients of triplet–triplet transitions*

Basically two techniques employing flash photolysis or pulse radiolysis are available. As noted below, these methods both involve assumptions. However, different assumptions apply in each case so that whenever possible the triplet molar absorption coefficients should be measured by both methods (Milder and Kliger 1985; Heihoff *et al.* 1990).

Singlet depletion method This technique uses flash photolysis excitation and involves comparing the loss of ground state absorption with the gain in triplet absorption obtained. Since assumptions have to be made concerning the absorption of the triplet state in the region of ground state absorption where the depletion is being monitored, this method is subject to uncertainties.

A related method, known as the complete conversion method, can be used for molecules where the singlet–triplet quantum yield is relatively high, the triplet lifetime is reasonably long, and the excited singlet lifetime short compared to the duration of the exciting pulse. Increasing the photolysis flash intensity can result in the conversion of all gound state molecules into the triplet state. This allows the triplet absorption coefficients to be determined, since then the concentration of the triplet state is equal to the concentration of the original ground state. In the region of ground state absorption, the spectrum and absorption coefficients of the triplet are found by adding the ground state absorption to the difference spectrum.

Since high triplet quantum yields and good separations of triplet and ground state absorptions are by no means universal, these methods of determining triplet absorption coefficients are applicable only in certain cases, for example, porphyrins.

Energy transfer method This is the most generally applicable method and involves comparing the unknown triplet molar absorption coefficient with that of another triplet, or product of a triplet, of known molar absorption coefficient. Initially the method (Land 1968) employed pulse radiolysis for triplet excitation and, by means of triplet–triplet energy transfer, led to a comparison between the molar absorption coefficients of the benzophenone ketyl radical, $(C_6H_5)_2COH^{\cdot}$, which was established independently, and the triplet molar absorption coefficient of several aromatic compounds whose triplet levels are lower than that of benzophenone. Pulse radiolysis (or flash photolysis) of benzophenone in cyclohexane results in the formation of triplet benzophenone, which abstracts hydrogen from the solvent to form $(C_6H_5)_2COH^{\cdot}$ (Porter and Wilkinson 1961*b*). Employing relatively high concentrations of benzophenone, increasing amounts of an acceptor cause the benzophenone triplet to be quenched (resulting in less ketyl radical being formed), with the consequent appearance of triplet absorption of the acceptor

$$(C_6H_5)_2CO \rightsquigarrow {}^3(C_6H_5)_2CO$$

$$\text{or} \xrightarrow[h\nu]{}$$

$$^3(C_6H_5)_2CO \xrightarrow{C_6H_{12}} (C_6H_5)_2COH^{\cdot}$$

$$^3(C_6H_5)_2CO \xrightarrow{\text{acceptor}} (C_6H_5)_2CO + {}^3\text{acceptor}$$

The decrease in $(C_6H_5)_2COH^{\cdot}$ concentration can be equated to the concentration of acceptor triplet formed. Hence the ratio of the absorption coefficient of $(C_6H_5)_2COH^{\cdot}$ to that of the acceptor triplet can be found.

Once the triplet molar absorption coefficients of a number of polycyclic aromatic hydrocarbons, e.g. biphenyl, naphthalene, and anthracene, had been established against $(C_6H_5)_2COH^{\cdot}$, these hydrocarbons in turn were used as secondary standards (normally as donors), to compare with a large number of other triplet absorption coefficients. Exactly the same principle is involved, except that here the donor triplet was observed directly. Quite often, when the lifetimes of donor and acceptor triplet are long, it is possible to arrange that all donor triplets are quenched by energy transfer to acceptor and simply:

$$\frac{\varepsilon^{D}}{\varepsilon^{A}} = \frac{A^{D}}{A^{A}}$$

where ε^{D} and ε^{A} are the molar absorption coefficients of donor and acceptor triplet respectively, A^{D} is the maximum absorbance of the donor triplet in the absence of acceptor, and A^{A} is the maximum absorbance of the acceptor triplet when both donor and acceptor are present.

In real situations, however, the picture is often more complex. For example, the donor triplets may also decay by means other than energy transfer and the acceptor can decay during its formation. If both occur, three reactions have to be taken into account:

$$^{3}\mathrm{D} \xrightarrow{k_1} \mathrm{D} \text{ or products}$$

$$^{3}\mathrm{D} + \mathrm{A} \xrightarrow{k_q} \mathrm{D} + {}^{3}\mathrm{A}$$

$$^{3}\mathrm{A} \xrightarrow{k_3} \mathrm{A} \text{ or products}$$

The rate constants for decay of donor in the absence of acceptor (k_1) and in the presence of acceptor ($k_2 = k_1 + k_q$ [Q]) are measured as well as the maximum absorbance [A^{A} (maximum observed)]. Then the true total absorbance is given by:

$$A^{A}\ (\text{total}) = A^{A}\ (\text{maximum observed})\ \frac{k_2}{k_2 - k_1}$$

For the case where the rate constant of decay of acceptor triplet (k_3) is not negligible, it can be shown from classical theory (Capellos and Bielski 1972) that

$$A^{A}\ (\text{maximum observed}) = A^{A}\ (\text{total})\ \exp\left[-\frac{\ln k_2/k_3}{k_2/k_3 - 1}\right]$$

This relation can be used to obtain the A^{A} (total) from the experimentally determined ratios k_2/k_3 and A^{A} (maximum observed).

For all these types of determination it is necessary to make several basic assumptions and fulfil various conditions. These have been discussed in detail elsewhere (Land 1968), but, briefly, the most important are that:

(1) one triplet acceptor molecule is formed after interaction with one triplet donor molecule, and
(2) the corresponding singlet excited states also formed are too short-lived to interfere.

The band-widths of electronic transitions often change markedly from one solvent to another. According to the classical expression for oscillator strength (f):

$$f = \text{constant} \int \varepsilon(\nu)\, \mathrm{d}\nu$$

where $\int \varepsilon(\nu)\mathrm{d}\nu$ is the area under the absorption curve and f is independent of solvent. Thus by equating areas under the absorption curves of triplet absorptions in different solvents the corresponding triplet molar absorption coefficients may be compared (see, for example, Bensasson and Land 1971).

The molar absorption coefficient of benzophenone triplet has also been determined directly by absolute measurements of absorbed light energy via calorimetry (Hurley *et al.* 1983), the value obtained validating measurements made by the energy transfer technique. Errors that can arise in triplet absorption coefficients due to lack of overlap between the laser beam and the analysed volume in laser flash photolysis measurements have also been discussed (Bazin and Ebbesen 1983).

Many of the triplet molar absorption coefficients that have been measured earlier by the energy transfer technique have been tabulated (Bensasson and Land 1978*a*). More recently, a compilation of triplet–triplet molar absorption coefficients measured by a variety of techniques has been prepared (I. Carmichael and Hug 1986). The experimental conditions required to measure decadic molar absorption coefficients of transient species in general have also been reviewed by the Organic Chemistry Division Commission on Photochemistry of the International Union of Pure and Applied Chemistry (Bonneau *et al.* 1991). The Commission on Photochemistry concluded that 'triplet states are probably the most useful reference transients', and adopted a list of reference compounds extracted from the extensive least squares fit of Bensasson and Land (1971) by I. Carmichael *et al.* (1987).

3.1.2.2 *Molar absorption coefficients of singlet–singlet transitions*

These may be estimated by measuring the intensity of an excited singlet absorption and comparing this with the intensity of triplet absorption derived from it following intersystem crossing (see, for example, Amand and Bensasson 1974). The flash photolysis technique is preferable to pulse radiolysis for such measurements, since all triplets excited by light are formed via the excited singlet, this not being the case for ionizing irradiation. Excited singlet molar absorption coefficients (ε_S) may be calculated using the formula:

$$\varepsilon_S = \varepsilon_T \times \frac{A_S}{A_T} \times \phi_T$$

where A_S and A_T are the absorbances of excited singlet and derived triplet, respectively, ε_T is the known molar absorption coefficient of the triplet, and ϕ_T is the known intersystem crossing quantum yield (see below).

3.1.2.3 *Singlet–triplet intersystem crossing quantum yields*

The first method of measuring ϕ_T employing flash photolysis involved the direct comparison of the number of quanta absorbed by a solution with the number of triplet states formed after conventional flash photolysis, the triplet molar absorption coefficient being known (Bowers and Porter 1967). Because of the difficulties in accurately measuring numbers of quanta absorbed from the continuum of flash lamps, and in allowing for intensity variations due to progressive light absorption through the reaction vessel, such direct determinations have not been employed for many systems. Subsequently, two methods of determination of ϕ_T have been developed and applied to a large number of molecules.

Heavy atom fluorescence quenching method This method (Medinger and Wilkinson 1965) compares the change in fluorescence intensity of a molecule in solution caused by adding a given concentration of heavy atom quencher, with the increase in triplet state absorption resulting from such addition of quencher observed on flash photolysis. A number of heavy atom quenchers have been used, including various alkyl halides, xenon (Horrocks and Wilkinson 1968), and dimethyl mercury (Vander Donckt and Lietaer 1972). It can be easily shown that

$$\frac{F^0}{F} = \left(\frac{A_T F^0}{A^0 F} - 1\right)\phi_T + 1$$

where F and F^0 are the fluorescence intensities in the presence and absence of the quencher, and A_T and A^0 are the initial absorbances of the triplet state following flash excitation in the presence and absence of quencher. This technique is most readily applied to fluorescent molecules whose singlet lifetime is relatively long.

Comparative method using laser flash photolysis This method makes use of the fact that actinometry is simpler with monochromatic pulses emitted by lasers as compared to broad-band polychromatic emissions from 'conventional' gas-filled flash lamps. The comparative method was introduced by Richards and Thomas (1970) for the determination of the triplet quantum yield of *N,N,N′,N′*-tetramethyl-*para*-phenylene diamine (TMPD). The energy input was measured by excitation of a solution of anthracene, used as standard, which had the same absorbance as that of TMPD at the wavelength of the exciting laser. The limitations of this simple method have been discussed by Amand and Bensasson (1975) who applied it to the determination of ϕ_T for 17 aromatic compounds. Furthermore, the necessity to use a low number of photons and a low fraction of conversion to the triplet state was discussed and illustrated by an application to the determination of the triplet quantum yield of retinal by Bensasson *et al.* (1978*b*).

Description of the method: The principle of the technique is that of relative actinometry. The concentration of triplet formed in a solution of a substance X, with unknown ϕ_T^X but known triplet molar absorbance coefficient, ε_T^X, is compared with the concentration of triplet formed in a solution of some standard, St, with known ϕ_T^{St} and ε_T^{St}.

To ensure that equal numbers of singlet excited states are formed in both solutions, the ground state absorbances (A) at the laser excitation wavelength are chosen to be low and identical

$$A_G^X = A_G^{St}$$

Optically thin solutions should be used. In an optically thin solution irradiated with an incident intensity I_0, the absorbance A must be low enough to ensure that the intensity absorbed I_{abs}, measured in einsteins cm^{-2}, $I_{abs} = I_0 (1 - 10^{-A}) \simeq 2.3.I_0A$. Usually, the solutions should have $A < 0.05$ in 1 mm, corresponding to the fact that the optical detection of the transient species is often performed in the first millimetre of the excited volume along the laser path. In addition, in order to ensure that this equality ($A_G^X = A_G^{St}$) remains valid during the flash, the ground state depletion should be the lowest possible for both solutions commensurate with a reasonable signal-to-noise ratio. The maximum absorbances of the triplets of unknown, A_T^X, and standard, A_T^{St}, should be determined at a time such that the concentration of excited singlet states (S_1) is zero and that of the triplet states (T_1) is a maximum. This is particularly important for molecules with rather long S_1 lifetimes relative to the laser pulse duration. These absorbances are then plotted as a function of laser intensity measured in relative arbitrary values by a monitoring system via a beam splitter. The slopes of these plots are straight lines for low laser intensities (see examples shown for retinal by Bensasson *et al.* (1978*b*) or for indole by Klein *et al.* (1981)). These slopes are transient triplet absorbance values normalized for the number of photons absorbed, called respectively α_T^X and α_T^{St}. It is easy to show that under such conditions

$$\phi_T^X = \phi_T^{St} \times \frac{\alpha_T^X}{\alpha_T^{St}} \times \frac{\varepsilon_T^{St}}{\varepsilon_T^X}$$

A determination of these slopes has the advantage of checking the region of linearity between the transient absorbance and the laser intensity.

In certain cases, the ground state molar absorption coefficient and the triplet quantum yield of X are much lower than those of the standard. Under these conditions it may be necessary to use a higher ground state absorbance for the unknown A_G than for the standard A_G^{St} and an appropriate correction should be made:

$$\phi_T^X = \phi_T^{St} \times \frac{\alpha_T^X}{\alpha_T^{St}} \times \frac{\varepsilon_T^{St}}{\varepsilon_T^X} \times \frac{A_G^{St}}{A_G^X}$$

It is important in these time-resolved experiments to ensure that the molecules in their first excited singlet or triplet states, or their photoproducts, do not absorb significant amounts of exciting light compared with the ground state absorption of the starting materials, since this would lead to erroneous actinometry (Speiser *et al.* 1973).

Actinometry: the flux of photons emitted during a laser pulse can be calculated from transient absorbance measurements. If the fraction (η) of ground state molecules excited to the triplet state is: $\eta = [T]/[G]_0$, and if η is $\ll 1$, the relation between I_0 and η is given (Bensasson *et al.* 1978b) by:

$$I_0 = -\log_{10}(1-\eta)/1000\ \phi_T \varepsilon_G$$

where the factor of 1000 arises from the normal units for ε ($dm^3\ mol^{-1}\ cm^{-1}$) and for I_0 ($mol\ cm^{-2}$))

$$I_0 = \eta/2.303 \times 1000\ \phi_T \varepsilon_G$$

$$= [T]/2303\ \phi_T \varepsilon_G [G]_0$$

This expression allows us to calculate the number of photons emitted during the pulse duration from the measurement of the fraction of molecules converted to the triplet when the following parameters are known: ϕ_T, ε_T at λ of observation, and ε_G at λ excitation.

This last expression has been applied by Klein *et al.* (1981) to the calculation of the number of photons emitted in their pulsed laser experiments dealing with naphthalene and anthracene triplets. They have shown that the triplet absorbances are proportional to the laser intensity up to $I_0 = 5 \times 10^{15}$ photons pulse^{-1} at 265 nm. This means that the pulse should not exceed ~ 4 mJ at 265 nm, as 1 Joule of light energy $= 5 \times 10^{15}\ \lambda$ (nm) photons. This laser intensity corresponds to a fraction η of ground state molecules excited to the triplet state

$$\eta = I_0 \times 2.303 \times 2000 \times \varepsilon_G \times \phi_T$$

Thus, for example, with $\varepsilon_G = 1000$ and $5000\ dm^{-3}\ mol^{-1}\ cm^{-1}$ and $\phi_T = 0.71$ and 0.75 for anthracene and naphthalene, respectively, at 265 nm, $\eta = 1.4$ per cent for anthracene and $\eta = 7$ per cent for naphthalene. These low numbers of photons and low conversions are also essential to minimize the consequences of a lack of perfect overlap between the excited and analysed volume (Bazin and Ebbesen 1983), of the neglect of the ground state depletion, and of photoselection effects (Razi Naqvi 1983).

Choice of standard molecules: The common UV laser excitation wavelengths used today are 266, 355, and 532 nm (neodymium–YAG); 347 and 694 nm (ruby); and 337 nm (nitrogen). Earlier work used neodymium–glass lasers with excitation wavelengths at 265, 353, and 530 nm.

For 266 nm excitation, the usual triplet quantum yield standard is naphthalene with $\phi_T = 0.75$, measured by Amand and Bensasson (1975) via the method of fluorescence quenching by xenon, and ε_T (414 nm) = 24 500 dm^3 mol^{-1} cm^{-1} in cyclohexane (Bensasson and Land 1971). For 355 and 347 nm, the usual reference is anthracene with $\phi_T = 0.71$ measured by Amand and Bensasson (1975) by the comparative method using, as standard, naphthalene with $\phi_T = 0.75$, determined with a 265 nm excitation via fluorescence quenching by xenon, and ε_T (422 nm) = 64 700 dm^3 mol^{-1} cm^{-1} in cyclohexane (Bensasson and Land 1971). For 530 or 532 nm a useful standard is $Ru(bpy)_3^{2+}$ with $\phi_T = 1$ and ε_T (360 nm) = 10 000 dm^3 mol^{-1} cm^{-1} (Rougée *et al.* 1982). For 694 nm, methylene blue $\phi_T = 0.58$ (McVie *et al.* 1978), ε_T (370 nm) = 7200 dm^3 mol^{-1} cm^{-1} (Wildes *et al.* 1977) has been used. Many of the ϕ_T values measured by this technique have been tabulated by Amand and Bensasson (1975) and Bensasson and Land (1978*a*).

A photoreversible fulgicide, commercially available under the name of Aberchrome-540, has also been proposed for laser actinometry (Wintgens *et al.* 1988). In this case the photoproduct is stable.

3.1.2.4 *Triplet energy levels*

Triplet states readily transfer their excitation energy to any molecule also present in the solution having a triplet of lower energy. Such transfer processes are useful for estimating triplet energy levels, particularly for compounds which do not phosphoresce or whose $S_0 \rightarrow T_1$ transitions cannot be readily enhanced by heavy atom effects. The procedure is as follows: the compound with an unknown triplet level is excited in the presence and absence of various compounds whose triplet levels are already known. If the unknown triplet is quenched, then its triplet level lies above that of the standard; if it is not quenched, then its level lies below that of the standard. In this way it is often possible to fix upper and lower limits to the lowest triplet levels of such compounds. The method is based on the finding that where the triplet levels of two compounds differ by more than 8–12 kJ mol^{-1}, then triplet energy will be transferred from the compound with the higher triplet level to the compound with the lower triplet level with a rate constant approaching the diffusion-controlled limit (Backstrom and Sandros 1958; Porter and Wilkinson 1961*a*). However, the technique is rather imprecise for molecules in which the energy transfer step involves a change in geometry of the acceptor. In this case a very small change in the rate of energy transfer occurs over a quite wide range of donor energies. This effect is particularly evident with polyenes (Saltiel and Hammond 1963; Sykes and Truscott 1970; Herkstroeter 1975*b*).

When the triplet levels of donor (D) and acceptor (A) are close, it is sometimes possible to observe an equilibrium between the two triplet excited states:

$$D^T + A \rightleftharpoons D + A^T$$

An equilibrium constant K may be defined by

$$K = \frac{[A^T][D]}{[A][D^T]}$$

and, provided a true equilibrium is attained, then the standard free energy change in the process $\Delta G^{\ominus}$ is given by:

$$\Delta G^{\ominus} = -2.303\ RT \log K$$

The assumption is then made that $\Delta G^{\ominus}$ is equivalent to the difference in the energies of the triplet states, ΔE_T. If one of the triplet levels is known accurately, the other can thus be obtained. This method has, for example, been employed for a number of aromatic hydrocarbon triplet states (Kira and Thomas 1974), cyclopentadiene and tetramethylbutadiene (Gorman *et al.* 1981), azulene (Gorman *et al.* 1984*a*), and the 1,2,4,5-tetracyanobenzene–benzene complex (Craig *et al.* 1984). Other possible methods of determining E_T include time-resolved thermal lensing (TRTL), the effect of temperature on energy transfer rates, and magnetic optoacoustic spectroscopy.

3.1.3 *Physico-chemical parameters of radicals*

3.1.3.1. *Molar absorption coefficients*

As with excited state intermediates, the corresponding molar absorption coefficients are important for measuring radical yields and for determining second-order dimerization or disproportionation rate constants. Because pulse radiolysis, under a variety of conditions, can often lead to high yields of radicals whose concentrations can be easily estimated, it is usually the more suitable technique for measuring radical molar absorption coefficients.

Radicals formed via one-electron reduction of solutes in water can very often be quantitatively prepared by reduction with e^-_{aq}, following pulse radiolysis. Since the molar absorption coefficient of the hydrated electron is well established ($\varepsilon^{720} = 1.85 \times 10^4\ dm^3\ mol^{-1}\ cm^{-1}$; Fielden and Hart 1967), a comparison of the initial absorption intensity of the hydrated electron at its maximum at 720 nm with the absorption intensity of the solute radical anion formed by reaction of the solute with e^-_{aq}, can lead to a determination of the radical molar absorption coefficient. Since oxidizing $OH^{\cdot}$ is also formed as a primary product of water radiolysis (see Section 1.2.4.1), steps must be taken to modify its effect, e.g. by conversion into further reducing $CO_2^{\cdot-}$ (by reaction with $HCOO^-$) or $CH_2OH^{\cdot}$ (by reaction with CH_3OH). For $10^{-1}\ mol\ dm^{-3}$ $HCOO^-$, the total reducing radical yield ($e^-_{aq} + CO_2^{\cdot-}$) is sometimes taken to be $G = 6.5$ (Land and Swallow 1971), this being higher than the total radical yield for pure water ($G = 6.0$), given in Section 1.2.4.1, because such high solute concentrations interfere with the radiation primary act.

When a solute is insoluble in water but soluble in alcohol the situation is simpler, since for dilute solutions the predominant radiation-induced radicals are both reducing: e^-_{RCHOH} and $R\dot{C}HOH$. In argon-flushed methanol the total scavengeable radical yield is $G = 5.6$ (D. W. Johnson and Salmon 1977).

Radicals formed by one-electron oxidation of solutes in water can often best be prepared by reaction with $OH^{\cdot}$ using nitrous oxide-saturated solutions, or with milder oxidizing agents such as $(SCN)_2^{\cdot-}$, $N_3^{\cdot}$ or $Br_2^{\cdot-}$ (see Section 3.1.1.2). The $(SCN)_2^{\cdot-}$ radical is also often used for ionizing radiation dosimetry. For N_2O-saturated 10^{-2} mol dm^{-3} thiocyanate solutions Schuler *et al.* (1980) give $G(OH^{\cdot}) = 6.13$ and a molar absorption coefficient of $(SCN)_2^{\cdot-}$ at 472 nm of 7580 dm^3 mol^{-1} cm^{-1}.

Approximate estimates of molar absorption coefficients of single-electron oxidized or reduced solutes may also be gained in non-polar solvents, for example hexane, using G values of $\sim$ 0.1–0.2 that have been measured for radiolytic free ion yields (Dawe and Land 1975).

When a triplet state with a known molar absorption coefficient can be quantitatively converted to a corresponding radical, e.g. a quinone triplet reduced by an alcohol (Amouyal and Bensasson 1976) or by an amine (Amouyal and Bensasson 1977), the molar absorption coefficient of the radicals can sometimes be measured by comparing the transient absorbances of the derived radicals with the transient absorbances of the precursor triplet state.

3.1.3.2 *Acidity constants*

An important property of a radical is its acidity constant. This may be readily determined since the absorption spectrum of the radical is usually different in the two states of protonation. Flash photolytic or pulse radiolytic generation of the radical is carried out under different conditions of solvent acidity, preferably using buffers. In the case of flash photolytic generation, the buffer should not absorb the photolysing light, and, in the case of pulse radiolytic generation, the buffers should not interfere with the reactions leading to the radical under study. A plot of absorbance due to the radical at a wavelength where the molar absorption coefficient differs for the two states of protonation against pH then leads to the pK_a required. As an example, Fig. 3.5 shows such data for the adriamycin semiquinone (Land *et al.* 1985) (for structures, see Chapter 11). Figure 3.5(a) shows the difference between the absorption spectra of semiquinone and parent adriamycin measured at pH 1.1 and 9.1 after completion of pulse radiolytic one-electron reduction of adriamycin by e^-_{aq} and $CO_2^{\cdot-}/\dot{C}O_2H$, but before significant radical–radical reactions had taken place. The variation with pH of the absorption change at 475 nm, ΔA, is shown in Fig 3.5(b). Since the parent absorbance does not change in this pH region, the variation must be due to a dissociation of the semiquinone. The points were fitted to the equation:

$$\Delta A = \frac{\Delta A_1}{1 + 10^{pH - pK_a}} + \frac{\Delta A_2}{1 + 10^{pK_a - pH}}$$

where ΔA_1, and ΔA_2 are the changes in absorbance well before and well

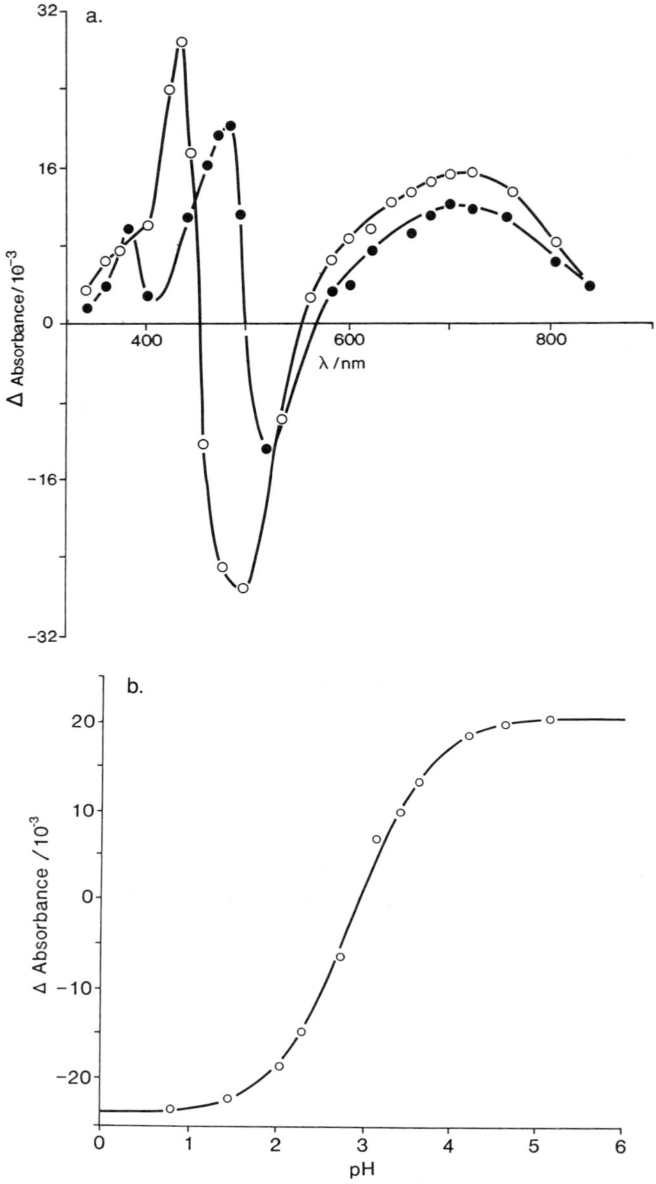

Fig. 3.5(a) Difference between absorption spectra of semiquinone and parent adriamycin measured 10 μs after the pulse, [adriamycin] = 6.3×10^{-5} mol dm^{-3}, optical path length 1 cm, dose 5.5 Gy. (○) pH 1.1 (1 mol dm^{-3} formic acid + H_2SO_4); (●) pH 9.1 (10^{-1} mol dm^{-3} formate + borate buffer). (b) Variation with pH of change in absorbance at 475 nm produced by the pulse [adriamycin] = 5.6×10^{-5} mol dm^{-3}. The solid line is a computed best fit with pK_a = 2.9 (taken from Land *et al.* 1985).

beyond the pK_a of the semiquinone, respectively. A good fit was obtained with $pK_a = 2.9 \pm 0.05$.

3.1.3.3 *Electron transfer: reduction potentials*

Radicals may decay via processes other than disproportionation or dimerization. For example, radicals may give up their unpaired electron to other solutes present. To study electron transfer from a solute A to another solute B in aqueous solution, a mixture of A and B is subjected to pulse radiolysis in the presence of suitable OH· scavengers. From a knowledge of the reactivities of A and B individually with e^-_{aq}, the relative concentrations of A and B are chosen so that all primary e^-_{aq} reacts initially with A. Electron transfer from A to B is then followed by monitoring the change in transient absorption spectrum of the solution from that of $A^{\cdot-}$ to that of $B^{\cdot-}$. Pseudo-unimolecular kinetics are observed if the concentration of $A^{\cdot-}$ is chosen to be small compared with the concentration of B. The equivalent process following one-electron oxidation by, for example, $N_3^{\cdot}$ can also be used to study electron transfer processes. A good example involves aromatic containing peptides and proteins (see Chapter 6).

When the reducing power of the radical species donating the electron is comparable to that of the species formed by acceptance of the electron (i.e. the one-electron reduction potentials of A and B are similar), the electron transfer from A to B does not go to completion. In such a situation, the equilibrium constant, *K*, for the reversible reaction:

$$A^{\cdot-} + B \underset{k_b}{\overset{k_f}{\rightleftharpoons}} A + B^{\cdot-}$$

may be obtained. Provided the rates of the forward and backward reactions are high, equilibrium can be achieved before the radicals disappear by other reactions. With species having differing absorption spectra, optical measurements can be used to measure *K*.

The equilibrium constant may be expressed in terms of the standard one-electron reduction potentials of A and B:

$$E^{\ominus}(B/B^{\cdot-}) - E^{\ominus}(A/A^{\cdot-}) = 2.303\frac{RT}{F}\log K$$

When the potentials are expressed in mV, the numerical value of $2.303\frac{RT}{F}$ at 25 °C is 59.

Figure 3.6 gives typical pulse radiolysis data from which one-electron reduction potentials may be deduced (Wardman and Clarke 1976). This figure shows the absorption spectra measured 50 μs after pulse radiolysis of a nitrogen-saturated solution containing isopropanol, benzyl viologen (BV^{2}), and duroquinone (DQ) (pH 7.2). The dots are the pulse radiolysis observations and the solid line is the known absorption spectrum of the benzyl viologen radical

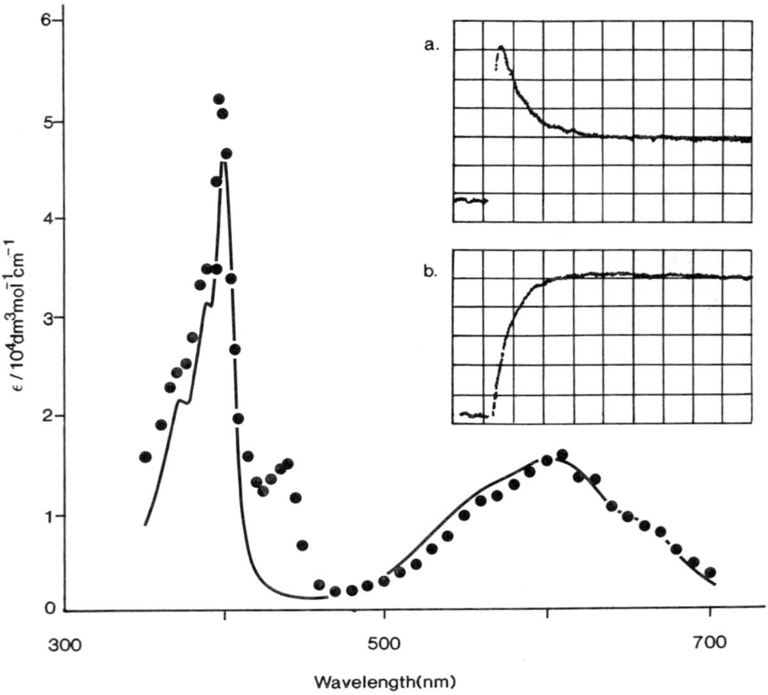

Fig. 3.6 Absorption spectrum (●) measured 50 μs after pulse radiolysis of a solution containing isopropanol, benzyl viologen (2×10^{-3} mol dm^{-3}), and duroquinone (4.4×10^{-5} mol dm^{-3}). Solid line: absorption spectrum of benzyl viologen radical cation. Inset (a): oscillogram showing decay of initial absorption at 600 nm following the pulse; horizontal displacements, time (1 division = 5 μs); vertically upward displacements, increasing absorption (1 division = 1.3 per cent absorption). Inset (b) building of absorption at 440 nm after the pulse; horizontal displacements, time (1 division = 5 μs); vertically upward displacements, increasing absorption (1 division = 0.6 per cent absorption) (taken from Wardman and Clarke 1976).

cation, $BV^{\cdot +}$, normalized to give the same relative absorbance at 600 nm as the pulse radiolysis experiment. The absorption at 600 nm rapidly decays to an equilibrium value, and the spectrum at 50 μs shows an absorption in the 350–470 nm region in addition to $BV^{\cdot +}$. This other absorption with maximum at 440 nm is due to the duroquinone radical anion, $DQ^{\cdot -}$. At 440 nm, $\varepsilon(DQ^{\cdot -}) \simeq 4.5\ \varepsilon(BV^{\cdot +})$ and the absorption at this wavelength increases after the pulse to an equilibrium value, at a rate similar to the decrease observed at 600 nm. From such measurements, together with experiments with BV^{2+} and DQ separately, an equilibrium constant for the reaction

$$BV^{\cdot +} + DQ \rightleftharpoons BV^{2+} + DQ^{\cdot -}$$

was obtained, leading to a ΔE value of 110 mV. Using $E^1(BV^{2+}/BV^{\cdot+}) = -354$ mV, an $E^1(DQ/DQ^{\cdot-})$ of -244 mV was thus calculated.

In certain cases an independent, although less accurate, estimate of K may also be obtained by measurement of the rate of approach to equilibrium. If, as is usual, $[A^{\cdot-}]$ and $[B^{\cdot-}] \ll [A]$ and $[B]$, then the first-order rate constant characterizing the rate of approach to equilibrium is given by:

$$k_{obs}(s^{-1}) = k_f[B] + k_b[A]$$

The ratio intercept:slope of a plot of $k_{obs}/[B]$ against $[A]/[B]$ then gives $k_f/k_b (= K)$.

Observed one-electron reduction potentials vary with pH whenever a hydrogen ion is involved. For example, when a single ionization of a semiquinone $Q^{\cdot-}$ needs to be considered, $E(Q/Q^{\cdot-})$ depends on pH according to

$$E(Q/Q^{\cdot-}) = E(Q/Q^{\cdot-})_7 + 59 \log \frac{K_i + [H^+]}{K_i + 10^7}$$

where K_i is the ionization constant. $E(Q/Q^{\cdot-})$ is therefore independent of pH above that corresponding to the pK_a of the semiquinone ($[H^-] \ll K_i$), but rises with decreasing pH below that pK. In the pH-independent region the redox process is the simple addition of an electron, whereas in the acid region it is $Q + H^+ + e^- \rightarrow QH^{\cdot}$.

The one-electron reduction potential $E(A/A^{\cdot-})$ is related to the two-electron reduction potential $E(A/A^{2-})$ by:

$$E(A/A^{\cdot-}) + E(A^{\cdot-}/A^{2-}) = 2E\,(A/A^{2-})$$

where $E(A^{\cdot-}/A^{2-})$ is the one-electron reduction potential of the radical $A^{\cdot-}$, i.e. the second one-electron reduction potential of A. Examples of the pulse radiolytic determinations of first one-electron, second one-electron, and two-electron reduction potentials, and their variations with pH, are given for quinone antitumour agents and models in Chapter 11.

Many of the one-electron reduction potentials that have been measured by pulse radiolysis have recently been assessed and tabulated (Wardman 1989).

3.2 Other monitoring techniques

3.2.1 *Luminescence*

The most common methods of measuring excited state lifetimes involve monitoring the time dependence of the sample emission with excitation being either pulsed or modulated. Usually the detection is *via* photomultipliers unless the emission is sufficiently intense to allow use of a photodiode.

For luminescence lifetimes of a few hundred nanoseconds, pulsed lasers allow direct measurement of the luminescence decay pulse-by-pulse (of course, signal averaging is also available). For shorter-lived signals the most popular method used is single photon counting and an additional method is the 'phase shift' technique. For fluorescence lifetimes in the region of a few picoseconds the streak camera may under certain circumstances be another possibility.

While not a direct monitoring technique, singlet lifetimes can also be estimated indirectly via the Strickler–Berg equation, together with the relationship between the fluorescent lifetime, τ_f, the radiative lifetime, τ_O, and the fluorescence quantum yield, ϕ_f, i.e. $\tau_f = \phi_f \tau_O$ (see Section 2.1.6). Strickler and Berg (1962) have shown that the intrinsic radiative lifetime, τ_0 is given by

$$1/\tau_0 = 2.88 \times 10^{-9} n^2 \nu_{nm}^2 A$$

where ν_{nm} is the wavenumber of the transition in cm^{-1}, approximated to ν_{max}, and which is taken as identical to the absorption maximum, n is the refractive index of the solvent, and A is the integrated absorption coefficient where the integration is performed over the whole absorption band: $\int(\nu)\,d\nu$ expressed in $dm^3\,mol^{-1}\,cm^{-2}$. With the assumption of a Lorentzian absorption profile, A can be approximated as $\pi\varepsilon_{max}\,\Gamma/2$ where ε_{max} is the decadic molar absorption at the maximum of the absorption profile which has a full-width-half-maximum of Γ in cm^{-1}. $1/\tau_0$ can then be expressed as $= (\nu_{max}^2 n^2 \varepsilon_{max} \Gamma)/2.21 \times 10^8$.

3.2.1.1 *Single photon counting (SPC)*

The SPC time-correlated method remains one of the most widely used techniques for measuring short luminescence (fluorescence) lifetimes. While other techniques tend to require rather high emission intensities SPC requires low intensities. The major components, shown in Fig. 3.7, are an excitation pulse (flash lamp or laser), a time-to-amplitude converter (TAC), and a multichannel analyser (MCA). Also start and stop photomultipliers (PMT) are used to start and stop the voltage ramp of the TAC.

In operation the firing of the lamp or laser generates a start pulse to the TAC and the detection of a stop pulse by the PMT stops the TAC. Under appropriate conditions (low emission intensity) the probability of a stop pulse occurring at a specified time will be directly proportional to the probability of the photon being emitted by the sample at that time. The MCA is

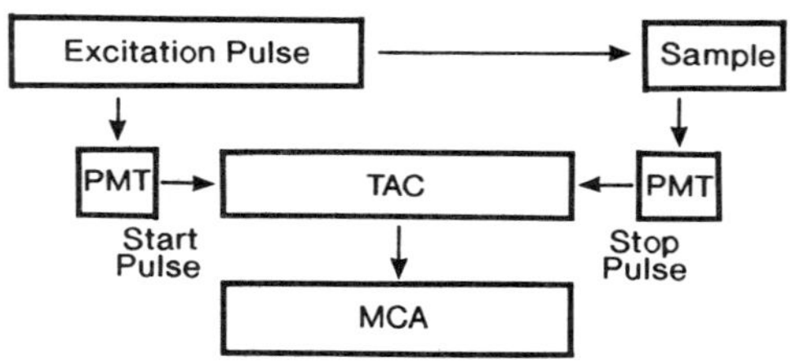

Fig. 3.7 Simplified scheme for single photon counting.

used in the pulse height analysis mode and so the MCA digitizes the height of the input pulse and increases by one the contents of the memory channel corresponding to this digital value. That is, the MCA converts the voltage to a time channel and, summing over many pulses, the MCA builds up a probability histogram of counts versus time channels. (For more details, see Demas 1983; O'Connor and Phillips 1984).

The SPC method has several advantages, for example, it is very sensitive and the response is not limited to the pulse width of the photomultiplier because it is the leading edge which does the triggering. Furthermore, the dynamic range and linearity of SPC is very large. There are relatively few drawbacks to SPC although it may be necessary to collect data for up to several hours for very weak fluorescence. In addition, problems associated with the computer analysis routine must always be considered (Cundall and Dale 1983).

An interesting extension to the infra-red region has been to the SPC detection of singlet oxygen via its 1270 nm (phosphorescence) with an appropriate red-sensitive photomultiplier as attempted by Yu *et al.* (1989).

3.2.1.2 *Phase shift*

In this technique, the sample is excited with a sinusoidally varying light source. The sample emission will also be sinusoidal but will lag behind the excitation. The longer the emission lifetime (τ), the more the delay and the greater the phase shift between the two waveforms. If the phase shift is δ, then $\delta = \tan^{-1}(2\pi\gamma\tau)$ where γ is the modulation frequency.

This technique has serious drawbacks if complex kinetics arise because only a single shift is measured. The problem is at least partially overcome by taking measurements at more than one frequency (see, for example, Lakowicz 1981; Lakowicz *et al.* 1988).

3.2.1.3 *Optical detection of magnetic resonance (ODMR)*

The triplet state is really composed of three triplet (magnetic) sub-levels. Under normal conditions this is of no significance but it is the basis of the ODMR technique. Using very low (cryogenic) temperatures and microwave excitation to change the relative populations of the triplet sub-levels in a magnetic field it is possible to detect corresponding changes in the phosphorescence. Such changes are very sensitive to the local environment of the chromophore and thus the ODMR method can give information of such micro-environments.

3.2.2 **Photoacoustic spectroscopy and photothermal methods**

A photoacoustic spectrum is the action spectrum of heat dissipated by an electronically excited system. The advantage offered by this method, is its applicability to heterogeneous, opaque or scattering biological systems *in vitro* or *in vivo*, for example carotenoproteins on shells, photosynthesizing leaves, and intact skin. General reviews on these techniques have appeared (Cahan *et al.* 1980; Rosencwaig 1980; T.A. Moore 1983; Peters and Snyder 1988; Braslavsky and Heihoff 1989).

Rockley and Devlin (1977) and Bernstein *et al.* (1982) were among the first to use photoacoustic spectroscopy (PAS) techniques for the observation of excited state absorption. Yee and Kliger (1983) introduced a differential time resolved PAS method which avoided the large PAS initial signal by ground state molecules. This PAS technique has an advantage over conventional laser flash photolysis absorption spectrophotometry in being generally more sensitive, especially for highly fluorescent species (Kliger 1989).

Two methods have been developed for the investigation of the time profile of the heat dissipated after absorption of a laser pulse (Braslavsky 1985, 1987; Braslavsky and Heihoff 1989). The first, called laser-induced optoacoustic spectroscopy (LIOAS), monitors, via a microphone (piezoelectric transducer), the acoustic wave associated with the microthermal properties of the sample. The second, called time-resolved thermal lensing (TRTL), monitors the refractive index change caused by the heat dissipated by the solute. The time resolution of both methods is in the range 30 ns to 1 μs. Lifetimes, thermodynamic data (e.g. the enthalpies of various intermediates) as well as quantum yields, relative or absolute, can be inferred from LIOAS and TRTL methods. A useful example is the determination of the absolute quantum yield of singlet oxygen production from porphyrins (Redmond *et al.* 1987).

3.2.3 *Diffuse reflectance in solids*

Transient spectrophotometry is normally performed using transparent samples. For highly scattering media, the photoacoustic detector is not always satisfactory. An additional method for observing transient species produced in non-transparent, heterogeneous, highly scattering substances is the diffuse reflectance technique. This method measures changes in the level of diffusely reflected monitoring light following excitation of a sample by a laser flash (Kessler and Wilkinson 1981; Wilkinson and Willsher 1984), or a high energy radiation pulse (Wilkinson *et al.* 1984). Yee and Kliger (1988) have introduced a spectrograph and an optical multichannel analyser into a diffuse reflectance laser flash photolysis set-up to detect time-resolved absorption spectra of opaque samples with high spectral resolution. They were able to show that in *n*-alkane crystalline Shpol'skii matrices triplet spectra are sharpened by about a factor of two.

3.2.4 *Rayleigh and Raman scattering*

When a molecule is irradiated with monochromatic light most of the scattered light is at the incident wavelength. This is known as elastic (Rayleigh) scattering. A small fraction of the light is modified in frequency by an inelastic (Raman) scattering.

3.2.4.1 *Rayleigh scattering*

The intensity of light scattered, I_θ, with unchanged wavelength, λ, through an angle, θ, depends upon the incident intensity, I_0, the observational angle θ, the distance, r_S, between the scattering volume and the detector, and the polarizability of the scatterer, α, via the equation:

$$I_\theta = \frac{I_0}{r_S^2} \cdot \frac{8\pi^4\alpha^2 (1 + \cos^2\theta)}{\lambda^4}$$

Rayleigh theory, applied to solutions of macromolecules, links I_θ to the number of scattering particles per unit volume and to their average molecular weights. Beck *et al.* (1974) have applied this theory to determine the rate constants for reactions producing scissions of polymer molecules. The same group (Lindenau *et al.* 1976) has followed the kinetics of DNA scission by hydroxyl radicals. A further example of the application of this theory is the measurement of dimer formation in the radiation-induced inactivation of the enzyme lysozyme by Hashimoto *et al.* (1981).

3.2.4.2 *Resonance Raman spectroscopy*

The Raman effect results from the inelastic scattering of light. In this inelastic process, the incident photon interacting with a molecule changes its energy by stimulating a vibrational and/or rotational transition in the molecule. The frequency of the inelastically scattered photon is modified by discrete amounts related to the gain or loss of vibrational or rotational energy of the molecule. Two sets of spectroscopic lines arise in Raman spectra, the so-called Stokes and anti-Stokes lines. The Stokes lines are much more intense and correspond to vibrations induced from the ground vibrational states, whereas the anti-Stokes (higher energy) correspond to transitions originating at an upper level. Thus Raman scattering, like infra-red spectroscopy, yields information about the nuclear motions of the scattering molecule. However, the intensity of the Raman lines is low and in order to improve this method the frequency of the laser source which induces the Raman scattering is selected to be at or near the electronic transition of a chromophoric group. This results in a resonance enhancement of the scattering from the vibrations coupled to the electronic excitation with the following important consequences, particularly for precious biological samples difficult to obtain in large quantities:

1. Scattering from molecules or part of the molecules not in resonance is weak.
2. Resonance enhancement increases the intensity by 10^6.
3. Sample concentrations as low as 10^{-6} mol dm^{-3} can be examined in volumes as small as a few microlitres.

Resonance Raman spectroscopy has been applied to ground state biological molecules, for example: porphyrins and haemoproteins (Asher and Sauer

1976), vitamin B12 (Mayer *et al.* 1973), chlorophyll (Lutz 1977), nucleic acids (Chinsky *et al.* 1978), and visual pigments (Mathies *et al.* 1977). The resonance enhancement makes possible time-resolved resonance Raman (TR^3) spectroscopy leading to information about the vibrational spectra of excited molecules and free radicals and hence their structures. The theory of resonance Raman spectroscopy has been described (Warshel and Dauber 1977, Mathies 1979; Long 1989). The technique has been developed in a number of laboratories to transients produced by flash photolysis and pulse radiolysis for a variety of systems, including CO–haemoglobin during its picosecond photodissociation (Terner *et al.* 1980), semiquinone and phenoxyl radicals (Tripathi and Schuler 1983; A. W. Parker *et al.* 1992), excited carotenoids and polyenes (Dallinger *et al.* 1979; Wilbrandt and Jensen 1981), and the purple membrane of *Halobacterium halobium* (El Sayed 1982).

3.2.5 *Infra-red spectroscopy*

The direct detection of transient organic free radical absorptions in solution by time-resolved infra-red (TRIR) spectroscopy between 1600 and 2000 cm^{-1} has been applied by Neville *et al.* (1989, 1991) to kinetic and spectral studies of β-scission of alkoxyl radicals:

$$[C_6H_5C(CH_3)_2O]_2 \xrightarrow{h\nu} 2\ C_6H_5C(CH_3)_2O^{\cdot} \rightarrow 2\ C_6H_5COCH_3 + CH_3^{\cdot}$$

as well as to the decay of acyl radicals $R\dot{C}{=}O$ generated by UV laser flash photolysis. Conventional laser flash photolysis using visible and UV spectrophotometry cannot unambiguously detect acyl radicals which have, as for example the benzoyl radical, very weak UV absorption. TRIR spectroscopy has also been applied to the study of carboxyhaem proteins in the region of the carbonyl stretching frequency (1900–2000 cm^{-1}). In this way J.N. Moore *et al.* (1988) have determined iron–carbonyl bond geometries of carboxymyoglobin and carboxyhaemoglobin in solution via picosecond TRIR spectroscopy.

Time-resolved Fourier transform infra-red (FTIR) techniques have also allowed the study of laser-induced photofragmentation dynamics in the gas-phase with vibrational and rotational details of the excited products (Leone 1989).

3.2.6 *Conductivity*

The formation and decay of charged species can be followed via the conductance of a solution (for a review of the various methods, see Asmus and Janata 1982*a*). The principle is to apply a voltage across a pair of electrodes suspended in a solution, and to measure the change in current flowing through the cell caused by a laser or high energy radiation pulse. The change in ion concentration is determined by measuring the change in voltage $\Delta V_L(t)$ across a load resistor R_L so that:

$$\Delta V_L(t) = \frac{V_B, R_L}{k_z 10^3} \sum_i \Delta c_i z_i l_i$$

where V_B is the voltage applied across the voltage divider consisting of the conductivity cell and the load resistor, Δc_i the change in concentration of the *i*th charged species, z_i its net charge, l_i the specific conductance of the species involved, and k_z the cell constant. The smallest detectable change in ion concentration is typically 1×10^{-7} mol dm^{-3} if caused by a proton and an anion. The time resolution is typically a few nanoseconds for aqueous systems (Janata and Schuler 1980) and 100 ps for solvents of low polarity (Beck 1979). Conductivity measurements have, for example, been applied to the study of deprotonation (Janata and Schuler 1980), electron reactions (Asmus and Janata 1982*a*), and ligand exchange in organic systems. As this method detects the sum of all ions present, it is necessary to have already investigated the system using another monitoring technique such as optical absorption spectroscopy, polarography or ESR.

3.2.7 *Polarography*

The redox properties of a number of free radicals have been investigated by polarography using pulse radiolysis (for review, see Asmus and Janata 1982*b*). Pulse radiolysis usually has the advantage of producing a more homogeneous distribution of radicals than flash photolysis, and, linked to polarography, leads to a better time resolution than other electrochemical methods such as cyclic voltammetry. The pulse radiolysis–polarography technique has been developed at the Hahn–Meitner Institute in Berlin (Henglein 1976). Similar flash photolysis experiments have also been performed (Berg and Reissman 1970; Perone and Drew 1971; Barker and Bozlan 1974) although this technique is less well developed.

For the pulse radiolysis set-up a time resolution of about 20 μs has been attained. The equipment at Berlin consists of a reaction cell made up of a rectangular quartz vessel containing a hanging drop mercury electrode as the working electrode, a mercury pool as the counter electrode, and a saturated calomel electrode (SCE) as the reference electrode. This set-up allows the simultaneous determination of optical absorption and polarographic current. At a given electrode potential, the polarographic current is recorded as a function of time after the accelerator pulse. Polarograms are constructed from a number of experiments conducted at different potentials by plotting the current at a given time after the pulse against the potential. In this manner the redox potentials of the appropriate couples can be measured.

3.2.8 *Magnetic resonance*

Time resolved magnetic resonance methods are spectroscopic techniques related to transitions between energy levels whose spacing is dependent on a magnetic field applied to the sample. They are concerned with the spin of unpaired electrons (Electron Spin Resonance spectroscopy, ESR) or the spin of nuclei having a magnetic moment (Nuclear Magnetic Resonance spectroscopy, NMR). These techniques coupled with pulse radiolysis and flash photolysis have not reached the time resolution of optical spectrophotometry but they can identify the structure of a transient radical and give information on its history. Chemical reactions conducted in stationary magnetic fields after a laser flash or a pulse of electrons produce spin-polarized products in systems not at equilibrium. Spin polarization of electrons in radicals or of nuclei in molecules can be detected by magnetic resonance during and after the reaction. A review on these techniques known as Chemically Induced Dynamic Magnetic Polarization (CIDMP), including both electron and nuclear spin polarization studies has been presented by Muus *et al.* (1977) and Salikhov *et al.* (1984). Such CIDMP methods have played an important role in the understanding of the primary molecular processes in photosynthesis (Hoff 1984, 1989).

3.2.8.1 *Flash photolysis ESR and CIDEP*

In conventional ESR experiments the absorption is observed while sweeping a strong magnetic field B. In time-resolved ESR techniques, there are two possible approaches: (i) the time-dependence of the signal can be detected at a particular B-value in the spectrum or (ii) the complete spectrum of the radicals is constructed by observing the signal at a given time after the photolytic flash and at different magnetic field values. An interesting phenomenon which enriches the flash-photolysis ESR experiments is the Chemically Induced Dynamic Electron Polarization (CIDEP) which makes possible the identification of the multiplicity of the precursor state producing the radicals and often of its nature, $n\pi^*$ or $\pi\pi^*$.

Almost all radicals observed in the first few microseconds after their formation show spectra with intensities and phases, different from those shown at thermal equilibrium, and the system is considered to be spin polarized. This is due to relative populations which are not described by the Maxwell–Boltzmann distribution. Intersystem crossing (ISC) occurs at different rates into the different sublevel triplet states. ISC populates preferentially one of the triplet sublevels, which are not degenerate as they become separated in the magnetic field B. This preferential population gives a spin polarized triplet. Reaction of this polarized triplet with a quencher produces a pair of spin-polarized radicals (Fig. 3.8):

$$^3M^* + Q \rightarrow R^{1\cdot}(\uparrow) + R^{2\cdot}(\uparrow)$$

which are in the same phase, either absorption or emission, depending on the

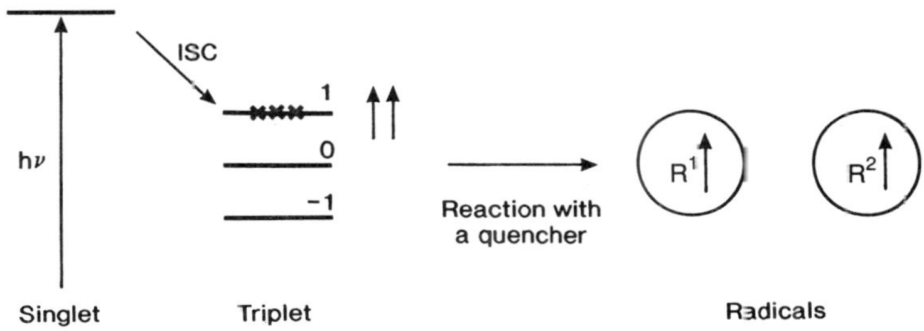

Fig. 3.8 Reaction of a polarized triplet sub-level with a quencher leading to spin-polarized radicals (adapted from McLauchlan 1985).

symmetries of the states involved and on the nature of the triplet, $^3n\pi$ or $^3\pi\pi$. This process is called the triplet mechanism (TM).

In other cases the spectra of the radicals are not in phase, half are in emission (E) and half in absorption (A). This process is more common, being due to magnetic and exchange interactions between the radicals directly after their formation or during a diffusional re-encounter, and is called the geminate radical pair mechanism (RPM). From a triplet precursor an emission is observed before the absorption (E/A pattern) with a preponderance of E or of A marked with an asterik. In contrast, from a singlet precursor, the pattern is inverted with an absorption being obtained before the emission (A/E). Time resolved ESR and CIDEP studies in blocked photosynthetic centres provided evidence that the primary charge separation proceeds from an excited singlet and not a triplet state (Hoff 1984). From triplet precursors, both TM and geminate RPM effects are often obtained in the same spectrum as shown in Fig. 3.9 where the overlapping spectra produced by 1,4-dichlorophthalazine reacting with 2,6-di-*t*-butylphenol are displayed. The sharp lines of the phenoxyl radical are superimposed on the unresolved broad spectrum of the heterocyclic radical (Fig. 3.9(a)). The presence of both TM and RPM mechanisms are analysed in Fig. 3.9(b)).

For a full description of Chemically Induced Dynamic Electron Polarization, see Hore *et al.* (1979*a*, and *b*).

3.2.8.2 *Flash photolysis NMR and CIDNP*

In laser flash photolysis NMR, a phenomenon occurs called Chemically Induced Dynamic Nuclear Polarization (CIDNP). It is related, as CIDEP, to a non-equilibrium population of the spins under study. It leads to enhanced NMR absorption or emission although the chemical shifts and the number of peaks of the NMR spectra of the species are exactly the same as in normal NMR spectra.

An example, taken from Hore (1985), illustrates the nuclear spin sorting process leading to CIDNP (Fig. 3.10). A molecule R^1H-R^2 is dissociated via its photo-excited triplet state into a pair of radicals

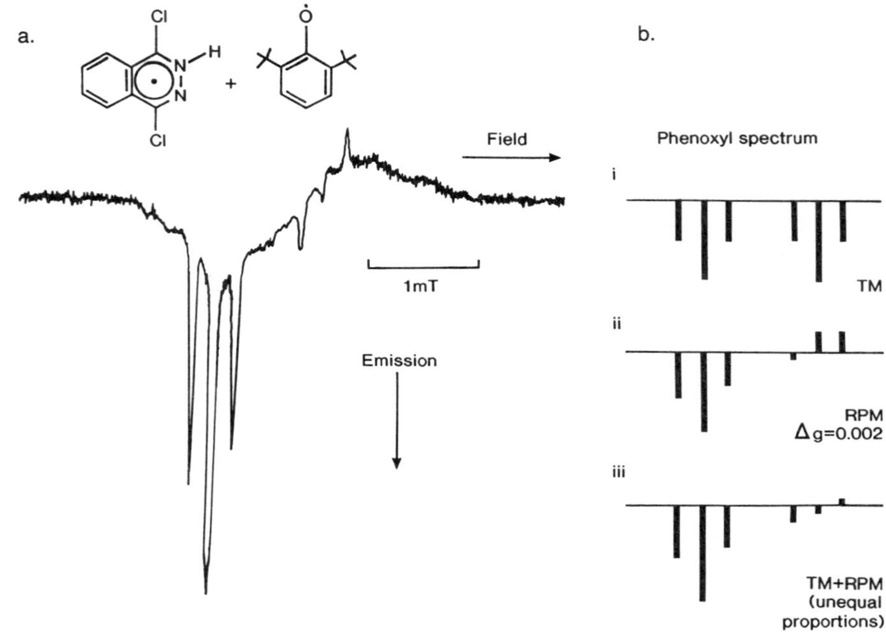

Fig. 3.9 Time integration spectra of radicals that show triplet mechanism (TM) and radical pair mechanism (RPM) effects. The sharp lines in experimental spectrum (a) are due to the phenoxyl radical whose unpolarized spectrum would be a doublet of triplets identical to, but opposite in phase from, the stick spectrum shown for TM-polarization in trace (i). Because the radicals are dissimilar, the RPM-polarization in the phenoxyl is of the E*/A type (ii). The final theoretical spectrum is obtained by adding (i) and (ii) to yield (iii), which is seen to reproduce the experimental behaviour (taken from McLaughlan 1985).

$$^{3}[\mathrm{R^{1}H^{\cdot}}(\uparrow)\ \ldots\ \mathrm{R^{2\cdot}}(\uparrow)]$$

in which the H atom is the only magnetic nucleus. This radical pair is in an unreactive spin correlated triplet because the formation of a bond requires the spins to be antiparallel. Thus this pair can either diffuse apart out of its solvent cage to become two independent free radicals able to react and give R^1HX and R^2Y (the escape products), or it can intersystem cross (ISC) to the singlet state:

$$^{1}[\mathrm{R^{1}H^{\cdot}}(\uparrow)\ \ldots\ \mathrm{R^{2\cdot}}(\downarrow)]$$

In this state the spins are antiparallel and can form diamagnetic products by recombination (Fig. 3.10). If ISC is faster for radical pairs containing the $R^1H^{\cdot}$ proton in its β spin state, these pairs will have a higher probability of recombining inside the cage in a fast geminate reaction. In contrast, the pairs

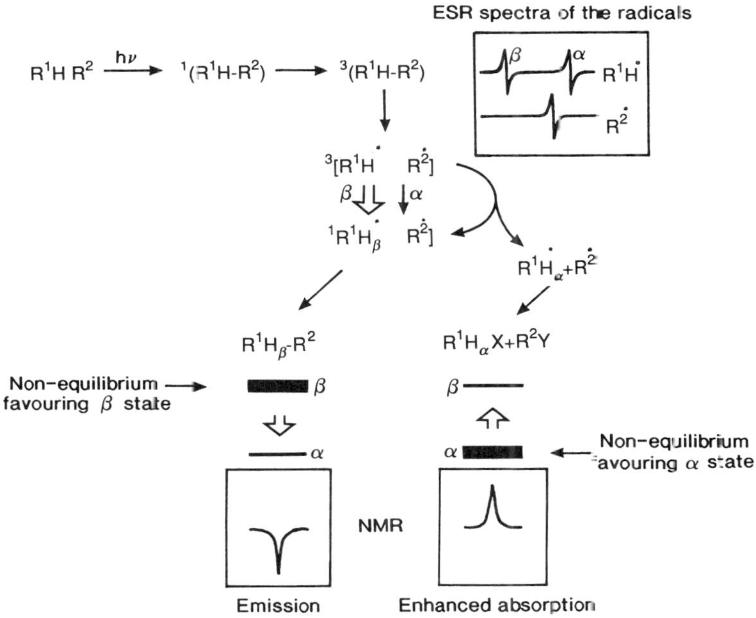

Fig. 3.10 Schematic representation of the nuclear spin sorting process leading to CIDNP (adapted from Hore 1985).

containing the $R^1H\cdot$ proton in its α state will have a greater probability of leaving the cage and producing the escape products, via a slower process. Via this spin sorting process, a non-equilibrium population of the nuclear spin states of the products is created, resulting in emission (E) for $R^1H_\beta R^2$ and enhanced absorption (A) for $R^1H_\alpha X$. In CIDNP studies, enhancement in NMR absorption spectra can be as high as a factor of 10–100. The time of the CIDNP process is related to the intersystem crossing time (1–10 ns). However, the polarization persists for a time governed by the nuclear spin lattice relaxation time of the products, often 1–10 s. The time resolution of laser flash NMR has been extended to the submicrosecond region (Hore 1985).

3.2.9 *Microwave conductivity*

The time-resolved microwave conductivity (TRMC) method provides a quantitative measurement of a change in the microwave conductivity (dielectric loss) of materials when subjected to a short pulse of irradiation delivered either from pulse radiolysis (Infelta *et al.* 1977) or flash photolysis (De Haas and Warman 1982) sources. The change in the high-frequency conductivity of a medium can be related to various parameters, including:

(1) charge distribution, which is able to confirm the occurrence of a photoinduced charge separation;

(2) dipole moments of excited singlet or triplet states or of radicals and rotational relaxation times;

(3) formation and decay kinetics of intermediates, including charge separation and charge recombination.

Using the TRMC technique, Warman *et al.* (1986, 1991) have studied charge separation in synthesized dyads in which an electron-donor moiety is separated from an acceptor by a rigid, saturated hydrocarbon framework of variable length, from 0.5 to 1.3 nm. They found charge separation to occur on a subnanosecond timescale, with close to unit quantum efficiency. They observed lifetimes of the charge separation reaching several hundred nanoseconds, with dipole moments approaching 70 Debye units. They noted very large solvent effects, with an exponential dependence of recombination times with the number of sigma bonds of the insulating bridge between donor and acceptor. Such studies simulate the primary steps in photosynthesis which involve very rapid electron transfer between donors and acceptors embedded in a membrane.

3.2.10 *Novel time-resolved techniques*

Other techniques developed during the last decade include, for example:

(1) *fast electrochemical methods*, interconnected with kinetic results obtained independently by pulse radiolysis (these methods have now been extended to response times of the order of a microsecond since the application of ultramicroelectrodes to cyclic voltammetry and spectroelectrochemistry; Andrieux *et al.* 1990);

(2) *time-resolved photo-thermal beam deflection spectroscopy* (Poston and Harris 1990), which is a photocalorimetric method related to the thermal lens method previously described, see Section 3.2.2;

(3) *time-resolved-circular dichroism*, a method sensitive to molecular structures, which has been applied to the photodecomposition of carboxymyoglobin by laser flash photolysis (Kliger 1989); and

(4) *time-resolved X-ray absorption measurements*, using synchrotron generated intense pulses of X-rays with durations of 10^{-10} s, which have been applied to monitor the recombination of CO with myoglobin after laser flash photolysis of carboxymyoglobin at room temperature (Mills *et al.* 1984).

4

Activated forms of oxygen

'[Oxygen] might not be so proper for us in the usual healthy state of the body: for, as a candle burns out much faster in dephlogisticated than in common air, so we might, as may be said, live out too fast, and the animal powers be too soon exhausted in this pure kind air'

Joseph Priestley (1775)

4.1 **Introduction**

Elements present in sedimentary rocks, 3–4 thousand million years old, are in their reduced forms, such as Fe^{2+}. The oxidation of iron and other elements is observed in rocks laid down later, 2 to 3 thousand million years ago, indicating the period of oxygen formation which probably occurred via the decomposition of water by photosynthetic systems. An ozone layer, produced from this oxygen in the upper atmosphere, protected the surface of our planet from UV radiation and allowed evolution towards more complex aerobic forms of life.

During the last century, Louis Pasteur (1861, 1863) discovered that oxygen can be lethal to some microorganisms and proposed the distinction between aerobes and anaerobes. Later, Paul Bert (1871, 1878) extended Pasteur's observation by showing that oxygen can also be lethal to aerobes when the oxygen pressure is higher than normal. These findings were unfortunately ignored between 1940 and 1954, when elevated concentrations of oxygen were used in incubators for premature babies, causing a number of disorders including irreversible damage to their retina (retrolental fibroplasia).

The first step in the understanding of the molecular mechanisms of oxygen toxicity is marked by an article in Science (1954) by Rebecca Gerschman, Daniel Gilbert, and their collaborators. It reviewed 'pertinent material leading to the hypothesis that oxygen poisoning and radiation injury have one common basis of action, possibly through the formation of oxidising free radicals, such as . . . $OH^{\cdot}$, $HO_2^{\cdot}$ and $RO_2^{\cdot}$'. The second important step in the investigation of oxygen toxicity was the finding by J. M. McCord and Fridovich (1969) of an enzymatic function of a protein containing both copper and zinc, known alternatively as erythrocuprein, hepatocuprein or cerebrocuprein. This function was the catalysis of the dismutation of the superoxide anion:

$$2O_2^{\cdot-} \xrightarrow{2H^+} H_2O_2 + O_2$$

Fridovich and coworkers subsequently reported superoxide dismutase activity for two other proteins, one containing iron and the other manganese. This discovery triggered a long controversy on the superoxide theory of oxygen toxicity. It drew a surge of attention, from many disciplines including radiation chemistry, on the molecular mechanisms of disorders linked to oxygen free radicals and to singlet molecular oxygen.

It has been possible to identify a number of activated forms of oxygen—singlet molecular oxygen, superoxide anion radicals, hydroxyl radicals, peroxy-radicals, hydrogen peroxide, nitric oxide—via direct detection *in vivo* and, more generally, via *in vitro* model systems.

In this chapter, we will consider the chemical, radiolytic, photochemical, and enzymatic methods of generating these species, the normal pathways of their formation *in vivo*, the methods used for their detection, their chemical properties *in vitro*, and how they induce damage in the cellular environment.

4.2 Production of the activated oxygen species

4.2.1 *Singlet oxygen* ($^1O_2$)

The oxygen molecule has a triplet multiplicity ($2S + 1 = 3$) in its ground state, ($^3\Sigma_g^-$) (see Section 2.1.3). In the gas phase its two lowest electronic excited

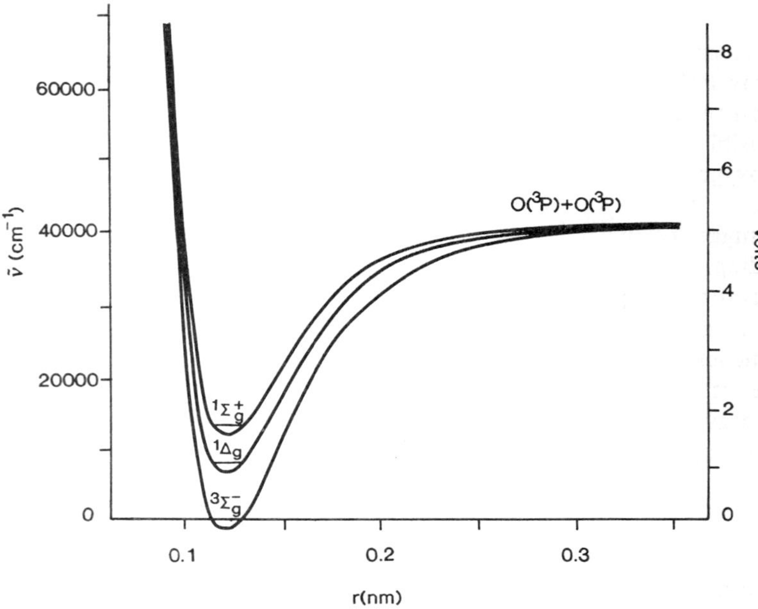

Fig. 4.1 Potential energy curves, including zero point energies, for three electronic states of molecular oxygen (taken from Herzberg 1950).

states, with term symbols $^1\Delta_g$ and $^1\Sigma_g^+$, are located, respectively, at near 0.98 eV (94.5 kJ mol^{-1}, 7882.1 cm^{-1}, 1268.7 nm) and 1.63 eV (157.2 kJ mol^{-1}, 13 121 cm^{-1}, 762.1 nm). The experimental potential energy curves for these three states (Fig. 4.1) show a very similar internuclear separation at equilibrium.

4.2.1.1. *Production* in vitro

Foner and Hudson (1956) described a method for producing singlet oxygen O_2 ($^1\Delta_g$), abbreviated as $^1O_2$, by subjecting gaseous oxygen O_2 ($^3\Sigma_g^-$), abbreviated as O_2, to an electric discharge. This technique has been used for the peroxidation of organic compounds (Corey and Taylor 1964).

A simple way to prepare singlet oxygen chemically is to mix hypochlorite ion with hydrogen peroxide:

$$OCl^- + H_2O_2 \rightarrow Cl^- + {}^1O_2 + H_2O$$

The weak red luminescence observed during such a reaction by Blanchetière (1913) and Mallet (1927) was finally interpreted by Arnold *et al.* (1965). These authors assigned the red luminescence at 634 and 703 nm to a double molecule or 'colliding' pair ('dimol'). These dimol emission bands arise, respectively, from a (0, 0) transition and a (0, 1) transition, the latter to a vibrationally excited ground state:

$$^1O_2 + {}^1O_2 \rightarrow \text{'Dimol'} \rightarrow 2\,O_2 + h\nu \;(634 \text{ and } 704\,\text{nm})$$

More recently, other chemical sources of singlet oxygen have been proposed (Murray 1979; Aubry 1985).

Electronically excited states of a photosensitizer P, (^1P, ^3P), produced by UV, visible or high energy radiation, are able to transfer their energy to ground state oxygen and convert it to the $^1O_2$ state (see review by Saltiel and Atwater 1988).

As suggested by Iu and Ogilby (1987) photosensitizers might be considered according to their energy levels and lifetimes.

Thus for sensitizers whose (i) singlet and triplet states may both be quenched by O_2 to give $^1O_2$, (ii) singlet excited state lifetime is long enough to ensure that quenching by oxygen can be a major deactivation channel, and (iii) energy difference ΔE (S_1–T_1) as well as the energy of their lowest triplet $E(T_1)$ exceed the energy of singlet oxygen, 7882 cm^{-1} (McLean *et al.* 1990), the following reactions apply:

$$^1P + O_2 \rightarrow {}^3P + {}^1O_2$$

and

$$^3P + O_2 \rightarrow {}^1P + {}^1O_2$$

Quantum yields of $^1O_2$ formation by sensitizers of this category, such as 9,10-dicyanoanthracene, can thus reach 2 (Dobrowolski *et al.* 1983).

In non-polar organic solvents, the interaction of excited singlet states of aromatic hydrocarbons with O_2 can give the triplet state of the hydrocarbon with higher efficiency and is called 'oxygen-induced intersystem crossing' (Potashnik *et al.* 1971):

$$^1P \xrightarrow{O_2} {}^3P$$

However for some sensitizers singlet and triplet states may both be quenched by O_2, but only triplet state quenching results in $^1O_2$. Mesoporphyrin IX dimethyl ester is an example of this situation. Although its singlet state lifetime (19 ns) is long enough to ensure that at 5×10^{-3} mol dm^{-3} oxygen approximately 60 per cent of the singlet state molecules are quenched, the ΔE (S_1–T_1) is too low (= 3270 cm^{-1}) and thus excludes the formation of $^1O_2$ (Iu and Ogilby 1987).

Finally, of course, there are sensitizers whose singlet states are not quenched because they are too short-lived. This is the case for acridine. Although its ΔE (S_1–T_1) (10 300 cm^{-1}) is large enough, its singlet state lifetime (< 50 ps) is too short to produce $^1O_2$ (Iu and Ogilby 1987) under normal conditions.

The quenching rate constants of singlet excited states of aromatic hydrocarbons by oxygen were shown to be usually above 10^{10} $dm^3\,mol^{-1}\,s^{-1}$. With few exceptions, it was found that the rate constant for energy transfer from triplet excited states to produce $^1O_2$ is normally 1/9 that of the diffusion controlled rate in the solvent (L.K. Patterson *et al.* 1970), as expected theoretically when spin-statistical factors are taken into account (Kearns 1971; Birks 1975; Saltiel and Atwater 1988).

Because the ΔE (S_1–T_1) value is often insufficient and because S_1 states usually have short lifetimes, singlet oxygen formation occurs most readily via the lowest triplet excited state of the sensitizer. Singlet oxygen formation is in competition with three other reactions: formation of superoxide anion, enhanced triplet decay, and formation of products, with the following scheme (which does not take into account second-order triplet-triplet annihilation important at high triplet concentration):

		Rate constant
$^3P \rightarrow P$	intrinsic decay	k_T
$^3P + P \rightarrow P + P$	quenching by ground state	k_{TG}
$^3P + O_2 \rightarrow P + {}^1O_2$	energy transfer	k_1
$^3P + O_2 \rightarrow P^{\cdot +} + O_2^{\cdot -}$	electron transfer	k_2
$^3P + O_2 \rightarrow P + O_2$	enhancement of intersystem crossing	k_3

Thus, defining $k_0 = k_T + k_{TG}[P]$, and ϕ_T as the triplet quantum yield of the photosensitizer, the quantum efficiency of $O_2(^1\Delta_g)$ production is:

$$\phi_\Delta = \phi_T \times \frac{k_1}{(k_1 + k_2 + k_3)} \times \frac{(k_1 + k_2 + k_3)[O_2]}{(k_1 + k_2 + k_3)[O_2] + k_0}$$

The ratio $k_1/(k_1 + k_2 + k_3)$, i.e. the fraction of triplet photosensitizer molecules quenched by oxygen and leading to $^1O_2$, has been called S_Δ by Gorman *et al.* (1978). The ratio of the sum of the triplet decays involving oxygen over the overall triplet decay has been called S_Q by Heihoff *et al.* (1990). This term is the only term in ϕ_Δ related to the concentration of O_2. Thus the expression of quantum efficiency of singlet oxygen becomes:

$$\phi_\Delta = \phi_T S_Q S_\Delta$$

This equation is simplified into $\phi_\Delta = \phi_T S_\Delta$ for

$$k_0 \ll (k_1 + k_2 + k_3)[O_2]$$

yielding $S_Q = 1$. This simplification is warranted in a number of cases with long-lived photosensitizer triplet states and with low concentration of the sensitizer ground state and of the sensitizer triplet excited state. However this simplification cannot be considered as a general rule as is sometimes assumed. Examples of applying the appropriate correction are given by Knox *et al.* (1988) and Heihoff *et al.* (1990).

For aromatic $\pi\pi^*$ triplets, the fraction of triplet molecules leading to singlet oxygen, S_Δ, is high = 0.7–1. In contrast, for the more polarizable $n\pi^*$ triplet states, S_Δ is lower = 0.25–0.35 (Redmond and Braslavsky 1988).

Several more recent studies of the mechanism of oxygen quenching of triplets invoke charge transfer interactions (McLean and Truscott 1990; McGarvey *et al.* 1992; McLean and Rodgers 1992).

Another source of singlet oxygen is the thermal decomposition of aromatic transannular peroxides (see structure given in Section 4.4.1.3). Such endoperoxides of aromatic compounds including rubrene and 9,10-diphenylanthracene (Wasserman *et al.* 1972), formed by their interaction with singlet oxygen, are stable indefinitely when stored in the solid state at 0–5 °C. In aprotic solvents at 80 °C, they are dissociated at an appreciable rate and reform $^1O_2$.

4.2.1.2 *Biological and environmental production*

Singlet oxygen monomol emission at 1270 nm has been observed directly during enzymatic reactions *in vitro* (Duran *et al.* 1984; Kanofsky 1984; Khan 1984). As noted by Kanofsky (1989), this biochemical production of singlet oxygen 'should not be uncritically extrapolated to living systems'. However, in the presence of intact human eosinophils, Kanofsky *et al.* (1988) have detected the phosphorescence characteristic of singlet oxygen.

Formation of singlet oxygen can also proceed via energy transfer from excited carbonyl molecules produced in the dark (Cilento and Adam 1988), by a number of pathways including:

(1) dioxetane decomposition:

$$\begin{array}{c} >C-C< \\ |\quad\;\;| \\ O-O \end{array} \rightarrow\; >C{=}O + {}^3(>C{=}O)$$

$${}^3(>C{=}O) + O_2 \rightarrow\; >C{=}O + {}^1O_2$$

(2) recombination of two secondary peroxyl radicals (Kellogg 1969):

$$\begin{array}{c} >COO^{\cdot} \\ | \\ H \end{array} + \begin{array}{c} >COO^{\cdot} \\ | \\ H \end{array} \rightarrow \begin{array}{c} >COH \\ | \\ H \end{array} + {}^3(>C{=}O) + O_2$$

$${}^3(>C{=}O) + O_2 \rightarrow\; >C{=}O + {}^1O_2$$

The peroxyl radical recombination can also directly release O_2 in a $^1\Delta_g$ state via a Russell mechanism (Russell 1957; Howard and Ingold 1968) (for more details see Section 4.3.4.1).

Singlet oxygen has been detected in natural waters (oceans, seas, lakes, and rivers) and is mainly generated via sunlight initiated photochemistry (Zepp *et al.* 1977). It can also be present in planetary atmospheres (Wayne 1991).

4.2.2 ***The superoxide anion radical and its protonated form, ($O_2^{\cdot-}/HO_2^{\cdot}$)***

The $O_2^{\cdot-}/HO_2^{\cdot}$ system can be generated by radiolysis and photolysis, although many biochemists often use enzymatic generation via the xanthine/xanthine oxidase method in aqueous solutions, or the KO_2–crown ether reagent (see below) in aprotic solvents, because they have no access to a high intensity light source or a high energy radiation source.

4.2.2.1 *Radiolytic production in aqueous solution*

As reported in Section 1.2.4.1, radiolysis of water with low LET radiation produces, the solvated electron, e_{aq}^{-} ($G = 2.7$), the hydrogen atom $H^{\cdot}$ ($G = 0.55$), and the hydroxyl radical $OH^{\cdot}$ ($G = 2.7$). In the presence of formate (0.1 mol dm^{-3}) $H^{\cdot}$ and $OH^{\cdot}$ radicals form $CO_2^{\cdot-}$, and this radical can transfer its electron to O_2 (concentration 2.6×10^{-4} mol dm^{-3} in aerated water):

$$CO_2^{\cdot-} + O_2 \rightarrow CO_2 + O_2^{\cdot-} \quad k = 2.4 \times 10^9 \, dm^3 \, mol^{-1} \, s^{-1}$$

Alcohols, such as ethanol, methanol, and 2-propanol in alkaline aqueous solutions can be used to replace formate for the scavenging of $H^{\cdot}$ and $OH^{\cdot}$ radicals (Bielski and Arudi 1983). Moreover, the solvated electron and $H^{\cdot}$ also react very efficiently with O_2 (both with $k = 2 \times 10^{10} dm^3 \, mol^{-1} \, s^{-1}$) to give $O_2^{\cdot-}/HO_2^{\cdot}$. The superoxide anion radical is in equilibrium with its protonated form, its conjugate acid, the perhydroxyl radical:

$$HO_2^{\cdot} \rightleftharpoons O_2^{\cdot-} + H^+, \; pK_a = 4.8 \text{ (Behar } et\ al. \text{ 1970)}$$

At pH 6.8 the ratio of the concentrations $[O_2^{\cdot-}]/[HO_2^{\cdot}]$ is 100/1, and thus at physiological pH, the equilibrium is towards $O_2^{\cdot-}$. However protons are present in specific cellular microenvironments, usually at the surfaces of membranes such as the interiors of lysosomes and phagosomes, where the pH can be as low as 5 (see also Section 12.5). In such microenvironments the superoxide anion is protonated and may attack polyunsaturated fatty acids (see Section 4.4.3.1) (Bielski *et al.* 1983).

4.2.2.2 *Photochemical production*

Two techniques have been proposed:

1. Photolysis of hydrogen peroxide in aqueous solution generates $OH^{\cdot}$ radicals, with a quantum yield of 0.82. The hydroxyl radicals react with the hydrogen peroxide to produce the $HO_2^{\cdot}/O_2^{\cdot-}$ system, with a rate constant $k = 1.7 \times 10^7 \, dm^3 \, mol^{-1} \, s^{-1}$ (Baxendale and Khan 1969; Nadezhdin and Dunford 1979):

$$H_2O_2 \xrightarrow{h\nu} 2\,OH^{\cdot}$$

$$OH^{\cdot} + H_2O_2 \rightarrow HO_2^{\cdot} + H_2O$$

2. The second method uses the quenching of flavin triplets by reductants (e.g. EDTA, ascorbic acid). The efficient quenching of the resulting flavosemiquinone anions by oxygen gives $O_2^{\cdot-}$ with $k = 3 \times 10^8 \, dm^3 \, mol^{-1} \, s^{-1}$ for lumiflavin (Vaish and Tollin 1971) and $k = 2.5 \times 10^8 \, dm^3 \, mol^{-1} \, s^{-1}$ for

lumiflavin-3 acetate (Faraggi *et al.* 1975). However, it should be emphasized that the flavin triplet can, at the same time, produce singlet oxygen. Similar formation of $O_2^{\cdot-}$ has been reported from the Rose Bengal radical anion (C. Lambert *et al.* 1990*a*), and *via* photooxidative quenching of excited states by methyl viologen (Mulazzani *et al.* 1988).

In natural waters, $O_2^{\cdot-}$ can be formed photochemically, possibly via photoionization of phenolic compounds. Additionally $O_2^{\cdot-}$ can arise from hydrated electron production by cosmic rays and by radiation emitted by radioactive ^{40}K (Swallow 1969). Petasne and Zika (1987) have emphasized that $O_2^{\cdot-}$ once formed in coastal sea waters can disproportionate to give hydrogen peroxide.

4.2.2.3 *Electrochemical production*

Superoxide anion can be prepared electrochemically by reduction of oxygen in strongly alkaline aqueous solutions and in aprotic media, such as dimethyl sulphoxide, dimethylformamide, acetonitrile, and pyridine (Sawyer *et al.* 1982*a*). Up to 10^{-2} mol dm^{-3} of $O_2^{\cdot-}$ can be produced by this technique and this has been useful in the study of its reactivity with substrates that are neither too reactive towards $O_2^{\cdot-}$ nor electroactive, such as alkyl halides (J. L. Roberts and Sawyer 1981).

4.2.2.4 *Generation* via *xanthine oxidase*

Since the first work of J. M. McCord and Fridovich (1968), biochemists often use the enzyme xanthine oxidase (XO) in presence of hypoxanthine or xanthine as a source of superoxide:

$$\text{hypoxanthine} + H_2O + 2O_2 \xrightarrow{(XO)} \text{xanthine} + 2O_2^{\cdot-} + 2H^+$$

$$\text{xanthine} + H_2O + 2O_2 \xrightarrow{(XO)} \text{uric acid} + 2O_2^{\cdot-} + 2H^+$$

In many physiological and biochemical studies it has been assumed that $O_2^{\cdot-}$ is the only radical species generated by the enzyme. It should be emphasized that these systems also generate H_2O_2 by dismutation of the superoxide anion (see Section 4.4) together with another oxidant more powerful than $O_2^{\cdot-}$ and H_2O_2. This other oxidant is able to generate ethylene gas by reaction with thioether and has characteristics similar to that of the hydroxyl radical (Halliwell and Gutteridge 1984). Subsequent studies have pointed to the essential presence of transition metals, such as iron, for the formation of this oxidizing species (Yamazaki and Piette 1990). Moreover, the ESR spin trapping technique has shown that both $O_2^{\cdot-}$ and $OH^{\cdot}$ radicals are generated by xanthine oxidase (Kuppusamy and Zweier 1989), although this is not unambiguous since $O_2^{\cdot-}$ may arise via a secondary process.

4.2.2.5 *The KO_2-crown ether reagent*

A convenient method for the preparation of solutions of superoxide anion was introduced by Valentine and Curtis (1975). Potassium superoxide, sparingly soluble in dry DMSO, can be dissolved up to a concentration of 0.15 mol dm^{-3} if solubilized with 0.30 mol dm^{-3} crown ether (dicyclohexyl-18-crown-6). This method has allowed many investigations of the reactions of $O_2^{\cdot-}$ in organic chemistry.

4.2.2.6 *Biological formation*

The one-electron reduction of O_2 to $O_2^{\cdot-}$ is thermodynamically unfavourable (see Section 4.4.2.) by 330 mV versus NHE (Ilan *et al.* 1976). This barrier is very important for living systems as it prevents their oxidative destruction. However, the detection of small amounts of $O_2^{\cdot-}$ in mitochondria indicated an electron leakage (< 5 per cent) from the normal pathway in the tetravalent reduction of oxygen via cytochrome oxidase (Fridovich 1978; Naqui *et al.* 1986; Nohl 1990). The production rate under physiological conditions has been estimated to be between 10^{-7} and 10^{-10} mol dm^{-3} s^{-1} (Chuaqui and Petkau 1987). Moreover, phagocytic cells, neutrophils, monocytes, macrophages, and eosinophils are also biological sources of $O_2^{\cdot-}$. These cells specialize in phagocytosis of foreign particles, including bacteria, in a plasma membrane vesicle. The phagocytosis mechanism involves the triggering of a 'respiratory burst', a multiplication of the rate of oxygen consumption by a factor of 10 or 20 accompanied by the formation of the superoxide radical by an oxidase complex (Tauber and Babior 1978; Markert *et al.* 1984) via:

$$NADPH + 2O_2 \rightarrow NADP^+ + H^+ + 2O_2^{\cdot-}$$

Superoxide anion can be generated by autoxidation of biomolecules such as haemoglobin, myoglobin, leukoflavins, and catecholamines, by a number of enzymes including xanthine oxidase and flavin dehydrogenases, by subcellular organelles such as mitochondria, chloroplasts, and microsomes, and by intact cells (Fridovich 1983). The role of ubiquinone (UQ) in mitochondrial $O_2^{\cdot-}$ formation has been reviewed and investigated by Nohl (1990). This author has shown via ESR that the ubisemiquinone remains stable in aerated aprotic systems, in agreement with thermodynamic considerations on one-electron redox potentials of UQ and O_2 in aprotic systems. However, Nohl has observed that traces of water cause the disappearance of $UQ^{\cdot-}$ and the formation of $O_2^{\cdot-}$. Pulse radiolysis studies have shown that the reaction $UQ^{\cdot-} + O_2 \rightarrow UQ + O_2^{\cdot-}$, in alkaline ethanol, is possible but slow with $k = 10^3$ dm^3 mol^{-1} s^{-1} (Bensasson and Land 1973).

4.2.3 *The hydroxyl radical ($OH^{\cdot}$)*

The very high reactivity of $OH^{\cdot}$ radicals makes it difficult to detect directly the generation of these species *in vivo*. Evidence of their presence is often deduced from the knowledge acquired from their chemical generation and their properties in model systems.

4.2.3.1 *Generation by the Fenton–Haber–Weiss reaction*

The reduction of hydrogen peroxide by iron chelates was observed by Fenton as long ago as 1894. Much later the hydroxylating properties of this reaction were discovered and attributed to the following:

$$Fe^{2+} + H_2O_2 \rightarrow Fe^{3+} + OH^- + OH^{\cdot}$$

Later, Haber and Willstäter (1931) and Haber and Weiss (1934) suggested the formation of hydroxyl radicals via a metal catalysed decomposition of H_2O_2, often written in the following overall form (see Sections 1.3 and 4.4.3.5):

$$O_2^{\cdot -} + H_2O_2 \rightarrow OH^{\cdot} + OH^- + O_2$$

This reaction, in the absence of a transition metal, has a rate constant too low to be measured in dimethylsulphoxide (Halliwell 1976). The rate constant was later estimated in aqueous solution to lie in the range $k = 2.25\ dm^3\ mol^{-1}\ s^{-1}$ (Ferradini *et al.* 1978), $< 10\ dm^3\ mol^{-1}\ s^{-1}$ (Czapski and Ilan 1978), and $0.13\ dm^3\ mol^{-1}\ s^{-1}$ (J. Weinstein and Bielski 1979). Metal ions and metal chelators (Fong *et al.* 1976; Van Hemmen and Meuling 1977; Borg *et al.* 1978; Halliwell 1978; Koppenol *et al.* 1978) lead to an acceleration of the process due to the consecutive reactions:

$$Cu^{2+} \text{ or } Fe^{3+} + O_2^{\cdot -} \rightarrow Cu^+ \text{ or } Fe^{2+} + O_2$$

$$Cu^+ \text{ or } Fe^{2+} + H_2O_2 \rightarrow Cu^{2+} \text{ or } Fe^{3+} + OH^{\cdot} + OH^-$$

$$\text{Net } O_2^{\cdot -} + H_2O_2 \rightarrow OH^{\cdot} + OH^- + O_2$$

The net reaction known as the Haber–Weiss reaction is also called a superoxide-driven Fenton reaction.

An excess of Fe^{2+} can reduce $OH^{\cdot}$ formation because of the following fast reaction (Stuglik and Zagorski 1981):

$$Fe^{2+} + OH^{\cdot} \rightarrow Fe^{3+} + OH^- \quad k = 3.2 \times 10^8\ dm^3\ mol^{-1}\ s^{-1}$$

The possibility of such Fenton–Haber–Weiss reactions for the *in vivo* formation of $OH^{\cdot}$ has been the topic of many debates related to the toxicity of $O_2^{\cdot -}$, a relatively unreactive radical (Groves 1980; Halliwell and Gutteridge 1985; Bielski and Cabelli 1991). Much evidence supports the existence of these metal-ion catalysed reactions giving $OH^{\cdot}$. The production of the hydroxyl radical in biological systems, such as phagocytes and hypoxanthine/xanthine oxidase mixtures, which are able to generate $O_2^{\cdot -}$, has been verified by many methods including the hydroxylation of various compounds (Grootveld and Halliwell 1986) or ESR trapping techniques (Finkelstein *et al.* 1980, 1981). As noted above, some results indicate that oxidizing species other than a free $OH^{\cdot}$ radical are produced and suggest the formation of high-valence iron species such as the ferryl ion, represented usually by $Fe^{II}(H_2O_2)$, $Fe^{II}O_2^{+}$, $Fe^{III}O^-$, or $Fe^{IV}(OH^-)_2$ in which the formal charge is represented by roman numerals with the oxidation state of iron being +4 (Walling 1975, 1982; Rahhal and Richter 1988; Rush *et al.* 1990). The extent to which $OH^{\cdot}$ free radicals, formed in a Fenton reaction, are not free but bound in two kinds of complexes, $Fe^{III}OH \ldots OH$ and $Fe^{IV}O$, depends upon the type of chelator used and upon H_2O_2 concentration, acording to Yamazaki and Piette (1991). In aqueous solutions, ferryl (Fe^{IV}) and perferryl (Fe^{V}) complexes have been produced and studied by pulse radiolysis (Bielski 1990). It should be noted that free Fe^{III} is extremely insoluble in water and *in vivo* it is always present in a chelated form.

4.2.3.2 *Sonolytic and radiolytic production*

Formation of hydroxyl radicals by ultrasound in aqueous solutions has been observed via chemical analysis of the reaction products (Henglein and Kormann 1985; Einhorn *et al.* 1989) or via spin trapping (Makino *et al.* 1983; Riesz *et al.* 1985).

As noted above in Section 3.1.1.2, a method of generating almost exclusively $OH^{\cdot}$ radicals *in vitro*, and thus allowing the study of their reactions with biomolecules, is to expose water to high energy radiation, in the presence of nitrous oxide. The latter converts the solvated electrons (G = 2.7) into extra oxidizing $OH^{\cdot}$ radicals (G = 2.7):

$$N_2O + e^-_{aq} \xrightarrow{H_2O} N_2 + OH^{\cdot} + OH^-$$

4.2.3.3 *Photochemical production*

As reported in Section 4.2.2.2, photolysis of hydrogen peroxide generates the hydroxyl radical with a high quantum yield but this radical rapidly reacts with H_2O_2 to give $HO_2^{\cdot}$.

The hydroxyl radical is photochemically produced in the atmosphere (Wayne 1991; Brune 1992; Thompson 1992) and in natural waters (Mopper and Zhou 1990). Its photoproduction in sea water has been demonstrated by flash photolysis (Zafiriou 1974). Nitrate ions in water, with a long-wavelength absorption band at 302 nm, are considered to be involved in this photoproduction (Zepp *et al.* 1987):

$$NO_3^- \xrightarrow{h\nu} NO_3^{-*} \rightarrow NO_2^{\cdot} + O^{\cdot-}$$

$$O^{\cdot-} + H_2O \rightarrow OH^{\cdot} + OH^-$$

4.2.3.4 *Electrochemical and stable solid chemical production*

The electrochemical reduction of Fe^{III}, Mn^{III}, Cu^{II} complexes in the presence of H_2O_2 in aqueous solutions produces $OH^{\cdot}$ radicals via a Fenton process (Sawyer *et al.* 1985).

A recent development has been the use of a solid solution of potassium peroxonitrite in potassium nitrate to produce $OH^{\cdot}$ radicals (King *et al.* 1992). Dissolution of this stable material in water is a very convenient route for $OH^{\cdot}$ generation.

4.2.3.5 *Biological production*

Many of the physical and chemical sources of hydroxyl radicals described above are able to produce these radicals *in vivo*: sonolysis using clinical equipment, shock waves used for destruction of kidney stones by lithotripters (Henglein *et al.* 1988), ionizing radiation used in radiotherapy, and mainly Fenton–Haber–Weiss reactions occurring via the $O_2^{\cdot -}$ leakage of the metabolism of oxygen (see Sections 4.2.3.1 and 4.2.5). Electrons leaking from mitochondria can be transferred from ubisemiquinone to H_2O_2 and produce $OH^{\cdot}$ radicals via a thermodynamically feasible reaction (Nohl 1990).

$$UQ^{\cdot -} + H_2O_2 \rightarrow UQ + OH^{\cdot} + OH^-$$

4.2.4 *The peroxyl ($RO_2^{\cdot}$) and the alkoxyl and phenoxyl ($RO^{\cdot}$) radicals*

4.2.4.1 *Radiolytic and photolytic production of peroxyl radicals*

Most carbon-centred radicals react with oxygen to give peroxyl radicals (also called peroxy radicals) with rate constants 0.5 to $5 \times 10^9\,dm^3\,mol^{-1}\,s^{-1}$:

$$R^{\cdot} + O_2 \rightarrow RO_2^{\cdot}$$

with the exception of some purine-derived radicals which appear quite unreactive (Willson 1970). In these investigations, pulse radiolysis has been a useful tool, $R^{\cdot}$ being produced via reaction of the parent molecule with $OH^{\cdot}$ in N_2O-saturated aqueous solutions (see Section 4.2.3.2). The reaction of radicals with oxygen has been reviewed by Neta *et al.* (1990).

When $R^{\cdot}$ is an aromatic or heterocyclic radical, it is possible to determine directly the rate constant of reaction with oxygen. Such $R^{\cdot}$ radicals have strong visible absorptions, whereas the corresponding peroxyl radicals absorb weakly. Thus it is possible to follow directly the addition of oxygen by measuring the pseudo-first-order rate constant of $R^{\cdot}$ decay as a function of oxygen concentration in solutions saturated with nitrous oxide–oxygen mixtures.

Aliphatic free radicals and their corresponding peroxyl radicals absorb mainly in the ultraviolet. However, it is possible to determine indirectly the rate constants of reaction with oxygen by using a competition method with a reference solute such as ferricyanide ion. In the case of methanol in nitrous oxide saturated aqueous solutions:

$$CH_3OH + OH^{\cdot} \rightarrow {}^{\cdot}CH_2OH + H_2O$$

$${}^{\cdot}CH_2OH + Fe(CN)_6^{3-} \xrightarrow{k_1} CH_2O + Fe(CN)_6^{4-} + H^+$$

In nitrous oxide–oxygen mixtures, the reduction of ferri- to ferro-cyanide is decreased, due to the competitive reaction

$${}^{\cdot}CH_2OH + O_2 \xrightarrow{k_2} {}^{\cdot}O_2CH_2OH$$

Taking A_0 and A as the ferricyanide absorption decrease, in the absence and presence oxygen, respectively, it is possible to write:

$$A_0/A = 1 + k_2[O_2]/k_1[Fe(CN)_6^{3-}]$$

With k_1 directly determined, k_2 is readily calculated.

The trichloromethylperoxyl radical, involved in carbon tetrachloride toxicity, can be prepared by pulse radiolysis in aqueous solutions, containing *t*-butanol, methanol or propanol and CCl_4 (Hiller *et al.* 1983; Huie *et al.* 1987)

$$e_{aq}^- + CCl_4 \rightarrow CCl_3^{\cdot} + Cl^-$$

$$CCl_3^{\cdot} + O_2 \rightarrow CCl_3O_2^{\cdot}$$

Moreover, laser flash photolysis techniques have been used for kinetic studies of the reactions of several free radicals with oxygen. The rate constants of these radical reactions are, for *t*-butyl: 4.9×10^9 (in cyclohexane), benzyl: 2.3×10^9 (in cyclohexane), and cyclohexadienyl: $1.6 \times 10^9\,dm^3\,mol^{-1}\,s^{-1}$ (in benzene) (Maillard *et al.* 1983).

4.2.4.2 *Chemical, photochemical, and radiolytic production of alkoxyl and phenoxyl radicals*

Catalysed electron transfer from complexed metals, M, in various oxidation states to peroxidic compounds produces alkoxyl radicals via an 'organic' Fenton reaction, where H_2O_2 is replaced by an organic hydroperoxide:

$$M^{(n-1)+}\text{-complex} + ROOH \rightarrow M^{n+}\text{-complex} + RO^{\cdot} + OH^-$$

The oxidized complex M^{n+} can also react with the peroxidic compound, including lipid peroxides, to give a peroxyl radical and the reduced metal complex:

$$M^{n+}\text{-complex} + ROOH \rightarrow M^{(n-1)+}\text{-complex} + RO_2^{\cdot} + H^+$$

It should be emphasized that the reaction of Fe^{2+} ions with lipid hydroperoxides ($1.5 \times 10^3\,dm^3\,mol^{-1}\,s^{-1}$) is over one order of magnitude faster than their reaction with H_2O_2 ($76\,dm^3\,mol^{-1}\,s^{-1}$) (Garnier-Suillerot *et al.* 1984).

In organic solvents, high yields of alkoxyl radicals can be obtained by flash photolysis. One of the most studied is the *t*-butoxyl radical (Wong *et al.* 1982). It is generated by flash photolysis of a 1:2 mixture of benzene and di-*t*-butyl peroxide (Paul *et al.* 1978):

$$(CH_3)_3CO{-}OC(CH_3)_3 \xrightarrow{h\nu} 2\,(CH_3)_3CO^{\cdot}$$

In polar solvents, alkoxyl radicals are relatively unstable. However, pulse radiolysis investigations in aqueous solution have shown that alkoxyl radicals can be formed by reaction of hydrated electrons with a number of different peroxides with high rate constants (Greenstock and Wiebe 1981; Saran *et al.* 1983):

$$e^-_{aq} + ROOH \rightarrow RO^{\cdot} + OH^- \quad k = 6\text{–}16 \times 10^9\ dm^3\ mol^{-1}\ s^{-1}$$

Phenoxyl radicals can be readily prepared by pulse radiolysis of the corresponding phenols. Addition of $OH^{\cdot}$ radical results in the formation of dihydroxycyclohexadienyl radical which can eliminate water leading to phenoxyl radical with a λ_{max} at $\approx$ 400 nm (Land and Ebert 1967; Chrysochoos 1968), while the phenol oxidation induced by $N_3^{\cdot}$ leads directly to phenoxyl radical (Alfassi and Schuler 1985). In the latter reaction, the neutral form of phenol is oxidized several orders of magnitude more slowly than is the phenoxide ion.

The spectral observation of a number of substituted phenoxyl radicals of biological interest (Steenken and Neta 1982) and of flavonoid aroxyl radicals (Bors *et al.* 1990) has been carried out by pulse radiolysis using various oxidizing radicals including $^{\cdot}CH_2CHO$ and $N_3^{\cdot}$. The molar absorbance coefficient and the λ_{max} of these substitute phenoxyl and aroxyl radicals vary strongly with their structure.

Flash photolysis of phenols also gives rise to phenoxyl radicals via photoejection of $H^{\cdot}$ atoms or photoejection of e^- followed by deprotonation of the corresponding radical cation (Land *et al.* 1961, Grossweiner *et al.* 1963). Phenoxyl radicals are important models of their biological derivatives such as tyrosinyl and tocopheryl radicals.

4.2.4.3 *Biological production of peroxyl and alkoxyl radicals*

In vivo, a catalysed electron transfer from complexed transition metals, M, in various oxidation states to peroxidic compounds should produce alkoxyl and peroxyl radicals as described above in Section 4.2.4.2. The possible formation of $RO_2^{\cdot}$ radicals *in vivo* is deduced from indirect evidence. They are then considered to be involved in the lipid (LH) peroxidation processes via a chain reaction:

$$LH + OH^{\cdot} \rightarrow L^{\cdot} + H_2O$$

$$L^{\cdot} + O_2 \rightarrow LO_2^{\cdot}$$

$$LO_2^{\cdot} + LH \rightarrow LOOH + L^{\cdot}$$

Bacteria and mammalian cells are more sensitive to ionizing radiation in the presence of oxygen (see Chapter 10). This oxygen effect is considered to be related to molecular processes in which $RO_2^{\cdot}$ radicals are involved.

Another example of biological involvement of peroxyl radicals is given by the hepatotoxic activity of CCl_4. Studies with spin traps showed that CCl_4 is metabolized *in vitro* and *in vivo* to $CCl_3^{\cdot}$. Pulse radiolysis studies demonstrated that the peroxyl radical, $CCl_3O_2^{\cdot}$, is much more reactive than $CCl_3^{\cdot}$ (Packer *et al.* 1978), and can be considered as the main damaging reactive species responsible for such hepatotoxicity (Slater 1987)

In natural waters, photo-oxidation of cumene (isopropylbenzene) and pyridine gives products specific of alkylperoxyl ($RO_2^{\cdot}$) and hydroxyl radicals (Mill *et al.* 1980).

4.2.5 *Metal ions in the human body and the production of oxyradicals*

In vivo, metal ion catalysts are able to transform $O_2^{\cdot -}$, H_2O_2, and ROOH into very reactive species such as $OH^{\cdot}$, ferryl complexes, $RO^{\cdot}$, and $RO_2^{\cdot}$, via Fenton–Haber–Weiss reactions described in Sections 4.2.3.1 and 4.2.4.2. As the metal compound binds to the target or in its vicinity, the subsequent reactions of H_2O_2 and ROOH take place at this site and the reactive radical is formed in the vicinity of the target and has no chance to diffuse away. Important evidence for the possibility of these reactions *in vivo* comes from experiments performed with metal-chelating agents, such as desferrioxamine, produced by Ciba–Geigy as its methane sulphonate (Desferal). Such chelating agents can decrease or suppress iron-dependent generation of $OH^{\cdot}$ from H_2O_2, inhibit lipid peroxidation, and decrease tissue injury in animal models (Halliwell and Gutteridge 1989). Different types of iron and copper, present *in vivo* and able to participate in radical reactions, have been reported: DNA-bound iron, iron and copper ions bound to albumin, protein-bound iron such as in ferritin, and Cu–SOD.

4.2.6 *Hydrogen peroxide and hydroperoxides* in vivo

Hydrogen peroxide has been observed in $O_2^{\cdot -}$-generating systems such as mitochondria and microsomes. It is probably formed via a dismutation reaction of $O_2^{\cdot -}$. Moreover, several enzymes, such as glycolate oxidase, D-amino acid oxidase, and urate oxidase, are able to produce H_2O_2 directly (Oshino *et al.* 1975). Production of hydroperoxides (ROOH) *in vivo* occurs via $RO_2^{\cdot}$ reduction. They can also be produced enzymatically by lipoxygenase and cyclo-oxygenase.

4.2.7 *The nitric oxide radical ($NO^{\cdot}$) and the nitrogen dioxide radical ($NO_2^{\cdot}$)*

Marletta *et al.* (1988) and Moncada *et al.* (1989) have observed that activated macrophages and neutrophils produce both $O_2^{\cdot -}$ and $NO^{\cdot}$. These species rapidly combine to give the peroxynitrite anion $ONOO^-$, which could be an additional cytotoxic agent produced by these cell types (Beckman 1990). The pK_a of peroxynitrite is 6.8 at 37 °C. Its protonated form dissociates into two radicals, $OH^{\cdot}$ and $NO_2^{\cdot}$ (Feigl 1988). The biosynthesis of nitric oxide occurs from L-arginine and is accompanied by the formation of NO_2^- and NO_3^- (Moncada *et al.* 1989).

Nitrogen dioxide, $NO_2^{\cdot}$, is also an air pollutant, produced by various combustion processes. Nitrogen dioxide can be formed by $OH^{\cdot}$ and e_{aq}^- reaction with nitrite and nitrate, respectively, in aqueous solution (Prutz *et al.* 1985*a*):

$$OH^{\cdot} + NO_2^- \rightarrow OH^- + NO_2^{\cdot}$$

$$e_{aq}^- + NO_3^- \rightarrow NO_3^{\cdot 2-}$$

$$NO_3^{\cdot 2-} + H_2O \rightarrow 2OH^- + NO_2^{\cdot}$$

Tobacco smoke contains $NO_2^{\cdot}$ and this can abstract hydrogen from linoleic and linolenic acid and trigger lipid peroxidation *in vitro* and *in vivo*. Interest in nitrogen oxides also arises from the potential of their involvement in the production of carcinogenic *N*-nitrosamines.

In recent years there has been a blossoming of interest in the role of $NO^{\cdot}$ in biological systems. Typical references are Saran *et al.* (1990), Kanner *et al.* (1991), Murphy *et al.* (1991), Beckwith *et al.* (1992), Bowry and Ingold (1992), and Snyder and Bredt (1992).

4.3 Methods of detection of activated forms of oxygen

4.3.1 *Detection of singlet oxygen*

There are two main methods, the spectroscopic method and the chemical method.

4.3.1.1 *Spectroscopic detection and radiative lifetime*

The first of these uses the unambiguous infra-red phosphorescence spectrum of the (0, 0) transition O_2 ($^1\Delta_g \rightarrow {}^3\Sigma_g^-$), which has a maximum at 1268 nm in the gas phase and is called singlet 'monomol' emission (Khan 1985). A small (~ 5–10 nm) spectral shift to longer wavelengths has been observed in aromatic solvents relative to non-aromatic ones: in benzene the wavelength maximum is located at 1279 nm (Bromberg and Foote 1989). Fast infra-red germanium detectors have allowed flash photolysis (e.g. Rodgers and Snowden 1982) and pulse radiolysis (Gorman *et al.* 1989; Land 1991) to be used for time-resolved detection of $^1O_2$.

The red 'dimol' emission, produced by a collision of two singlet oxygen molecules, has also been used to identify $^1O_2$, for example, in the case of the redox cycling of paraquat in isolated rat liver microsomes (Cadenas *et al.* 1983). However, it should be noted that other red luminescence processes overlap this 634/704 nm dimol spectral region.

Since the transition O_2 ($^1\Delta_g \rightarrow {}^3\Sigma_g^-$) is strongly forbidden, its radiative lifetime, τ_r, is long and can be nearly 1 h in the absence of collisions in the gas phase at low pressure, and 8 s in air. In solution, the non-radiative channel of deactivation (rate constant k_{nr}) is much larger than that of the radiative channel (rate constant k_r). The total rate constant of singlet oxygen deactivation is $k_\Delta = k_r + k_{nr}$, and the phosphorescence consequently has an extremely low quantum efficiency, $\phi_p = \phi_\Delta k_r/k_\Delta$. The value of k_r is now accepted to be solvent dependent and, for example, is increased by 25 times in going from trifluoroethanol to carbon disulphide. A reasonably good correlation between the solvent polarizability (related to the solvent refractive index) and the rate constant k_r of the radiative deactivation has been established (Scurlock and Ogilby 1987; Schmidt and Afshari 1990). The non-radiative decay of singlet oxygen will be discussed in Section 4.4.1.1.

The challenge of singlet oxygen direct detection in biological environments has been discussed by Gorman and Rodgers (1992). They suggest that singlet oxygen intermediacy in photodynamic reactions *in vivo* may only be unambiguously demonstrable by future experiments using IR detection and high-power continuous wave excitation.

4.3.1.2 *Chemical detection*

The chemical methods of singlet oxygen identification in chemical and biological systems use quenchers (Q) such as cholesterol, dimethylcyclohexene, 1,3-diphenylisobenzofuran (DPBF), *p*-nitrosodimethylaniline (RNO), 1,4-diazabicyclo[2.2.2]octane (DABCO), and sodium azide (NaN_3). Some reports of singlet oxygen detection have also used the increased lifetime of singlet oxygen in deuterated water. These various techniques should be used with caution and their relevance is discussed below.

The most specific of the singlet oxygen scavengers is cholesterol. The product of its reaction is a 5α-hydroperoxide, as distinct from the products of its radical oxidation, which are mixtures of 7α- and 7β-hydroperoxides (Foote 1979). However, the rate constant of singlet oxygen quenching by cholesterol is rather low: $k = 5.7 \times 10^4\,dm^3\,mol^{-1}\,s^{-1}$ in benzene (Vever-Bizet *et al.* 1989), so that this method is not useful for the detection of low concentrations of $^1O_2$. Dimethylcyclohexene has also been used as a specific scavenger since the products determined by gas chromatography after reduction are very different from those formed via free radical autoxidation (Foote 1979).

In addition, furans have been used as chemical traps of singlet oxygen. For instance, the bleaching of DPBF in non-aqueous solvents monitored at 415 nm, ($\varepsilon_{415} = 2.21 \times 10^4\,dm^3\,mol^{-1}\,cm^{-1}$) has often been used to detect singlet oxygen (Gorman *et al.* 1978; Bensasson *et al.* 1981). However, the furans are very easily oxidized by peracids, peroxyl radicals, hydrogen peroxide, and halogens (Foote 1979), so that this technique can lack specificity.

A useful detector of singlet oxygen in aqueous solutions has been suggested by Kraljic and El Moshni (1978). These authors observed that RNO does not react with singlet oxygen itself but is bleached (λ_{max} at 440 nm) by the trans-annular peroxide (AO_2) formed by reactions of imidazole (A) or histidine (A) with singlet oxygen. Thus, the system imidazole or histidine plus RNO has been proposed as a sensitive and selective test for the presence of singlet oxygen in aqueous solutions.

In general, the rates of several reactions must be considered in applying such chemical techniques.

	Rate constant
$^1P + Q \rightarrow$ products	k_1
$^3P + Q \rightarrow$ products	k_2
$^3P + O_2 \rightarrow P + {}^1O_2$	k_3
$^1O_2 + Q \rightarrow$ products	k_4
$^1O_2$ + biological target $\rightarrow$ products	k_5

For the above it is essential to arrange that the rate of the reaction of $^1O_2$ with the quencher ($k_4[Q]$) is much greater than k_5 [biological target], and also that the rates of quenching of ^1P and ^3P ($k_1[Q]$ and $k_2[Q]$) are unimportant compared to the reaction rate of the photosensitizer with O_2 ($k_3[O_2]$) to give $^1O_2$.

Typically, in the case of DABCO, it should be emphasized that this amine reacts efficiently with singlet excited states of sensitizers with $k_1 \approx 10^9\,dm^3\,mol^{-1}\,s^{-1}$ (Young and Martin 1972) so that $k_1[Q]$ would be $\sim 10^4\,s^{-1}$ for [Q] of $10^{-5}\,mol\,dm^{-3}$. The rate constant for the reaction of DABCO with singlet oxygen ($k_4 \approx 10^7\,dm^3\,mol^{-1}\,s^{-1}$ in many organic solvents) must be compared with that of the acceptor (k_5) under study.

In the case of azide, the rate constants (k_2) for the reaction with the triplet states of the sensitizers, methylene blue and pyrene, are 5×10^5 and $< 5 \times 10^2\,dm^3\,mol^{-1}\,s^{-1}$, respectively, while the rate constant for energy transfer to ground state oxygen (k_3) is $3 \times 10^9\,dm^3\,mol^{-1}\,s^{-1}$. Thus, under typical conditions ($0.3\,mol\,dm^{-3}\,N_3^-$ and $5 \times 10^{-3}\,mol\,dm^{-3}\,O_2$) the quenching of the triplet state sensitizer by ground state oxygen efficiently competes with that by azide (Hasty *et al.* 1972). This is not always the case. For instance, N_3^- can very efficiently quench the singlet excited state of the cationic sensitizer, tetrakis(*p*-*N*-methylpyridyl) porphyrin $(H_2TMPyP)^{4+}$. The rate constant of singlet oxygen quenching by azide, k_2, (determined by Hasty *et al.* (1972) as $2.2 \times 10^8\,dm^3\,mol^{-1}\,s^{-1}$), must be compared with that by the chemical or biological acceptor under study.

The fact that the intrinsic decay rate constant, k_d, of singlet oxygen is smaller in D_2O ($1.7 \times 10^4\,s^{-1}$, $\tau = 60\mu s$) than in H_2O ($2.5 \times 10^5\,s^{-1}$, $\tau = 4\,\mu s$) by more than an order of magnitude (see Section 4.4.1.1) has been used as a means to investigate a possible role of singlet oxygen in a biological system. However, an isotope effect will occur only if the reaction of singlet oxygen with a quencher, Q, obeys the inequality $k_4[Q] \ll k_d$. When the reverse is true, the yield of oxidized products in H_2O will be equal to that in D_2O, even if the reaction occurs via singlet oxygen. Furthermore, in any comparisons between D_2O with H_2O solutions, conclusions should take into account not only a modification of the singlet oxygen lifetime but also modifications of the photophysical properties of the sensitizer. For example, excited singlet and triplet lifetimes of a sensitizer and quantum yields of fluorescence and triplet state formation can be increased in D_2O relative to H_2O as observed in the case of sulphonated aluminium phthalocyanine (Beeby *et al.* 1992; Simpson *et al.* 1992).

4.3.2 *Detection of superoxide anion and its protonated form*

4.3.2.1 *Optical detection by pulse radiolysis*

Both superoxide anion and its protonated form have distinct absorption spectra (Bielski 1978) shown in Fig. 3.2. However, because of the presence of other absorbing species, the direct detection of $O_2^{\cdot -}$ cannot normally be used. The following reaction with tetranitromethane has been suggested (von Sonntag 1987*a* and references therein):

$$C(NO_2)_4 + O_2^{\cdot -} \rightarrow C(NO_2)_3^- + NO_2^{\cdot} + O_2$$

the nitroform anion being monitored at 350 nm ($\varepsilon = 15\,000\ dm^3\ mol^{-1}\ cm^{-1}$).

4.3.2.2 *ESR detection of superoxide anion*

Superoxide anions can also be detected by ESR spin trapping. This involves the addition of $O_2^{\cdot -}$ to diamagnetic compounds, such as nitrones, leading to relatively long-lived free radicals, nitroxide spin adducts, which are easier to detect than the short-lived $O_2^{\cdot -}$ (Finkelstein *et al.* 1980, 1981). There are many disadvantages in this technique:

1. The rate constant for spin trapping is low. For example, for 5,5-dimethyl-1-pyrroline-N-oxide (DMPO), one of the most often used nitrones, it is 10–30 $dm^3\ mol^{-1}\ s^{-1}$ for $O_2^{\cdot -}$ and $6.6 \times 10^3\ dm^3\ mol^{-1}\ s^{-1}$ for $HO_2^{\cdot}$. A high concentration of spin trap (0.06 to 0.2 $mol\ dm^{-3}$) is thus necessary.

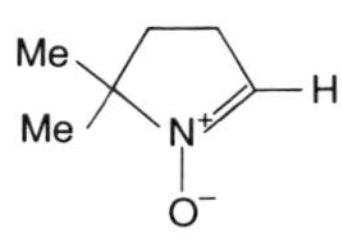

2. The $O_2^{\cdot -}$ spin adducts are usually unstable. In the case of DMPO, the spin adduct decays by a complex mechanism via uni- and bi-molecular kinetics (Buettner 1990).

4.3.2.3 *Spectrophotometric detection of superoxide anion and inhibition by superoxide dismutase (SOD)*

The reduction of ferricytochrome *c* to ferrocytochrome *c* is often employed to measure $O_2^{\cdot -}$ concentrations quantitatively. It is monitored by spectrophotometry using the change in molar absorption coefficient ($\Delta\varepsilon$ at 550 nm of 2.1×10^4 $dm^3\,mol^{-1}\,cm^{-1}$ and $\Delta\varepsilon$ at 418 nm of 7.0×10^4 $dm^3\,mol^{-1}\,cm^{-1}$). In tissue extracts, ferricytochrome *c* can be reduced by many other compounds, thus the $O_2^{\cdot -}$ concentration is evaluated from the (SOD)-inhibitable component of the total reduction. The cytochrome *c* must be purified by gel chromatography in order to avoid any contamination with SOD. Inhibition of a reaction by SOD is usually considered as an implication of $O_2^{\cdot -}$ in a reaction. The reaction with the metallo-enzyme SOD involves alternate reduction and oxidation of the copper, known as a 'ping pong' type mechanism, which has been investigated by pulse radiolysis (Klug-Roth and Rabani 1976; Goldstein *et al.* 1988 and references therein):

$$\text{SOD-Cu}^{2+} + O_2^{\cdot -} \rightarrow \text{SOD-Cu}^{+} + O_2$$

$$\text{SOD-Cu}^{+} + O_2^{\cdot -} \xrightarrow{2H^+} \text{SOD-Cu}^{2+} + H_2O_2$$

In certain cases, an excess of SOD can lead to oxidation of a biological substrate X by a displacement of the following equilibrium to the right:

$$O_2 + X \rightleftharpoons O_2^{\cdot -} + X^{\cdot +}$$

4.3.3 *Detection of the hydroxyl radical*

4.3.3.1 *Optical detection by pulse radiolysis*

The OH· radical spectrum in the ultraviolet, shown in Fig. 3.2, has been determined by pulse radiolysis (Nielsen *et al.* 1976; see also Czapski and Bielski 1993). The molar absorption coefficient at 188 nm is 540 $dm^3\,mol^{-1}\,cm^{-1}$. The extremely low value of this parameter, together with its rather inaccessible wavelength, make optical detection of OH· of little or no practical value. Consequently, in solution, it is usually monitored by the absorption of the product of its reactions.

4.3.3.2 *Spin trapping of OH· radicals*

As described above for the superoxide radical, the hydroxyl radical can also be trapped by DMPO as a long-lived spin adduct and detected by ESR (Finkelstein *et al.* 1980, 1981; Kuppusamy and Zweier 1989). Most of these trapping studies are qualitative but some investigators have reported kinetic studies (Yamazaki *et al.* 1990; Yamazaki and Plette 1991).

4.3.3.3 *Chemical detection*

The most common chemical methods are:

(1) conversion of methional and analogues to ethene gas;

(2) aromatic hydroxylation; and

(3) deoxyribose degradation.

Ethene production from reaction of hydroxyl radical with methional or analogues is readily measured by gas chromatography (GC). The rate constant for the reaction of OH· with methional, measured by pulse radiolysis, is $8.2 \times 10^9\,dm^3\,mol^{-1}\,s^{-1}$ (Bors *et al.* 1976). The scavenger chosen is usually 2-keto-4-methylthiobutyric acid (KMB), as it has a higher water solubility and a higher reactivity than methional. Furthermore, methional is avoided for its fetid odour. The radical generating system is incubated with KMB and a small sample of the headspace gas is used for quantitative analysis of ethene by GC. However, the method is not specific to the hydroxyl radical as it has been shown that alkoxyl and alkylperoxyl radicals can also produce ethene from methional (Pryor and Tang 1978). Thus the radical generating system should be further investigated by other techniques to fully identify the radicals present. KMB is not a satisfactory scavenger of radical processes *in vivo* because of its toxicity at high concentrations.

The presence of hydroxyl radicals can be identified and assayed by the formation of aromatic hydroxylated products using high-performance liquid chromatography (Grootveld and Halliwell 1986; Karam *et al.* 1991).

Damage of deoxyribose by hydroxyl radicals forms products which can react with thiobarbituric acid (TBA) in acidic conditions, and lead to the formation of a pink product with an absorption maximum at 532–535 nm (Halliwell and Gutteridge 1981).

Some drawbacks of chemical detection methods have been reported by Winterbourn (1987).

4.3.4 *Detection of peroxyl and alkoxyl radicals*

The presence of peroxyl (or dioxyl, Koppenol 1990*a*) and of alkoxyl radicals in a system can be inferred from indirect evidence via chemiluminescence or direct evidence via spin-trapping and ESR spectroscopy.

4.3.4.1 *Chemiluminescence*

Triplet carbonyls and singlet oxygen can be formed via self reaction of peroxyl radicals ($RO_2^{\cdot}$) by a concerted reaction (Kellogg 1969; Nakano *et al.* 1976) leading in two steps to the formation of singlet oxygen:

$$2{>}CHO_2^{\cdot} \rightarrow {>}CHOH + {}^3({>}C{=}O) + O_2$$

$${}^3({>}C{=}O) + O_2 \rightarrow {>}C{=}O + {}^1O_2$$

These two sequential reactions are rationalized in a Russell diagram (Fig. 4.2) showing the geometry of the transition state, a tetraoxide intermediate, and the products of the reaction in the solvent cage (Russell 1957; Kellogg 1969). The reaction exothermicity (480–630 kJ mol^{-1}) is localized in the carbonyl fragment which is in its excited triplet state, thus satisfying the spin selection rules. This carbonyl triplet excited state produces singlet oxygen by energy transfer.

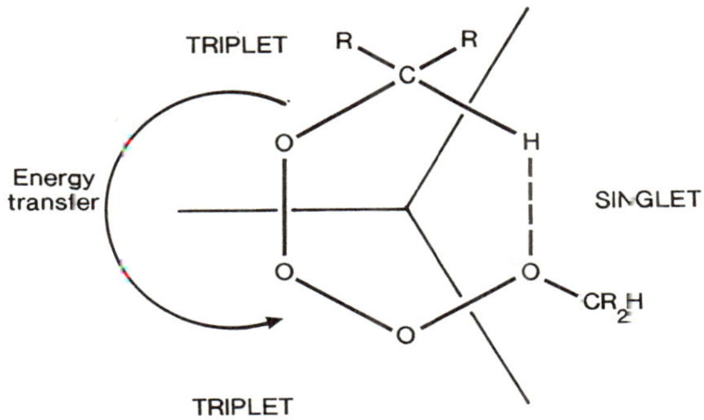

Fig. 4.2 'Russell diagram' showing the probable geometry of the transition state, the position of bond cleavage, and the energy transfer process from the incipient carbonyl to oxygen in a ROO· + ROO· reaction (adapted from Kellogg 1969).

The disproportionation of alkoxyl radicals can also be a source of triplet ketone (Phillips *et al.* 1967) via a Russell mechanism:

$$2{>}\dot{C}HO \rightarrow {>}CHOH + {}^3({>}C = O)$$

This chemiluminescence method of detection can be applied to tissues (Murphy and Sies 1990), but lacks specificity and must be complemented by ESR.

4.3.4.2 *ESR detection of peroxyl and alkoxyl radicals*

Detection of peroxyl radicals by spin trapping followed by ESR spectroscopy is difficult as their spin adducts appear to be much less stable than the hydroperoxyl adduct (Janzen 1980). However, ESR detection of peroxyl and alkoxyl radicals following reaction of hydroperoxides with haem-proteins has been possible (M. J. Davies 1988) using the spin trap DMPO.

4.3.5 *Detection of lipid peroxidation*

The detection of lipid peroxidation is considered to be evidence of free radical reactions occurring via a radical chain reaction propagated via peroxyl radicals as described in Section 4.2.4.3. Lipid peroxidation destroys the integrity of the biological membranes and also membranal enzymes and proteins. It is a highly destructive process involved in toxicology, ageing, and many human diseases.

Three main biochemical methods are being utilized for analysis of lipid peroxidation. These methods can all lead to artifactual results and need special precautions, which have been reviewed by Kappus (1985) and Halliwell and Gutteridge (1989). In addition, low level chemiluminescence is also used as an assay for lipid peroxidation.

4.3.5.1 *Conjugated dienes*

During lipid peroxidation, conjugated dienes are formed from the reaction of the lipid radical with oxygen (Figs 4.3 and 4.4). Both *cis* and *trans* isomers are produced. They can be assayed by their UV absorption maxima at 236 and 233 nm with molar absorption coefficients $\varepsilon = 26\,000$ and $28\,000\ \mathrm{dm^3\,mol^{-1}\,cm^{-1}}$, respectively, the method being applicable for studies of peroxidized lipids formed either *in vivo* or *in vitro*. It is not a very sensitive method and it is difficult in the case of *in vivo* investigations to obtain control samples. However, the method can identify which organ or organelle has suffered lipid peroxidation.

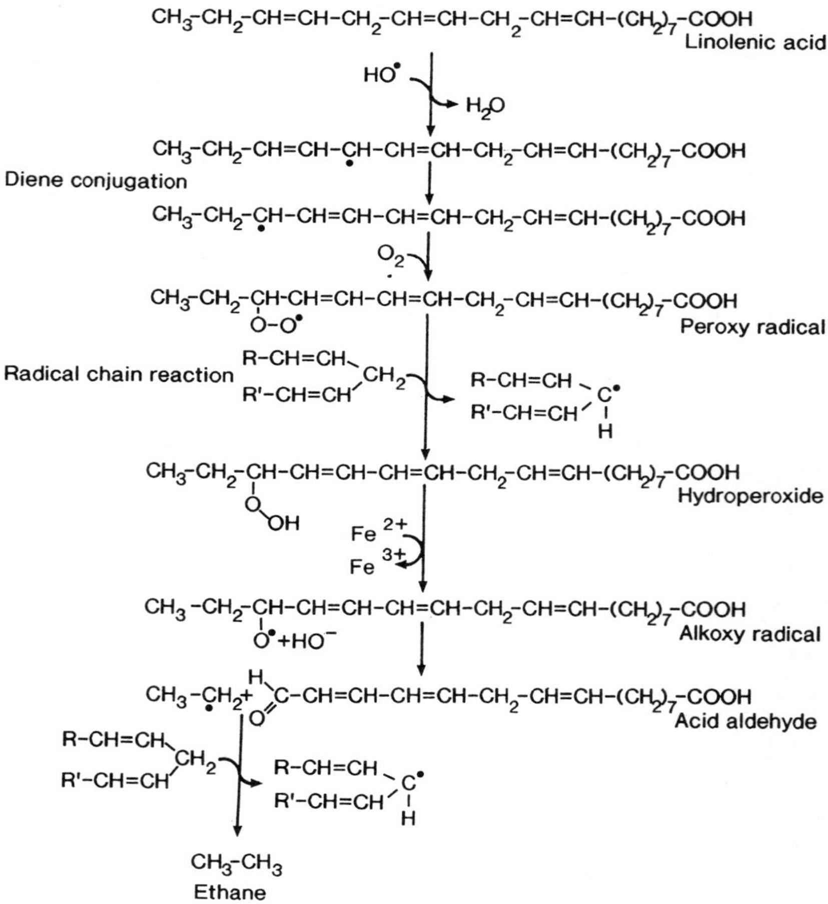

Fig. 4.3 Scheme representing the formation of ethane during lipid peroxidation induced by hydroxyl radicals. The steric configuration is not shown (from Kappus 1985).

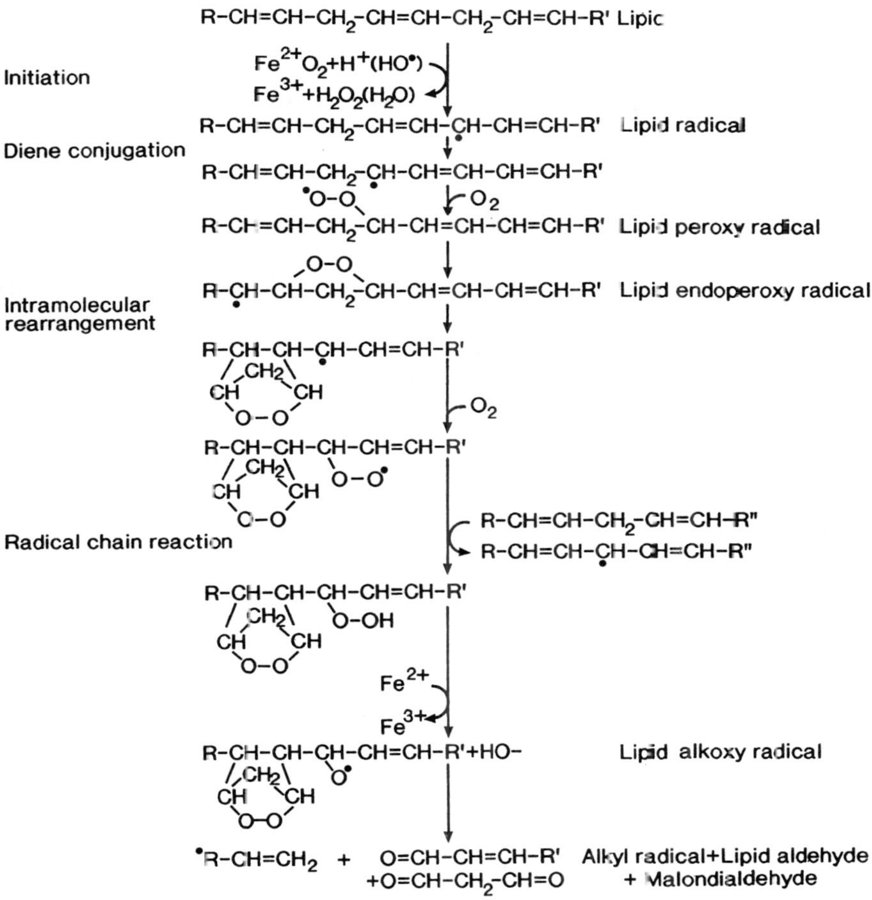

Fig. 4.4 Scheme representing the formation of malondialdehyde during lipid peroxidation induced by ferrous–oxygen complexes or by hydroxyl radicals (from Kappus 1985).

4.3.5.2 *Malondialdehyde*

Malondialdehyde (MDA), as a product, can be a useful indicator of *in vitro* and *in vivo* lipid peroxidation (Fig. 4.4). It reacts with thiobarbituric acid (TBA) and its concentration can be determined from the visible absorbance of its reaction products. Under acidic conditions, one molecule of MDA reacts with two molecules of TBA with the elimination of two molecules of water to produce a Schiff's base with an absorption maximum at 532–535 nm.

4.3.5.3 *Alkanes*

Alkanes are formed during lipid peroxidation via the series of reactions depicted in Fig. 4.3. The collecting of exhaled hydrocarbons, mainly ethane, is considered as the best non-invasive method to monitor lipid peroxidation *in vivo*. In their CCl_4-toxicity investigations, Lindstrom and Anders (1978) have observed a good correlation between ethane exhalation and hepatic levels of conjugated dienes. Pretreatment of mice by an antioxidant, such as vitamin E, suppresses ethane exhalation.

However, alkanes are minor products of peroxidation, less important than malondialdehyde, and are metabolized by the liver. Furthermore, they can be produced by oxidation of other substrates or by gut bacteria.

4.3.5.4 *Low level chemiluminescence*

Low level chemiluminescence of tissue fractions has been associated with lipid peroxidation (Cilento 1982; Cadenas *et al.* 1984; Murphy and Sies 1990). It is considered that chemiluminescence might arise as described in Section 4.3.4.1 via (i) the self reaction of peroxyl or alkoxyl radicals and (ii) the exergonic thermolytic decomposition of dioxetanes. In both cases the luminescent molecules would be mainly carbonyl triplets (380–450 nm), usually considered as poor light emitters, and singlet oxygen.

Another chemiluminescence pathway occurring both *in vitro* and *in vivo* was proposed some time ago by Linschitz (1961). This was based on an electron exchange process observed in the porphyrin-catalysed decomposition of tetralin peroxide. This process, revived by many groups including Koo *et al.* (1978) and Kaklika *et al.* (1979), has been called 'chemically initiated electron-exchange luminescence' or 'CIEEL' (Koo and Schuster 1978). The reaction usually proceeds via a dark catalysis of a dioxetanone decomposition by a fluorescer F of low oxidation potential:

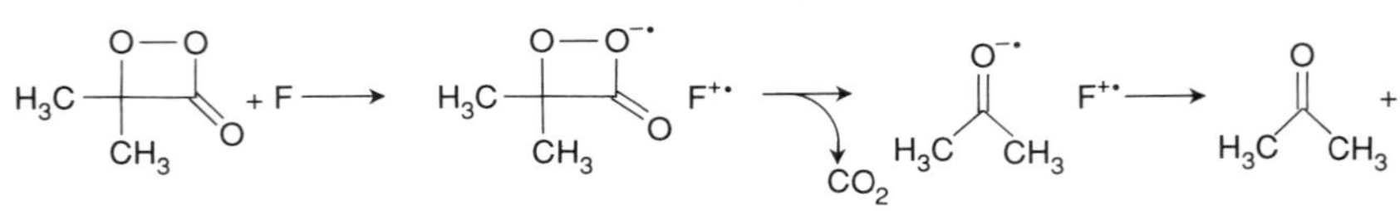

This complex process, involving two successive electron transfers, has been thoroughly discussed by Wilson (1985). A spectrum of mechanisms between a complete electron transfer and a charge transfer complex between the fluorescer and the dioxetane has been proposed (Catalani and Wilson 1989; McCapra 1990). It should be emphasized that dioxetanones are peroxides involved in the bioluminescence of fireflies (Shimomura 1982).

More sensitive photodetectors which are becoming available should open new pathways in the investigation of oxidative luminescent processes in tissues and organs.

4.4 Chemical properties of the activated forms of oxygen

4.4.1 *Reactions of* $^1O_2$

The $^1O_2$ reaction modes and products have been reviewed, for example by Wasserman and Murray (1979) and by Frimer (1985). A compilation of rate constants for its reactions has been published (Wilkinson and Brummer 1981, with a new edition in preparation). In this section some typical examples of the main reactions of singlet oxygen are presented.

Singlet oxygen interacts with other molecules in four ways: it can transfer its excitation energy, returning to the ground state while the interacting molecules enter (i) a vibrational or (ii) an electronic excited state. The third and fourth reactions are chemical combination usually with compounds containing carbon–carbon double bonds or else formation of a charge transfer intermediate which can lead to $O_2^{\cdot-}$.

4.4.1.1 *Intermolecular electronic-to-vibrational energy transfer*

In solution, the radiationless route of singlet oxygen physical deactivation is solvent dependent. Its second-order deactivation rate constant is composed additively of increments k_{xy} corresponding to energy-accepting oscillators of xy atom pairs, including C—H, C—D, O—H, O—D. . . . or for various atomic groupings including $—CH_3$, $>CH_2$. Such increments were determined by Hurst and Schuster (1983), Rodgers (1983*b*), and Schmidt and Brauer (1987). They allow a calculation of the singlet oxygen lifetime τ_Δ and the corresponding second-order rate constant k_d for a given solvent molecule X_nY_m, via the following equation:

$$k_d = (\tau_\Delta[S])^{-1} = \sum_{xy} N_{xy} k_{xy}$$

with [S] the molar concentration of the solvent and N_{xy} the number of particular xy pairs occurring in the interacting molecule.

Additivity values have also been determined using solute molecules: an increment of $570\,dm^3\,mol^{-1}\,s^{-1}$ per $—CH_2—$ has been found for the quenching of $^1O_2$ by saturated fatty acids in C_6D_6 and CD_3OD (Vever-Bizet *et al.* 1989) to be compared with $813\,dm^3\,mol^{-1}\,s^{-1}$ per $—CH_2—$ found for neat *n*-alkanes and *n*-alcohols (Rodgers 1983*b*). A strong exponential correlation was found between k_{xy} and the highest fundamental vibrational energy of the respective oscillator (Schmidt and Brauer 1987). This explains the important difference between the lifetimes of $^1O_2$ in H_2O (4 μs) and D_2O (60 μs) and, of course, between all hydrogenated and perdeuterated solvents (Table 4.1).

Table 4.1 Singlet oxygen lifetimes, $\tau_{\Delta}(\mu s)$ ($\pm$ 10 per cent), determined by time-resolved luminescence measurements

Solvent	Lifetime
Benzene	30
Benzene-d_6	630
Chlorobenzene	46
Chlorobenzene-d_5	1200
Hexafluorobenzene	3900
Toluene	27
Toluene-d_8	320
Acetone	50
Acetone-d_6	750
Acetonitrile	62
Acetonitrile-d_3	520
Carbon disulphide	1400
Water	4
Water-d_2	60
SDS (0.01 mol dm^{-3})/water-d_2	57
CTAB (0.01 mol dm^{-3})/water-d_2	56
Triton X-100 ($\sim$0.03 mol dm^{-3})/water-d_2	32

4.4.1.2 *Intermolecular electronic-to-electronic energy transfer*

For an efficient energy transfer to a molecule A in its singlet ground state:

$$^1O_2 + A \rightarrow {}^3O_2 + {}^3A$$

the triplet energy level of A must be lower than the energy level of $^1O_2$.

Polyenes are the best known among the molecules able to perform this type of $^1O_2$ quenching. An efficient quenching occurs for molecules with a number of conjugated double bonds (n) greater than 9. This is consistent with an estimation of their triplet energy level as discussed in detail in Chapter 7.

Azomethine dyes quench singlet oxygen with rate constants as high as $6 \times 10^9\ dm^3\ mol^{-1}\ s^{-1}$. Their triplet energy levels have been estimated to be $\leq 87\ kJ\ mol^{-1}$ by Herkstroeter (1975*a*). Nickel(II) complexes, used as photostabilizers, are also able to quench singlet oxygen by an energy transfer mechanism (Evans 1980).

4.4.1.3 *Addition to C=C double bonds*

G.O. Schenck and Ziegler (1944) observed the formation of a transannular 1,4-peroxide via a photosensitized oxygenation of α-terpinene, using chlorophyll as sensitizer. This was the first synthesis of ascaridole, a constituent of chenopodium (Mexican tea).

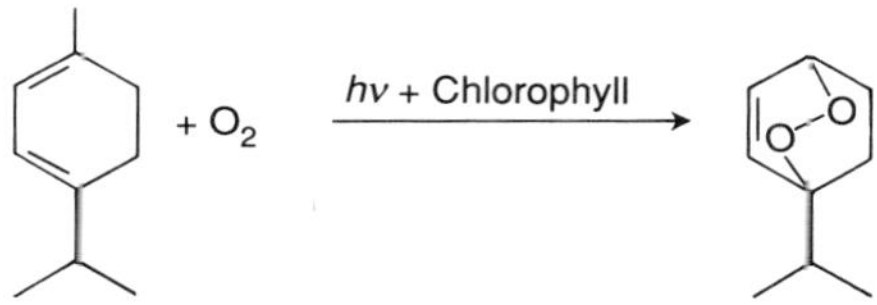

α-Terpinene Ascaridole

This reaction is now considered as the natural pathway of ascaridole biogenesis. Transannular peroxides, previously termed epidioxides or photo-oxides, are now called endoperoxides (G.O. Schenck 1949). The prefix endo means *internal* from the greek word 'endon' = within. The first discovery of a photochemical transannular addition of oxygen to an aromatic hydrocarbon was established for rubrene by Moureu *et al.* (1926) and Clar and John (1930):

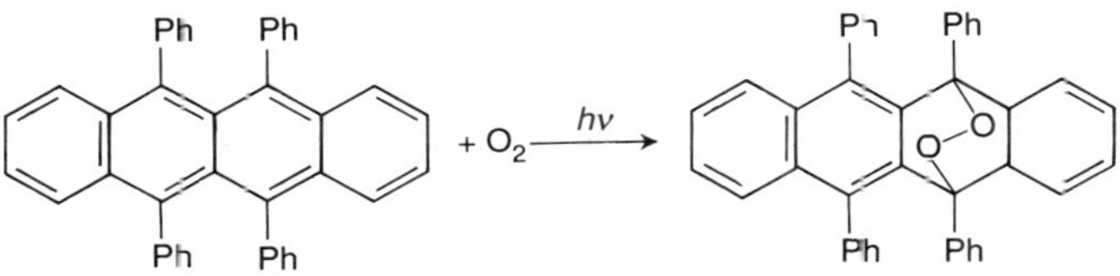

Rubrene

This reaction has now been extended to numerous polynuclear hydrocarbons such as anthracene.

The photoaddition of $^1O_2$ to conjugated dienes (containing two double bonds separated by a single bond) can occur via a cycloaddition reaction:

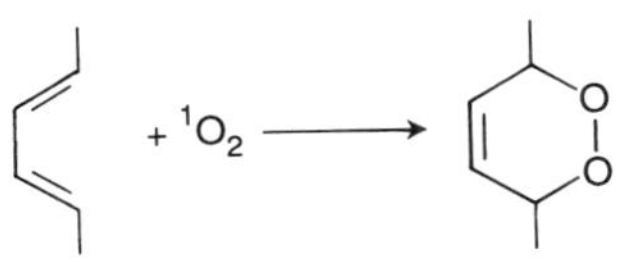

This is related to the classical thermal Diels–Alder reaction (Denny and Nickon 1973). The products of this photoaddition are also called endoperoxides.

With electron-rich-donating atoms, such as N or S, adjacent to a double bond, singlet oxygen can react to give an unstable dioxetane which decomposes thermally to give a carbonyl triplet (Schulte-Herbrüggen and Cadenas 1985):

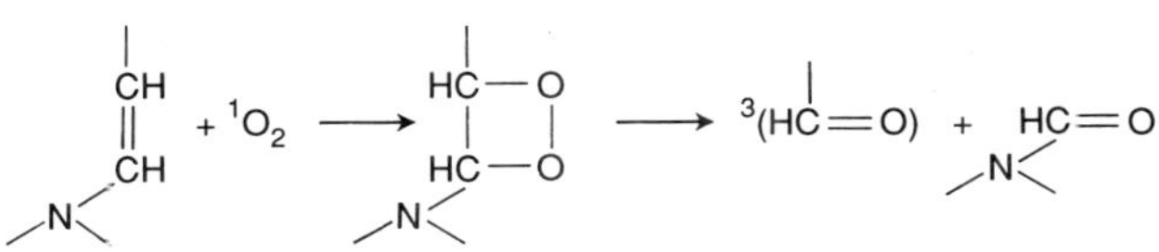

Dioxetanes are involved in a number of bioluminescence processes occurring via electron transfer (Section 4.3.5.4).

Singlet oxygen reacts with alkyl-substituted olefins and can produce allylic hydroperoxides with a shifted double bond as required by an ene-type reaction:

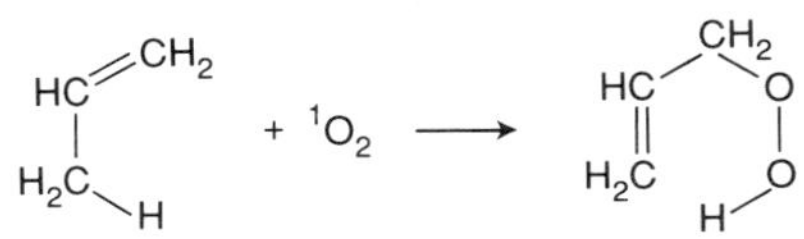

This reaction, discovered by G.O. Schenck (1948), is of great importance in synthesis as there is no loss of optical activity in the oxidation of optically active olefins (Stephenson 1980). There has been some use of this ene reaction in the perfumery industry to synthesize rose oxide. The corresponding specific reaction with cholesterol was described earlier (Section 4.3.1.2). Other related reactions involving $^1O_2$ are discussed in Chapters 5 and 6.

4.4.1.4 *Charge transfer quenching*

The reduction potential E_7 for the couple $[O_2(^1\Delta_g)/O_2^{\cdot -}]$ is given by:

$$E_7\,[O_2\,(^1\Delta_g)/O_2^{\cdot -}] = E_7[^3O_2/O_2^{\cdot -}] + \Delta E_{00}\,[O_2\,(^1\Delta_g)]$$

with $E_7\,[^3O_2/O_2^{\cdot -}] = -0.330$ V (NHE) as given by Ilan *et al.* (1974) and ΔE_{00} $[O_2(^1\Delta_g)] = +0.977$ V (Herzberg 1950), so that $E_7\,[O_2(^1\Delta_g)/O_2^{\cdot -}] = +0.647$ V (NHE). This shows that in comparison with ground state oxygen, singlet oxygen is a more efficient electron acceptor.

A correlation between the logarithm of the rate constant of singlet oxygen quenching and the ionization potential of the quenchers, either amines (Young and Martin 1972) or sulphides (Young and Brewer 1979) or phenols (Scurlock *et al.* 1990), suggests a charge transfer mechanism. This is related to the Rehm and Weller (1970) mechanism of fluorescence quenching involving a collision complex, followed by the formation of a charge transfer complex:

$$^1O_2 + A \rightleftarrows [^1O_2 \ldots A] \rightleftarrows [O_2^{\cdot -} \ldots A^{\cdot +}] \rightarrow O_2 + A$$

With a polar solvent and an amine of low ionization potential, a complete electron transfer occurs to form a radical cation and the superoxide anion, as demonstrated in the case of *N,N,N′,N′*-tetramethyl-*p*-phenylenediamine (Manring and Foote 1982). The $O_2^{\cdot -}$ formation has also been observed for the mitochondrial component, nicotinamide adenine dinucleotide phosphate and its analogues (G. Peters and Rodgers 1981) and for *N,N*-dimethylaniline and other electron donors (Saito *et al.* 1983).

Time resolved studies of $^1O_2$ quenching by thiols (Rougee *et al.* 1988) and phenols (Scurlock *et al.* 1990) in aqueous solution at different pH values show that the reaction occurs via a charge transfer from the thiolate, RS^-, and the phenolate, ϕO^-, respectively. In the case of singlet oxygen quenching by the azide ion, N_3^-, already discussed in Section 4.3.1, the evidence of a charge transfer complex is implied by the identification of the azide radical, $N_3^{\cdot}$, using the spin trapping technique (Harbour and Issler 1982). It is tempting to suggest that the quenching of singlet oxygen by iodide ion, I^-, also occurs via a charge transfer complex (Rosenthal and Frimer 1976).

4.4.2 *Thermodynamics of oxygen reduction*

The reduction potential diagram below summarizes the relations between the various states of oxygen (see, for example, Sawyer 1990):

$$O_2 \xrightarrow{-0.33} O_2^{\cdot -} \xrightarrow{+0.89} H_2O_2 \xrightarrow{+0.38} OH^{\cdot} \xrightarrow{+2.3} H_2O$$

$$O_2 \xrightarrow{+0.28} H_2O_2 \xrightarrow{+1.35} H_2O$$

$$O_2 \xrightarrow{+0.815} H_2O$$

The numbers are the potentials in volts for the successive reduction of O_2, in aqueous solution, at pH 7 and with an oxygen pressure of 1 atmosphere (versus NHE).

Ground state oxygen is a rather strong oxidant (+0.815 V) taking into consideration its tetravalent overall reduction into H_2O. The two-electron reduction step leading to H_2O_2 has a potential equal to ½(0.89–0.33) = 0.28 V, and this is markedly smaller than the potential value for the two subsequent steps to water. Ground state oxygen is even less oxidizing for the first one-electron reduction step (−0.33 V). This barrier of −0.33 V is of the greatest chemical and biological importance as it prevents destructive oxidations which are thermodynamically favoured by the overall reduction potential O_2/H_2O = +0.815 V.

The reduction potential of the $O_2/O_2^{\cdot -}$ couple is −0.16 V if referred to a 1 mol dm^{-3} O_2 instead of the oxygen pressure of one atmosphere. This potential is lower in aprotic solutions and in DMF $E(O_2/O_2^{\cdot -}) = -0.84$ V versus SCE if one atmosphere pressure is the standard state, and −0.60 V if 1 mol dm^{-3} is the standard state (Sawyer *et al.* 1982*b*).

The reduction potentials are modified by pH as shown in Fig. 4.5. Furthermore, the acid–base properties of the three reduced species of O_2 are:

$$HO_2^{\cdot} \rightleftarrows O_2^{\cdot -} + H^+ \qquad pK_a = 4.8$$

$$H_2O_2 \rightleftarrows HO_2^- + H^+ \qquad pK_a = 11.7$$

$$OH^{\cdot} \rightleftarrows O^{\cdot -} + H^+ \qquad pK_a = 11.9$$

At pH 7, the superoxide anion can interact with a couple $S^{\cdot +}/S$ whose redox potential is between −0.33 and +0.89 V by two possible reactions:

$$O_2^{\cdot -} + S^{\cdot +} \rightarrow O_2 + S$$

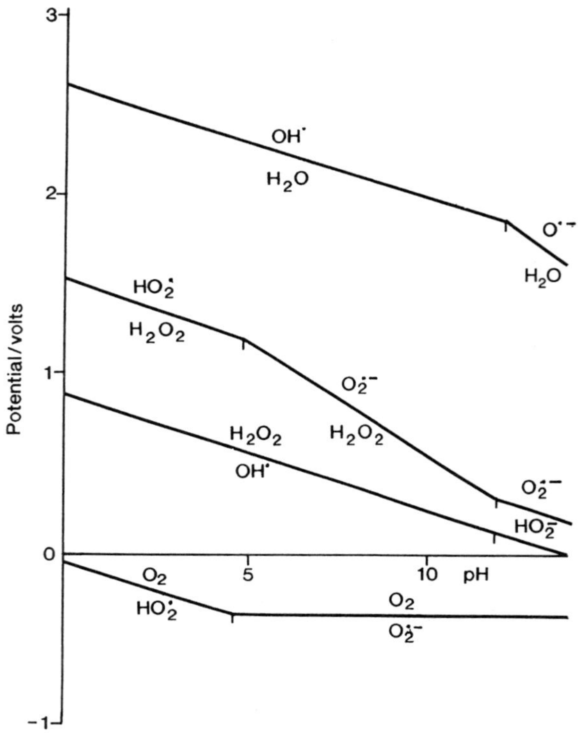

Fig. 4.5 Potential-against-pH diagram for species derived from oxygen.

$$O_2^{\cdot -} + S^{\cdot +} \xrightarrow{2H^+} H_2O_2 + S$$

The fact that these reactions are thermodynamically feasible does not mean that they are kinetically possible. It is essential therefore to determine the rate constants in each case. One such example is the Fenton–Haber–Weiss reaction which only occurs in presence of redox active metal ions as described in Section 4.2.3.1. Another example is:

$$O_2^{\cdot -} + O_2^{\cdot -} \xrightarrow{2H^+} H_2O_2 + O_2$$

which only occurs extremely slowly, probably due to poor overlap of the orbitals involved (Koppenol and Butler 1977).

4.4.3 *Reactions of $O_2^{\cdot -}$/$HO_2^{\cdot}$*

In aqueous solution and other protic solvents, the negative charge of $O_2^{\cdot -}$ and its high energy of solvation make it rather inert. For example, it does not react with negatively charged phospholipids, nucleic acids, and other polyanions. It is in equilibrium with its protonated form, the hydroperoxyl radical $HO_2^{\cdot}$ with a pK_a of 4.8. Thus, in physiological conditions, nearly 1 per cent of $O_2^{\cdot -}$ is in the protonated form, which is much more reactive (Gebicki and Bielski 1981; Bielski *et al.* 1983) and can lead to dismutation reactions. In organic and aprotic solvents, superoxide anion is not in equilibrium with its protonated form, thus dismutation does not occur. Reactions in these environments are of great interest due to their similarity with the hydrophobic phospholipid bilayer of membranes. Some examples of reactions in hydrophilic and hydrophobic media are described below.

4.4.3.1 *Protonation and disproportionation in aqueous solution*

The kinetics for the protonation and disproportionation of $O_2^{\cdot -}$ have been determined by pulse radiolysis in water and reviewed by Bielski *et al.* (1985):

$$HO_2^{\cdot} \rightleftarrows H^+ + O_2^{\cdot -} \quad pK_a = 4.8 \text{ in water}$$

$$HO_2^{\cdot} + O_2^{\cdot -} \xrightarrow{H^+} H_2O_2 + O_2 \; (k = 8.5 \times 10^7 \, dm^3 \, mol^{-1} \, s^{-1})$$

$$HO_2^{\cdot} + O_2^{\cdot -} \xrightarrow{H_2O} H_2O_2 + O_2 + OH^- \; (k = 9.7 \times 10^7 \, dm^3 \, mol^{-1} \, s^{-1})$$

$$HO_2^{\cdot} + HO_2^{\cdot} \rightarrow H_2O_2 + O_2 \; (k = 8.3 \times 10^5 \, dm^3 \, mol^{-1} \, s^{-1})$$

$$O_2^{\cdot -} + O_2^{\cdot -} \xrightarrow{2H^+} H_2O_2 + O_2 \; (k < 0.3 \, dm^3 \, mol^{-1} \, s^{-1})$$

It should be emphasized that, in the close proximity of membranes, the pH has been reported to be lower than the physiological pH of 7.4, and thus $HO_2^{\cdot}$ can attack a double-allylic H atom of linolenic, linoleic, and arachidonic fatty acids and produce peroxides with a reaction exothermicity of $\approx 42 \, kJ \, mol^{-1}$ (Bielski *et al.* 1983).

On the basis of thermodynamic considerations, singlet oxygen could be formed via dismutation of $O_2^{\cdot -}$ (Koppenol 1978). The experimental evidence for this formation has remained controversial (Aubry *et al.* 1981). However, Sugimoto *et al.* (1988) have obtained evidence for the transient formation of hydrogen tetraoxide (HOOOOH) from the dimerization of $HO_2^{\cdot}$ and its concerted homolytic dissociation to singlet oxygen and hydrogen peroxide.

4.4.3.2 $O_2^{\cdot -}$ *as a reductant*

The superoxide anion, $O_2^{\cdot -}$, even at low pH reduces many metal cations, such as Fe^{III} and Cu^{II}, free, chelated or in proteins. These reductions involving Fenton–Haber–Weiss reactions (see Section 4.2.3.1) lead to the formation of oxidizing species, including $OH^{\cdot}$ radical and the ferryl ion. These are considered as the main cause of $O_2^{\cdot -}$ toxicity *in vivo*. The superoxide anion reduces some quinones, depending upon the one-electron reduction potential of the quinone, leading to the formation of semiquinone radicals (Patel and Willson 1973).

As mentioned in Section 4.3.2.3, pulse radiolysis has demonstrated the mechanism of the catalytic dismutation of $O_2^{\cdot-}$ by an alternate reduction and oxidation of the copper in the metalloenzyme SOD. The rate constant of the reaction of $O_2^{\cdot-}$ with SOD–Cu^{II} is $1.6 \times 10^9\ dm^3\ mol^{-1}\ s^{-1}$. Similar reactions occur for many other Cu^{II} complexes.

The superoxide anion can also react in aqueous solution with organic halides, such as bromouridine, via a dissociative electron transfer:

$$O_2^{\cdot-} + BrU \rightarrow U^{\cdot} + Br^- + O_2$$

whereas in aprotic solvents, it reacts as a nucleophile.

4.4.3.3 *Nucleophilic reactions*

In aprotic media, $O_2^{\cdot-}$ is a good nucleophile and can produce hydroperoxides, peroxides, alcohols, or carbonyl compounds (Frimer 1983). For instance, with halides RX it can give dialkylperoxides via the following scheme:

$$RX + O_2^{\cdot-} \rightarrow RO_2^{\cdot} + X^-$$

$$RO_2^{\cdot} + O_2^{\cdot-} \rightarrow ROO^- + O_2$$

$$ROO^- + RX \rightarrow ROOR + X^-$$

4.4.3.4 *Chain processes involving $O_2^{\cdot-}/HO_2^{\cdot}$*

Chain processes initiated by $O_2^{\cdot-}/HO_2^{\cdot}$ which result in multiple damage per initial radical and might have deleterious effects *in vivo* have been proposed: the chain oxidation of linoleic acid by $HO_2^{\cdot}$ (Gebicki and Bielski 1981), the addition of $O_2^{\cdot-}$ to ketomalonate leading to decarboxylation in oxygenated aqueous solutions (Schuchmann *et al.* 1991), and the oxidation of a thiol, dithiothreitol (Zhang *et al.* 1991).

4.4.3.5 *The Czapski hypothesis of $O_2^{\cdot-}$ toxicity*

The toxicity of $O_2^{\cdot-}$ is usually attributed to a reduction of a metal ion M^{n+} bound to a biological target, leading to the production of the highly deleterious hydroxyl radical via a Fenton–Haber–Weiss reaction with $OH^{\cdot}$ reacting at the site of its formation:

$$O_2^{\cdot-} + \text{biol-}M^{n+} \rightarrow O_2 + \text{biol-}M^{(n-1)+}$$

$$\text{biol-}M^{(n-1)+} + H_2O_2 \rightarrow (\text{biol-}M^{n+} \ldots OH^{\cdot}) + OH^-$$

$$(\text{biol-}M^{n+} \ldots OH^{\cdot}) \rightarrow \text{biological damage}$$

However, as emphasized by Czapski *et al.* (1988), many other reducing agents are present in cells at much higher concentrations than superoxide, such as glutathione, vitamin C, and NADPH. These authors have thus presented the following hypothesis: the uniqueness of $O_2^{\cdot-}$ toxicity may arise from its ability to oxidize rather than reduce metal compounds

$$O_2^{\cdot-} + M^{n+} + 2H^+ \rightarrow O_2 + M^{(n+1)} + H_2O_2$$

The oxidized metals, Fe^{IV}, Cu^{III} or Mn^{III}, are highly oxidizing species able to attack the biological target. If bound to a biological molecule, the oxidized metal may cause an alteration of the binding site via an intramolecular electron transfer.

4.4.4 *Reactions of OH·*

There are three main types of reactions: electron transfer, H abstraction, and addition to double bonds, all with high rate constants in the range 10^9–10^{10} $dm^3\,mol^{-1}\,s^{-1}$. The kinetic data of these reactions have been critically reviewed by Buxton *et al.* (1988).

4.4.4.1 *Electron transfer*

The hydroxyl radical, OH·, is the most reactive oxidizing free radical. At 25 °C its standard reduction potential is 2.31 V (Koppenol 1989) at pH 7, and even higher in acidic solutions as shown in Fig. 4.5. It oxidizes metal ions, M^{n+}, to higher oxidation states via a hydrated hydroxide ion:

$$OH^{\cdot} + M^{n+} \rightarrow (M\,OH)^{n+} \xrightarrow{H^+} M^{n+1} + H_2O$$

In the case of the oxidation of Fe^{2+} to Fe^{3+}, the rate constant of reaction is $3 \times 10^8\,dm^3\,mol^{-1}\,s^{-1}$. It can also oxidize organic compounds such as chlorpromazine to give its radical cation (Asmus *et al.* 1979).

4.4.4.2 *Hydrogen abstraction*

The OH· radical reacts with most C—H bonds with rather high rate constants via H atom abstraction to give water and an organic radical. As the O—H bond energy in water (ΔH (O—H) = 495 kJ mol^{-1}) is higher than the C—H bond energy (ΔH (C—H) = 385 to 410 kJ mol^{-1}), the abstraction of carbon-bound H atoms is always exothermic and the relative yield of abstraction of oxygen-bound H atoms is very low.

4.4.4.3 *Addition to double bonds*

Hydroxyl radicals add to double bonds and give secondary radicals which, in the presence of O_2, lead to the formation of peroxyl radicals. For example, in the case of thymine, uracil, and cytosine, the addition occurs mainly at C5 with only a minor reaction at C6. Addition of oxygen leads to the formation of peroxyl radicals which can abstract an H atom from a sugar moiety in nucleosides. Oxygen addition then leads to peroxyl radicals of the sugar moiety which undergo a sequence of reactions, producing finally strand breakage (see Chapter 5).

4.4.4.4 *Reactions of $O^{\cdot -}$*

In strongly alkaline solution, far removed from biological conditions, OH· is converted to its conjugate $O^{\cdot -}$:

$$\overset{OH^-}{OH^{\cdot} \rightleftarrows O^{\cdot -} + H_2O}$$

with $pK_a = 11.9$. This basic form of the radical has properties different from those of $OH^{\cdot}$. Electron transfer with inorganic ions is much slower for $O^{\cdot -}$. With an aromatic molecule carrying an aliphatic chain, $O^{\cdot -}$ abstracts an $H^{\cdot}$ atom from the aliphatic chain whereas $OH^{\cdot}$ adds preferentially to the aromatic ring (Neta *et al.* 1972).

4.4.5 *Reactions of peroxyl ($RO_2^{\cdot}$) and alkoxyl, phenoxyl, and aroxyl ($RO^{\cdot}$) radicals*

With the exception of some purine-derived radicals, all carbon-centred free radicals react with oxygen to give peroxyl radicals, which play an important role in living systems as ultimate agents in oxygen toxicity. Peroxyl radicals derived from most biological molecules have been investigated and their unimolecular and bimolecular reactions surveyed (von Sonntag 1987*a*; Neta *et al.* 1990; von Sonntag and Schuchmann 1991).

One of the most important unimolecular reactions of some peroxyl radicals is the elimination of $O_2^{\cdot -}/HO_2^{\cdot}$.

$$\overset{\displaystyle OO^{\cdot} \atop \displaystyle |}{>COH} \rightarrow >C{=}O + O_2^{\cdot -} + H^+$$

In the case of peroxylglycine radicals, the $O_2^{\cdot -}$ elimination rate constant is $1.5 \times 10^5\ s^{-1}$ at pH 7.9 (Abramovitch and Rabani 1976).

Peroxyl radicals undergo termination reactions with each other to form intermediate tetraoxides and to produce an excited carbonyl compound, an alcohol, and $^1O_2$, according to a Russell mechanism (as described above in Section 4.3.4.1). The yields of singlet oxygen via the Russell mechanism range from 3.4 to 6 per cent for primary and from 3.9 to 14 per cent for secondary alkylperoxyl terminations (Niu and Mendenhall 1990). The decomposition of the tetraoxide to give molecular oxygen and two oxyl radicals is another possibility (Ingold 1969):

$$2{>}CHOO^{\cdot} \rightarrow {>}CHOOOOCH{<} \rightarrow O_2 + 2\,{>}CHO^{\cdot}$$

The rate constants of the above reactions were determined by pulse radiolysis. In particular, it was shown that peroxyl radicals of linoleate and linolenate systems disappear essentially by a second-order process with a rate constant $2k = 10^7\ dm^3\ mol^{-1}\ s^{-1}$ (Hasegawa and Patterson 1978). Poor electron donors, such as linoleic acid (LH) react very slowly with peroxyl radicals. Thus Gebicki and Bielski (1981) reported a rate constant of $36\ dm^3\ mol^{-1}\ s^{-1}$ for the reaction:

$$LOO^{\cdot} + LH \rightarrow LOOH + L^{\cdot}$$

The chain reaction of lipid peroxidation can be blocked by a reaction of $LOO^{\cdot}$ with α-tocopherol for which $k = 8 \times 10^4\ dm^3\ mol^{-1}\ s^{-1}$ (L.K. Patterson 1981). The role of β-carotene as a radical ($RO_2^{\cdot}$) trapping molecule has been discussed by Burton and Ingold (1984). These workers reported the sensitivity of this process to the oxygen partial pressure. A somewhat analogous addition of $O_2^{\cdot -}$ (generated via pulse radiolysis) to certain carotenoids has recently been suggested by Conn *et al.* (1992), see Chapter 7.

In contrast with aliphatic counterparts, the α-halogenated peroxyl radicals are more reactive: for example, $k\,(CCl_3O_2^{\cdot} + \text{linoleate}) = 0.39 \times 10^7\ dm^3\ mol^{-1}\ s^{-1}$ and $k\,(CCl_3O_2^{\cdot} + \text{vitamin E}) = 5 \times 10^8\ dm^3\ mol^{-1}\ s^{-1}$ (Willson 1985). This is due to their greater redox potential: $E_7 = 1.3$ V for $CCl_3O_2^{\cdot}$ (extracted from Jovanovic and Simic 1988) and 1.05 V for $CH_3O_2^{\cdot}$ (Jovanovic *et al.* 1992).

Most of the studies on alkoxyl radicals have involved the *t*-butoxyl radical generated via laser flash photolysis of di-*t*-butyl peroxide (Scaiano 1987). The reactions investigated were β-cleavage and hydrogen abstraction:

$$(CH_3)_3COOC(CH_3)_3 \xrightarrow{h\nu} 2\,(CH_3)_3CO^{\cdot}$$

$$(CH_3)_3CO^{\cdot} \longrightarrow CH_3COCH_3 + CH_3^{\cdot}\ (\beta\text{-cleavage})$$

$$(CH_3)_3CO^{\cdot} - RH\,(\text{solvent}) \longrightarrow (CH_3)_3COH + R^{\cdot}$$

$$(CH_3)_3CO^{\cdot} + XH\,(\text{substrate}) \longrightarrow (CH_3)_3COH + X^{\cdot}$$

The β-cleavage is faster for $PhCH_2CH_2O^{\cdot}$ ($k = 2.3 \times 10^7\ s^{-1}$) than for $(CH_3)_3CO^{\cdot}$ ($k = 2.9 \times 10^4\ s^{-1}$), this being due to the greater stability of the benzyl radical over the methyl radical produced by the cleavage. The rate constant of H abstraction from O—H bonds is faster in non-polar than in hydrogen bonding solvents. For example, *t*-butoxyl radicals abstract $H^{\cdot}$ from phenol with rate constants of 3.3×10^8, 2.2×10^7, and $4.7 \times 10^6\ dm^3\ mol^{-1}\ s^{-1}$ in benzene, methanol, and pyridine, respectively (Scaiano 1987).

The reactions of phenoxyl radicals and other aroxyl (aromatic oxyl) radicals, originating from aromatic hydroxy derivatives such as flavonoids, have been studied by pulse radiolysis. These radicals do not react with oxygen, $k < 10^2\ dm^3\ mol^{-1}\ s^{-1}$ (Simic 1990). The higher the stability of the phenoxyl or aroxyl radical, the better are the antioxidant properties of the parent compound. A very reactive aroxyl radical formed after scavenging another radical would not interrupt a radical chain reaction. For example, the α-tocopheryl radical ($EO^{\cdot}$) formed from an efficient antioxidant, vitamin E (EOH), is rather long lived, due to delocalization of the unpaired electron. The rate of formation of $EO^{\cdot}$ in the reaction of vitamin E with the halogenated peroxyl radical $CCl_3O_2^{\cdot}$ is $5 \times 10^8\ dm^3\ mol^{-1}\ s^{-1}$. This radical decays very slowly with $k\,(EO^{\cdot} + EO^{\cdot}) = 3.5 \times 10^2\ dm^3\ mol^{-1}\ s^{-1}$ (Simic 1990) and can be repaired by vitamin C with a rate constant of $1.55 \times 10^6 \, dm^3 \, mol^{-1} \, s^{-1}$ to give an ascorbyl radical which can itself be repaired by NADH (Packer *et al.* 1979):

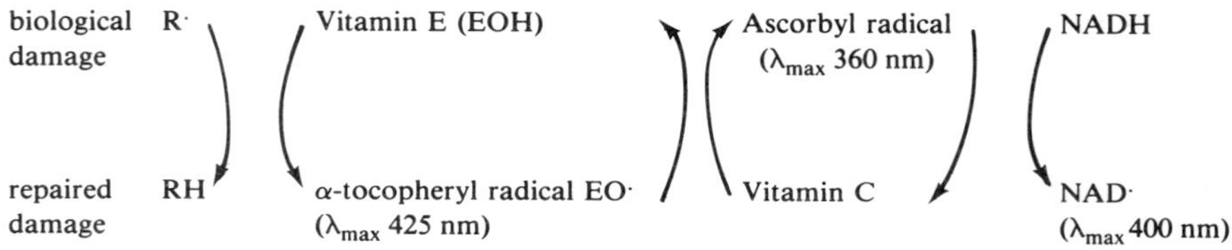

The flavonoids kaempferol and quercetin are as efficient radical scavengers as vitamin E, and have high rate constants of reactions with linoleic acid peroxyl radicals ($> 10^8 \, dm^3 \, mol^{-1} \, s^{-1}$). However, only the quercetin radical decays slowly enough to make this flavonoid an excellent antioxidant ($2k = 3.4 \times 10^6 \, dm^3 \, mol^{-1} \, s^{-1}$) in contrast with kaempferol whose aroxyl radical decays with $2k = 1.4 \times 10^8 \, dm^3 \, mol^{-1} \, s^{-1}$.

At the end of this section on oxygen centred radicals, it is interesting to compare their reactivities. Their rate constants of reaction follow the order OH· > alkoxyl > peroxyl > phenoxyl. From bond-dissociation energies, Koppenol (1990*b*) has derived the reduction potentials versus NHE, at neutral pH, of typical alkoxyl and alkylperoxyl radicals. These have provided a quantitative basis for the empirical observation that alkoxyl radicals are more reactive than alkylperoxyl radicals towards hydrogen abstraction of a doubly allylic hydrogen atom present in polyunsaturated fatty acids (PUFA). Moreover, calculations have indicated that an alkylperoxyl radical may not abstract a non-allylic hydrogen from a hydrocarbon. These thermodynamic considerations constitute a useful basis for the understanding of lipid peroxidation (Simic *et al.* 1992).

4.5 **Biomedical properties of the activated forms of oxygen**

As referred to in the introduction of this chapter, it has been recognized since Joseph Priestley (1775) that oxygen could be both beneficial and deleterious to life. The favourable aspect for aerobic life is the reduction of oxygen to water which leads to the efficient production of adenosine triphosphate (ATP), the source of energy in biological systems. However, increasing evidence suggests that oxygen metabolites, including $^1O_2$, $O_2^{\cdot -}$, OH·, RO·, and $RO_2^{\cdot}$ are involved in many diseases. Potential sources, detection, and reactions of these oxygen metabolites have been described in previous sections of this chapter. Unambiguous evidence of their participation in tissue damage is based upon direct detection by spin-trapping and Electron Spin Resonance. Indirect evidence is based upon:

(1) activation of lipid peroxidation, associated with protein and DNA modifications;
(2) the stimulation of peroxidation by catalytic metal ions and by the metabolism of toxic chemicals; and
(3) the decrease of the oxidative damage by sequestration of metal ions or by non-enzymatic and enzymatic antioxidants.

The role of oxygen-centred radicals, peroxides, and singlet oxygen in human diseases has been investigated for endogenous and exogenous damage in a large number of cases (Halliwell and Gutteridge 1989). Some examples of free radical toxicity are given below.

4.5.1 *Endogenous oxyradical toxicity*

Endogenous oxyradicals are involved in pathological processes including inflammation, cardiovascular diseases, cancer, and ageing. Enzymatic and non-enzymatic defence mechanisms can intercept and detoxify these oxyradicals and their products. Superoxide dismutases (SOD), copper–zinc enzymes present in all eukaryotic cells such as yeasts, plants, and animals, enhance the dismutation rate of $O_2^{\cdot -}$, producing H_2O_2, by almost four orders of magnitude (see Section 4.4.3.2). In turn, catalases, which are haem (Fe^{III}–protoporphyrin) enzymes, catalyse the dismutation: $2H_2O_2 \rightarrow 2H_2O + O_2$. Glutathione peroxidase, a seleno-enzyme, also catalyses the destruction of H_2O_2 at the expense of the oxidation of glutathione, GSH, to GSSG: $H_2O_2 + 2\,GSH \rightarrow 2\,H_2O + GSSG$. Another enzyme, glutathione reductase, reforms GSH. The second line of antioxidant defence includes non-enzymatic molecules as diverse as vitamins E and C, β-carotene, glutathione, urate, and flavonoids.

Phagocytosis is associated with the formation of $O_2^{\cdot -}$, H_2O_2, $OH^{\cdot}$, and $^1O_2$, which are not only bactericidal molecules but also mediators of inflammation. Such inflammation is suppressed in pharmacological models by SOD and catalase, allopurinol, and free radical scavengers (Flohé *et al.* 1985; Weiss 1986).

In cardiovascular injury, ESR measurements have shown a burst of oxygen free radicals at the moment of reperfusion following oxygen deprivation (ischaemia). SOD eliminates this burst and reduces the infarct size (Weisfeldt *et al.* 1988). Electrophysiological changes and arhythmias in perfused rat hearts have been produced by Rose Bengal photoactivation. The protective effect of histidine suggests a prominent role of $^1O_2$ in this process (Bernier *et al.* 1991).

Healthy humans excrete in the urine products of oxidative damage of DNA and RNA such as thymine glycol, thymidine glycol, 5-hydroxymethyluracil, 5-hydroxymethyldeoxyuridine, and 8-hydroxydeoxyguanosine which are derived from repair of oxidized DNA (Ames 1989; Simic 1992). These oxidized products can also be generated *in vitro* by radiation and photosensitization reactions via activated forms of oxygen. These modifications of DNA can lead to endogeneous carcinogenesis when the repair mechanisms and the cellular antioxidant defences are overwhelmed (Cerutti 1985; Loeb 1989).

The phenomenon of ageing and the maximum lifespan expectancy are related to oxidative damage. As an example, the rate of $O_2^{\cdot -}$ and H_2O_2 mitochondrial production from mouse, rat, rabbit, guinea pig, pig, cow, drosophila, and housefly has been shown to be inversely related to maximum life span potential (Sohal *et al.* 1990).

4.5.2 *Free radicals and xenobiotics*

Pathological conditions due to oxygen radicals generated by scores of xenobiotics are actively studied and reported in many reviews (see Halliwell and Gutteridge 1989, 1990). We will briefly examine the cases of paraquat, cigarette smoke, quinones, and polycyclic hydrocarbons.

The bipyridyl herbicide, paraquat (1,1′-dimethyl-4,4′-dipyridylium ion with each nitrogen positively charged, also called methyl viologen or MV^{2+}) is manufactured as a salt with Cl^- ion by Imperial Chemical Industries. Its toxicity to plants, bacteria, animals, and humans is due to the generation of $O_2^{\cdot -}$ via MV^{2+} reduction to $MV^{\cdot +}$, followed by $MV^{\cdot +} + O_2 \rightarrow MV^{2+} + O_2^{\cdot -}$. This mechanism has been investigated by pulse radiolysis (Farrington *et al.* 1973, 1978). It has also been proposed that the synthetic narcotic 1-methyl-4-phenyl-1,2,3,6-tetrahydropyridine (MPTP), following oxidation by monoamine oxidase to the 1-methyl-4-phenyl-pyridinium ion (MPP^+), gives rise to a radical closely related to $MV^{\cdot +}$. This radical $MPP^{\cdot}$ can also generate $O_2^{\cdot -}$ (Chacon *et al.* 1987, 1989) which may contribute to chemically induced Parkinson's disease.

The induction of SOD synthesis and the increased damage by metal ions have confirmed the existence of toxic reactions involving $O_2^{\cdot -}$ *in vivo*.

A great variety of other oxidants such as $NO^{\cdot}$, $NO_2^{\cdot}$, $ROO^{\cdot}$, H_2O_2, present in cigarette smoke, and of semiquinone radicals present in cigarette tar, are considered as promoters of lung cancer in smokers (Halliwell and Gutteridge 1989).

Redox cycling of quinones can generate large quantities of $O_2^{\cdot -}$ via the one-electron reduction of a quinone to its semiquinone radical:

$$Q^{\cdot -} + O_2 \rightleftarrows Q + O_2^{\cdot -} \quad \text{equilibrium constant } K$$

The position of the equilibrium is dependent upon the reduction potential of the quinone and the relative concentration of Q and $O_2^{\cdot -}$: $E(O_2/O_2^{\cdot -}) - E(Q/Q^{\cdot -}) = RT/F(\ln K)$. This equilibrium can be pulled to the right by removal of $O_2^{\cdot -}$ even with a low O_2 concentration and high $E(Q/Q_2^{\cdot -})$. It can also be modified by the environment as oxygen is more soluble in organic solvents than in water (2.3×10^{-3} mol dm^{-3} and 2.7×10^{-4} mol dm^{-3} for air saturated cyclohexane and water, respectively, at 25 °C) and because reduction potentials are changed by the dielectric constant. Rate constants for reaction of semiquinones (including chemotherapeutic agents) with oxygen have been plotted by Wardman (1990) as a Marcus function yielding the reorganizational parameter $\lambda = 77\ \text{kJ mol}^{-1}$ (see Section 2.4.3). The formation of O_2^- and other activated oxygen species may contribute to the actions of clinically used quinone anti-cancer drugs on DNA (Gaudiano and Koch 1991). Some other effects of the antitumour quinones, such as cardiotoxicity and skin toxicity, are also related to oxygen radical formation. These drugs are discussed in more detail in Chapter 11.

Szent-Györgi *et al.* (1960) have correlated the carcinogenic activity of polycyclic hydrocarbons with the magnitude of ESR signals formed by reaction of these compounds with molecular iodine. These ESR signals are attributed to cationic free radicals (R. P. Mason 1982), which are considered to be one of the electrophilic species binding covalently to DNA and triggering the carcinogenic process (Rogan *et al.* 1988; Cavalieri and Rogan 1990). Other examples of similar investigations by spin-trapping methods of oxygen centred radicals *in vitro* and *in vivo* are described by Janzen and Bolton (1990). Some difficulties in correlating free radical formation and cellular injury have been noted by Rosen and Halpern (1990).

4.5.3 *Chemopreventative agents against oxyradicals*

Mortality from cancer and from ischaemic heart disease has been inversely correlated to the presence of antioxidants in plasma (Gey *et al.* 1987; Gey and Puska 1989). Epidemiological knowledge has led to the organization of 18 intervention trials in progress using cancer chemopreventative drugs involving natural antioxidants and anti-inflammatory agents including β-carotene, retinol/retinoic acid, vitamins C and E, 4-hydroxyphenyl retinamide, piroxicam, and aspirin (Bertram *et al.* 1987; Boone *et al.* 1990). Supplementation of antioxidants in the diet at high levels is questioned by evidence that it might depress the natural enzymatic antioxidant levels. This compensatory theory is also related to the failure of antioxidant-supplemented diets in significantly increasing longevity in animals (Cutler 1984). These intervention trials in progress should answer this question.

5

Pyrimidines, purines, and nucleic acids

'DNA is the beginning and the end.'

Michael G. Simic

5.1 Occurrence and importance

All cells contain both deoxyribonucleic acid (DNA) and ribonucleic acid (RNA). Most of the DNA is located in the nucleus and in the mitochondria, whereas most of the RNA is in the cytoplasm, and more precisely in subcellular organelles involved in protein synthesis. DNA contains purine bases, adenine (A) and guanine (G), and pyrimidine bases, thymine (T) and cytosine (C). The RNA structure differs from that of DNA by two features: the sugar unit is ribose and not deoxyribose and thymine is replaced by uracil (U). DNA stores genetic information in all prokaryotic and eukaryotic cells using a code in which the 'letters' are the four bases of DNA. There are two genetic functions of DNA:

1. DNA replicates itself and transmits its information to daughter cells. In the replication, the two strands of helical polynucleotide chains coiled around a common axis unwind and separate (Watson and Crick 1953*a*, *b*). Each chain then acts as a template for the formation of a new chain.
2. DNA is able to express its information via RNA synthesis from a DNA template. This process called 'transcription' is followed by another process called 'translation' in which certain RNA molecules are templates for the synthesis of proteins. In this last process, the sequence of the four bases in RNA is translated into a sequence of the 20 amino acids of a protein, with a group of three bases coding for one amino acid.

Nucleic acids are the major molecular targets for deleterious effects, including mutagenicity, carcinogenicity, and lethality, produced by ionizing and ultraviolet radiation. At a time when genetic studies were little developed, mutation effects by UV and ionizing radiation had already been observed. For example, stable heritable variations in colony morphology, cellular morphology, and pathogenicity were noted by Henri (1914) among the surviving cells of *B.anthracis* cultures irradiated by UV. Mutagenic effects of X-rays in *Drosophila* were observed by H. J. Muller (1927). The action spectrum for killing of bacteria (Gates 1930) and the action spectrum for mutation in corn (Stadler and Uber 1942) showed a maximum and a shape very similar to that of the absorption spectrum of nucleic acids, indicating DNA as the molecule responsible for these effects. The fact that UV and ionizing radiation can be delivered more simply than chemical mutagenic substances triggered extremely fruitful interdisciplinary studies involving genetics, photochemistry, and radiation chemistry of DNA, both *in vivo* and *in vitro*. The transformations induced by UV and ionizing radiation in nucleic acids and related compounds are of great importance not only for their relevance to mutagenesis, carcinogenesis, and radiotherapy but also since the discovery of the ozone hole over Antartica for their relevance to environmental photobiology (Wayne 1991).

Nucleotides also play many other roles in biology. For example adenosine di- and tri-phosphates (ADP, ATP) are the energy pool of enzymatic reactions and muscle action, 8-aza-adenosine is a potent antibiotic, adenosine derivatives such as puromycin and kinetin are an antibiotic and a plant growth hormone, respectively. The three-dimensional structures of all these molecules are essential features of their biochemical functions.

5.2 Structure

Nucleotides, the monomeric units of nucleic acids, are composed of a phosphate (p or P), a sugar, and a purine or pyrimidine base. The unit consisting of a sugar and a base is a nucleoside. Figure 5.1 shows the structures of the normal bases and sugars and their abbreviations. In ribonucleic acid (RNA) the sugar is ribose, so the four nucleosides are termed rA, rU, rC, and rG. Unusual nucleosides can occur in RNA, such as 4-thiouridine. The four nucleotides of RNA are termed AMP, UMP, CMP, and GMP. In deoxyribonucleic acid (DNA) the sugar is 2′-deoxyribose, so the four nucleosides are termed dAdo, dThd, dCyd, and dGuo. All known DNA are made up from the four nucleotides, dAMP, dTMP, dCMP, and dGMP.

The polynucleotide chains of DNA and RNA are shown in Fig. 5.2. Most DNA consist of two single strands of the same length allowing full base pairing by hydrogen bonds, A with T and C with G, by two and three hydrogen bonds, respectively. The two single strands of DNA are twisted about the same axis to form a double helix. The classical Watson–Crick right-handed double helix, called B-DNA, contains about 10 residues per turn, with the planes of the bases perpendicular to the helix axis. Two other double helices have also been observed: another right-handed double helix, called A-DNA, containing about 11 residues per turn, where the planes of the base pairs are tilted 20 degrees away from the perpendicular to the helix axis, and a left-handed double helix, called Z-DNA, containing about 12 residues per turn. An important feature of these three structures is the presence of two kinds of grooves, called the major and minor groove. In the vicinity of the three different helices, water is organized with distinct hydration patterns (Palecek 1991). This water has very different properties compared with ordinary water (Ebert 1980) and it appears not to be capable of solvating electrons (van Lith *et al.* 1986).

The DNA length varies from micrometres in viruses to several centimetres in mammals. In human somatic cells, each of the 46 chromosomes contain a single DNA duplex molecule about 4 cm long. This 2 m total length of DNA surrounded by polypeptides, called histones, is packed into a nucleus 0.5 μm in diameter in a highly ordered structure and stores all the genetic information.

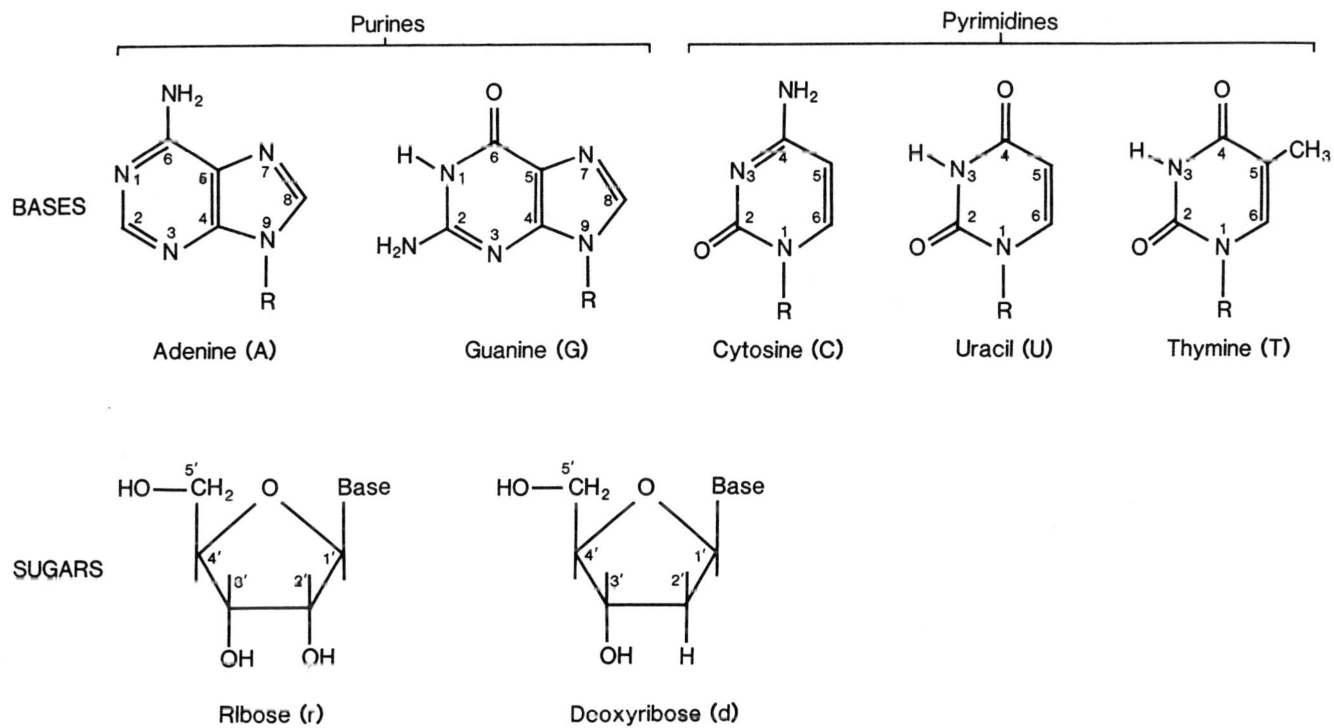

Fig. 5.1 Structures of the normal base and sugar constituents of nucleic acids.

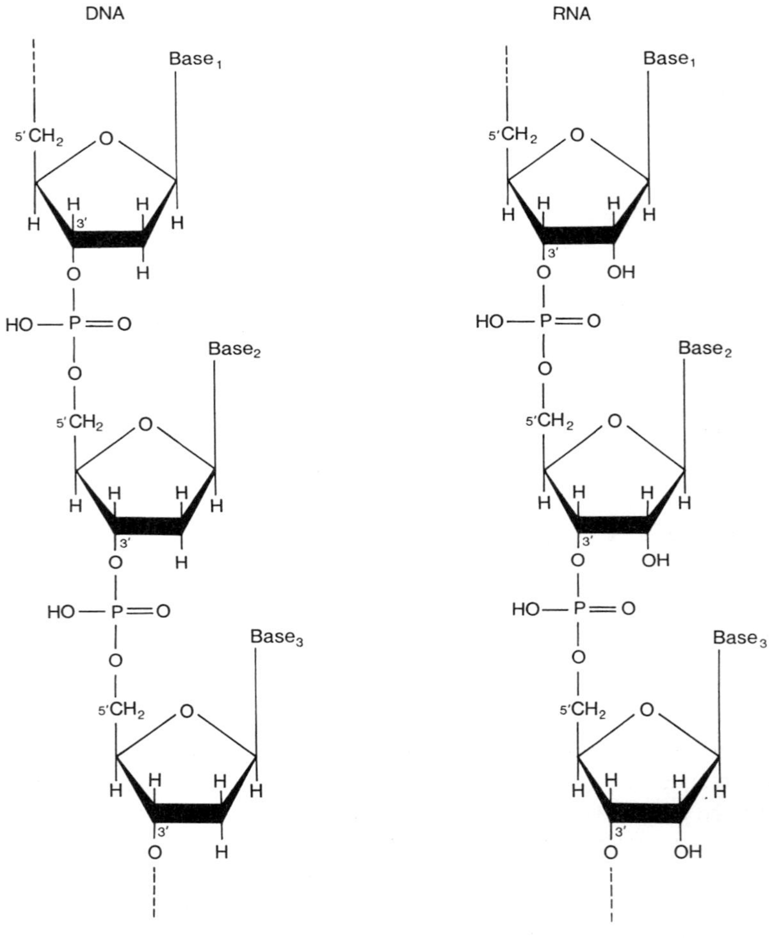

Fig. 5.2 Polynucleotide chains of DNA and RNA.

5.3 Ground state absorption spectra

The strong near UV absorption of all nucleic acids is due to the purine and pyrimidine bases. The absorption spectra of the bases are shown in Fig. 5.3. The absorption bands located at 260–270 nm are due primarily to $\pi\pi^*$ transitions with hidden weaker $n\pi^*$ transitions not well located. Substitution by ribose results in a small bathochromic shift. The free bases are not perfect models for nucleic acids due to the absence of the glycosidic C—N bond. For this reason *N*-9- methyl purine and *N*-1- methyl pyrimidine are better models, with absorption spectra very similar to those of nucleosides. Going from

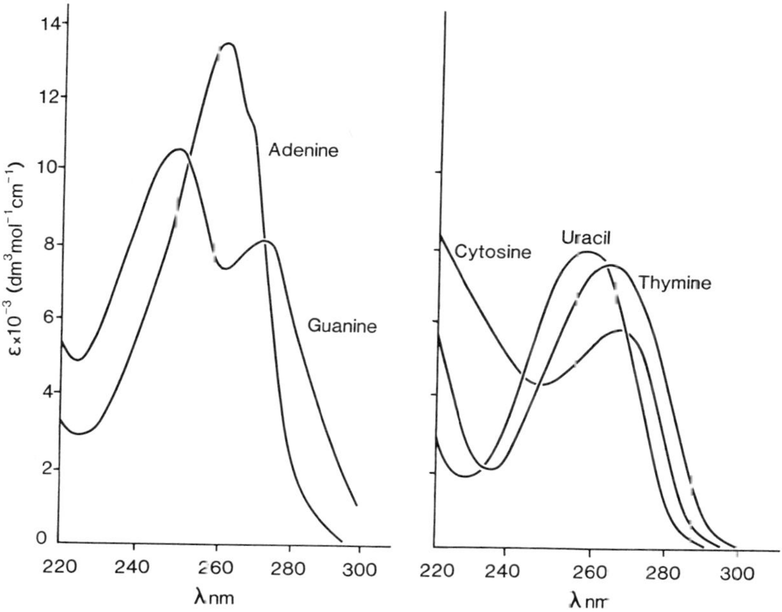

Fig. 5.3 Absorption spectra of nucleic acid purine and pyrimidine bases.

ribonucleosides to deoxyribonucleosides, or to nucleotides, does not modify the absorption spectrum. For pyrimidines and their corresponding nucleosides the 260 nm band is due to one $\pi\pi^*$ transition polarized in the plane of the molecule. For purines and their corresponding nucleosides two $\pi\pi^*$ transitions polarized at right angles occur around 260–270 nm, the second transition appearing as a shoulder in the case of adenine, adenosine or dAMP and almost as a separate peak in the case of guanine, guanosine or dGMP (Figs 5.3 and 5.4). All bases, nucleosides, and nucleotides have a second absorption band around 200 nm, due to one or more $\pi\pi^*$ transitions more intense than the band at 260–270 nm. The absorption spectra are pH sensitive (Fig. 5.4).

In the case of DNA the first absorption band observed at 260 nm is qualitatively that corresponding to the sum of the constituent mononucleotides. The sugar phosphate backbone makes no significant contribution to the UV absorption spectra at wavelengths greater than 200 nm. However, the intact double helix has a weaker overall absorption intensity than the single strand, indicating that the nucleic acid bases strongly interact with each other. Upon increasing the temperature, the UV absorbance increases, indicating less stacking interactions.

Extensive reviews (Daniels 1976; Hauswirth and Daniels 1976) deal with the electronic transitions of the ground state of the bases, the mononucleosides and

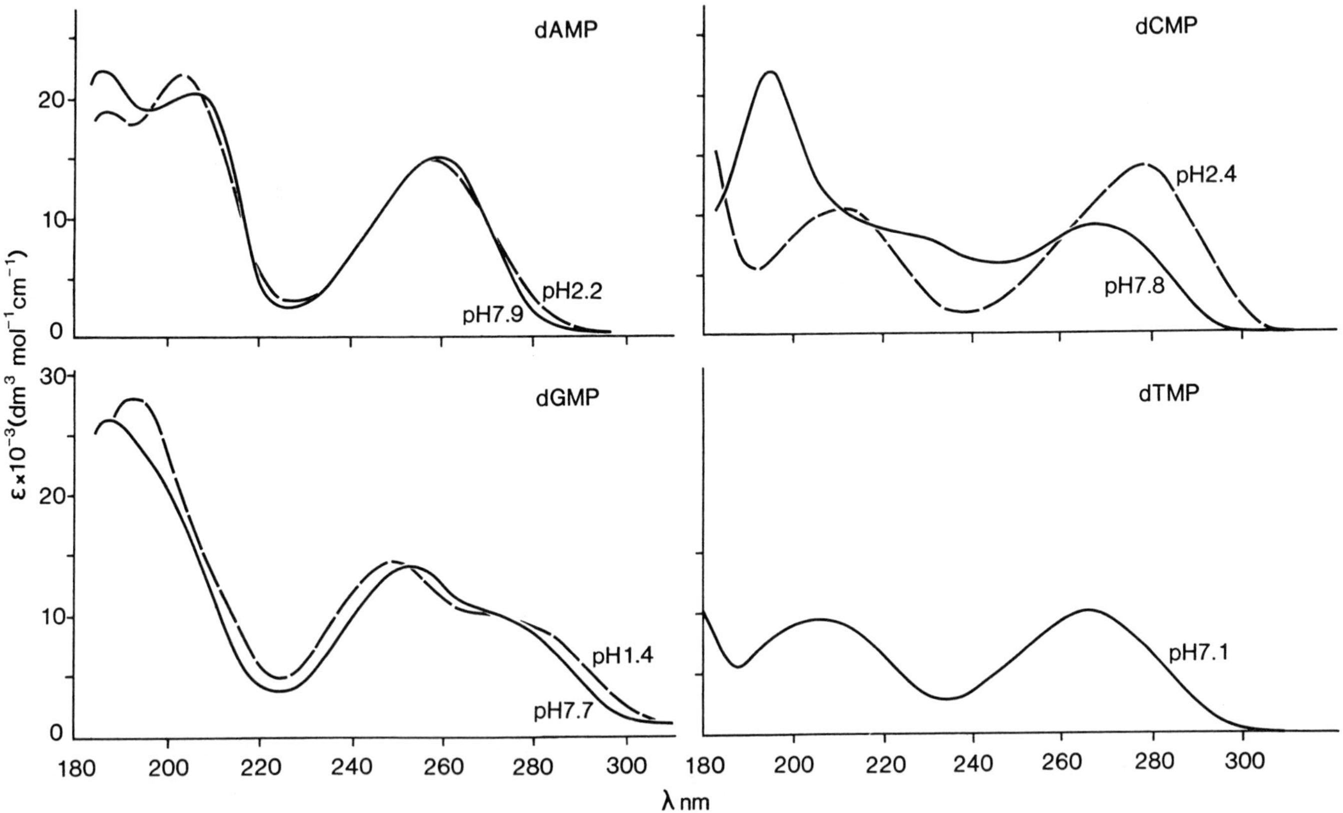

Fig. 5.4 Absorption spectra of four deoxynucleotides (taken from Voet *et al.* 1963).

mononucleotides, of DNA and RNA. These reviews also consider the pertubation of the transitions by the base stacking effects in dinucleotides, oligonucleotides, single-stranded and double-stranded polynucleotides, and finally DNA and RNA.

In the rare nucleoside, 4-thiouridine, substitution of the oxygen atom in position 4 of the uracil ring by sulphur produces a shift of the main $\pi\pi^*$ from 260 to 330 nm. This interesting property made this minor nucleoside, found in position 8 and 9 of the *Escherichia Coli* tRNAs, a selective photoactive probe of tRNA photochemistry (see review by Favre 1990).

5.4 Fluorescence of monomers, polynucleotides, and DNA

The spectra, quantum yields, and lifetimes of the fluorescence of the monomers, di- and poly-nucleotides, and DNA have been determined mainly in mixtures of water and ethylene (or propylene) glycol or of ethanol and methanol transparent glasses at 77 K and in neutral aqueous solution at room temperature, 300 K.

The emission maxima of the monomers are in the region 310 to 350 nm with slight shifts, usually to the red, when going from the bases to the nucleosides, nucleotides, and polynucleotides at 77 K or 300 K. Polynucleotides and DNA also show a long wavelength emission in the 400 nm region (Daniels 1976, 1983).

The nature of the lowest singlet excited state ($n\pi^*$ or $\pi\pi^*$) responsible for the fluorescence of these compounds is still under discussion. This lowest state can be modified by the environment as shown in the case of thymine. This molecule at 77 K has a relatively long fluorescence lifetime ($\tau_F \approx 1$ ns) and a relatively low fluorescence quantum yield ($\phi_F = 0.004$) in a polar aprotic solvent such as 2-methyltetrahydrofuran, whereas in a polar protic solvent such as ethanol–methanol (3:1), it has a shorter lifetime ($\tau_F < 0.5$ ns) and a higher ϕ_F value of 0.035 (R.S. Becker and Kogan 1980). These fluorescence property changes of thymine indicate that the lowest excited singlet corresponds to an $n\pi^*$ transition in aprotic solvents and to a $\pi\pi^*$ transition in protic solvents.

The fluorescence quantum yields of some 30 DNA components and derivatives, investigated or reinvestigated by Görner (1990*a*) at 77 K in different environments, are weak, lying in the 10^{-3}–10^{-2} range. For native DNA at 77 K, in ethylene glycol–water mixtures at pH 7, ϕ_F is ≤ 0.008 (Görner 1990*a*). At room temperature, the ϕ_F of the DNA components are in the 10^{-5} and 10^{-4} range, weaker than at 77 K (see review by Cadet and Vigny 1990). The energies of the S_1 states of the bases have been estimated from the intersection of the absorption and emission spectra. They are in the order: 425 kJ mol^{-1} for A and U, 417 kJ mol^{-1} for C, 412 kJ mol^{-1} for T, and 399 kJ mol^{-1} for G at 300 K (Daniels and Hauswirth 1971). The singlet energy levels are rather similar for the corresponding nucleosides and nucleotides: 425 kJ

mol^{-1} for AMP, 422 kJ mol^{-1} for UMP, 407 kJ mol^{-1} for TMP, 406 kJ mol^{-1} for CMP, and 404 kJ mol^{-1} for GMP at 300 K (Cadet and Vigny 1990).

At room temperature, the singlet lifetimes of the nucleic acid monomers are in the picosecond range. They have been estimated via the Strickler and Berg equation (see Section 3.2.1) and the relationship between the fluorescence lifetime, τ_F, the radiative lifetime, τ_0, and the fluorescence quantum yield, ϕ_F, which is $\tau_F = \phi_F \times \tau_0$ (Daniels and Hauswirth 1971; Vigny and Duquesne 1976). They have also been measured via energy transfer to Eu^{III} ions (Guéron *et al.* 1974), via high intensity picosecond laser flash photolysis (Nikogosyan and Letokhov 1983), and measured directly via synchrotron radiation time-correlated photon counting (Ballini *et al.* 1988) and ps single-photon spectroscopy (Rigler and Claesens 1986).

For some di- and poly-nucleotides, Ballini *et al.* (1988) found a fast fluorescence (<100 ps) around 300 nm where the mononucleotides also emit. In addition, they observed a new emission in the 380–420 nm region with lifetimes in the nanosecond range, which might be due to excimers. However, no excimer was observed in double-stranded calf thymus DNA at room temperature by Georghiou *et al.* (1985). Using a streak camera and 266 nm ps laser pulses, these authors detected a major fluorescence decay time component (≈90 per cent) of 10 ps and a second decay time component of 65 ps.

Singlet energy transfer between nucleic acid bases and dyes in intercalation complexes does not extend beyond two base pairs from the dye molecule with no exciton migration along the chain (Rayner *et al.* 1980).

5.5 **Phosphorescence and triplet energy levels of nucleic acid components and DNA**

Phosphorescence of some 30 DNA components and derivatives at 77 K in different environments have been investigated or reinvestigated and discussed by Görner (1990*a*). Phosphorescence spectra of purines at 77 K show vibrational structure, for example in the case of adenine at 368, 385, 405, and 423 nm, in contrast with the unstructured spectra of cytosine, uracil, and thymine (Daniels 1976; Görner 1990*a*).

The $\pi\pi^*$ nature of the phosphorescent state for the purine and pyrimidine bases has been demonstrated via low temperature studies from out-of-plane polarization of the emission and from ESR data (Daniels 1976).

Triplet energy levels of nucleotides determined from the blue edge of the phosphorescence spectra at 77 K are in the order: 333 kJ mol^{-1} for CMP, 328 kJ mol^{-1} for UMP, 325 kJ mol^{-1} for GMP, 319 kJ mol^{-1} for AMP, and 314 kJ mol^{-1} for TMP (G. J. Fisher and Johns 1976). As a consequence, TMP is the ultimate energy sink and an important target of photodamage in DNA.

The phosphorescence quantum yields (ϕ_P) at 77 K are in the range of 10^{-3}–10^{-2}, usually weaker for pyrimidine than for purine derivatives. For native DNA at 77 K, in ethylene glycol–water mixtures at a pH of 7, ϕ_P is ≤ 0.02. (Görner 1990*a*).

The phosphorescence decays for many pyrimidine and purine derivatives, measured in glasses at 77 K (Görner 1990*a*) exhibit a monoexponential decay and lie between 0.3 s for orotic acid and 2 s for adenine in aerated or non-aerated ethylene glycol–water glasses at 77 K, while at room temperature triplet–triplet absorption studies show triplet lifetimes in the microsecond range in deaerated solutions. For native and denatured DNA, the phosphorescence lifetime at 77 K is 0.7 and 0.5 s, respectively (Görner 1990*a*).

5.6 Triplet absorption studies of pyrimidines and of their nucleosides and nucleotides

5.6.1 *Absorption spectra*

The triplet–triplet absorption spectra of thymine, uracil, thymidine, uridine, thymidine monophosphate, and uridine monophosphate have been studied by laser flash spectroscopy (265 nm) at room temperature between 300 and 450 nm (Salet and Bensasson 1975; Salet *et al.* 1979) in a range of solvents. The identification of the observed transients as triplet states was proven via sensitization of β-carotene and retinol triplets in acetonitrile or ethanol. In water, these spectra showed a maximum between 350 and 420 nm with shifts to the red when going from the pyrimidines (~350 nm for thymine and 375 nm for uracil) to nucleosides and nucleotides (400–420 nm). The thymine and uracil triplet spectra in aqueous solution obtained by Whillans and Johns (1971) using conventional flash photolysis showed spectra with no structure and maxima located at 350 nm and 390 nm for thymine and uracil respectively. Later studies by the same authors (1972) report, rather surprisingly, several peaks and shoulders at intervals of 30 nm. Hayon (1969) reported transient absorption spectra for uracil and thymine obtained by pulse radiolysis in acetonitrile, dimethyl formamide (DMF), and isopropanol solvents which he assigned to triplet–triplet absorption. These spectra are very different from those reported by Salet and Bensasson (1975) and Whillans and Johns (1971), and are probably due to radicals, since triplet states are in fact generated with only a very low yield by pulse radiolysis in polar solvents (see Section 1.2.4.2).

Görner (1990*b*) has complemented these studies by investigating transients of uracil and a series of 17 related pyrimidines obtained using 248 nm, 20 ns excimer laser pulses in acetonitrile, ethanol, and aqueous solutions. The spectra obtained in a wider wavelength range show generally

(1) a maximum around 290–310 nm;

(2) a second band or a shoulder at 360–420 nm, this second band being similar to the triplets observed in the previous studies of Salet and Bensasson and Whillans and Johns; and

(3) a band peaking at 700 nm due to the solvated electron.

5.6.2 *Lifetimes*

The results of Szabo *et al.* (1970), P. J. Wagner and Bucheck (1970), Whillans and Johns (1971), Salet and Bensasson (1975), and Salet *et al.* (1979) concerning kinetic parameters of the triplet decay, at room temperature, are in reasonable agreement. The triplet state of these compounds is quenched by ground state molecules according to

$$k_{obs} = k_0 + k_s \, [M]$$

where k_0 is the rate constant of the unimolecular decay in the absence of ground state quenching, k_s is the rate constant of the triplet quenching by the ground state, and [M] the concentration of the molecules studied. The k_0 values lie in the region 10^5 s^{-1} and k_s around 10^9 dm^3 mol^{-1} s^{-1} (see Bensasson *et al.* 1983). Görner (1990*b*) obtained, for uracil and uracil derivatives, lifetimes varying from 0.5 to 4 μs, not corrected for self-quenching, at typical concentrations of 0.3×10^{-3} mol dm^{-3} in aqueous solutions.

Dimerization of thymine and thymine derivatives has been shown to occur via the triplet state using triplet sensitizers such as acetone, acetophenone, and propiophenone. Thus, the dimers formed by direct excitation are probably due to triplet reaction with the ground state (G. J. Fisher and Johns, 1976).

5.6.3 *Molar absorption coefficients*

The molar absorption coefficients, ε_T, of the pyrimidine triplet absorption have been obtained in acetonitrile or ethanol by the energy transfer technique (see Section 3.1.2.1) using, as standard, retinol triplet (Salet and Bensasson 1975). In order to obtain an estimate of ε_T in water, it was assumed that the oscillator strength of the triplet transition is independent of the solvent (Bensasson and Land 1971; Lavalette *et al.* 1971). The ε_T values at 340 nm thus estimated are in acetonitrile and water, respectively, for thymine $\approx$4800 dm^3 mol^{-1} cm^{-1} and 5200 dm^3 mol^{-1} cm^{-1} and for uracil $\approx$2500 dm^3 mol^{-1} cm^{-1} and 2400 dm^3 mol^{-1} cm^{-1}. These values are estimated with the assumption that the pyrimidine triplet quenching only occurs via energy transfer (Heihoff *et al.* 1990). It should be emphasized that Milder and Kliger (1985) have measured ε_T for uracil in acetonitrile by the singlet-depletion method and obtained a value which agrees reasonably well with the ε_T determined via energy transfer to retinol by Salet and Bensasson (1975). In the case of 4-thiouridine, no such agreement was obtained on account of simultaneous electron and energy transfer between the triplet of the base and retinol (Heihoff *et al.* 1990). Using the ground state bleaching method, Milder and Kliger (1985) have estimated an $\varepsilon_T \approx$ 2400 dm^3 mol^{-1} cm^{-1} at 540 nm for 4-thiouridine and $\approx$2500 dm^3 mol^{-1} cm^{-1} at 590 nm for 1,3-dimethyl-4-thiouracil in water.

5.6.4 *Intersystem crossing yields*

Estimates of the values of ϕ_T for thymine (T), uracil (U), their nucleosides and nucleotides have been obtained by the comparative method (Section 3.1.2.3) in different solvents (Salet and Bensasson 1975; Salet *et al.* 1979). The ϕ_T values are low and markedly smaller in water (10^{-2}–10^{-3}) than in ethanol or acetonitrile (0.2–0.04). For T and U, they are in reasonable agreement with the results of G. J. Fisher *et al.* (1974) obtained indirectly from the yields of the photoadditions of T and U triplets to cysteine, and with the results of Lamola and Eisinger (1971) obtained for TMP and UMP by the europium ion energy transfer method using acetophenone as a standard.

Using the extinction coefficients estimated by Salet and Bensasson (1975), Görner (1990*b*) has determined the ϕ_T of uracil and a large series of related pyrimidines. Taking account of the low ε_T and ϕ_T values, most of the reported ϕ_T values are in reasonable agreement.

For the rare nucleoside, 4-thiouridine, a high intersystem crossing yield was observed in water, 0.7–0.9 (Milder and Kliger 1985; Heihoff *et al.* 1990), in contrast with all the other components of nucleic acids.

5.6.5 *Ordering of nπ* and ππ* excited states in pyrimidines*

A rigorous assignment of the lowest transitions in the first absorption band of thymine, uracil, and their corresponding nucleosides and nucleotides, has not yet been made. However, to suggest an ordering of the $^{1,3}n\pi^*$ and $^{1,3}\pi\pi^*$ excited states, Salet *et al.* (1979) used

(1) the El Sayed rule which states that intersystem crossing transitions between singlet and triplet $n\pi^*$ states and between singlet and triplet $\pi\pi^*$ states are forbidden, whereas transitions from singlet $n\pi^*$ to triplet $\pi\pi^*$ and from singlet $\pi\pi^*$ to triplet $n\pi^*$ are allowed (El Sayed 1963; R.S. Becker 1969);

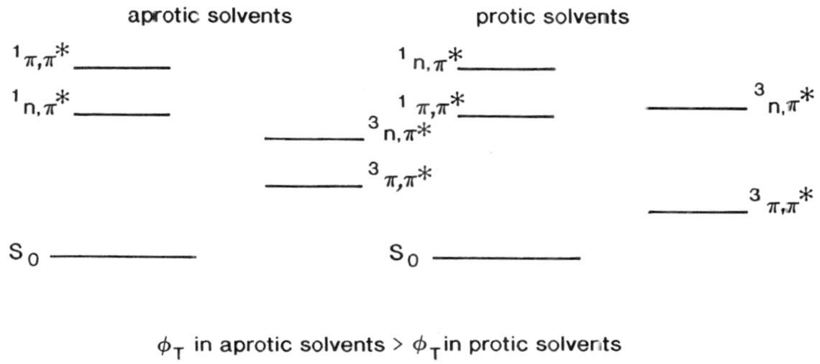

Fig. 5.5 Ordering of excited states of pyrimidines and of their corresponding nucleosides and nucleotides with the assumption of n, π^* being the lowest in aprotic solvents.

(2) the decrease of ϕ_T in going from non-H-bonding solvents to H-bonding solvents; and

(3) the quantum yields of fluorescence, ϕ_F, and phosphorescence, ϕ_P, and the lifetimes of fluorescence, τ_F, and of phosphorescence, τ_P.

Two possible orderings were proposed (Salet *et al.* 1979). Experimental data on the fluorescence of thymine (R. S. Becker and Kogan 1980) would rather favour the possibility illustrated in Fig. 5.5. This situation is similar to that described for retinals (see Section 7.2.2.2).

5.7 Triplet absorption studies of purines

The purines form part of the light absorbing chromophores of nucleic acids, but, as mentioned above, the light energy which they absorb is transferred to pyrimidines. For this reason, photochemical information on purines is rare (Elad 1976). Conventional flash photolysis of purine, studied as a model compound for guanine and adenine, revealed an absorption band in ethanol between 400 and 450 nm assigned to the triplet state (Connolly 1969). Subsequently, Arce *et al.* (1980) reported flash photolysis experiments in acetonitrile and water. The short-lived absorption ($t_{1/2} \sim 20\ \mu s$) in the region 310–460 nm is removed by triplet quenchers such as Mn^{II}, Co^{II}, and oxygen. The similarity of this transient absorption with the low temperature spectrum obtained during continuous UV irradiation of purine in a glass at 77 K allowed Arce *et al.* (1980) to assign these bands to the purine triplet. In aqueous solution, these workers also observed the presence of e^-_{aq} due to a photoionization of the purines. For 2-aminopurine a triplet absorption has been reported by Szabo *et al.* (1975).

5.8 Excited state reactions of pyrimidines, purines, and derivatives in solution and in DNA

This section will only describe the photoproducts observed by conventional monophotonic far-UV photochemistry (with low intensity or fluence rate $\approx 10^{-1}$–$10\ W\,m^{-2}$ with low fluences $<0.1\ kJ\ m^{-2}$). The dominant process resulting from exposure of nucleic acids and related compounds to vacuum-UV and to multi-photon laser UV (10^9–$10^{13}\ W\,m^{-2}$) is the formation of radical cations. Thus, this non-conventional photochemistry, showing similarities with ionizing radiation chemistry, will be reviewed in Section 5.9 which deals with radicals of nucleic acids.

5.8.1 *Pyrimidines*

The main products in the photochemistry of DNA in aqueous solutions arise from pyrimidines, as identified by UV irradiation of pyrimidine monomers in aqueous solution. They are

(1) the cyclobutane dipyrimidines Pyr <> Pyr;

(2) the hydrates; and

(3) the pyrimidine photoadducts or 'spore photoproducts'.

Typical structures are shown in Fig. 5.6(a)–(c) (for reviews see G. J. Fisher and Johns 1976; Cadet and Vigny 1990). These three classes of pyrimidine photoproducts are also the three main types of DNA base photoproducts within cells (Jagger 1976).

The formation of cyclobutane dipyrimidines is the major photo-induced alteration in DNA (Setlow 1966) and RNA (Ottensmeyer and Whitmore 1968). Furthermore, it has been shown theoretically by molecular orbital calculations (Mantione and Pullman 1964), and experimentally by sensitization (Rahn and Patrick 1976), that pyrimidine cyclobutane photodimerization involves the triplet state of these molecules. However, dimerization in the solid state and in aggregates might also occur via a singlet excited state mechanism (G. J. Fisher and Johns 1976). The increase of ϕ_T of pyrimidines when going from protic to aprotic solvents infers important consequences on the photochemistry of

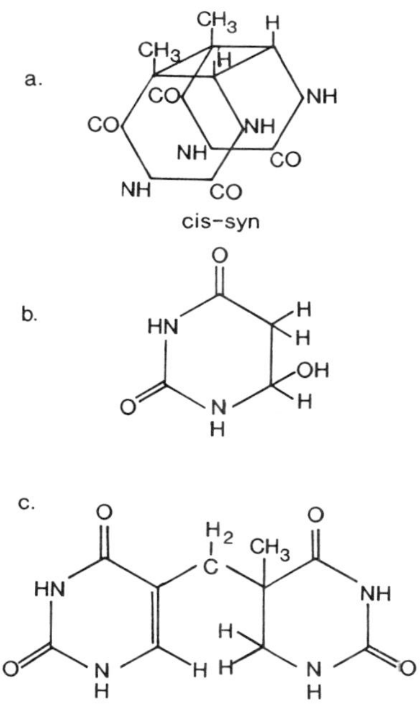

Fig. 5.6 Typical structures of three classes of pyrimidine photoproducts: (a) *cis-syn* cyclobutane thymine dimer, (b) uracil hydrate, and (c) spore product (5-thyminyl-5, 6-dihydrothymine).

nucleic acids depending on the micro-environment of the molecular components. In particular, the increase of the photo-formation of pyrimidine dimers in one strand on nuclear DNA during the S phase of the cell cycle (Chanet *et al.* 1975) might be due to the disruption of H-bonding between paired thymine and adenine. The quantum yield of dimerization of pyrimidines is very low, for example for thymine it is 4.7×10^{-4}. Four possible stereo configurations (*cis-syn*, *cis-anti*, *trans-syn*, *trans-anti*) of cyclobutadithymidines have been isolated and identified mainly by IR and NMR spectroscopies (Cadet and Vigny 1990).

The monomerization of thymine dimers Thy <> Thy can be induced by hydrated electrons and has been observed by flash photolysis and pulse radiolysis (Santus *et al.* 1972). This mode of pyrimidine dimer splitting by hydrated electrons might be involved in the enzymatic mechanism of photoreactivation, via a charge-transfer complex.

Photosensitization of pyrimidine dimer splitting by a covalently bound indole has also been described (see, for example, Van Camp *et al.* 1987). More recently, Kim *et al.* (1990) have shown a marked solvent dependency of this process with the ϕ for dimer splitting ranging from 0.06 in water to 0.41 in a non-polar environment. These authors interpret their results in terms of back transfer from an intermediate charge-separated species being in the Marcus inverted region. Further, it is suggested that photolyases may have evolved in which a low polarity active site retards back transfer and thus contributes to the overall efficiency of the repair mechanism associated with enzymatic dimer splitting.

The mechanism of photosplitting of the pyrimidine dimers by the *E.coli* photolyase has been studied by flash photolysis. In particular, the involvement of reduced flavin, $FADH_2$, as the main chromophore of the photosensitized reaction has been shown (Jorns *et al.* 1987). It is assumed that $FADH_2$ or $FADH^{\cdot}$ can act as electron donor to the C(6) of the 5′-site of Thy <> Thy (Sancar 1987). Doublet-quartet intersystem crossing of flavin radical in photolyase has been proposed as a major relaxation process (Okamura *et al.* 1989).

The second type of photoproducts are the pyrimidine photohydrates which are produced via the singlet excited state since triplet quenchers have no effect on the quantum yields of photohydration and since photohydrates are not detected in solutions photosensitized by acetone, acetophenone or benzophenone. The quantum yields of formation are very low, for example 10^{-2} for 3′-CMP at pH 7.1, and 3×10^{-6} at pH 6 for thymine (Cadet and Vigny 1990).

Other types of DNA pyrimidine photoproducts are the pyrimidine(6–4) pyrimidone photoadduct and the 5,6-dihydro-5(α-thyminyl)thymine, the so-called 'spore photoproducts'. The nature of the photoexcited intermediate of these photoproducts is unknown. The mutation potential of pyrimidine(6–4)pyrimidone adducts is being actively investigated. The results obtained by Cleaver *et al.* (1988) provide evidence that far-UV mutagenesis in a *xeroderma pigmentosum* revertant, that repairs only Pyr(6–4)Pyo, is due to these photoadducts and not to Pyr <> Pyr.

5.8.2 *Purines*

Purines are far less photoreactive than pyrimidines (Elad 1976). Quantum yields of the photodegradation of various purine components in frozen aqueous solutions (pH 5.6) at 77 K are below 10^{-4} and 0.3×10^{-4} for adenine and guanine nucleosides (or nucleotides), respectively (see Cadet and Vigny 1990). For prolonged UV irradiation of DNA, Errera (1952) has observed some modifications of adenine but the amount of damage produced *in vivo* in the pyrimidines is far into the 'overkill range' before photodamage of purine can be detected. The nature of the excited states of purines involved in their photochemistry is not well known. Alcohols (Connolly and Linschitz 1968; Ben Ishai *et al.* 1973), amines (Lorberbaum *et al.* 1976), and ethers (Jerumanis and Martel 1970) photoadd to purines. Flash photolysis studies of these systems with time resolution in the nanosecond range should show any possible role of the purine triplet state in these photoadditions.

Irradiation of adenine, guanine, and their nucleoside and nucleotide derivatives in aqueous solutions at 77 K can lead to photoionization through a biphotonic process via a triplet state intermediate (Hélène *et al.* 1966; Arce and Rodriguez 1985, 1986; Arce and Pacheco 1986).

5.8.3 *DNA strand breaks and DNA–DNA and DNA–protein crosslinks via monophotonic excitation*

Monophotonic irradiation by UV is rather inefficient at producing DNA single-strand breaks (SSB) compared to that of Pyr <> Pyr dimer formation in bacterial (M. J. Peak and Peak 1982) and human (M. J. Peak *et al.* 1987) cells. The overall formation of the photoproducts and the relative distribution of *trans-syn* and *cis-syn* isomers of Pyr <> Pyr are dependent on the secondary structure of DNA (Patrick and Rahn 1976; Rahn and Patrick 1976; Ye *et al.* 1990). The SSB production can be enhanced in the presence of oxygen or of photosensitizers such as benzophenone (see Section 5.10). The SSB photolesions are presently considered as minor and not correlated to mutagenesis (M. J. Peak and Peak 1982). In contrast with UV, ionizing radiation is very effective in producing SSB (Hutchinson 1985).

DNA-DNA crosslinks have been observed in dry or tightly packed DNA as in sperm heads, but are likely to represent no biological importance to normal wet cells (K. C. Smith 1966). Photosensitizers, such as furocoumarins which intercalate between the stacked base pairs, can efficiently lead to interchain crosslinks (see Chapter 9).

DNA and protein bind covalently via UV, visible, and ionizing irradiation

(K.C. Smith 1976a; Shetlar *et al.* 1985). These crosslinks can be used to map contact points between a protein and DNA in irradiated cells (M.M. Becker and Wang 1984). They have been related to ageing and carcinogenesis (Chapter 9).

5.9 **Radicals**

Much of our knowledge of the free radicals of pyrimidines, purines, nucleosides, nucleotides, and DNA has stemmed

(1) from radiolysis studies (von Sonntag 1987*a*), in particular, pulse radiolysis (von Sonntag 1987*b*),
(2) from laser-pulse-induced photolysis using high-intensity multi-photonic irradiation involving sequential biphotonic excitations in two-steps at 266 nm and 248 nm (Croke *et al.* 1988, Nikogosyan 1990; Wala *et al.* 1990), monophotonic excitations at 193 nm (Candeias and Steenken 1992, Görner *et al.* 1992), or biphotonic excitation in one step with 532 nm light (Hefetz *et al.* 1992).

The radiolysis investigations have usually been performed with dilute aqueous solutions, the radicals being produced by reactions with the transient species of water radiolysis: OH·, H·, and e^-_{aq}. Originally, identification of the base intermediates, generated by pulse radiolysis, was only possible by optical absorption spectrophotometry, but more recently conductivity measurements have enhanced progress. Radiation-induced strand breakage and crosslinking in solutions of DNA itself have been studied by light scattering methods. ESR studies in the solid state have also been invaluable in the positive identification of radicals derived from the DNA bases.

5.9.1 ***Pyrimidine radicals produced by OH·, H·, and*** e^-_{aq}

The rate constants for reaction of OH· with the pyrimidines, their nucleosides and nucleotides are of the order of 5×10^9 dm^3 mol^{-1} s^{-1}. In its reaction with thymine, OH· can undergo three processes: addition at C5 [I], addition at C6 [II] or hydrogen abstraction from the methyl group [III]. Little OH· attack occurs at the sugar or phosphate groups in the corresponding nucleosides and nucleotides.

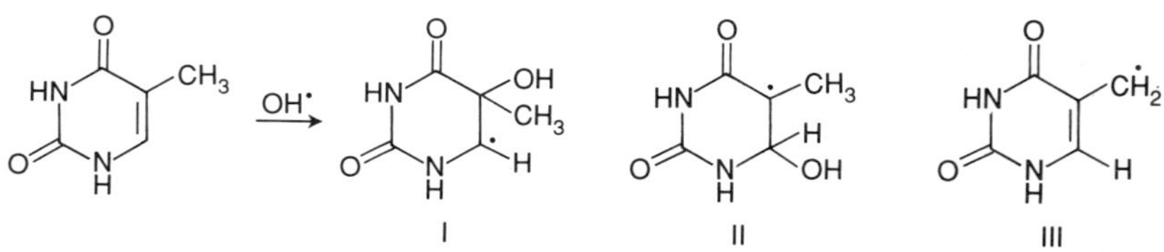

These three radicals can be distinguished by their different redox properties, radical I being reducing, radical II being oxidizing, and radical III being neither a strong reductant nor a strong oxidant (Fujita and Steenken 1981). Radical I accounts for ~65 per cent of the reactions of OH· and is monitored by its ability to reduce tetranitromethane to $C(NO_2)_3^-$ ($k \sim 10^9$ dm^3 mol^{-1} s^{-1}):

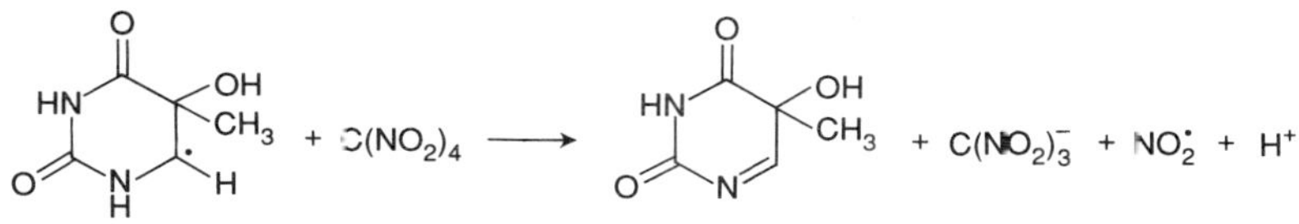

Radical II accounts for ~25 per cent of OH· reactions and is able to oxidize *N,N,N′,N′*-tetramethylphenylenediamine (TMPD) to yield the characteristic absorbing cation TMPD·$^+$ ($k \sim 10^9$ dm^3 mol^{-1} s^{-1}):

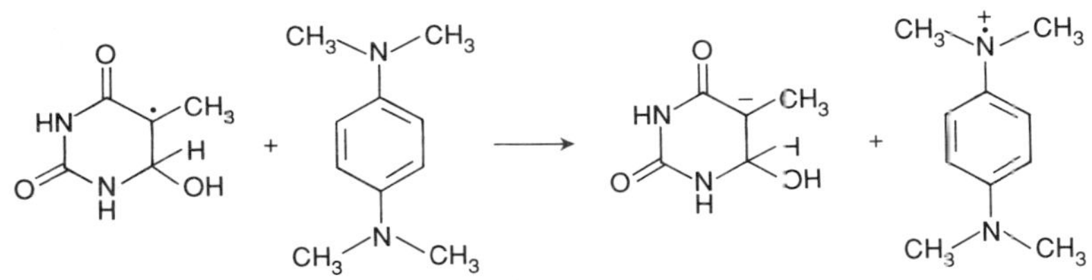

The remaining 10 per cent of the OH· abstract H· from the methyl group of thymine to form radical III which, in the presence of oxygen, forms a peroxyl radical which can also oxidize TMPD to TMPD·$^+$. The reducing radical I can undergo a base-catalysed dehydration reaction to yield another radical which is again able to oxidize TMPD to TMPD·$^+$. Radicals analogous to I and II are also formed from uracil and cytosine.

The rate constants for the reaction of H· with pyrimidines ($k \sim 5 \times 10^8$ dm^3 mol^{-1} s^{-1}) are somewhat lower than the corresponding reactions of OH·. The main reactions of H· are again at the 5 and 6 double bond positions, the relative proportions of such attack being strongly influenced by any substituents. In the case of thymine, the proportion of C5 to C6 is 40 per cent to 60 per cent, with negligible H· abstraction from the methyl group. As with the corresponding OH· adducts, the C5 radicals are reducing and their yield can be monitored by their rapid reaction with $C(NO_2)_4$.

The pyrimidines react with the solvated electron with a diffusion-controlled rate constant ($k \sim 10^{10}$ dm^3 mol^{-1} s^{-1}), the presence of a negatively charged phosphate group in nucleotides lowering the rate constant to some extent. The corresponding pyrimidine radical anions have ketyl character and are subsequently protonated by H^+ or water. Whilst the cytosine radical anion reacts very rapidly with water itself ($t_{1/2}$ = 210 ns), the thymine and uracil electron adducts are longer lived with respect to protonation by water, but react at diffusion controlled rates with H^+. Protonation occurs mainly on the O at the (4) position, more slowly at C6 but not at all at C5:

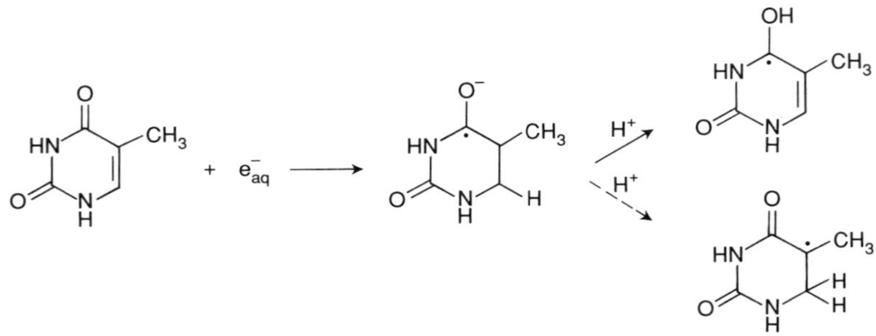

The electron adducts and their oxygen protonated forms are reducing radicals which can readily transfer electrons to electron-affinic radiosensitizers (see Chapter 10).

5.9.2 *Purine radicals produced by OH·, H·, and e^-_{aq}*

Progress in elucidating the identity of purine-derived radicals has been considerably slower than in the case of the pyrimidines. The purine case is more complicated but a coherent pattern is now emerging (Steenken 1989). Attempts have been made to extend to the purines those methods that proved successful with pyrimidines, for example, to classify the various conceivable purine OH· adducts in terms of their redox properties. In such systems, however, the different mesomeric structures are sometimes suggestive of both easy oxidizability and reducibility. Such 'redox ambivalence' can make the assignments of radical structures on the basis of redox titration data rather ambiguous.

For deoxyguanine derivatives three main sites of OH· attack have been considered (O'Neill 1983):

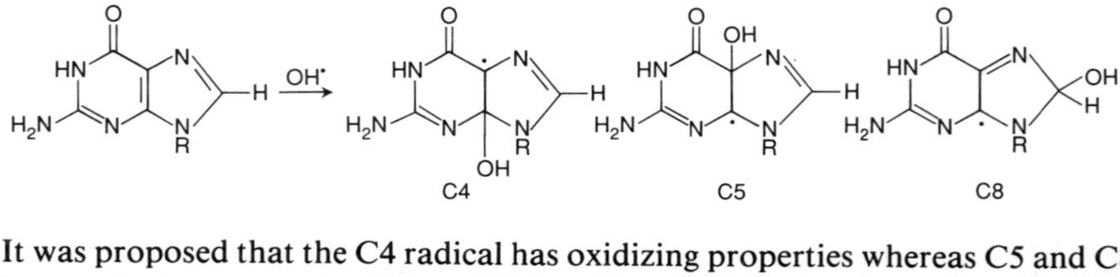

It was proposed that the C4 radical has oxidizing properties whereas C5 and C8 have reducing properties. The oxidizing C4 radical has been shown to be identical with that produced on one-electron oxidation of deoxyguanosine with $Br_2^{\cdot-}$ or $SO_4^{\cdot-}$ identified with deoxyguanosine which has lost H· (written as G(−H·)):

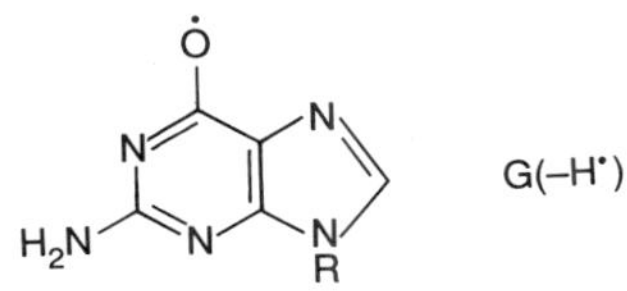

Thus the OH· adduct to C4 must rapidly eliminate a molecule of water. The G(−H·) radical has pK_a values at 3.9 for protonation and at 10.8 for deprotonation. It is able to oxidize ionized thiols such as glutathione and cysteine. The reducing C5 radical can undergo dehydration to yield G(−H·) radical, a reaction which thus results in redox inversion. The C8 reducing radical leads to several stable products including 8-hydroxyguanine derivatives, their 2-electron reduced counterparts, 8-hydroxy-7,8-dihydroguanine compounds and 5-formamido-2,6-diamino pyrimidines.

Addition of OH· to 6-substituted purines such as adenine leads to two different OH· adducts, C4 and C8; for example

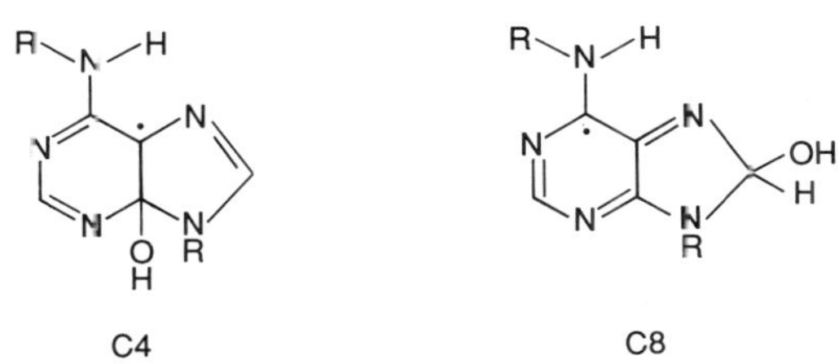

The weakly oxidizing C4 radical can dehydrate into a strongly oxidizing neutral N-centred radical:

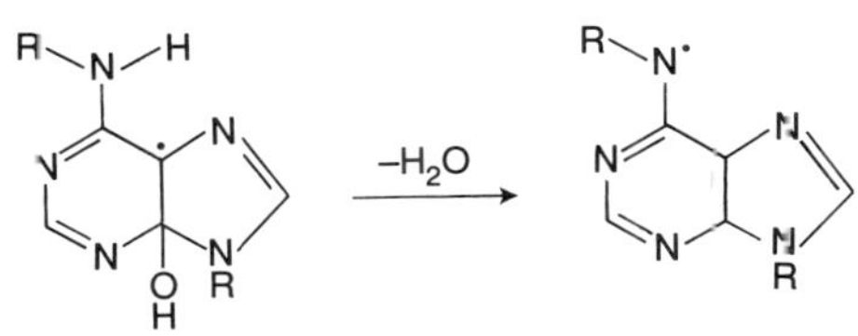

The C8 radical is strongly reducing and can also undergo ring opening ultimately forming a 5-formamido-6-amino pyrimidine. This radical can

also be reduced, thus demonstrating redox ambivalence. The long-known first-order changes in absorbance observed on pulse radiolysis of adenine, deoxyadenosine, and its nucleotides are assigned to the ring opening and dehydration reactions.

The rate constants for reactions of purines with H atoms are lower than those for pyrimidines. Adenine is almost three times less reactive with H· (1.4×10^8 $dm^3\ mol^{-1}\ s^{-1}$) than thymine (3.8×10^8 $dm^3\ mol^{-1}\ s^{-1}$) in neutral solution (Scholes and Simic 1968) and six times less in acid solutions (0.83×10^8 dm^3 $mol^{-1}\ s^{-1}$ compared with 5.1×10^8 $dm^3\ mol^{-1}\ s^{-1}$; Neta *et al.* 1971). The site of attack of H·, investigated by ESR in powder samples, is C8 and C2 in adenine and C8 in guanine derivatives (Holmes *et al.* 1967; Bernhard 1981).

The purines react very rapidly with e^-_{aq} ($k \sim 10^{10}$ $dm^3\ mol^{-1}\ s^{-1}$), although such rates decrease on introduction of a phosphate group, due to electrostastic repulsion between the negatively charged reactant, e^-_{aq} and the nucleotide. Since nitrogen has a greater electron affinity than carbon, it is reasonable to assume that the negative charge in the radical anion resides mainly on the nitrogens. Using conductance and optical detection with nanosecond time resolution, Visscher *et al.* (1987, 1988) showed that such anions are rapidly protonated ($k > 1.4 \times 10^8\ s^{-1}$) to give the N protonated radicals (ANH):

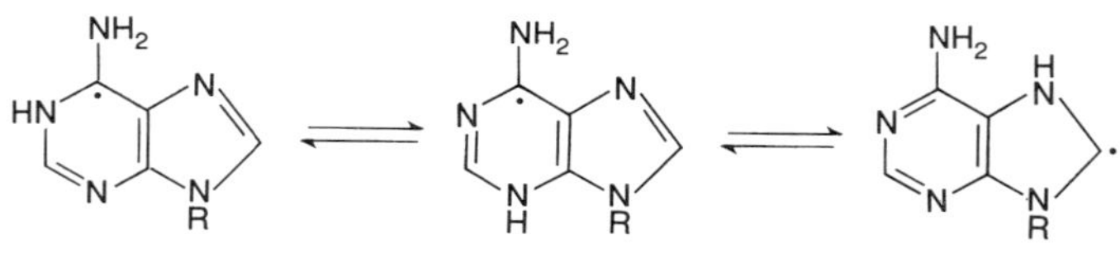

In terms of redox behaviour, these N protonated radicals are strongly reducing, but all are converted via catalysis by OH^-, HPO_4^{2-}, and $H_2PO_4^-$ into the C protonated radicals:

The pK_a value for ANH was roughly estimated to be 12, which compares with 3.5 for the corresponding value for the parent compound, adenosine. This large change in basicity has implications regarding the radiation chemistry of DNA.

Two adenine dinucleotides, nicotinamide adenine dinucleotide (NAD^+) and flavin adenine dinucleotide (FAD), are important electron transport agents in biological systems.

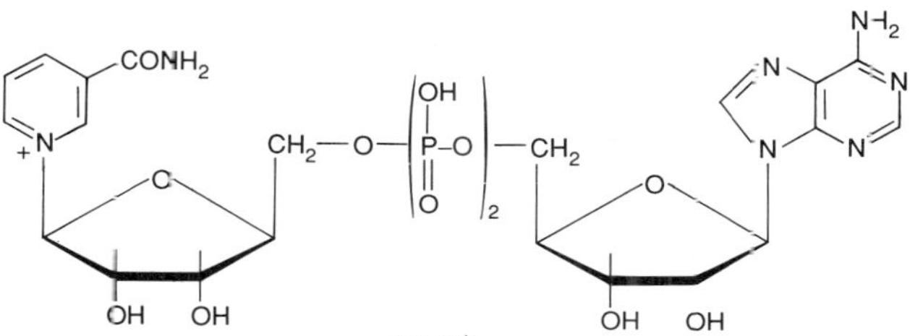

NAD$^+$

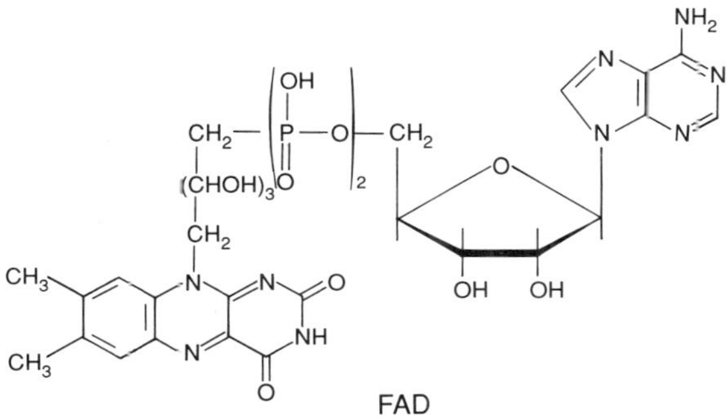

FAD

In both cases it is the prosthetic group which is responsible for the redox function. NAD$^+$ and FAD are quantitatively reduced to the intermediate NAD$^\cdot$ and to the FAD semiquinone by e_{aq}^- and $CO_2^{\cdot -}$, any electron which might be scavenged by the adenine moiety being rapidly channelled to the nicotinamide and flavin groups, respectively. In the absence of other reactants, the NAD$^\cdot$ radical decays to yield a dimer, whereas the flavosemiquinone disproportionates. The extensive pulse radiolysis and flash photolysis studies of the properties of the one-electron reduced form of NAD$^+$ and FAD have been summarized earlier (Bensasson *et al.* 1983; von Sonntag 1987*b*).

5.9.3 *DNA radicals produced by OH$^\cdot$, H$^\cdot$, and e_{aq}^-*

DNA constituents, the pyrimidines and purines are found to have high reactivity ($k \sim 10^9$–10^{10} dm^3 mol^{-1} s^{-1}) towards e_{aq}^- and OH$^\cdot$, whereas the ribose phosphate moiety is somewhat less reactive towards OH$^\cdot$ ($k \sim 10^9$ dm^3 mol^{-1} s^{-1}) and almost unreactive towards e_{aq}^- ($k < 10^7$ dm^3 mol^{-1} s^{-1}) due to the high negative charge of neighbouring phosphate groups. An e_{aq}^- produced near to DNA will therefore react only with a pyrimidine or purine base, whereas OH$^\cdot$ may react also with a sugar unit, usually leading to DNA strand breaks.

As with the separate DNA bases, $OH^{\cdot}$ radicals react with nucleic acids mainly by addition to the double bond of the bases, the percentage of hydrogen abstraction by $OH^{\cdot}$ depending upon the structure of the DNA. Double stranded DNA shows less hydrogen abstraction than single stranded DNA. The distribution of the addition of $OH^{\cdot}$ to the bases of DNA has been found to correspond to the reactivity of the $OH^{\cdot}$ radicals with the respective base monomers. The spectrum obtained after $OH^{\cdot}$-attack corresponds closely to that of a weighted-average spectrum of nucleotides.

There is little attack by the $OH^{\cdot}$ and $H^{\cdot}$ radicals at the sugar moiety. In the absence of oxygen, strand breakage and base release are considered to be caused by the base $OH^{\cdot}$-adduct as demonstrated by ESR in the case of poly(U). The strand breakage mechanism occurs after the following consecutive steps: the $OH^{\cdot}$-adduct base radicals (6-yl-5-OH and 5-yl-6-OH) abstract an H atom from the sugar moiety forming sugar radicals causing a strand break and a base release (Hildenbrand and Schulte-Frohlinde 1989). In the presence of oxygen, sugar peroxyl radicals abstract H atoms from the sugar moiety causing the single strand breaks (SSB) and the base release. The yield of SSB formation by $OH^{\cdot}$ radicals is the same in the presence and absence of oxygen (Schulte-Frohlinde *et al.* 1990). This shows that the oxygen enhancement effect (OER) is not related to SSB.

The polynucleotide most investigated as a model of DNA is poly(U). Table 5.1 shows that $OH^{\cdot}$ radicals are more reactive than $H^{\cdot}$ and e^-_{aq} for strand breakage and base release in poly(U) (von Sonntag 1987*b*).

5.9.4 *Radicals of nucleic acids and their components produced by high intensity laser pulses*

Photoionization of nucleic acids and their constituents in aqueous solutions by powerful nanosecond and picosecond UV laser pulses has been detected via the observation of the absorption of the solvated electron, e^-_{aq}, between 500 and 800 nm peaking at 720 nm. Several pathways have been observed for this photoionization: sequential biphotonic with 248 and 266 nm pulses, monophotonic with 193 nm pulses or two-photon simultaneous absorption with 532 nm.

In the case of biphotonic photoionization, the second light quantum is absorbed by the excited singlet and triplet states of the bases with 248 and

Table 5.1 Efficiencies (in per cent) of $OH^{\cdot}$, $H^{\cdot}$, and e^-_{aq} for inducing strand breakage and release of unaltered uracil in poly(U)

	$OH^{\cdot}$	$H^{\cdot}$	e^-_{aq}
Strand breakage	41	19	0
Base release	51	28	3

266 nm ns laser-pulses (5 MW cm^{-2}), whereas it is absorbed mainly by singlet excited state excitations with 248 and 266 nm ps pulses (4 GW cm^{-2}) (Zavilgelsky *et al.* 1984; Nikogosyan 1990; Wala *et al.* 1990). The ratio of both processes depends on the yield, lifetime, and absorption coefficient of the excited state involved and on the rate of excitation. This type of two-step excitation by photons of ≤ 4.9 eV ($\lambda \geq 248$ nm) populates high-lying electronic states ≈ 6–10 eV and induces biphotonic photoionization of nucleic acid bases, whose gas-phase ionization potentials are ≥ 8 eV. Monophotonic ionization of purines and pyrimidines has been produced by the 193 nm pulses of an argon fluoride excimer laser (Candeias and Steenken 1992; Görner *et al.* 1992) since the photon energy available of 6.4 eV adds to the solvation energies of the ions formed (≤ 3.5 eV) to give a total of ≤ 9.9 eV. The production of hydrated electrons follows a square dependence with intensity at low intensity, goes over to a linear dependence with intensity, before reaching saturation at higher intensities (Nikogosyan 1990).

Photoionization is considered as being at the origin of single strand breaks observed in polynucleotides and DNA. The kinetics of single strand breaks in deoxygenated aqueous solutions of poly(U), following laser pulses, have been investigated using conductivity as the analysing method (Opitz and Schulte-Frohlinde 1987). The rate constants for SSB formation by laser photoionization are identical to those due to OH-adduct radical formation after pulse radiolysis, both measured by conductimetric detection. To explain these kinetic data it has been suggested that in laser photoionization the radical cation of the base reacts with water forming OH-adduct radicals with a faster rate than the succeeding H-abstraction from the sugar. In this hypothesis, the rate-determining steps of SSB formation by radiolytic OH· initiation or by laser-induced radical cations are identical.

Croke *et al.* (1988) have shown that the cleavage of DNA by high intensity UV laser irradiation occurs at the phosphodiester bond lying immediately 3′ (see Fig. 5.2) to guanine residues. These authors proposed that a two-photon excitation of DNA leads to photoionization of a base, followed by intramolecular hole migration to guanine, the base with the lowest ionization potential (see Section 5.9.6); this is followed by a phosphodiester bond cleavage 3′ to the guanine.

Water excitation by high-intensity laser pulses (266 nm, 30 ps, 10–10^4 GW m^{-2}) leads to the formation of solvated electrons, hydroxyl radicals, and hydrogen atoms (Nikogosyan and Letokhov 1983; Oraevsky and Nikogosyan 1985). This is an example of simultaneous absorption of two photons (2×4.66 eV) above the ionization threshold (≈ 6.5 eV) of water. Such biphotonic water photolysis is very similar to water radiolysis. In such experimental conditions, very efficient protective effects by hydroxyl radical scavengers, such as *tert*-butanol, showed that OH· radicals play a major role in the decomposition of thymine and demonstrated the 'indirect' character of this solvent biphotonic photolysis.

Another two-photon laser photochemistry of DNA at wavelengths for which there is no monophotonic transition has been observed with 28 ps, 532 nm laser pulses of 1–8 GW cm^{-2} (Hefetz *et al.* 1992). The biphotonic photolysis yielded cyclobutylpyrimidine dimers with a two-photon cross section of 0.5×10^{-52} cm^2 $photon^{-1}$ s per nucleotide. Experiments performed using focused 532 nm, 28 ps pulses with intensity of 360 TW cm^{-2}, above the threshold of dielectric breakdown for water, resulted in breaks in DNA strands but no cyclobutylpyrimidine dimers. The free radical scavenger, mannitol, partially inhibited formation of the strand breaks, indicating that mechanical processes initiated by the shock wave of the dielectric breakdown also caused DNA strand breaks.

5.9.5 *Intramolecular radical transfer reactions*

Primary effects of ionizing radiation on duplex DNA result in the formation of cationic and anionic radicals. Several lines of investigation, including MO calculations, photoelectron spectroscopy, electrochemistry, ESR studies, and pulse radiolysis, established that the positive holes are trapped by guanine (G) and the electrons stabilized on thymine (T).

The electron transfer processes in DNA can be predicted from the one electron redox potentials of each base. The ionization potentials of nucleic acid bases measured by electron spectroscopy (Hush and Cheung 1975) increase in the order: G (8.24 eV), A (8.44 eV), C (8.94 eV), T (9.14 eV), U (9.50 eV). The aqueous phase oxidation potentials versus NHE at pH 7 are in the order G (1.53 V), A (1.63 V), T (1.73 V), and C (1.88 V) measured electrochemically (Kittler *et al.* 1980), and the one electron redox potentials at pH 13, E^1_{13}, versus NHE measured by pulse radiolysis (Jovanovic and Simic 1986) are in the order G (0.63 V), A (0.75 V), T (0.79 V), C (0.81 V), U (0.88 V). Thus, guanine is the most easily oxidized of the nucleic acid bases and might be considered as a 'positive hole' sink in DNA.

Pulse radiolysis measurements also showed that the DNA-guanyl radical (E^1_7 (G) = 1.04 V) can rapidly oxidize the aromatic amino acids in glycyltryptophan and tyrosine methyl ester with rate constants $k = 3.6 \times 10^7$ and 1.7×10^8 mol^{-1} dm^3 s^{-1}, respectively (Jovanovic and Simic 1989). Steenken *et al.* (1992) have recalculated the oxidation potential of G and found a value slightly higher, $E^1_7 \geq 1.17$ V, using the data of Jovanovic and Simic (1986) but taking the value 0.75 V versus NHE for the reference compound tryptophan at pH 13 (Merenyi *et al.* 1988). The dependence on redox potential of the rate constants for oxidation of reductants by the deoxyguanosine radical at pH 7 follows a Marcus relation (O'Neill and Chapman 1985), suggesting an outer-sphere electron transfer mechanism. ESR studies have confirmed that the positive charge in DNA migrates to G (Sevilla *et al.* 1991). These ESR investigations have also indicated that, at low temperature, cytosine is the major site for electron attachment in DNA (75 per cent $C^{\cdot -}$ and 25 per cent $T^{\cdot -}$). However, at higher temperatures, an equilibrium exists between $C^{\cdot -}$ and $T^{\cdot -}$, and an irreversible protonation at C6 of the thymine anion shifts the reduction path to TH·:

$$C^{\cdot -} \rightleftharpoons T^{\cdot -} \rightarrow TH^{\cdot}$$

A similar conclusion has been reached from pulse radiolysis studies (Steenken *et al.* 1992). Moreover, these pulse radiolysis experiments have shown U, T, C, and their nucleosides and nucleotides to have very similar reduction potentials, $E^1_8 \approx -1.1$ V versus NHE at pH 8.

In conclusion, the damage produced in DNA by ionizing radiation or by high-intensity laser pulses can be considered as migrating to specific sites: the positive charge being localized on the guanine and the negative one on the thymine.

5.9.6 *Double-strand breaks*

Double-strand breaks (DSB) are generated via either a single ionization event directly producing one DSB or the formation of two random single-strand breaks formed at a critical distance (h) on two opposite strands. In the absence of oxygen, this critical distance has been determined to be 40–70 base pairs for DSB induced by 248 nm laser pulses (Bothe *et al.* 1990). Siddiqi and Bothe (1987) found a similar value $h \approx 43$ measured with OH· radicals as initiators. In addition, they observed that a single OH· radical can lead to DSB by a radical transfer mechanism with a probability between 1 and 3 per cent. The quantum yields of DSB formed by laser-induced photoionization are lower than the quantum yields of single-strand breaks ϕ_{SSB}. For example, in argon-saturated solutions $\phi_{DSB} = 0.2 \times 10^{-5}$, whereas $\phi_{SSB} = 5 \times 10^{-5}$ for a total laser energy $E_{tot} = 5$ J (at a laser pulse energy of 100 mJ) with 248 nm, 20 ns pulses (Schulte-Frohlinde *et al.* 1990).

In the hierarchy of DNA lesions, unrepaired double-strand breaks, although rarely produced, appear to be the major type of DNA damage resulting in reproductive cell death. Single-strand breaks tend to be repaired (McWilliams *et al.* 1983). A typical base damage, such as the conversion of thymine to thymine glycol, is only 33 per cent lethal (Hariharan *et al.* 1977). In contrast, there is increasing evidence that reproductive cell death can result from unrepaired double-strand breaks (Lloyd 1985). This evidence is derived from studies showing that with increasing LET, both the ratio of double-strand breaks to single-strand breaks and the relative biological effectiveness increase (Frankenberg 1984; Frankenberg-Schwager *et al.* 1984; Kiefer 1985). Another strong line of evidence results from studies with Auger emitting radionuclides (such as iodo (^{125}I)-desoxyuridine) covalently bound to DNA. These radionuclides are extremely cytotoxic and produce DSB with very high efficiency (one or more per disintegration); a single DSB is sufficient to produce mutation (Little 1991; Little *et al.* 1991). This conclusion is also related to the observation that most radiation-induced mutations involve large-scale molecular events such as deletions, chromosomal rearrangements, and long tracks of recombination, events associated with DNA breaks rather than base damage (Little 1991).

5.10 **Photoactive compounds and DNA**

5.10.1 ***The triplet state of aromatic probes in the study of dynamics and structure of DNA***

Complexation and intercalation of small molecules, including drugs, mutagens, and carcinogens to DNA have been studied by a large number of techniques. We shall describe some examples of the application of the triplet state of ligands used as probes of the dynamics of DNA molecules in the ns, μs, and ms ranges (Geacintov and Brenner 1989).

The fluorescence decay of benzo(*a*)pyrene and benzo(*e*)pyrene bound to DNA is multiexponential, and thus indicates a heterogeneity of binding sites; this is not surprising on account of the presence of 10 distinct intercalative binding sites in DNA. Whereas this fluorescence decay is in a range of tens of nanoseconds, the triplet decay of the same ligands is monoexponential over 2–3 lifetimes in the ms range (Geacintov *et al.* 1976). These results suggest that there is little or no exchange between binding sites in the tens of nanosecond time scale while on the millisecond range, the molecules have time to move from site to site, this change averaging the triplet lifetimes.

The triplet decay of a molecule can be monitored by following the recovery of the partially bleached ground state or by monitoring changes in the $S_1 \rightarrow S_0$ fluorescence, using a second light source (Corin *et al.* 1987). This method has led to the possibility of investigating the relative internal flexibility, bending stiffness, and rotational motion of DNA, poly(dA).(dT), and poly(dG).(dC) via methylene blue triplet used as probe (Hagerman 1988; Allison *et al.* 1989).

Photo-CIDNP effects (see Section 3.2.8.2) are produced via the formation of a radical pair generated from the triplet of a photo-excited dye (D), usually a flavin and the substrate (AH). The radical pair can be generated by either an electron transfer or a hydrogen-atom abstraction:

$$^3D + AH \rightarrow \overline{^3D^{\cdot -}\ AH^{\cdot +}}$$

$$^3D + AH \rightarrow {}^3DH^{\cdot}\ A^{\cdot}$$

Mechanistic studies of photo-CIDNP effects for pyrimidine, purines, dinucleotides, and oligonucleotides have been reported and have emphasized the importance of guanine (G) (E. F. McCord *et al.* 1984*a*; Stob *et al.* 1989). It has also been shown that CIDNP effects generated by the triplet of riboflavin reacting with G could be detected in loop structures and for bulged-out nucleotides in DNA or RNA (E. F. McCord *et al.* 1984*b*). Attempts to observe CIDNP signals for double-stranded nucleic acids might have been unsuccessful on account of the use of a bulky riboflavin as probe. With 3-(carboxy-methyl) lumiflavin, signals have been detected for oligonucleotide duplexes showing an interaction of the excited probe with central guanosine residues (Katahira *et al.* 1991). These signals disappear when the drug chloromycin binds to the minor groove and prevents the access of the flavin to the base, this access being necessary for the photoreaction to take place. The fact that the drug bleomycin cuts DNA at the same site where the CIDNP signals were observed, indicates that a peculiar structural feature is certainly related to the high affinity of the drug and the flavin for the central guanosines of the duplex (Katahira *et al.* 1991). Although the photo-CIDNP technique has not been applied to nucleic acids as much as to proteins (Section 6.6.4), the situation will probably change as the above preliminary examples have allowed the use of photo-CIDNP for structural studies of nucleic acids.

5.10.2 *Photochemical effects of photosensitizers*

Endogenous biomolecules, including 4-thiouridine, chlorophylls, porphyrins, bilirubin, flavins, and exogenous compounds including numerous drugs and food additives can produce 'photodynamic' reactions. These photo-oxidative modifications involve three main biological targets: nucleic acids, membrane lipids, and proteins. Nucleic acids are considered the main targets for ionizing radiation, but their damage by photosensitizers is not thought to be the ultimate cause of cell death, probably due to efficient enzymatic repair. We shall describe some examples of Type I photoreactions, involving electron transfer or hydrogen abstraction and Type II photoreactions, involving singlet oxygen reactions.

5.10.2.1 *Hydrogen abstraction and electron transfer*

Purine nucleic acid components, and particularly guanine, as free base, nucleoside or nucleotide, are preferentially damaged by photosensitized oxidation (Kittler and Löber 1977). Benzophenone- or riboflavin-mediated photooxidation of 3′,5′-di-*O*-acetyl-2′-deoxyguanosine produced four main decomposition products. Two of them have been identified as the α and β anomers of 3,5-di-*O*-acetyl-2-deoxy-D-*erythro*-pentose (Cadet *et al.* 1986). Guanine radical intermediates from the Type I photoreactions and from the hydroxyl reaction give rise to identical decomposition products. In this reaction, a radical cation of the guanine derivative may deprotonate and/or react with a water molecule:

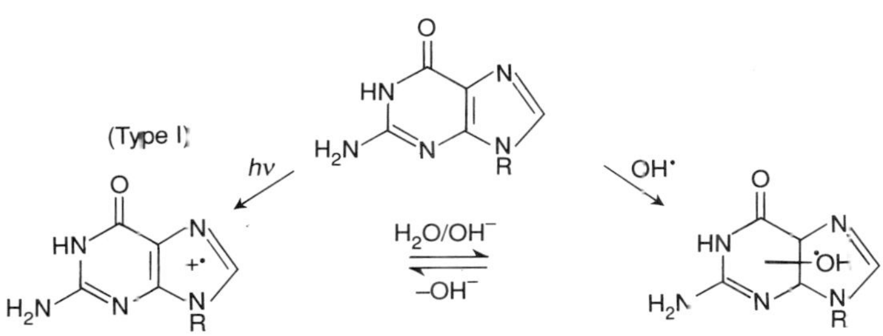

Other examples of Type I photosensitization have been observed with ketones, pterins, and menadione (2-methyl-1,4-naphthoquinone). Ketone sensitizers can induce chain breakage resulting from attack on the deoxyribose residues by $OH^{\cdot}$ radicals and also the formation of thymine dimers via triplet sensitization (Charlier *et al.* 1972). Confirmation of a predominant Type I mechanism in the acetophenone- and benzophenone-mediated photo-oxidation of both nucleosides dAdo and dGuo has been obtained by the analysis of the final products (Cadet *et al.* 1986). The reduction of the triplet state of the naturally occuring pterins by guanosine and xanthine has been studied by flash photolysis; the bimolecular rate constant of the triplet pterin quenching was ≈ 3.5 and 1.4×10^9 dm^3 mol^{-1} s^{-1}, respectively, whereas adenosine did not show any reactivity (Chahidi *et al.* 1981). A menadione (MQ) photosensitized oxidation of pyrimidine nucleosides has been investigated by flash photolysis (G. J. Fisher and Land 1983). The triplet state of MQ was quenched by thymine, uracil, 6-methyluracil, and orotic acid with rate constants nearly diffusion-controlled to give the pyrimidine radical cation and the semiquinone-MQ-radical anion. Another variety of electron transfer reaction is worth mentioning: the electron transfer by quenching of the triplet state of thymine and uracil by a series of electron acceptors studied by flash photolysis which allowed Kemp *et al.* (1985) to determine an intrinsic barrier to electron transfer of $\Delta G^{\neq}(0) = 6.7$ kJ mol^{-1} in terms of Rehm and Weller and Marcus descriptions (see Section 2.4).

5.10.2.2 *Singlet oxygen reactions*

The rate constant of deactivation of $^1O_2$ by guanine derivatives ($\approx 5 \times 10^6$ dm^3 mol^{-1} s^{-1}) is much greater than that observed for all the other nucleic acid components ($\approx 10^4$–10^5 dm^3 mol^{-1} s^{-1}) (Rougée and Bensasson 1986). This is consistent with the fact that singlet oxygen reacts predominantly with guanine in DNA (Kawanishi *et al.* 1986). As a result of the dGuo oxidation four products have been identified: the two diastereoisomers of 4,8-dihydro-4-hydroxy-8-oxoguanine, cyanuric acid, and 8-oxo-dGuo. Lee and Rodgers (1987), using flash photolysis and rose bengal as photosensitizer, confirmed that dGuo can be considered as the only nucleotide reacting with $^1O_2$ ($k = 5.3 \times 10^6$ dm^3 mol^{-1} s^{-1} at pH 7) and that DNA presents a lower reactivity ($k = 5.1 \times 10^5$ dm^3 mol^{-1} s^{-1}) probably due to a greater difficulty in penetrating the macromolecular matrix.

The possibility for singlet oxygen to induce true single-strand breaks is still controversial. Nieuwint *et al.* (1985) and Lafleur *et al.* (1987) in plasmid DNA, Decuyper-Debergh (1988) in double-stranded phage DNA, Epe *et al.* (1988) in PM2 DNA have not detected single-strand breaks by singlet oxygen. In these studies $^1O_2$ was produced by thermal decomposition of an endoperoxide or by the separated-surface-sensitizer technique which avoids reactions with the photosensitizer (Midden and Wang 1983). In contrast to the previous investigations, it has been shown that singlet oxygen can break both plasmid and bacteriophage DNA (Di Mascio *et al.* 1989) and pBR322 DNA (Blazek *et al.*

1989; Devasagayam *et al.* 1991). Piette (1991) concludes from an analysis of these contradictory findings that the guanine residues oxidized by singlet oxygen can probably rearrange and lead to SSB formation when alkali is present and that the yield of true SSB formation is either very low or non-existent. This conclusion is supported by a study showing that the yield of SSB is 17 times lower than the yield of guanine oxidized by singlet oxygen (Schneider *et al.* 1990).

DNA is not thought to be the main target of singlet oxygen toxicity in eukaryotic cells (see, for example, Dubbelman *et al.* 1988 and Section 12.4). However, some evidence of the cytotoxic, mutagenic, and genotoxic effects of DNA damage by singlet oxygen has been provided for yeast, bacteria, and viruses (Epe *et al.* 1989; E. Muller *et al.* 1990; Piette 1991). In particular, the lesions of deoxyguanosine induced by singlet oxygen block DNA synthesis (Ribeiro *et al.* 1992). It should be emphasized that oxidative stress by triggered polymorphonuclear cells or hydrogen peroxide can reactivate human immunodeficiency virus (Legrand-Poels *et al.* 1990) and that chemical carcinogens and depurination can activate protooncogen c-Ha-Ras 1; thus it is essential to study the result of guanine oxidation by singlet oxygen on these processes (Piette 1991).

The determination of the relative importance of the Type I and Type II mechanisms in the photo-oxidation of 3′,5′-di-*O*-Ac-dGuo has been reviewed by Cadet and Vigny (1990) for many photosensitizers. For example it is 45 per cent for Type I and 55 per cent for Type II in the case of methylene blue, 65 and 35 per cent in the case of riboflavin, 53 and 47 per cent in the case of haematoporphyrin, 50 per cent each in the case of 3-carbethoxypsoralen, and 95 per cent and 5 per cent in the case of 5-methoxypsoralen.

In water, the diffusion length of singlet oxygen (δ) within one lifetime (τ_Δ) of its intrinsic decay is $(6D\tau_\Delta)^{1/2}$, D being the diffusion coefficient of oxygen in water ($\approx 2.4\ 10^{-9}\ \mathrm{m^2\ s^{-1}}$) and τ_Δ in water being $\approx 4\ \mu\mathrm{s}$. This leads to $\delta \approx 250\ \mathrm{nm} = 0.25\ \mu\mathrm{m}$. *In vivo*, assuming that τ_Δ is $<0.6\ \mu\mathrm{s}$ because of the singlet oxygen reactions, δ is $<0.037\ \mu\mathrm{m}$.

5.10.2.3 *DNA photosensitized damage at specific sites*

In Section 5.1, we have seen that DNA expresses its information via RNA synthesis from a DNA template, and that this process called 'transcription' is followed by another process called 'translation' in which certain RNA molecules are templates for the synthesis of proteins. Oligonucleotides can be targeted to a specific sequence of the DNA double helix and inhibit transcription (this is called the antigene strategy) or targeted to a messenger RNA and inhibit translation (this is called the antisense strategy). Both strategies should be able to artificially control gene expression and in particular inhibit the expression of oncogenes in tumour cells. The minimum size of an oligonucleotide in order to recognize once a single given sequence in a genome is rather short. It can be calculated assuming a random distribution of the bases (Hélène

1987). For the human genome, which contains 4×10^9 base pairs, a minimum length of 15 to 19 bases is required to find the complementary sequence only once. DNA cleavage or crosslinking at specific sites has been obtained using antisense or antigene oligonucleotides attached to photosensitizers, including azido derivatives, proflavine, ruthenium complexes, psoralens, porphyrins, and ellipticines (Le Doan *et al.* 1991, Takasugi *et al.* 1991, and references therein). Active investigations are being developed in order to evaluate these strategies *in vivo*.

5.11 **Concluding remarks**

Despite many studies by photochemical and radiolytic methods which established, during the last 20 years, the major photophysical parameters and identified the main photolytic and radiolytic products of the nucleic acids and of their components, there are still a number of controversies including for example the consequences of the interaction of DNA with singlet oxygen (Piette 1991). Nevertheless, much useful progress has been made, for example:

(1) the identification of the main products, base damage, and strand breaks, resulting from the reactions of DNA with radicals such as e^-_{aq}, and $H^{\cdot}$, and $OH^{\cdot}$, which are fundemental to the chemotherapy and radiotherapy of tumours,

(2) the understanding of the mechanism of the thymine dimer repair via photomediated reduction.

A recent major interest has been the application of ultra-high intensity lasers, which have allowed the radicals, usually associated with the pulse radiolysis technique, to be generated in a laser flash photolysis experiment. A range of mono-, bi-, and multi-photonic effects can be observed, leading to cyclobutylpyrimidine dimers and strand breaks in DNA and to dielectric breakdown in the solvent at the highest intensities (Nikogosyan 1990; Görner *et al.* 1992; Hefetz *et al.* 1992). This laser technique which can mimic γ-irradiation should facilitate the studies of high-lying electronic states of DNA and might be useful for the investigation of microirradiation damage to organelles in cells.

Aerobic life leads to the generation of oxidative species, including $O_2^{\cdot -}$, $OH^{\cdot}$, H_2O_2, and $^1O_2$. These oxidants alter DNA during normal metabolism, produce damage very similar to that observed in radiolytic or photolytic studies and have implications for carcinogenesis and ageing processes (Adelman *et al.* 1988; J.R. Wagner *et al.* 1992). There has also been an increasing interest in the effects of mutagens, of environmental toxicants, and dietary effects on DNA and hence on ageing and carcinogenesis (see, for example, Simic 1991; Simic and Bergtold 1991). A greater understanding of the fundamental radiolytic, photolytic, chemical, and metabolic oxidative reactions involving DNA should certainly assist medicine in advances against such hazards.

6

Amino acids, peptides, and proteins

'Proteins are the workhorses of life.'

H. Frauenfelder

6.1 Introduction and importance

Increasing interest in the photo- and radiation chemistry of proteins is due to:

(1) the effects of irradiation on food proteins (see, for example, Swallow 1991);

(2) the increasing recognition of radical induced biochemical injury via reactions with proteins (Willson *et al.* 1985);

(3) the recognition of the role of electron transfer reactions involving intramolecular processes in proteins (see, for example, McLendon 1988; Prutz *et al.* 1989) and the application of Marcus theory to such electron transfer processes (see, for example, Bertrand 1991, Faraggi and Klapper 1991 and references given in Section 6.8);

(4) studies of two photon absorption processes (see, for example, Nikogosyan and Görner 1992);

(5) a more detailed understanding of protein conformations and dynamics probed by tryptophan triplet (Geacintov and Brenner 1989);

(6) photo-deligandation in haem proteins, see, for example, Hochstrasser and Johnson 1988; Frauenfelder *et al.* 1991;

(7) the role of a tyrosyl residue in re-reducing the oxidized chlorophyll special pairs in photosystem 2 of photosynthesis (P_{680}—see Chapter 7) and subsequently oxidizing the manganese cluster (see, for example, Mathis *et al.* 1989); and

(8) crosslinking of DNA to proteins (see, for example, Simic and Dizdaroglu 1985)

Nearly half the dry weight of living material is composed of protein (from 'proteios' meaning, in Greek, of the first rank). Many thousands of protein structures have now been established and the role of proteins and enzymes extends to nearly all biological processes.

The 20 naturally occurring amino acids are linked via peptide bonds to produce dipeptides, tripeptides . . . proteins. Such bonds have a *trans* conformation with delocalization of electrons so that there is considerable resistance to rotation about the C—N bond.

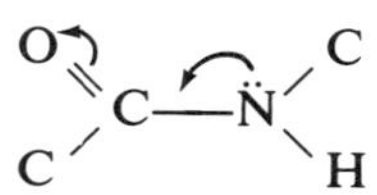

Four levels of protein structure can be distinguished. The primary structure is the sequential order of the amino acid residues. The secondary structure describes the ordered helical arrangement formed by these residues. The tertiary structure concerns the folding of this helical strand, while the quaternary structure deals with the non-covalent associations of independent tertiary structural units. For example, the protein of haemoglobin is composed of four tertiary sub-units. Examples of the most common polypeptide secondary structures are the alpha helix, the beta sheet, and the random chain formed from an alpha helix when the hydrogen bonds are disrupted.

6.2 Ground state absorption

The peptide bond exhibits a longest wavelength weak absorption band at 210–220 nm with a molar absorption coefficient of about 100 $dm^3\ mol^{-1}\ cm^{-1}$. This weak $n\pi^*$ transition usually appears as a small shoulder on the long wavelength tail of the strong $\pi\pi^*$ band near 190 nm ($\varepsilon \sim$ 4–7000 $dm^3\ mol^{-1}\ cm^{-1}$).

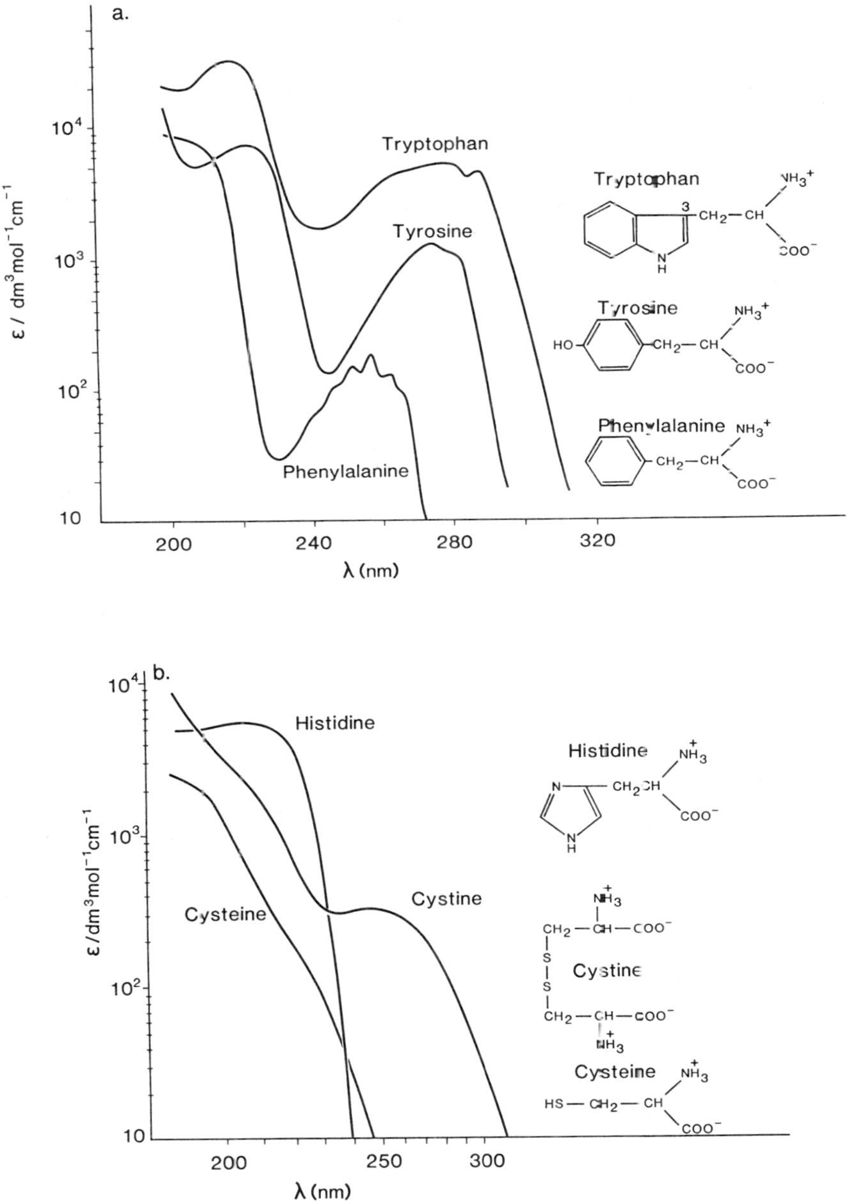

Fig. 6.1 (a) Absorption spectra of tryptophan, tyrosine, and phenylalanine at pH 6. (b) Absorption spectra of histidine, cystine, and cysteine at neutral pH.

The photochemistry of proteins is dominated by the role of the three amino acids with aromatic side chains, tryptophan, tyrosine, and phenylalanine, by the two sulphur containing amino acids, cysteine and cystine, and by histidine which contains an imidazole side chain. The structures and ground state absorption spectra of these six amino acids are given in Fig. 6.1.

Tryptophan, with an indolic side chain, is the amino acid with the longest wavelength ground state absorption (in the near ultraviolet) and it absorbs solar radiation reaching Earth (for example, at $\lambda = 290$ nm $\varepsilon = 3935$ dm^3 mol^{-1} cm^{-1} in neutral solution; Fasman 1976). It thus plays an important role in such phenomena as enzyme photo-inactivation (McLaren and Hidalgo-Salvatierra, 1964) and lens protein damage (Zigman 1977; Dillon and Atherton 1990). Other amino acids are also important with respect to their excited states and radicals and particularly because of their role in electron transfer.

The disulphide group of cystine can be split by ultraviolet and, since the S—S bonds are important for protein tertiary structure, the effects of UV absorption by cystine can be substantial.

6.3 Singlet-oxygen interactions

There have been two topics of most interest with respect to the singlet oxygen–amino acid ground state interaction. Firstly, the ability of amino acids to react with singlet oxygen and the type of reaction (chemical or physical) involved, and secondly the oxidative conversion of tryptophan to *N*-formylkynurenine (NFK) and the subsequent photosensitizing properties of the NFK.

In an early study I.B.C. Matheson and Lee (1979) generated singlet oxygen in D_2O by direct irradiation of the dissolved oxygen with the 1060 nm Nd/YAG laser line and measured the chemical reaction rate constants (k) with several amino acids. For histidine the rate constant followed the histidine pK_a curve with $k = 1 \times 10^8$ dm^3 mol^{-1}s^{-1} at pD ≥ 8 and about 0.5×10^7 dm^3 mol^{-1} s^{-1} at low pD. The corresponding values (at pD 8.4) for tryptophan, methionine, tyrosine, and alanine were 3, 1.6, 0.8 and 0.7×10^7 dm^3 mol^{-1} s^{-1}, respectively. By comparison with the total quenching rate constant (chemical reaction plus physical quenching, I.B.C. Matheson *et al.* 1975) the quenching of singlet oxygen by the amino acids was shown to be virtually all due to a chemical reaction except for tryptophan. For this indolic amino acid the chemical and physical rate constants were reported as 3×10^7 and 2×10^7 dm^3 mol^{-1} s^{-1}, respectively. However, such results are not unambiguous since, for example, the chemical reaction products can vary with the experimental conditions and the photosensitizer used (Gorman *et al.* 1979). Somewhat more recently Lindig and Rodgers (1981) have reported a rate constant of 6.1×10^7 dm^3 mol^{-1} s^{-1} for singlet oxygen quenching by tryptophan in D_2O, with a very similar value (5.6×10^7 dm^3 mol^{-1} s^{-1}) being reported by Rougée and Bensasson (1986) and also by Yoshimura and Ohno (1988), 7×10^7 dm^3

$mol^{-1}\ s^{-1}$. The latter workers studied the lumiflavin-sensitized photo-oxidation of tryptophan and found that a Type I reaction of the sensitizer to produce radicals was much more important than the Type II reaction involving singlet oxygen (see sections 4.2.11, 4.4.1, and 9.4.3).

M.C. Palumbo *et al.* (1990) have studied the interaction of singlet oxygen with a range of indolic derivatives and have attempted to distinguish between physical and reactive (chemical) quenching. Indoles substituted in the 3 position gave significant chemical reaction rates whereas the parent, indole, and those derivatives substituted in other nuclear positions gave mainly physical interactions. Palumbo *et al.* discuss this behaviour in terms of exciplex formation between singlet oxygen and the indolic compound.

With respect to the interaction of amino acids with singlet oxygen, the behaviour of histidine is of interest because it is a means of indicating if enzymic activity loss during photo-irradiation is due to a histidine/singlet oxygen reaction. An example of this is the rose bengal photosensitized inactivation of enolase (Westhead 1965).

There is also some interest in the wide range of photo-products that can arise from cystine (RSSR) photo-oxidation. It should be noted that while the molar absorption coefficient of cystine is much lower than that of tryptophan or tyrosine in the near UV (see Fig. 6.1), the disulphide bridge can have great significance in protein structure so that cystine photolysis can lead to quantum yields for enzymic inactivation much greater than that due to inactivation by tryptophan or tyrosine photolysis (see, for example, Creed 1984*c*). While there is little quantative data, Foote and Peters (1971) have proposed that the RSSR reaction with singlet oxygen proceeds via a zwitterion:

$$\mathrm{RSSR} + {}^1\mathrm{O}_2 \rightarrow \mathrm{RS}^+(\mathrm{O}_2)^-\mathrm{SR}$$

$$\mathrm{RS}^+(\mathrm{O}_2)^-\mathrm{SR} + \mathrm{RSSR} \rightarrow 2\ \mathrm{RSS(O)R}$$

in which a second RSSR molecule completes the reaction.

Of considerable commercial interest is the yellowing and photodegradation of wool and the response of wool to differing UV wavelengths. In this respect the recent results of G. J. Smith (1992) are of interest. Thus, oxygen was found to react with that triplet species in wool keratin which was produced by radiation at 350 nm and phosphoresced at 500 nm. As a result of this interaction singlet oxygen was detected from its 1270 nm luminescence. However, irradiation at 265 nm produces the triplet of tryptophan but no singlet oxygen. A possible role of *N*-formylkynurenine was discussed. It was suggested that singlet oxygen 'bleaches' any yellowing at the longer wavelengths of wool irradiation. This could explain the increased yellowing of wool at wavelengths shorter than about 320 nm.

While it is accepted that in dye sensitized systems singlet oxygen reacts with tryptophan to form NFK there is some debate as to the precise mechanism of the reaction. It seems likely that the excited oxygen attacks the indole ring in the 3 position to yield a zwitterionic form as an initial step rather than a dioxetane intermediate (see Creed 1984*a*). The mechanism of the direct photo-oxidation of tryptophan to NFK is not clear and does not simply involve singlet oxygen (or $O_2^{\cdot -}$) but may involve a radical reaction between the tryptophan lowest triplet and (ground state) oxygen.

NFK is not only the major tryptophan photo-oxidation product but also its primary metabolite. It absorbs more strongly at longer wavelengths than tryptophan and this is one of the most important endogenous photosensitizers (see Section 9.4.3). It is a powerful photodynamic agent with $\phi_T \sim 0.2$ (Pileni *et al.* 1978, 1979), and its photosensitizing properties have been investigated by a number of workers. Pileni *et al.* have used 265 and 353 nm laser flash photolysis to study NFK and report that the NFK (S_1) state reacts with water, via an exciplex, probably to produce the hydroxyl radical $OH^{\cdot}$, while the NFK (T_1) state reacts with many biological compounds including vitamins, amino acids, and nucleic acid bases. The semi-reduced NFK thus formed can, in turn, reduce substrates such as cytochrome *c* and oxygen.

6.4 Singlet excited states

Most interest in the singlet excited states and fluorescence properties have concerned tryptophan and, as a simpler model system, indole. Figure 6.1 shows the absorption spectrum of tryptophan with bands at ~220 nm and also 280 nm and 288 nm with a shoulder at 265 nm. It should be noted that even at wavelengths as long as 300 nm the molar absorption coefficient is ~500 $dm^3\ mol^{-1}\ cm^{-1}$ and that the light absorption by proteins at wavelengths beyond 290 nm is thus due mainly to the typtophan residues. The absorption bands of tryptophan in the 250–300 nm region consist of two electronic transitions (sometimes written with the term symbols ${}^1A \rightarrow {}^1L_a$ and ${}^1A \rightarrow {}^1L_b$). As reviewed by Creed (1984*a*) the ${}^1A \rightarrow {}^1L_a$ transition is much more susceptible to solvent shifts than the ${}^1A \rightarrow {}^1L_b$ transition. One of the possible explanations of this is that the ${}^1L_a$ state forms an exciplex with polar solvent molecules.

For tyrosine and phenylalanine the transitions are that of a substituted benzene system with the lowest energy transitions described as being to the ${}^1L_b$ state (sometimes written as the ${}^1B_{2u}$ state). The first excited singlet levels are such that energy transfer will be from phenylalanine → tyrosine → tryptophan and this, of course, is why in most proteins the fluorescence is dominated by that of tryptophan.

All three aromatic amino acids show a broad, structureless fluorescence with emission maxima at ~280, 300 and 350 nm for phenylalanine, tyrosine, and tryptophan, respectively, and the effects of pH on the fluorescence intensity from these amino acids were reviewed by R.S. Becker (1969) and Santus *et al.* (1980) and are related, at least in part, to the $-NH_2/-NH_3^+$ equilibrium. The ϕ_F value for indole has been reported as 0.35 (Klein *et al.* 1981), 0.2 for tryptophan, 0.14 for tyrosine, and 0.04 for phenylalanine (see, for example,

Cantor and Schimmel 1980 and, for indolic compounds, Santus *et al.* 1980)

There has been some interest in the interaction and distribution of indoles, including tryptophan, with micelles and polyelectrolytes (see, for example, Encinas and Lissi 1986; Zimerman *et al.* 1990). Based on fluorescence studies, Zimerman *et al.* noted a strong interaction between sodium poly(styrene sulphonate) and a range of indoles including tryptamine, indole-3-propionic acid, and indole-3-butyric acid, but no such interaction was measureable for tryptophan itself, probably due to the complexity of the system which cannot be described by a simple equilibrium.

The fluorescence studies of Encinas and Lissi using the detergents, CTAB, SDS, and TX-100, were carried out to mimic biological membranes. This work was based on the quenching of pyrene fluorescence by tryptophan and has allowed a quantitative estimation of the effects of the micellar charge on the distribution of the tryptophan zwitterion between the micellar and aqueous phases. Recently Silva *et al.* (1991) have extended the micellar work to the formation of the product between typtophan and riboflavin (the simultaneous presence of these two compounds has been related to hepatoxic and cytotoxic effects; Farrel *et al.*, 1982; Vileisis *et al.* 1982). Under anaerobic conditions, Silva *et al.* found that the formation of adducts was inhibited in acid solution by SDS and by CTAB. However, the oxidation of tryptophan photoinduced by riboflavin was reported to be much faster in basic solutions.

There has been much debate concerning the number of fluorescence lifetimes that tryptophan exhibits with single, double, and triple components reported. Furthermore, various effects of pH, buffer composition, ionic strength, solvation, and artifacts due to photo-decomposition (see, for example, Gudgin *et al.* 1983; Leismann *et al.* 1983) have been reported. The situation is further complicated by more than one conformer arising, each with a different fluorescence lifetime (Petrich *et al.* 1983).

Tyrosine, on the other hand, generally exhibits a single fluorescence lifetime of ~4 ± 1 ns with −COOH substituted derivatives of tyrosine showing double exponential decays suggesting the presence of more than one conformer (Creed 1984*b* and references therein).

6.5 Triplet states

6.5.1 *Absorption spectra*

The triplet photophysical properties of indole, tryptophan, and its derivatives; of phenol, tyrosine, and its derivatives; and of phenylalanine and its derivatives were reviewed earlier by Bensasson *et al.* (1983) and Creed (1984*a*, *b*, *c*).

Flash photolysis has been used to study the triplet excited states of amino acids for about 20 years. In general, flash photolysis leads not only to triplet generation but also to photo-ionization yielding the corresponding radical cation (or, depending on pH, the deprotonated neutral radical). There is often substantial overlap between the triplet-triplet absorption spectrum and that due to the radicals. Nevertheless, using a combination of laser flash photolysis with and without oxygen (to remove the triplet—see Chapter 3) and pulse radiolysis (to generate the radicals), it has been possible to characterize the absorption spectra and reactivity of both the lowest triplet state and of the radicals. Figure 6.2 shows some typical triplet–triplet spectra for tryptophan, tyrosine, and phenylalanine in aqueous solution.

The ϕ_T for indole in water was reported by Klein *et al.* (1981) as 0.23, and a slightly lower value of 0.18 was given earlier by Volkert *et al.* (1977) for tryptophan. Somewhat more recently Bazin *et al.* (1983) reported the effects of varying the excitation wavelength on the ϕ_T of tryptophan and found values

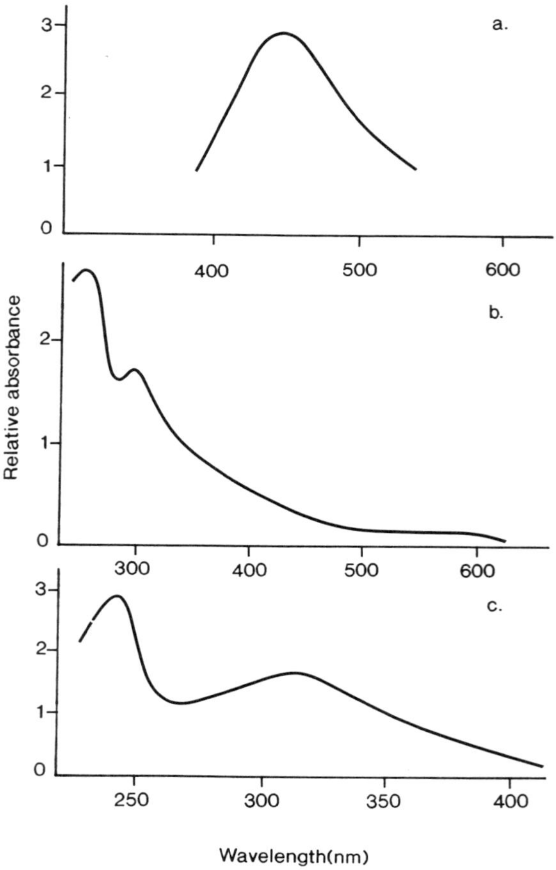

Fig. 6.2 Absorption spectra of triplet states of (a) tryptophan, (b) tyrosine, and (c) phenylalanine (Adapted with permission from Bent and Hayon, *J. Am. Chem. Soc.* **97**, 2599–606, 2606–12 and 2612–9, Copyright 1975, American Chemical Society.)

of 0.26, 0.32, and 0.30 at excitation wavelengths of 280, 292, and 300 nm, respectively. That is, a triplet quantum yield of 0.29 ± 0.03, essentially constant for excitation over the tryptophan near UV band. However, Previtali (1984) has measured the ϕ_T for indole as a function of laser intensity and obtained 0.55 ± 0.05. This was based on an assumed monophotonic photoionization quantum yield (ϕ_i) of 0.15, but see Section 6.6.1. Very recently, also using data based on a range of laser intensities and a theoretical fitting technique, Nikogosyan and Görner (1992) have estimated ϕ_T values of 0.21, 0.50, and 0.40 for 248 nm excitation of tryptophan, tyrosine, and phenylalanine, respectively, in aqueous solution.

The lifetimes of the triplet states of indole, tryptophan, and several tryptophan derivatives have been previously reviewed (Bensasson *et al.* 1983), and lie in the range 8–20 μs. The triplet lifetimes of tyrosine (10 μs after correction for ground state quenching) and several of its derivatives are in the same range, while the corresponding values for phenylalanine are 2–3 μs depending on pH and compounds of related structures have triplet lifetimes from 1–5 μs (Bent and Hayon 1975*a*,*b*, *c*).

The triplet energy levels of tryptophan, tyrosine, and phenylalanine were established from the early phosphorescence work of McGlynn *et al.* (1969) as 297, 342, and 345 kJ mol^{-1}, respectively.

6.5.2 *Reactivity*

Both triplet–triplet energy transfer and electron transfer reactions have been identified as reactions of the tryptophan lowest triplet state. Thus, with anthracene an energy transfer process is observed and with oxygen both energy and electron transfer have been suggested (see, for example, Creed 1984*a*).

Both tryptophan and tyrosine undergo electron transfer reactions with disulphides such as lipoate:

$$^{3}\text{TrpH} + \text{RSSR} \rightarrow \text{TrpH}^{\cdot+} + \text{RSSR}^{\cdot+}$$

$$^{3}\text{TyrOH} + \text{RSSR} \rightarrow \text{TyrO}^{\cdot} + \text{H}^{+} + \text{RSSR}^{\cdot-}$$

such reactions are of biological importance because RSSR$^{\cdot-}$ dissociates according to the following equilibrium (Hoffman and Hayon 1972).

$$\text{RSSR}^{\cdot-} \rightleftharpoons \text{RS}^{\cdot} + \text{RS}^{-}$$

Intramolecular electron transfer of this type involving a neighbouring —SS— group has been reported for cystinylbistyrosine. This molecule has a rather short triplet lifetime (~40 ns), the decay of which leads to an absorption at λ_{max} ~ 420 nm characteristic of the RSSR$^{\cdot-}$ anion (Bent and Hayon 1975a).

Many other molecules of biological relevance, including oxygen, quench tyrosine triplet via electron transfer. Typically, the rate constants are 4.8 × 10^{9} dm^{3} mol^{-1} s^{-1} for oxygen, 3.8 and 7.7 × 10^{8} dm^{3} mol^{-1} s^{-1} for histidine at pH 3.8 and 7.7 respectively, and 5.2 × 10^{8} dm^{3} mol^{-1} s^{-1} for cysteine (for a more complete listing, see Bensasson *et al.* 1983). Typically, such reactions lead to the phenoxyl radical (TyrO·) and the superoxide radical anion:

$$^3\text{TyrOH} + O_2 \rightarrow \text{TyrO}^{\cdot} + H^+ + O_2^{\cdot -}$$

The quenching of phenylalanine triplet by oxygen has a rate constant similar to that for tryptophan and tyrosine triplets ($\sim 3 \times 10^9$ dm^3 mol^{-1} $^{-1}$ at pH 7.5). However, as noted above, the triplet lifetime of phenylalanine is rather shorter than that of tryptophan and tyrosine . This is due to an intramolecular dissociation (Bent and Hayon 1975*b*)

$$NH_3^+ - CH(CH_2C_6H_5)COO^- \rightarrow C_6H_5CH_2^{\cdot} + [NH_3^+ - CH\ COO^-]^{\cdot}$$

6.6 **Radical generation**

6.6.1 *One and two photon absorption*

There have been many measurements of the photoionization of tryptophan, tyrosine, and phenylalanine over the last 30 years or so. A wide range of values have been given for ϕ_i with the mechanism of ionization being reported by some groups as monophotonic and some as biphotonic, while others give data in conflict with their own previous work! The early studies have been reviewed by Creed (1984*a*, *b*, *c*) and are summarized by Nikogosyan and Görner (1992). Bazin *et al.* (1983) used a YAG pumped dye laser system to measure ϕ_i for tryptophan in water and reported a value of 0.075 ± 0.02 over the wavelengths 265–300 nm, and that the process was monophotonic. Subsequently, Previtali (1984), using laser flash photolysis of indole in the presence of the amino acids, glycine, proline, and hydroxyproline suggests that the photoionization is a mixed mono- and bi-photonic process.

Nikogosyan and Görner (1992) have used KrF and ArF excimer lasers, at 248 and 193 nm, respectively, to study the ϕ_i values for the aromatic amino acids as a function of laser intensity, *I*. In this detailed study it was found that $\phi_i = 0.05$, 0.03, and 0.01 for tryptophan, tyrosine, and phenylalanine, respectively, as $I \rightarrow 0$ for 248 nm excitation. Figure 6.3 shows the effect of increasing *I* on the observed values of ϕ_i. These data show that 248 nm ionization is, at least partially, two-step in nature. Such a biphotonic mechanism was also suggested in the early work of Bent and Hayon (1975*a*, *b*, *c*) and Lachish *et al.* (1976). Nikogosyan and Görner conclude that most of the hydrated electrons in their work are formed via the route: $S_O \rightarrow S_1 \rightarrow T_1 \rightarrow T_n \rightarrow$ ionization. Of course, the marked dependency of ϕ_i on *I* does explain, at least to some extent, the wide range of values given previously since often the irradiation intensity and excitation wavelength were not comparable. In the low intensity region ($I < 5 \times 10^{10}$ W m^{-2}) Nikogosyan and Görner report changed slopes of 1.3, 1.6, and 1.7 for tryptophan, tyrosine, and phenylalanine respectively (data not shown in Fig. 6.3). These authors therefore suggest that there is a

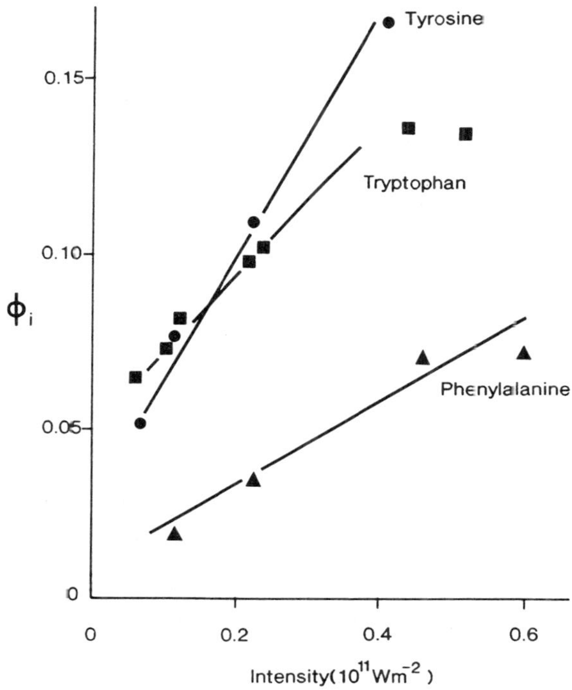

Fig. 6.3 Variations in apparent ϕ_i for tryptophan, tyrosine, and phenylalanine as a function of high laser intensity at 248 nm (taken from Nikogosyan and Gorner 1992).

monophotonic contribution to the 248 nm excitation and that this is most marked for tryptophan.

The ϕ_i data obtained from 193 nm excitation (0.32, 0.13, and 0.07 for Trp, Tyr, and Phe respectively) indicated no dependence on I so that, in this case, contrary to the 248 nm irradiation data, the solvated electron arises via a monophotonic process.

The UV photolysis of aromatic amino acids and some related di- and tripeptides have been studied by Khoroshilova *et al.* (1990). Using 266 nm laser excitation these workers noted significant peptide bond scission (up to 15 per cent) suggesting that such processes must be taken into account during high intensity (two photon) laser studies of protein inactivation.

An interesting use of photoionization in proteins is suggested by the recent work of Greenaway *et al.* (1991). These workers reacted pyridoxal 5′-phosphate with free amino groups (e.g as in lysine) and reduced the so-formed Schiff base with, for example, sodium borohydride. The adducts so produced were photoionized with a 308 nm excimer laser. Thus, using this technique, it is possible for flash excitation, outside the region of absorption by aromatic residues of proteins, to produce electrons at a specific site, leading to significant biological effects.

Finally, in this section, it should be noted that with tyrosine and its derivatives, radicals can arise other than via photoionization. This is due to a rupture of the O—H bond giving TyrO·, H^+, and the solvated electron (or TyrO· and H·). A similar reaction has been reported for dopa (Land *et al.* 1986, see Section 8.2.1).

6.6.2 *Generation with oxidizing radicals*

Pulse radiolysis is often used to generate oxidizing radicals such as OH·, $Br_2^{\cdot -}$, $(SCN)_2^{\cdot -}$, $SeO_3^{\cdot -}$, $NO_2^{\cdot}$, and the azide radical $N_3^{\cdot}$. The reactions of several of these radicals with amino acids were reviewed earlier (Bensasson *et al.* 1983) and are only summarized here together with some of the more recent studies.

The rate constants for the reaction of these oxidizing inorganic radicals with tryptophan lie in the range 1–10 × 10^8 dm^3 mol^{-1} s^{-1} (see Table 6.1). Typically, the reactions involved are:

$$Br_2^{\cdot -} + TrpH \rightarrow 2Br^- + TrpH^{\cdot +}$$

followed by deprotonation of $TrpH^{\cdot +}$ reflecting a pK_a of ~4.2 (Bent and Hayon 1975*c*). The absorption spectra of amino acid radicals produced by both light and ionizing radiation are similar. Figure 6.4 shows the $TrpH^{\cdot +}$ and Trp· spectra.

For tyrosine, the radicals arise from the loss of the phenolic hydrogen to give

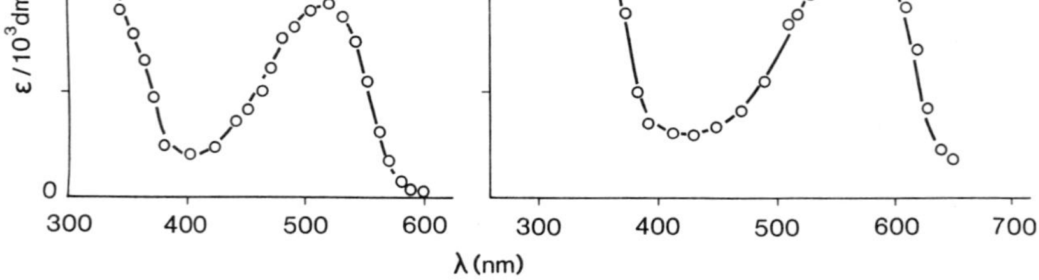

Fig. 6.4 Absorption spectra of the product of oxidation of tryptophan by $(SCN)_2^{\cdot -}$ at pH 6.7 (Trp·) and pH 3.4 ($TrpH^{\cdot +}$) ([Tryptophan] = 3 × 10^{-3} mol dm^{-3}, [SCN^-] = 5 × 10^{-2} mol dm^{-3}, N_2O-saturated aqueous solutions, spectra measured 30 μs after 3.5 Gy pulse) (taken from Posener *et al.* 1976).

Table 6.1 Bimolecular rate constants (10^7 dm^3 $mol^{-1}s^{-1}$) for reactions of radicals with some amino acids at room temperature[a]

Radical	Tryptophan	Tyrosine	Histidine	Phenylalanine	Cysteine	Methionine	Reference[b]
$Br_2^{\cdot -}$	77	2.0 (pH 7.5)	1.5 (pH 7.6)	<0.1	18 (pH 6.6)	1.1 (pH 7.3)	a
$(SCN)_2^{\cdot -}$	27	0.5	<0.1	<0.1	5 (pH 6.6)	0.2	a
$I_2^{\cdot -}$	<0.1	<0.1	<0.1	<0.1	11 (pH 6.8)	0.1	a
$CO_3^{\cdot -}$ (pH 11.2 ± 0.3)	44	29	0.7	<0.1	27	12	a
$Cl_2^{\cdot -}$ (pH 2)	260	27	1.4	0.6	85	0.7	a
$N_3^{\cdot}$	410 (pH 7.4)	10 (pH 6.5)	<0.1 (pH 7.8)	<0.1 (pH 7.8)	1.4 (pH 6.3)	<0.1 (pH 6.5)	b
	450 (pH 11.4)	360 (pH 11.8)	1.7 (pH 11.2)	—	270 (pH 10.5)	<0.5 (pH 11.4)	b

[a] pH 7.0 ± 0.2 except where otherwise stated.
[b] (a), Adams *et al*. (1972*a*); (b), Land and Prütz (1979).

TyrO· species. The corresponding absorption spectrum is given in Fig. 6.5. The rate constants for the reaction of tyrosine with the oxidizing radicals are less than for tryptophan with typical values being given in Table 6.1.

Most inorganic radicals do not oxidize phenylalanine but OH· does react ($k \sim 6.6 \times 10^9$ dm^3 mol^{-1} s^{-1}) to give a transient species absorbing at 318 nm (Klapper and Faraggi 1979).

The oxidizing radicals $N_3^{\cdot}$, $SeO_2^{\cdot}$, and OH· can lead to a somewhat more complex situation than simple one-electron oxidation (Butler *et al.* 1984):

$$N_3^{\cdot} + TrpH \rightarrow N_3^- + TrpH^{\cdot +}$$

$$TrpH^{\cdot +} + N_3^- \rightarrow HN_3 + Trp^{\cdot}$$

with the rate constant for the second of these reactions being 2×10^8 dm^3 mol^{-1} s^{-1}. Phosphate ions and hydroxyl ions were also found to rapidly deprotonate $TrpH^{\cdot +}$. Furthermore, at high azide concentrations the azide radical reacts with its parent ion to form $N_6^{\cdot -}$

$$N_3^{\cdot} + N_3^- = N_6^{\cdot -}$$

The redox potential of tryptophan and tyrosyl radicals (see Fig. 6.8) are such

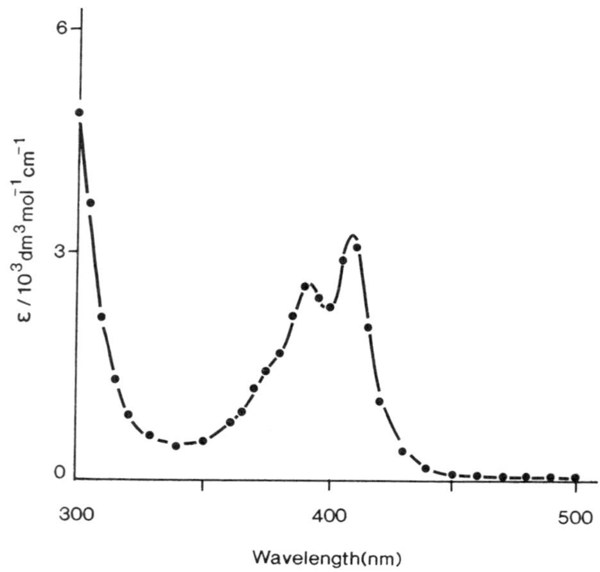

Fig. 6.5 Absorption spectrum of product of oxidation of tyrosine by $N_3^{\cdot}$ at pH 7.1 ([Tyrosine] = 1×10^{-3} mol dm^{-3}, [N_3^-] = 5×10^{-2} mol dm^{-3}, N_2O-saturated aqueous solution, spectrum measured 48 μs after 2.7 Gy pulse; Land, unpublished). Spectrum normalized to ε_{405} nm = 3200 dm^3 mol^{-1} cm^{-1} (Bansal and Fessenden 1976).

that both $N_3^{\cdot}$ and $N_6^{\cdot-}$ are thermodynamically capable of oxidizing tryptophan and tyrosine although, at azide concentrations below 10^{-1} mol dm^{-3}, the reactive species was shown to be uncharged (Butler *et al.* 1984), thus the reaction entirely involves $N_3^{\cdot}$.

It has also been suggested that $N_3^{\cdot}$ adds to tryptophan and tyrosine (Singh *et al.* 1982). However, these claims are associated with the use of high doses and alternative explanations based on different states of protonation of products formed from the decay of Trp$^{\cdot}$ and, in the case of tyrosine, the formation of biphenols (Butler *et al.* 1984), have been given.

The mechanisms of oxidation by the azide radical can be via electron transfer:

$$N_3^{\cdot} + TrpH \rightarrow N_3^- + TrpH^{\cdot+}$$

or H atom transfer

$$N_3^{\cdot} + TrpH \rightarrow HN_3 + Trp^{\cdot}$$

and this has been investigated by Solar *et al.* (1991). Using a pulse radiolysis comparative technique with H_2O and D_2O, these workers interpret their data as suggesting that electron transfer occurs in all cases.

Selenium dioxide radicals are in acid–base equilibrium (Tamba and Badiello 1985):

$$H_2SeO_3^{\cdot+} \rightleftharpoons HSeO_3^{\cdot} \rightleftharpoons SeO_3^{\cdot-}$$

with the corresponding pK_a values being 3.9 and 7.4. Pulse radiolysis studies have led to the following order for redox potentials:

$$Br_2^{\cdot-} > HSeO_3^{\cdot} > (SCN)_2^{\cdot-} > SeO_3^{\cdot-} > I_2^{\cdot-}$$

so that the redox potential values are 0.92 V for $HSeO_3^{\cdot}$ and 0.66 V for $SeO_3^{\cdot-}$. Thus, the acid form of the radical is a more powerful oxidant than the base, following the general trend for such acid–base couples (Hayon and Simic 1974). The reaction of selenite radicals with tyrosine is mainly pH-independent in the range 5.6–11.2 (Tamba and Badiello 1985). This is in marked contrast to the increase in reactivity seen by several other groups for tyrosine oxidation by other oxidizing radicals in this pH range. Such an increase in rate is related to the deprotonation of TyrOH (p$K_a \sim 10$) and the increased ease of oxidation of TyrO$^-$ by the oxidizing radicals. Tamba and Badiello suggest that the results with the selenium dioxide radicals may be due to a balancing of the tyrosine deprotonation and the acid–base forms of the radical.

Hydroxyl radicals are not usually used to generate one-electron oxidized species because of other competitive processes such as addition reactions. However, a characterization of the radiation-induced products of OH$^{\cdot}$ reactions with amino acids is worthwhile to help understand the effects of ionizing radiation on proteins. Thus Karam *et al.* (1984), using a ^{60}Co γ-source and high performance liquid chromatography have shown that as with the early studies of tyrosine, the tyrosyl peptides (L-tyr-L-tyr and L-tyr-L-tyr-L-tyr) undergo crosslinking *via* C—C bonds following OH· addition, but also that these peptides link via ether groups.

Shen *et al.* (1987) have studied the reaction of OH· and several other oxidizing radicals with indole and a series of methyl substituted indoles. For all radicals except OH·, a quantitative conversion to the indolyl radical was obtained, while for the OH· reaction about 50 per cent of indolyl radicals were obtained with the remainder of the OH· reaction being assigned to addition products.

Imidazole, the functional group present in histidyl residues, is also an important protein site for OH· attack with rate constants about 3–9 × 10^9 dm^3 mol^{-1} s^{-1} being reported (Rao *et al.* 1975). The structures of the radicals so produced may involve OH· addition at the C2 or C5 positions (Samuni and Neta 1973).

RSH compounds such as cysteine react rapidly with $Br_2^{\cdot -}$ or OH· to form thiyl radicals.

$$RSH + Br_2^{\cdot -} \rightarrow RS^{\cdot} + 2Br^- + H^+$$

With disulphide, RSSR, oxidation to the corresponding radical cation $RSSR^{\cdot +}$ occurs. Subsequent reactions of $RSSR^{\cdot +}$ can be complex and include S—S cleavage, disproportionation and water addition (see, for example, Bensasson *et al.* 1983).

6.6.3 *Generation with reducing radicals*

As with the previous section this topic was reviewed earlier (Bensasson *et al.* 1983) and only a summary is presented here together with some more recent results. Also, the reader is referred to von Sonntag (1987*a*) for a more recent review.

The most important reducing reactions with proteins involve hydrated electron addition to aromatic amino acids, sulphur containing amino acids, and to histidine.

Indole and tryptophan derivatives add e_{aq}^- with rate constants of 2–5 × 10^8 dm^3 mol^{-1} s^{-1} to yield electron adducts such as $TrpH^{\cdot -}$ with λ_{max} at ~340 and 510 nm. Similar rate constants have been obtained for electron addition to tyrosine and, as with tryptophan, hydrogen atom adducts can also be obtained under suitable conditions such as low pH.

The reaction of e_{aq}^- with phenylalanine (k ~ 1.6 × 10^8 dm^3 s^{-1}) can lead to both addition to the aromatic ring and de-amination (Mittal and Hayon 1974).

Inactivation of enzymes can follow from disulphide bond reduction and this topic and other aspects of sulphur reduction in proteins has attracted considerable attention (von Sonntag 1987*a*). Typically, rate constants for electron addition to disulphides are ~2–30 × 10^9 dm^3 mol^{-1} s^{-1} and the resulting anion radical has a broad absorption maximum at 410 nm. However, the absorption spectra of disulphide radical anions do show substantial variations of λ_{max}

and molar absorption coefficient ($\sim$4–9 $\times$ 10^3 dm^3 mol^{-1} cm^{-1}, from 410–425 nm) as described by von Sonntag (1990). At low pH protonation can occur to yield $RSSRH^{\cdot}$, which like $RSSR^{\cdot -}$ itself, can dissociate

$$RSSR^{\cdot -} + H^+ \rightarrow RSSRH^{\cdot} \rightarrow RS^{\cdot} + RSH$$

Favaudon *et al.* (1990) have investigated the $CO_2^{\cdot -}$ radical induced cleavage of S—S bonds in proteins. Such a process offers a method of probing disulphide bond reactivity in proteins. The $CO_2/CO_2^{\cdot -}$ couple is -1.85 V (Surdhar *et al.* 1989), and $CO_2^{\cdot -}$ electron donation to proteins has been suggested to be selective for S—S bonds (G.E. Adams *et al.* 1972*b*). However, $CO_2^{\cdot -}$ can also react with cystine groups (Faraggi and Klapper 1988) and can also oxidize tyrosine:

$$CO_2^{\cdot -} + TyrOH \rightarrow HCO_2^- + TyrO^{\cdot}$$

The resulting phenoxyl radical is a good oxidant (DeFelippis *et al.* 1989) and can react with reducing radicals such as $RSSR^{\cdot -}$

$$RSSR^{\cdot -} + TyrO^{\cdot} \xrightarrow{H^+} RSSR + TyrOH$$

Under acid conditions, however, Favaudon *et al.* (1990) did find that $CO_2^{\cdot -}$ reacted with a range of proteins to form the protonated disulphide as the main species, and its major decay path was formation of thiyl radicals. At pH 8 the deprotonated $RSSR^{\cdot -}$ is formed, this assignment being based on its 400–425 nm absorption (see, for example, Akhlaq and von Sonntag 1987). This species decays through competing inter- and intra-molecular routes, including disproportionation, protein–protein cross-linking, electron transfer with tyrosyl residues, and reaction with sulphydryl groups. Thus, as noted by Favaudon *et al.* the fate of the oxygen-centred phenoxyl radicals in proteins should take account of sulphur radical electron transfer reactions such as

$$Tyr - O^{\bullet} + RSSR \xrightarrow{H^+} Tyr - OH + R{-}S{-}S{-}R^{\cdot +}$$

The reaction of e_{aq}^- with sulphydryl groups has also been reported (see, for example, M.Z. Hoffman and Hayon 1973) and probably leads to dissociation reactions such as

$$RCH_2SH + e_{aq}^- \rightarrow R\dot{C}H_2 + SH^-$$

The protonated histidyl residue and the carbonyl moiety of the peptide bond are the two other important sites of hydrated electron addition in proteins. In both cases the initial electron adduct is unstable often decaying in a complex manner. For the peptide bond one of the reactions involves deamination to yield ammonia. In addition, the rate constant for the reaction between e_{aq}^- and polyglycine and polyalanine increases linearly with chain length (Tal and Faraggi 1975) indicating that the carbonyl group can be significant in the reactivity of proteins towards e_{aq}^-.

6.6.4 *Generation by probe triplets*

The photo-CIDNP method described in Section 3.2.8.2 can give information about the surface structure of proteins, their conformation and their interaction with ligands. Visible excitation of a dye D (usually a flavin derivative) to its triplet ^3D in the presence of proteins (AH) can lead to radical pair formation via electron transfer in the case of tryptophan, the methylated lysines, and methionine, or hydrogen atom-abstraction in the case of histidine and tyrosine (Stob and Kaptein 1989):

$$^3D + AH \rightarrow {}^3\overline{D^{\cdot -}\,AH^{\cdot -}}$$

$$^3D + AH \rightarrow {}^3\overline{DH^{\cdot}A^{\cdot}}$$

The NMR absorption or emission lines are enhanced specifically for accessible amino acids (AH), and thus these reactions can be used to probe the external part of the protein. For example, in the case of platelet factor 4 (PF4), a secretory peptide known to play an important role in coagulation, photo-CIDNP NMR experiments have revealed that Tyr-60 is accessible to the flavin dye in the monomer state but not in the tetramer state, suggesting that the association process either (i) induces a new monomer sub-unit conformation which makes Tyr-60 inaccessible to the solvent or (ii) buries Tyr-60 within the tetramer structure (K. H. Mayo and Chen 1989).

In the case of apocytochrome *c*, the haem-free precursor of the mitochondrial cytochrome *c*, photo-CIDNP enhancements of histidine, tryptophan, and tyrosine are reduced in the presence of the membrane model SDS. These results indicate a lesser accessibility of the residues to the flavin dye in the micellar system and have been interpreted as indicating a protein-micelle interaction burying the residues (Snel *et al.* 1991).

Photo-CIDNP is also well suited for the study of Type I photosensitized reactions between a photosensitizer such as a porphyrin and a biomolecule target such as tyrosine (Le Nouen *et al.* 1990).

6.7 **Radical reactivity**

Radicals produced by both one-electron oxidation and reduction undergo subsequent processes. Thus, oxidation of tryptophan to produce the corresponding radical cation, is followed by dissociation reflecting a pK_a of about 4.2.

$$\text{Trp H}^{\cdot +} \rightleftharpoons \text{Trp}^{\cdot} + H^+$$

The 'repair' of these indolyl radicals can be achieved by electron donors (Jovanovic and Simic 1985*a*), such as ascorbate, *N,N,N′,N′*-tetramethyl-*p*-phenylene-diamine-hydrochloride, phenolic antioxidants, and the Vitamin E analogue, Trolox. However, sulphydryl compounds which are good H· donors are rather unreactive, as is oxygen.

In non-polar solvents oxidation can lead to indolyl radicals, I·, and to N—H bond rupture followed by H· addition to the parent molecule to yield $IH_2^{\cdot}$.

Tyrosyl phenoxy radicals decay with second-order kinetics ($2k = 4.7 \times 10^8$ $dm^3\ mol^{-1}\ s^{-1}$) with the formation of 2,2′ dimers which then enolize to produce a 2,2′ biphenolic compound (Prütz *et al.* 1983). The results of Hashimoto *et al.* (1982) and Seki and Schnabel (1982) show that the formation of such biphenolic species is a rather common and efficient process in tyrosyl peptides and proteins. Both cysteine and oxygen substantially preclude biphenol formation and 'repair' the protein by converting TyrO· back to tyrosine:

$$\mathrm{TyrO^{\cdot} + CysH \rightarrow TyrOH + Cys^{\cdot}}$$

In the case of oxygen, the repair is via the superoxide radical anion:

$$e^-_{aq} + O_2 \rightarrow O_2^{\cdot -}$$

$$\mathrm{TyrO^{\cdot}} + O_2^{\cdot -} \xrightarrow{H^+} \mathrm{TyrOH} + O_2$$

More recently Cudina and Josimovic (1987) have used γ and pulse radiolysis to study the interaction of oxygen and tyrosine radicals. The second-order rate constants reported are $k(\mathrm{TyrO^{\cdot}} + O_2) \leq 2 \times 10^5$ $dm^3\ mol^{-1}s^{-1}$ and $k(\mathrm{TyrO^{\cdot}} + O_2^{\cdot -}) = 1.7 \times 10^9$ $dm^3\ mol^{-1}\ s^{-1}$. In addition, tyrosine itself was found to react with $HO_2^{\cdot}$ to yield dopa.

As noted above, disulphide radicals readily undergo dissociative, often complex, reactions leading to S—S cleavage:

$$\mathrm{RSSR^{\cdot +} + OH^- \rightarrow RSH + RSO^{\cdot}}$$

$$\mathrm{RSSR^{\cdot -} \rightleftharpoons RS^{\cdot} + RS^-}$$

6.8 Electron migration within proteins

Long range energy transfer in many systems has been studied for over 40 years following its early description by Förster (1948). On the other hand long range electron transfer (LRET) was not accepted as an important phenomena until much later (see reviews by Isied *et al.* 1992; Winkler and Gray 1992). This was despite some rather direct experimental evidence arising from pulse radiolysis studies (see, for example, Bensasson *et al.* 1983), some of these are discussed below (Section 6.8.1). However, in the last decade LRET has become accepted and there has been an explosion of interest in several systems exhibiting this effect, e.g photosynthetic reaction centres (see Section 7.7) and respiratory electron transfer chains, with particular interest in LRET in proteins. One of the proteins most studied has been cytochrome *c* because the X-ray crystallographic structure of several cytochrome *c* systems are available.

At least three factors are now known to affect the rate constants for LRET in biological systems and these have been much investigated recently from both a theoretical and experimental viewpoint:

1. Electronic coupling of the redox centres and its dependence on the separation distances.
2. Nuclear factors including molecular vibrations of the redox reactants and rearrangement and polarization of the surrounding medium.
3. The accessibility of different protein conformations, with the possibility that this factor can lead to directional electron transfer. As pointed out by Sutin and Brunschwig (1990), for example, the effects of this third factor, can be very marked in the Marcus inverted region (see Section 2.4.4)

In the last few years much of our understanding of LRET in proteins has come from extending the Marcus theory from its original description of small inorganic molecules in solution to redox centres with large separations in proteins, see, for example, Bertrand (1991) and references therein, and to experiments on model systems. Some of the most elegant model systems are based on proteins structurally modified by genetic engineering (site directed mutagenesis) and often involve a ruthenium modified cytochrome *c* system (see, for example, Bechtold *et al.* 1986; McLendon *et al.* 1987; Karas *et al.* 1988).

LRET has become an interdisciplinary area spanning theoretical physics to modern biochemical techniques and this range of study has been reviewed by Bertrand (1991), Kuki (1991), B.M. Hoffman *et al.* (1991), Therien *et al.* (1991); Mauk (1991); McLendon (1991), and Sykes (1991).

6.8.1 *Radical transformations involving peptide functions*

Pulse radiolysis investigations on simple amino acids and peptide systems have shown that a variety of radical transformations can be initiated both by one-electron reducing as well as one-electron oxidizing species (Prütz, 1987). The transformations occur in particular between aromatic and sulphur-containing functions and some of these reactions have also been unambiguously detected in peptides and proteins. The mechanism of such transformations is not clear and can be electron transfer through bond or through space or H atom transfer.

6.8.2 *Transformations initiated by one-electron reducing species*

Some radical transformations inititated under reducing conditions are illustrated in Fig. 6.6(a), the corresponding radical structures being given in Fig. 6.7. Where appropriate, the notation used in the latter figure for different radical species involved indicates the main site of the free electron. The peptide bond carbonyls of aliphatic amino acids are reactive towards the hydrated electron and can result in fast deamination of the protonated terminal amino group (Simic and Hayon 1971, Faraggi and Tal 1975). Cyclic peptides (diketopiperazines), models for a protein peptide bond, form more stable

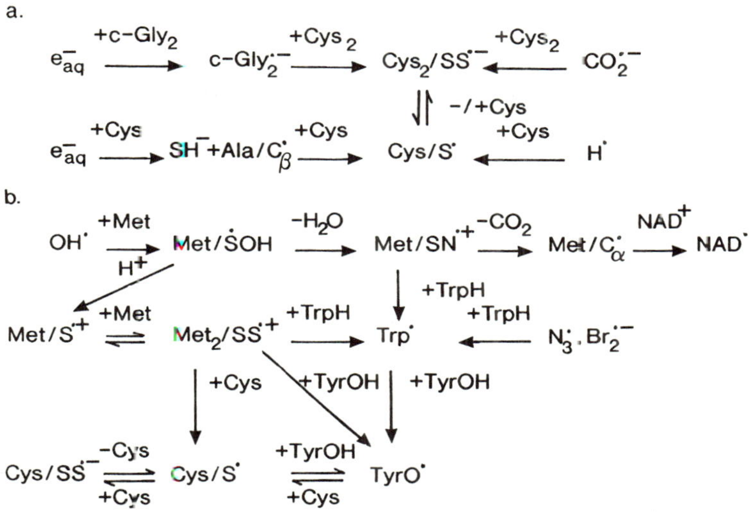

Fig. 6.6 Scheme showing amino acid radical transformations initiated by one-electron (a) reductants and (b) oxidants (adapted from Prütz 1987).

carbonyl electron adducts which can transfer electrons to thiols (Hayon and Simic 1971). One-electron reduction of a cysteine residue results in the rapid dissociation of the transient into SH^- and β-carbon alanyl radicals which eventually oxidize cysteine to the thiyl radical Cys/S· (Wilkening *et al* 1968).

The thiyl radical Cys/S· which predominates at low thiol concentrations due to the equilibrium:

$$\mathrm{Cys/S^{\cdot} + Cys \rightleftharpoons Cys_2/SS^{\cdot -} + H^+}$$

is a species capable of oxidizing a tyrosine residue (Prütz *et al.* 1986):

$$\mathrm{Cys/S^{\cdot} + TyrOH \rightleftharpoons Cys + TyrO^{\cdot}}$$

At pH 9.15 an equilibrium constant $K = 20$ was obtained, leading to a reduction potential difference E(Cys/S·, Cys)$-E$(TyrO·, TyrOH) of 80 mV. This equilibrium represents an important crossover point from reductive to oxidative radical transitions.

6.8.3 *Transformations initiated by one-electron oxidizing species*

A selection of radical transformations initiated under oxidizing conditions is illustrated in Fig. 6.6(b), the corresponding radical structures being given in Fig. 6.7. The OH· adduct at the sulphur function of methionine, Met/SOH·, can convert into various other methionyl species depending upon the pH and peptide structure. In neutral solution, water elimination results in the

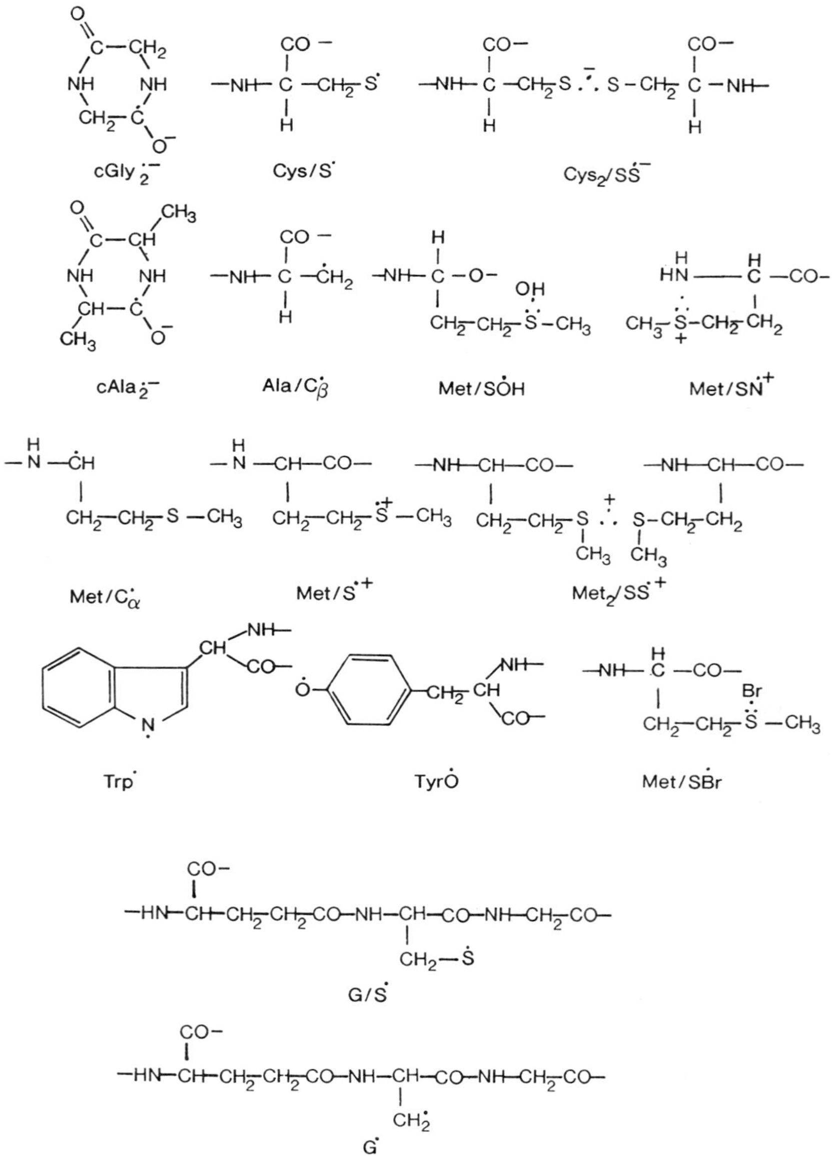

Fig. 6.7 Structures of some radicals involved in amino acid radical transformations.

formation of the three-electron bonded Met/SN$^{\cdot +}$ radical cation (Asmus *et al.* 1985; Bobrowski and Holcman 1987) which rapidly decarboxylates to form Met/$C_{\alpha}^{\cdot}$ which can reduce NAD^+ to NAD$^{\cdot}$ (Hiller and Asmus 1983).

Decarboxylation is prevented in the methionyl peptide Met-Gly, and the corresponding more stable Met-Gly/SN$^{\cdot +}$ is a strong oxidant which can oxidise tryptophan to Trp$^{\cdot}$ (Prütz *et al.* 1986). This Met-Gly/SN$^{\cdot +}$ species has been found to equilibrate with azide:

$$\text{Met} - \text{Gly/SN}^{\cdot +} + N_3^- \underset{-H^+}{\overset{+H^+}{\rightleftharpoons}} \text{Met} - \text{Gly} + N_3^{\cdot}$$

the corresponding equilibrium constant K being 20 at pH 6.8. This constant was pH dependent, and in Fig. 6.8 is shown a plot of $E°$(Met/SN$^{\cdot +}$ + H^+, Met) against pH, together with the corresponding data for azide, tryptophan, cysteine, and tyrosine (Prütz *et al.* 1989).

At low pH, methionyl peptides of the type Gly-Met, eliminate water from the OH$^{\cdot}$ adduct leading to sulphur radical cations Met/S$^{\cdot +}$ which rapidly equilibrate with methionine to form the dimeric disulphide radical cation (Fig. 6.6(b)):

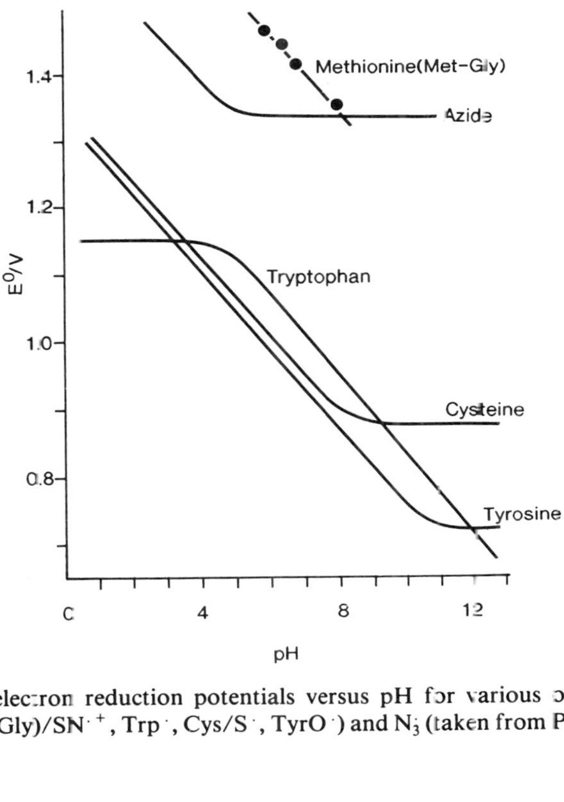

Fig. 6.8 One-electron reduction potentials versus pH for various oxidizing peptide radicals ((Met–Gly)/SN$^{\cdot +}$, Trp$^{\cdot}$, Cys/S$^{\cdot}$, TyrO$^{\cdot}$) and $N_3^{\cdot}$ (taken from Prütz *et al.* 1989).

$$Met/S^{\cdot +} + Met \rightleftharpoons Met_2/SS^{\cdot +}$$

This dimeric species was also found to be capable of the one-electron oxidation of tryptophan and tyrosine to form Trp· and TyrO·, respectively (Bobrowski and Lubis 1986).

As noted above, a major problem in identifying OH· induced radical transformations in proteins is the lack of specificity of the OH· radical. More selective one-electron oxidation of specific peptide units is possible with inorganic radicals such as $Br_2^{\cdot -}$ (Adams *et al.* 1972*b*) or $N_3^{\cdot}$ (Land and Prütz 1979). The $Br_2^{\cdot -}$ species appears to be a selective reagent for methionine and $Br_2^{\cdot -}$ attack upon the protein ribonuclease results initially in the formation of a three-electron bonded methionyl radical Met/SBr·. This methionyl radical decays intramolecularly to TyrO· the process probably involving the solvent exposed Met_{29} and the adjacent $TyrOH_{25}$ (Prütz *et al.*, 1985*b*).

The azide radical, on the other hand, selectively oxidizes trytophan to generate the indolyl radical Trp·. In peptides of the type TrpH—X—TyrOH efficient and complete intramolecular radical transfer between the indolyl and tyrosyl functions have been demonstrated by monitoring the first-order disappearance of the Trp· absorption peak at 510 nm, and the simultaneous appearance of TyrO· monitored at 405 nm. The clear separation and characteristic nature of the absorption bands of Trp· and TyrO· facilitate such studies, which have been carried out on a wide range of peptides and proteins.

Early studies of the transformation Trp· → TyrO· in peptides and proteins have been summarized by Bensasson *et al.* (1983). Erabutoxin *b*, a neurotoxin with 62 amino acids, contains just one tryptophan and one tyrosine unit, $TrpH_{25}$ and $TyrOH_{29}$. Despite the proximity of these functions, there was no evidence for a Trp· → TyrO· transfer unless the toxin was denatured by reduction of the four disulphide groups which stabilize the rigid structure (Prütz *et al.* 1982). It was therefore concluded that the rigid conformation of native erabutoxin *b* does not enable direct contact of the reactive groups, and that the peptide backbone does not provide a channel for the Trp· → TyrO· transfer.

Further evidence that the reaction requires direct contact between the two reacting centres is the very slow transfer found in the sterically restricted cyclo-D-TrpH—L-TyrOH peptide (Prütz *et al.* 1986). However, the L,L diastereoisomer shows a transfer about 30 times faster. These results point to an exchange Dexter mechanism requiring overlap of the corresponding electronic orbitals.

On the other hand, a series of TrpH/TyrOH peptides joined by various numbers of proline bridges, which hold the redox centres rigidly apart at a known distance, have been studied by Faraggi *et al.* (1989), DeFelippis *et al.* (1990), and Bobrowski *et al.* (1990) (see Table 6.2). The rate constant of the Trp· → TyrO· transfer were found to decrease exponentially with the distance between the TrpH and TyrOH, and the mechanism of transfer was suggested to be through-bond (DeFelippis *et al.* 1990). A similar long range electron transfer mechanism was proposed for Trp· → TyrO· in the enzyme lysozyme (M. Weinstein *et al.* 1991).

It may be seen from Fig. 6.8 that, whilst in the pH range 3–12, the one-electron reduction potential of the Trp· radical is more positive than that of TyrO·, at more extremes of pH the order of the reduction potentials is reversed. Accordingly, as has been found experimentally (Butler *et al.* 1986), below pH 3 and above pH 12 the transformation occurs in the opposite direction: TyrO· → Trp·. At pH 7, only small changes occur in redox potential among different tryptophan- and tyrosine-containing peptides (DeFelippis *et al.* 1991).

6.9 Tryptophan as a protein structural probe

As noted above the fluorescence of proteins is often dominated by that of tryptophan. Of course, if there is no tryptophan present, but tyrosine is present, the fluorescence will be dominated by that of tyrosine (see, for example, Longworth 1971, 1983). The fluorescence properties of amino acids are strongly dependent on their environment and that of tryptophan has been much exploited as a structural probe for proteins.

Conventional steady state fluorescence is only really useful if the protein contains a single tryptophan residue. However, since most proteins contain a number of such residues the fluorescence spectrum is a composite one. Fluorescence lifetime measurements, can give information on systems containing more than one tryptophan since different locations can give rise to different lifetimes (Cundall and Dale 1983). Studies of the emission spectrum, lifetime, and anisotropy have been used to give information on protein conformation, molecular dynamics, ligand binding, and separation distances. Much of this work has been previously reviewed and will not be discussed here (see, for example, Cundall and Dale 1983, Lakowicz 1983).

The short lifetime of the fluorescence excited state limits the time resolved aspects of this probe technique to processes with time constants of a few tens of nanoseconds or less. Conformational changes and molecular dynamics occurring over longer time scales can, however, be studied if the much longer-lived lowest triplet state is monitored (see, for example, Ghiron *et al.* 1988; Geacintov and Brenner 1989). Examples of the use of phosphorescence are given by Kai and Imakubo (1979), Horie and Vanderkooi (1982), and Vanderkooi *et al.* (1987*a*, *b*) but are limited to lifetimes in the millisecond time range; such long lifetimes may be due merely to inaccessibility to oxygen, or the rigidity of the local chromophore environment.

Strambini and Gonnelli (1985) have compared the triplet lifetime of free tryptophan and tryptophan in an enzyme in environments of varying viscosity. Their results suggest that the long triplet lifetime in proteins in fluid solutions is related to their further rigid microenvironment. Papp and Vanderkooi (1981) have reviewed the use of extrinsic quenchers such as oxygen, to probe protein structures, although such quenching data is often difficult to interpret (Geacintov and Brenner 1989). The use of UV laser flash photolysis allows the time resolution to be extended to tens of nanoseconds or shorter. However, the most convenient laser source is the 266 nm Nd YAG frequency quadrupled line. As shown in Fig. 6.1 this is not selective for tryptophan since this wavelength is also strongly absorbed by tyrosine (and, as shown in Fig. 5.3, by nucleic acid bases). Ghiron *et al.* (1988) have shown that this is not a serious drawback. Thus, for example the UV excitation of HSA, which contains a single tryptophan as well as many other 266 nm absorbing moieties, showed that the transients due to the non-indolyl absorbing species did not interfere with those due to the tryptophan. In a related experiment, Ghiron *et al.* studied the bacteriophage fd and showed that the tryptophan triplet absorption and phosphorescence were also readily identified and not significantly modified by the nucleic acid UV absorbers. A similar conclusion in terms of nearby disulphide groups was reached by Li and Galley (1989) and Li *et al.* (1989). Thus, it seems likely that the tryptophan triplet has a general applicability as a long-lived intrinsic protein 'reporter' group for future studies. Finally, it should be noted that tryptophan can also be used as a reporter group in conjunction with the ODMR technique, (see Section 3.2.1.3) based on the triplet sub-level splitting being sensitive to the microenvironment of the tryptophan residue (see, for example, Geacintov and Brenner 1989; Tsao *et al.* 1989).

6.10 **Protein dynamics**

In contrast with the reactions of small molecules, the reactions of biomacromolecules involve an understanding of their local or global folding and unfolding and of their three-dimensional fluctuations between conformational substates. These motions are studied by the new and growing field of 'protein dynamics'. The analysis of the protein structural changes during their reactions, over times that range from femto to milli-seconds, has been made possible by the advances of time-resolved techniques. Investigations of biochemical reactions controlled by dynamics of proteins include time-resolved studies of:

(1) photodeligandation and rebinding of CO or O_2 to myoglobin or to haemoglobin,
(2) control of electron transfer in photosynthetic centres;
(3) primary steps of vision; and
(4) fluorescence and molecular dynamics of tryptophan and tyrosine, hydrophobic residues which can be exposed to a polar solvent during the trypsin proteolysis of lysozymes.

Most areas of investigation on protein dynamics have been reviewed in a special issue of *Chemical Physics* (Eaton and Szabo 1991).

6.11 Concluding remarks

Photochemical studies of amino acids in simple environments were undertaken more than a decade ago. More recent work has concentrated on:

(1) protein macromolecules and protein dynamics;

(2) theoretical studies of electron transfer related to Marcus theory;

(3) two-photon absorption processes and

(4) electron transfer and related free radical cascades.

Of particular commercial interest (G. J. Smith 1992) is the linkage of an increased singlet oxygen yield on photolysis of wool to a reduction in the yellowing of the wool and the involvement of NFK as a photosensitizer in this system. The recent two-photon absorption studies of Nikogosyan and co-workers have allowed much of the previous literature inconsistencies on the quantum yield of photoionization to be rationalized, and have led to a better understanding of the consequences of two-photon absorptions at high laser intensities during laser inactivation of proteins. Radicals can also be generated photochemically via O—H bond cleavage, for example the quantum yield of ejected electrons from dopa is ~9 per cent compared to a 19 per cent quantum yield of dopasemiquinone (Land *et al.* 1986). Much of the recent interest in radicals concerns inter- and intra-molecular electron transfer reactions. Variations in distant electronic couplings might be one of the major parameters controlling the rates of electron transfer in proteins (Wuttke *et al.* 1992 and references therein). Estimates are available in the literature for the rate constants of the inter- and intra-molecular electron transfer reactions referred to in Fig. 6.6 under a variety of conditions, and for a number of related species, for example, radicals derived from the cysteine peptide glutathione (γ-Glu-Cys-Gly). Some of these rate constants are listed in Tables 6.2 and 6.3.

Table 6.2 Rate constants of Trp· → TyrO· in peptides at pH 7 (s^{-1})

n	TrpH-(Gly)$_n$-TyrOH[a]	TrpH-(Pro)$_n$-TyrOH[b]	TyrOH-(Pro)$_n$-TrpH[c]
0	7.3×10^4	7.7×10^4	6.7×10^4
1	5.1×10^4	2.6×10^4	7.2×10^3
2	2.4×10^4	4.9×10^3	3.2×10^3
3	3.3×10^4	1.5×10^3	2.0×10^3

[a] Prütz *et al.* (1981).
[b] Bobrowski *et al.* (1990).
[c] Faraggi *et al.* (1989).

Table 6.3 Rate constants of intermolecular radical reactions of some sulphur-centred peptide radicals

Transformation	pH	$k(\mathrm{dm^3\ mol^{-1}s^{-1}})$	Reference[a]
$\mathrm{Met/C_\alpha^{\cdot} \rightarrow Cys/S^{\cdot}}$	8.2	2.7×10^8	a
$\mathrm{Met/C_\alpha^{\cdot} \rightarrow (Cys\text{-}Gly)SS^{\cdot -}}$	8.1	$\sim 7 \times 10^6$	a
$\mathrm{(Met\text{-}Gly)SN^{\cdot +} \rightarrow Cys/S^{\cdot}}$	8.0	$\sim 8 \times 10^7$	a
$\mathrm{(Met\text{-}Gly)SN^{\cdot +} \rightarrow Trp^{\cdot}\text{-}Gly}$	6.8	4×10^7	b
$\mathrm{(Gly\text{-}Met)_2/SS^{\cdot +} \rightarrow Cys/S^{\cdot}}$	8.3	3.5×10^8	a
$\mathrm{(Gly\text{-}Met)_2/SS^{\cdot +} \rightarrow (Met\text{-}Gly)SN^{\cdot +}}$	8.2	6.0×10^7	a
$\mathrm{(Gly\text{-}Met\text{-}Gly)_2SS^{\cdot +} \rightarrow Cys/S^{\cdot}}$	8.2	1.7×10^8	a
$\mathrm{(Gly\text{-}Met)_2/SS^{\cdot +} \rightarrow Gly\text{-}TyrO^{\cdot}}$	8.9	9.0×10^7	a
$\mathrm{(Gly\text{-}Met)_2/SS^{\cdot +} \rightarrow Gly\text{-}TyrO^{\cdot}}$	7.5	2.3×10^7	a
$\mathrm{(Gly\text{-}Met)_2/SS^{\cdot +} \rightarrow Trp^{\cdot}\text{-}Gly}$	7.0	4.8×10^8	a
$\mathrm{Met_2/SS^{\cdot +} \rightarrow TyrO^{\cdot}}$	1	3.8×10^7	c
$\mathrm{Met_2/SS^{\cdot +} \rightarrow Trp^{\cdot}}$	1	3.8×10^8	c
$\mathrm{(Cys\text{-}Gly)/S^{\cdot} \rightarrow Gly\text{-}TyrO^{\cdot}}$	6.8	6.3×10^6	a
$\mathrm{Ala/C_\beta^{\cdot} \rightarrow Cys/S^{\cdot}}$	8.2	5.0×10^6	a
$\mathrm{G/S^{\cdot} \rightarrow Gly\text{-}TyrO^{\cdot}}$	8.1	5.8×10^6	a
$\mathrm{G^{\cdot} \rightarrow G/S^{\cdot}}$	8.0	7.2×10^6	a

[a] a, Prütz *et al.* (1989); b, Prütz *et al.* (1986); c, Bobrowski and Lubis (1986)

Several model intermolecular free radical cascades have been characterised by studying appropriate mixtures of amino acids and small peptides, including for example:

$$\mathrm{OH^{\cdot} \rightarrow Met/SN^{\cdot +} \rightarrow Met/C_\alpha^{\cdot} \rightarrow (Cys\text{-}Gly)_2/SS^{\cdot -} \rightarrow Cys\text{-}Gly/S^{\cdot} \rightleftharpoons TyrO^{\cdot}}$$ (Prütz *et al.* 1989)

$$\mathrm{e_{aq}^{-} \rightarrow cyclo\ Ala_2^{\cdot -} \rightarrow (Cys\text{-}Gly)_2SS^{\cdot -} \rightarrow Cys\text{-}Gly/S^{\cdot} \rightleftharpoons TyrO^{\bullet}}$$ (Prütz *et al.* 1989)

$$\mathrm{Br_2^{\cdot} \rightarrow MetSBr^{\cdot} \rightarrow Trp^{\cdot} \rightarrow TyrO^{\cdot}}$$ (Prütz *et al.* 1985b)

The oxidative and reductive peptide cascades eventually lead to cysteinyl and tyrosyl intermediates, which often themselves subsequently decay via terminating radical–radical disproportionation or crosslinking reactions, for example, giving rise to 2,2′-biphenols (Prutz *et al.* 1983). In the presence of antioxidants, such as α-tocopherol and ascorbic acid, both $\mathrm{Trp^{\cdot}}$ and $\mathrm{TyrO^{\cdot}}$, either free or in the enzyme lysozyme, can be repaired (Hoey and Butler 1984).

There has been recent interest in the photoswitchable binding of substrates to proteins as a means of regulating biotransformations by light (Willner *et al.* 1992).

7

Carotenoids

The yellow pages?

7.1 Introduction

The term carotenoid refers to a group of organic lipophilic pigments which range in colour from yellow to red. They are chemically related to β-carotene as C_{40} polyenes or as C_{20} polyenes such as vitamin A (retinol). Frequently the C_{20} polyenes are called retinoids. The basic skeleton structure of the C_{40} carotenoids is symmetrical, based on two C_{20} halves and all are derived from tetraterpenes consisting of eight C_5 isoprene units.

7.1.1 *Occurrence and structure*

Several hundred carotenoids or related molecules derived by hydrogenation, dehydrogenation, oxidation or degradation, have been isolated from natural sources over the last 100 years or so (Zeichmeister 1962, Isler *et al.* 1971). More recently, commercial products based on the chemical synthesis of β-carotene, apocarotenal, canthaxanthin, and other carotenoids have allowed the more widespread use of these naturally occurring pigments as food colorants, etc.

In nature, C_{40} carotenoids are found in photosynthetically active materials, such as leaf tissues and also in fruits, petals, pollen, and seeds and in animals. The feathers of many birds, for example, are coloured because of the carotenoid ingested with food and it is now known that the colouring of the feathers is essential to sexual behaviour and thus to the propagation of the species. If the carotenoid contains oxygen as a heteroatom it is called a xanthophyll; Fig. 7.1 gives the structures of several important carotenes and xanthophylls.

Vitamin A or retinol, a C_{20} carotenoid, is found mainly in liver, fish oils, and butter. Vitamin A is important to many body functions and low vitamin A levels are associated with increased risk of cancer, and both retinol and β-carotene have been suggested as preventing cancer. As the aldehyde retinal, C_{20} carotenoids are pivotal in vision and retinal also helps to protect the skin and mucous membranes. Of course, it is the carotenoids in plants that the body converts to vitamin A and vitamin A derivatives. A normal balanced diet will meet most people's needs and deficiency is rare in the advanced world. Nevertheless, the World Health Organization reports over 100 000 new cases of blindness caused by vitamin A deficiency each year.

As noted above, Fig. 7.1 gives the structures of some of the important C_{40} carotenoids. This figure also includes the structures of the following C_{20} carotenoids: retinol, 11-*cis* and all-*trans* retinal, and also of several polyenes of carbon chain length other than C_{40} and C_{20}, and some typical carotenoids arising in bacterial photosynthesis.

7.1.2 *Roles in photosynthesis and vision*

Carotenoids such as β-carotene and a range of bacteriophotosynthetic carotenoid pigments have been much studied for their role in photosynthesis and it is now well established that these pigments perform two major functions. If a photosynthetic organism which lacks carotenoids (e.g the R26 mutant of Rhodobacter (Rb) Sphaeroides) is irradiated in the presence of air, the photodynamic reaction via energy transfer from the bacteriochlorophyll (or chlorophyll) triplet state to produce singlet oxygen leads to self-sensitized destruction, via photo-oxidation, of the organism (see, for example, Cogdell and Frank 1987 and Section 4.2.1.1). Probably, the essential function of carotenoids is to intercept the chlorophyll triplet state, thus prevent the formation of singlet oxygen, and hence protect the organism from this damaging

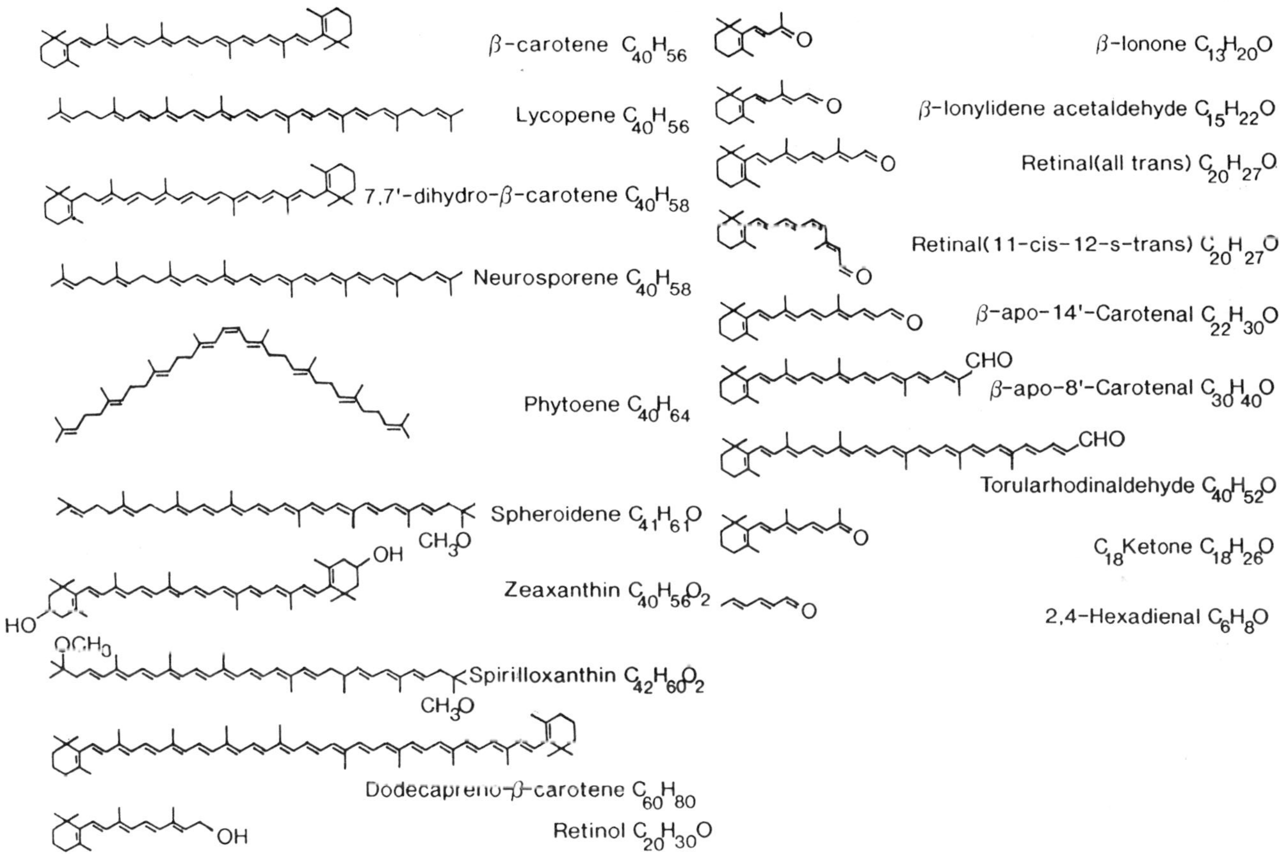

Fig. 7.1 Structures of carotenoids and analogues.

photo-oxidative reaction. The other essential function of carotenoids is as accessory light-harvesting pigments in the antenna system of photosynthetic organisms. They absorb light in the wavelength region where the chlorophylls only absorb weakly and transfer this energy efficiently, ultimately to the reaction centres of the system. More details concerning the function of carotenoids in photosynthesis are given below (Section 7.7).

The role of retinal in the primary stage of the vision process is associated with the 11-*cis* to all-*trans* isomerization of this polyene, while bound to the protein opsin. In fact, in the visual pigment, the 11-*cis*-retinal exists as the 11-*cis*, 12-s-*trans* isomer, and it is the conformational changes forced upon the visual pigment protein by the isomerization to the all-*trans* form that initiates the early stages in the visual process and which have been most studied by flash photolysis methods (see Schoenlein *et al.* 1991 and references therein).

7.1.3 *Use as colorants*

It must be emphasized that many of the carotenoids used in the commercial world are those which occur in nature. For example, carotenoid additions to farm animal and fish feedstuffs are often only to supplement or replace the carotenoid usually available in the wild.

The most widely used carotenoids as colorants are β-carotene, apocarotenal and canthaxanthin which are used in a range of fat-based foods such as margarine, edible oils, salad creams, and sauces as well as water-based products such as beverages and puddings. They are also used to colour confectionery products. Another typical example is crocetin (as the diglycoside) which is the widely used natural food colorant, saffron. Astaxanthin is added to the feed stock of salmon in fish farms to achieve the natural pink coloration of the wild fish (which have a ready source of this carotenoid). Similarly, egg yolks are coloured with carotene addition to farm chicken feedstuffs. Finally, it should be noted that β-carotene/canthaxanthin mixtures have been used to achieve a skin coloration somewhat akin to a sun-tan, but the use of canthaxanthin for this purpose is now discouraged due to possible accumulation in the eye.

While the C_{20} carotenoids such as Vitamin A are not colorants, they are also added to many foodstuffs such as margarine and dairy products as a fortification.

7.1.4 *Use in medicine*

Of course, Vitamin A deficiency treatment involves a clinical use of retinol, simply as a dietary supplement. The scavenging effects of carotenoids, with respect to photosensitizer triplets, singlet oxygen, and free radicals, may well imply beneficial therapeutic uses to such molecules. Typical of the discussions of anti-cancer properties of β-carotene, retinol, retinoic acid, and retinoids in general, are Ward *et al.* (1980), Schroder and Black (1980), Epstein and Grekin

(1981), Peto *et al.* (1981), Mathews-Roth and Krinsky (1984, 1987), Mathews-Roth (1986), and Ziegler *et al.* (1992).

Examples of the use of active metabolites of retinoids are the 'Roche Products' drugs etretinate and isotretinoin, derivatives of retinoic acid, which are used for a wide range of dermatological disorders such as psoriasis, eczema keratoticum, and morbus Darier. There are some possible question marks with respect to associated skin photosensitivity but this is probably only associated with high dosage of these retinoids (Ferguson and Johnson 1986, 1989). From a photochemical point of view, the use of carotenoids to treat a type of porphyric disease known as erythropoetic protoporphyria (EPP) is of interest (see Section 9.8). Here, the situation is somewhat analogous to the protective effect discussed below for photosynthesis, in that the carotenoid is thought to intercept the protoporphyrin triplet ^3PP to prevent the formation of the 'biologically damaging' singlet oxygen. Either or both of the following energy transfer processes may arise in the use of carotenoids to treat EPP:

$$^3PP + Car \rightarrow PP + {}^3Car$$

and/or

$$^3PP + O_2 \rightarrow PP + {}^1O_2$$

$$^1O_2 + Car \rightarrow O_2 + {}^3Car$$

Treatment of EPP with carotenoids is, of course, not a cure for the disease but simply reduces the skin photosensitivity which would arise via the processes:

$$^3PP + O_2 \rightarrow PP + {}^1O_2$$

$$^1O_2 - \text{biological substrate} \rightarrow \text{photo-damage}$$

The use of β-carotene alone can give a cosmetically unacceptable skin coloration and mixtures of β-carotene/canthaxanthin were previously used. However, as noted above, due to possible accumulation of canthaxanthin in the eye, it is not in current use.

7.2 Absorption and fluorescence spectra

Both the C_{40} and C_{20} series show absorptions with very high molar absorption coefficients from the ground state to an excited state termed $1\,^1B_u$. It is this strong ($S_0 \rightarrow S_2$) absorption which is responsible for the intense colours of these molecules. It is accepted that these transitions obscure a symmetry forbidden absorption to a lower-lying $2\,^1A_g$ (S_1) state, however the role of this lower-lying state is important, for example with respect to photosynthesis as discussed below (Section 7.7) and to the fluorescence of carotenoids. Figure 7.2 shows the absorption spectra of β-carotene and retinal in various isomeric forms. As can

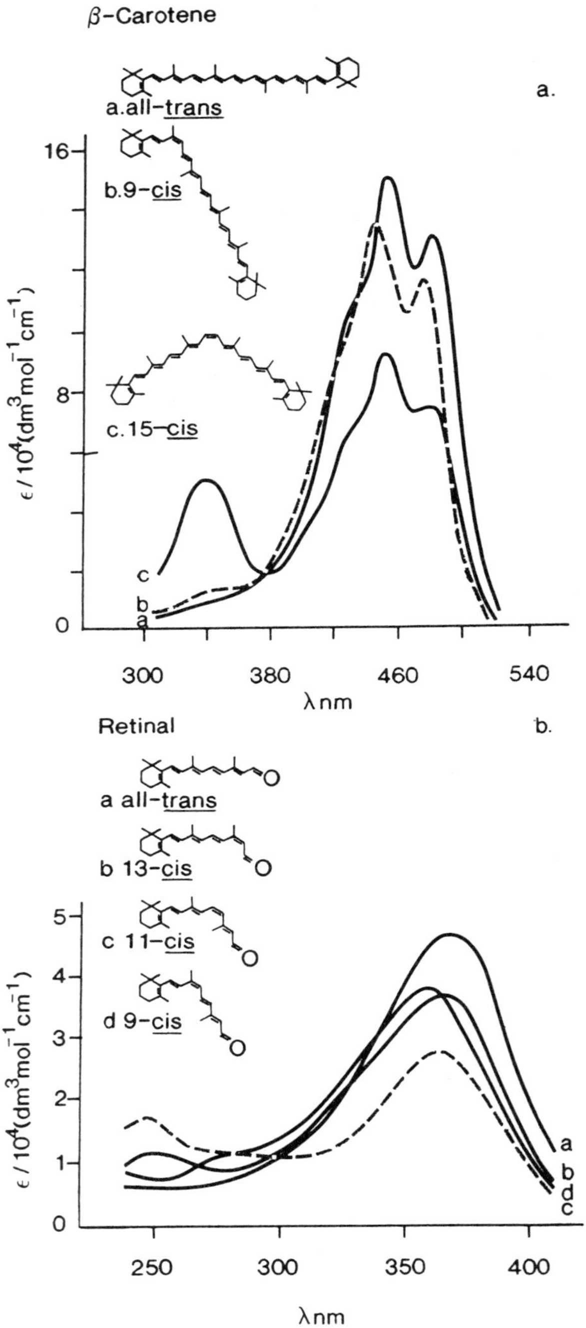

Fig. 7.2 Absorption spectra of isomeric (a) β-carotenes and (b) retinals in a hydrocarbon solvent.

be seen both groups show an additional '*cis*' peak for the various *cis*-isomers at shorter wavelengths than the main $1\,^1A_g \rightarrow 1\,^1B_u$ transition. The molar absorption coefficient of the main band is reduced for the *cis* isomers compared to the all-*trans* isomers.

In general, as would be expected, the molar absorption coefficient is also reduced and the wavelength of maximum absorption is shifted to the blue for the C_{20} (5/6 conjugated double bonds) compared to the C_{40} (11 conjugated double bonds). Bathochromic shift of the absorption bands in carotenoids with increasing numbers of conjugated double bonds is illustrated in Table 7.1 for phytoene, three conjugated double bonds, to dodecapreno-β-carotene, 19 conjugated double bonds. It should be noted, however, that aggregation of carotenoids usually leads to a blue shift in the absorption peaks.

Hirayama (1955) has proposed the following empirical formula to express the absorption maxima of polyene spectra:

$$(\lambda_{max})^2 = (A - B \times 0.920^N) \times 10^4 \text{ nm}^2$$

where A and B are dependent upon the solvent and the type of polyene studied; N is a proposed index defined by the number of conjugated double bonds, the number and types of substituents, and of spectroscopy effective structures: $N = n_1 + n_2 + n_3 \ldots$ For example, the increment for one conjugated double bond is $n = 1$, for a double bond in the β-end group it is $n = -0.8$, for a caboxylic acid it is $n = +0.3$. This empirical equation predicts the absorption maximum for infinite chain length at 608 nm or 16 500 cm^{-1} for polyene hydrocarbons, which is in agreement for polyene absorption spectra in ethanol.

The data in Table 7.1 are rather solvent dependent, and the $1^1A_g \rightarrow 1^1B_u$ transition undergoes a shift to longer wavelengths when the carotenoids are studied in non-polar solvents of increasing refractive index (Hudson *et al.* 1982; Anderson *et al.* 1991). The shift (of frequency) follows a linear dependence on the ratio $(n^2 - 1)/(n^2 + 2)$ where n is the refractive index of the solvent (see Section 2.1.7). This ratio is related to the solvent polarizability and Anderson

Table 7.1 Wavelength maxima in hexane of main ground state (λ) and triplet (λ_T) absorption as a function of number of conjugated double bonds (n)

Polyene	n	λ(nm)	λ_T(nm)
Phytoene	3	285	315
Retinal	5	360	450
7,7′-dihydro-β-carotene	8	405	480
Septapreno-β-carotene	9	410	495
β-Carotene	11	445	515
Decapreno-β-carotene	15	495	592
Dodecapreno-β-carotene	19	534	635

et al. (1991) have reported that β-carotene and spheroidene follow the relationship:

$$\tilde{\nu}_{solvent} = \tilde{\nu}_{gas} - K\left[\frac{n^2 - 1}{n^2 + 2}\right]$$

with K values near 11 750 cm^{-1}

Protonation of retinal Schiff base has been much studied as a model of the visual pigments, and leads to a marked bathochromic shift from ~360 nm for unprotonated base to ~450 nm when protonated (see for example, M.M. Fisher and Weiss 1974). More recently, Bobrowski and Das (1982) have used pulse radiolysis to study the rate constants of protonation of various retinyl molecules. Only lower limits (~10^7 dm^3 mol^{-1} s^{-1}) are reported for retinal and retinol with values ~10^9 dm^3 mol^{-1} s^{-1} being reported for Schiff bases. In the case of retinol and retinyl acetate, the initial H^+ adduct undergoes loss of water and acetic acid, respectively, producing the retinyl carbenium ion (λ_{max} 585 nm) which subsequently decays by a first order process with half-life ~10 μs.

7.2.1 *The 2^1A_g state*

The existence of the symmetry forbidden 2^1A_g states for linear polyenes was implied by a number of spectroscopic studies about 10–15 years ago (Hudson and Kohler 1974; Christensen and Kohler 1976; Holtom and McLain 1976; Fang *et al.* 1977, 1978) and for the C_{20} molecule, retinol, by Hudson and Kohler (1972), Christensen and Kohler (1974, 1976), Birge and Pierce (1979), and Birge (1981). The energy level of the 2^1A_g state of β-carotene is of particular importance to photosynthesis since it must be above that of the S_1 state of chlorophyll if it is involved in the antenna energy transfer reactions (see Section 7.7). However, the S_1 energy levels of the C_{40} carotenoids are not yet well established either experimentally or theoretically.

Three major approaches have been described to obtain the 2^1A_g energy levels: Raman spectroscopy, two-photon absorption spectroscopy, and an extrapolation of the fluorescence data of the shorter chain polyenes which, unlike the C_{40} carotenoids, fluoresce from the 2^1A_g level.

The early Raman work (Thrash *et al.* 1977) suggests a $S_2 - S_1$ energy gap for β-carotene of 3470 cm^{-1}, but more recent resonance Raman measurements suggest that the Raman technique is not useful in observing the $1^1A_g(S_0) \rightarrow 2^1A_g(S_1)$ transition (Watanabe *et al.* 1987).

The $1^1A_g \rightarrow 2^1A_g$ transition is not symmetry forbidden by two-photon spectroscopy and this has been applied with some success to short polyenes including retinal (Birge 1986). However, little or no progress has been made in applying these techniques to the (extremely inefficiently fluorescing) photosynthetic carotenoids.

Probably the most useful data to-date arise from the studies of Cosgrove *et al.* (1990), DeCoster *et al.* (1992), and Anderson *et al.* (1992) in which the $2^1A_g \rightarrow 1^1A_g$ fluorescence of shorter carotenoids ($n < 8$) has been extrapolated to the longer photosynthetic carotenoids. This leads to a $2^1A_g \rightarrow 1^1A_g$ gap of 5500–6500 cm^{-1} for β-carotene ($n = 11$) and sets the ^1A_g energy level of β-carotene at ~167 kJ mol^{-1} (~14 000 cm^{-1}).

Very recently, J.L. Haley *et al.* (1992) have encapsulated β-carotene in an asymmetric environment in a zeolite cavity. Reflectance spectroscopy on such a system shows an additional band at 700 ± 20 nm (14 300 cm^{-1}) in good agreement with the above extrapolated value.

7.2.2 *Fluorescence*

In general, for the short polyenes such as the retinals (C_{20}), a Stokes-shifted fluorescence from the S_1 (2^1A_g) state is observed following light absorption to the S_2 (1^1B_u) state. For the longer (e.g $n = 11$) carotenoids such an emission is not detected but an extremely weak (anti-Kasha) emission (see Section 2.1.6) is seen from the S_2 state (cf. azulene).

7.2.2.1 *C_{40} carotenoids*

Until quite recently the C_{40} carotenoids such as β-carotene were regarded as non-fluorescent. An early report of fluorescence from β-carotene was given by Cherry *et al.* (1968) although the possibility that this was due to an emission from a trace impurity was not eliminated. However, in recent years a weak carotene fluorescence has been confirmed by several groups (L.V. Haley and Koningstein 1983; Watanabe *et al.* 1987; Bondarev *et al.* 1988; Gillbro and Cogdell 1989; Shreve *et al.* 1991*a*, *b*).

Gillbro and Cogdell have reported fluorescence from all-*trans* β-carotene and from two bacterial carotenoids, rhodipin (extracted from *Chr. vinosum* strain D) and spheroidenone (extracted from *Rb. sphaeroides* strain 2.4.1 aerated to convert the spheroidene to spheroidenone). These workers chose carbon disulphide as the solvent because many other solvents give rise to interfering Raman lines. Quite good agreement between the fluorescence excitation spectra and the corresponding absorption spectra was noted which virtually eliminates the possibility that the emission was due to an impurity. A fluorescence quantum yield (ϕ_F) of ~3×10^{-5} was obtained which, taken with a calculated intrinsic radiative lifetime (τ_0) of ~1 ns leads to a calculated non-radiative lifetime (τ_f) of ~3×10^{-14} s, ($\tau_f = \tau_0\phi_F$). However, Gillbro and Cogdell also measured the carotenoid ground state (1^1A_g) recovery to be ~15 ps using a pump-probe technique. Clearly, since the rate of recovery of the ground state is much slower than the implied rate calculated for the non-radiative decay of the S_2 (1^1B_u) state, it is necessary to assume that the excited S_2 state is not decaying directly to the ground state. Rather similar results were obtained by Bondarev *et al.* in *n*-hexane as solvent who obtained $\phi_F \sim 4 \times 10^{-5}$ and hence $\tau_f \sim 4 \times 10^{-14}$ s at 77 K and ~8×10^{-14} s at 4.2 K. Such short lifetimes are

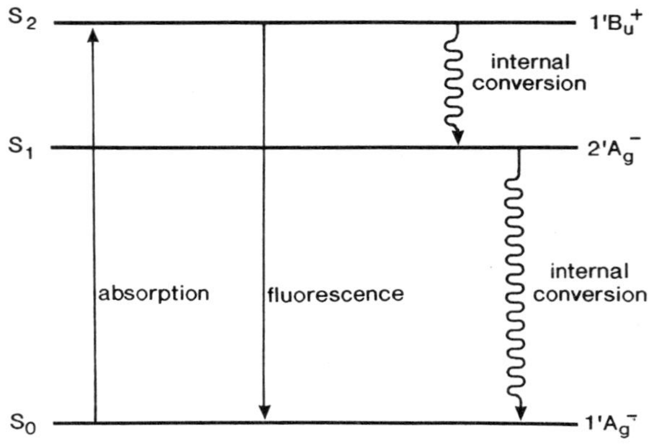

Fig. 7.3 Scheme of β-carotene energy levels.

characteristic of high electronic states in polyatomic molecules and the data are consistent with the energy level scheme given in Fig. 7.3. In this scheme the excited fluorescing state $1^1B_u^+$ decays by fast internal conversion to the forbidden ($2^1A_g^-$) state and it is the decay of the $2^1A_g^-$, state to the ground state which determines the kinetics of the ground state recovery.

The fluorescence emission detected in isopentane at 77 K, (Fig. 7.4), shows only a very small Stokes shift so it arises from the ^1B_u state ($1^1B_u^+ \rightarrow 1^1A_g^-$) rather than from the $2^1A_g^-$ state. Indeed, Bondarev *et al.* have searched for

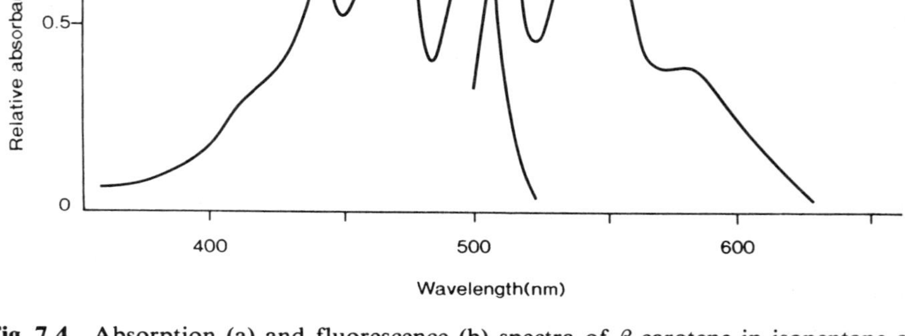

Fig. 7.4 Absorption (a) and fluorescence (b) spectra of β-carotene in isopentane at 77 °K (adapted from Bondarev *et al.* 1989).

fluorescence at longer wavelengths from the $2^1A_g^-$ state, but failed to observe any emission, implying $\phi_F < 10^{-5}$ for this state. Bondarev *et al.* also used a (6 ps) pulse excitation laser spectrometer to obtain the kinetics of the $2^1A_g^-$ decay of β-carotene and report a lifetime of ~10 ps which was independent of solvent viscosity and polarity. Wasielewski and Kispert (1986) used direct picosecond transient spectroscopy to obtain a lifetime of 8.4 ps for β-carotene with similar data reported for canthaxanthin and β-8′-apocarotenal. Recently, Shreve *et al.* (1991*a*, b) used a femtosecond system and report the $2^1A_g^-$ lifetime for β-carotene in carbon disulphide as 11 ps. These workers have also reported a direct measurement of the $1^1B_u^+$ state lifetime of β-carotene at 200–250 fs and that it decayed into the longer-lived $2^1A_g^-$ state. These direct measurements are in good agreement with the fluorescence and ground state recovery data.

7.2.2.2 *C_{20} carotenoids*

The fluorescence of retinal and retinol has been documented for some time. For retinal, fluorescence is exhibited in H-bonding solvents (Takemura *et al.* 1976, 1978) but not in hydrocarbon solvents. This has been related to a solvent induced excited state—switching between the $n\pi^*$ state and the forbidden $2^1A_g^-$ ($\pi\pi^*$) state with only the $2^1A_g^-$ state exhibiting fluorescence. Das and Becker (1979) have discussed such switching as a function of environment (solvent, temperature, etc) and the number of conjugated double bonds in the molecule. This switching of the excited state order not only affects ϕ_F but also the triplet quantum yield (ϕ_T) as discussed later.

The fluorescence of retinol has also been studied by several groups (see, for example, Christensen and Kohler 1974, 1976, Honig and Ebrey 1974; Becker *et al.* 1976; Becker 1988). Unlike the C_{40} carotenoids discussed above there is a substantial Stokes shift. It seems likely that this is due to the emission arising from the $2^1A_g^-$ state and this may also explain the anomalously long fluorescence lifetimes measured. Picosecond fluorescence lifetimes have been reported (Everaert and Rentzepis 1982) for the all-*trans* retinal Schiff base with *n*-butylamine (*N*-all-*trans*-retinylidine-*n*-butylamine) in hexane as a function of temperature (20 K to 298 K). Generally, two fluorescence decays were detected, for example, at 298 K, $\tau_1 = 22$ ps (major component) and $\tau_2 = 265$ ps. These results seem best interpreted in terms of two different conformers, probably existing in the ground state and first excited singlet state. It was suggested that a twisted form is the fast fluorescing state, while the slower process corresponds to the decay of a 'planar' conformer.

7.3 **Triplet state**

7.3.1 *Absorption spectra*

The triplet minus singlet absorption spectrum of β-carotene was first reported using flash photolysis and a variety of triplet sensitizers (Chessin *et al.* 1966).

This has since been studied by several groups using both flash photolysis (see, for example, Mathis 1969; Rodgers and Bates 1980) and pulse radiolysis (see, for example, Land *et al.* 1970, 1971; Almgren and Thomas 1980; Rodgers and Bates 1980). More recently such measurements have been extended to bacterial carotenoids (Cogdell *et al.* 1983; Borland *et al.* 1989 and also to the water soluble carotenoid, crocetin (I. B. C. Matheson and Rodgers, 1982; Craw and Lambert 1983).

The triplet absorption of all-*trans*-retinal, formed directly without a sensitizer, was reported somewhat earlier (Dawson and Abrahamson 1962) and subsequently, of all-*trans*-retinol using a sensitizer (Sykes and Truscott 1969). Typical triplet spectra of C_{40} and C_{20} carotenoids are given in Fig. 7.5). The triplet spectra of a number of other carotenoids are given by, for example, Truscott

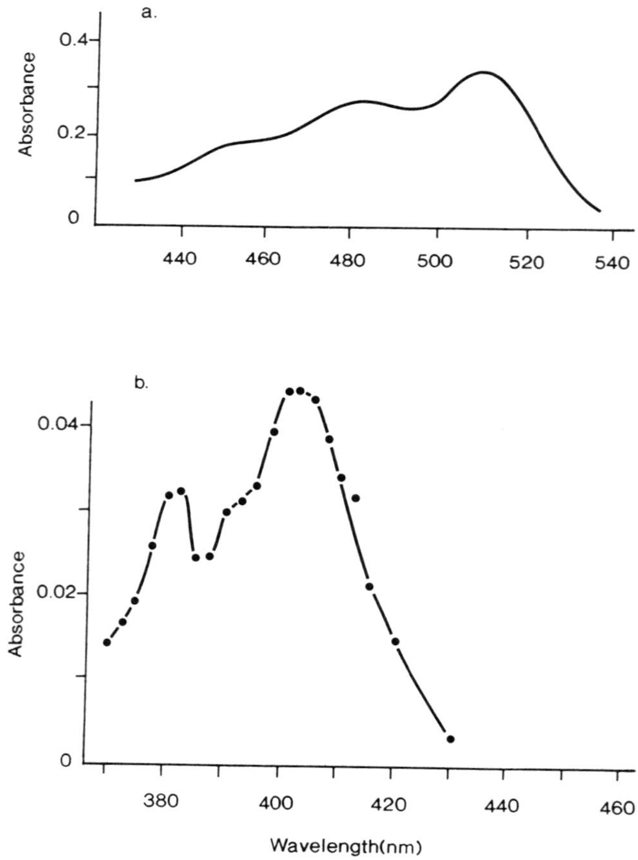

Fig. 7.5 Triplet absorption spectra of (a) all-*trans*-β-carotene and (b) all-*trans*-retinol in hexane.

et al. (1973) and Rodgers and Bates (1980), for several bacterial carotenoids by Bensasson *et al.* (1976), and various carbonyl containing polyenes (retinal analogues) by R.S. Becker *et al.* (1978), Das and Becker (1979), Bensasson *et al.* (1984*a*), and Chattopadhyay *et al.* (1985). In general, the same trends can be observed in these triplet–triplet spectra as noted above for the ground state singlet–singlet absorptions, that is, increasing absorption coefficients and bathochromic shifts with increasing number of conjugated double bonds (see Table 7.1). Quite a good linear plot can be obtained comparing the triplet–triplet absorption maxima to those of the ground state, singlet–singlet maxima (R.S. Becker *et al.* 1978) (see Section 7.7).

7.3.2 *Energy levels and interaction with oxygen*

There is some information on the lowest triplet energy levels of the C_{20} carotenoids, retinal and retinol. Thus Sykes and Truscott (1971) studied the rate of energy transfer from a series of triplet donors to retinol and estimated the triplet level of 155 ± 10 kJ mol^{-1}. Raubach and Guzzo (1971) used high pressure oxygen perturbation to observe directly an absorption ($S_0 \rightarrow T_1$) for retinal at 803 nm corresponding to a lowest triplet of ~150 kJ mol^{-1}. This latter measurement is clearly of a vertical (spectroscopic) energy level whereas

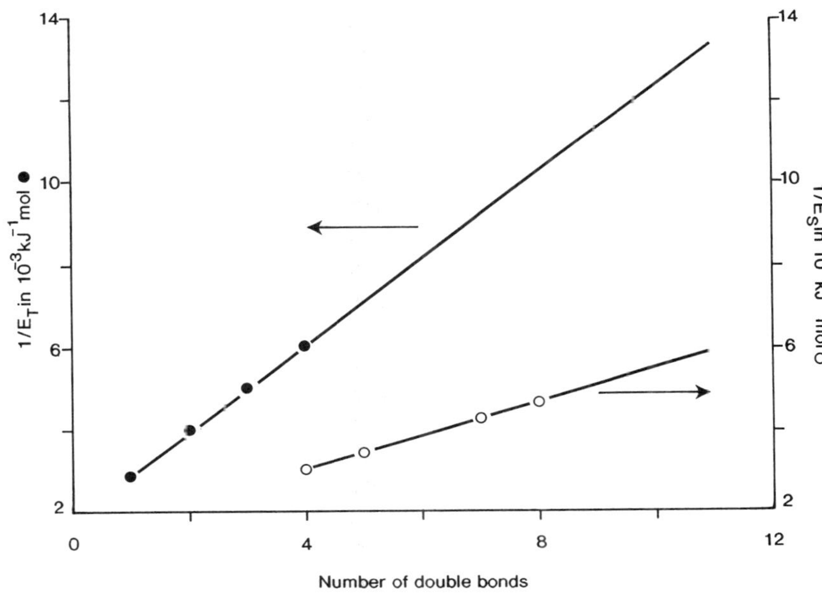

Fig. 7.6 Linear correlations for $^1/E_T$ and $^1/E_S$ versus the number of double bonds for carotenoids. E_T taken from Evans (1960, 1961), equation of line: $y = 1.8724 + 1.053\,x$, $R = 0.999$. E_S taken from Cosgrove *et al.* (1990). equation of line: $y = 1.3623 + 0.41907\,x$, $R = 0.999$.

those obtained by energy transfer may, at least in part, involve a non-vertical, twisted triplet. Nevertheless, perhaps rather surprisingly, similar values are reported for the two measurements (Gorman *et al.* 1984*b*).

For the C_{40} carotenoids, there are no equivalent data available to that given above for the retinoids. Extrapolation (Bensasson *et al.* 1976) of the early oxygen perturbation studies of Evans (1960, 1961) leads to an equation, $^1/E_T$ (in 10^{-3} kJ mol^{-1}) = 1.87 + 1.053n where n is the number of conjugated double bonds and hence to an estimate of 74 ± 10 kJ mol^{-1} for the β-carotene triplet energy level (E_T) (see Fig. 7.6). Taking into account the empirical data of Hirayama (1955) that suggests a subtraction should be made for double bonds which are not fully effective, as for instance the bond in the β-end group (the empirical increment of Hirayama is $n = -0.8$ for the double bonds at the β-end of the β-carotene) and that a correction should be made of +0.1 per alkyl group interacting with the polyene chain, $n = 11 - 1.6 + 0.6 = 10$, leads to an estimate of E_T of 81 kJ mol^{-1}.

Recently, there have been a number of studies of the β-carotene triplet energy level, a comparison of this level with that of singlet oxygen (94 kJ mol^{-1}), and the reaction of singlet oxygen with various carotenoids. The triplet states of carotenoids are well known to be quenched by ground state oxygen. This is not an energy transfer process but simply enhanced intersystem crossing associated with the triplet state character of ground state oxygen in the reaction complex with the carotenoid:

$$^3\text{Car} + {}^3\text{O}_2 \rightarrow [{}^3\text{Car} \ldots {}^3\text{O}_2] \rightarrow \text{Car} + {}^3\text{O}_2$$

Indeed, the early work of Foote (1976, and references therein) and the direct detection of the β-carotene triplet by flash photolysis following the quenching of singlet oxygen by β-carotene (Farmilo and Wilkinson 1973) unambiguously demonstrates energy transfer in the other direction, from singlet oxygen to β-carotene:

$$^1\text{O}_2 + \text{Car} \rightarrow {}^3\text{O}_2 + {}^3\text{Car}$$

Therefore the triplet energy level of β-carotene must be less than or approximately equal to that of singlet oxygen. Furthermore, the rate constant for the quenching of singlet oxygen by β-carotene (1.4×10^{10} dm^3 mol^{-1} s^{-1}) is only about half the diffusion controlled value (Wilkinson and Ho 1978, Rodgers and Bates 1980; Conn *et al.* 1991), while that of longer carotenoids such as dodecapreno-β-carotene is 2.3×10^{10} dm^3 mol^{-1} s^{-1} (Conn *et al.* 1991). This may suggest a similar energy level of triplet β-carotene and singlet oxygen. A similar conclusion was arrived at from earlier energy transfer studies (Land *et al.* 1970; Merkel and Kearns 1971; Herkstroeter 1975*b*).

In a recent study, Gorman *et al.* (1988) have measured the rate constants for singlet oxygen quenching by β-carotene in toluene as a function of temperature. At -10 °C a change from a diffusion limited to a pre-equilibrium limit was noted and, for the pre-equilibrium limited process at temperatures above −10 °C, the activation parameters correspond to a triplet energy of β-carotene being somewhat above that of singlet oxygen. On the other hand, Conn and McLean (unpublished) have measured the quenching of the triplet state of silicon naphthalocyanine (triplet energy level ~90 kJ mol^{-1}; Firey *et al.* 1988) by all-*trans* β-carotene and, allowing for steric factors, estimated the β-carotene triplet level at ~90 kJ mol^{-1}, i.e somewhat below singlet oxygen.

Another indirect measurement of the carotene triplet level can be made via the suggestion that it is approximately one-half the energy of the 2^1A_g state (Hudson *et al.* 1982). This arises from the valence bond description of the 2^1A_g state wavefunction in terms of two coupled triplets. From the data given above in Section 7.2.1, this leads to a lowest triplet level for all-*trans*-β-carotene of 92 kJ mol^{-1}, which is in good agreement with the values based on energy transfer rate constants given above.

A more direct method of obtaining the energy level of the lowest triplet state of β-carotene is via photo-acoustic spectroscopy. This technique is based on measuring the heat generated from the decay of a triplet and has been applied to β-carotene (Redmond and Lambert, unpublished). The errors inherent in the technique have not yet allowed a useful estimate of the energy level.

7.3.3 *Quantum yields*

The early work of Bensasson *et al.* (1975) established ϕ_T for several isomers of retinal (C_{20}). This work was extended by R.S. Becker *et al.* (1978) and Das and Becker (1979) who established the triplet spectra and ϕ_T values for a number

Table 7.2 Quantum yields of triplet formation (ϕ_T) of polyenes in hydrocarbon solvent as a function of number of conjugated double bonds (*n*)

Polyene	n^a	ϕ_T	Ref[b]
β-Ionone (C_{13})	2	0.10	a
β-Ionylidene acetaldehyde (C_{15})	3	0.42[c]	a
All-*trans*-C_{17}-polyenal	4	0.72	b
All-*trans*-retinol (C_{20})	5	0.03	c
All-*trans*-retinal (C_{20})	5	0.41	a
β-Apo-14-carotenal (C_{22})	6	0.41	a
All-*trans*-C_{24}-polyenal	7	0.018	c
β-Apo-8′-carotenal (C_{30})	9	3×10^{-3}	a
All-*trans*-β-carotene (C_{40})	11	1×10^{-3}	d
Torularhodinaldehyde (C_{40})	13	1×10^{-3}	a

[a] Excludes carbonyl group.
[b] a, Becker *et al.* (1978); b, Das and Becker (1979); c, Rosenfeld *et al.* (1976); d, Hashimoto *et al.* (1991).
[c] 0.27 ± 0.15 for four isomers see Bensasson *et al.* (1984*a*).

of carbonyl polyenes (C_{13}, C_{15}, C_{22}, C_{24}, C_{30}, and C_{40}) in hexane as solvent—see Table 7.2. These results established the importance of state ordering of excited states as pivotal to triplet state quantum yields and other parameters such as triplet lifetimes. They also showed that C_{15} aldehydes and C_{18} ketones (lower homologues of C_{20} retinal) were at the border with respect to the relative energies of the lowest excited states ($^{1,3}n\pi^*$ and $^{1,3}\pi\pi^*$). Consequently, Bensasson *et al.* (1984*a*) used the laser flash photolysis method to study several *cis* and di-*cis* isomers of the C_{15} aldehyde and C_{18} ketone. The triplet yields are reported in Table 7.2.

These and the early results in non-polar solvents have allowed a reasonably complete description of the relative positions of the $n\pi^*$ and $\pi\pi^*$ singlet and triplet levels in various groups of the polyenals and polyenones which account for virtually all the triplet parameters of such retinal related molecules. In polar solvents a modification of the state ordering of retinal $n\pi^*$ and $\pi\pi^*$ singlet and triplet states led to a large decrease of the ϕ_T values (Bensasson and Land 1978*b*).

The triplet quantum yields (ϕ_T) of all simple hydrocarbon carotenoids are very low, $\ll 0.1$, and direct excitation in the absence of a sensitizer is difficult. There are a few reports of the detection of a weak triplet absorption in some microheterogeneous environments, such as detergents but these may well be due to trace impurities. On the other hand, for the carbonyl containing systems the nature of the lowest singlet and triplet excited states ($n\pi^*$ or $\pi\pi^*$) can lead to significant ϕ_T values for molecules such as retinal (C_{20}) and β-apo-14′-carotenal (C_{22}) and to very low ϕ_T values (~ 0.003) for molecules such as β-apo-8′-carotenal (C_{30}). As with the fluorescence properties noted above, this is best understood in terms of a switching of the relative positions of the lowest excited $n\pi^*$ and $\pi\pi^*$ excited state.

A particularly interesting recent picosecond study of H. Hashimoto *et al.* (1991) showed that direct intersystem crossing from S_1 to T_1 can be observed for all-*trans*-β-carotene and three of its *cis* isomers. The ϕ_T values were estimated at $\sim 10^{-3}$ in all cases. Furthermore, the Raman spectra of S_1 and T_1 were obtained and showed that no isomerization takes place in S_1 (see 7.4 below).

The triplet states of retinoic acid and retinyl acetate have been studied by Rosenfeld *et al.* (1973) and Lo *et al.* (1982) using laser flash photolysis and, in the case of Lo *et al.*, also by pulse radiolysis. Both molecules exhibit a rather low ϕ_T, ~ 0.0013 and 0.025 for retinoic acid and retinyl acetate, respectively. Additional species are detectable following pulsed excitation of these molecules and these are discussed below.

7.3.4 *Energy transfer*

A study of triplet–triplet energy transfer involving two isomers of β-ionone (BI) (short-chain members of the retinal homologous series) as both an acceptor and donor molecule and of all-*trans* 2,4-hexadienal (HD) has been reported by Chattopadhyay *et al.* (1985). The second-order rate constants for the quenching of a series of sensitizers by BI was two to three times smaller than those by HD even though the (spectroscopic) triplet energies are estimated to be the same ($\sim 230\,kJ\,mol^{-1}$). This difference was attributed to steric effects associated with ground state distortion and alkyl-substitution in BI. Of particular interest was that the triplet of BI was not quenched by ferrocene or azulene (triplet energies $\sim 67\,kJ\,mol^{-1}$; Herksroeter 1975*b*) despite an exothermic energy gap of $\sim 60\,kJ\,mol^{-1}$. This can be understood (like that of olefin triplets in general), in terms of the BI triplet undergoing rapid and extensive geometric relaxation (distortion) to a form of much lower energy that cannot be quenched by ferrocene or azulene. This would clearly require a decrease in the vertical triplet energy of BI upon relaxation to be $\sim 65\,kJ\,mol^{-1}$.

Chattopadhyay *et al.* (1985) also reported triplet lifetimes of BI in a wide range of solvents including benzene, methanol and ethylene glycol, with values in the range 160 ± 30 ns. Similarly, short lifetimes have been reported for several acyclic conjugated diene and triene triplets (see, for example, Gorman and Hamblett 1983). These short triplet lifetimes were attributed to configurational distortion rather than to a particular excited state characteristic (e.g. $n\pi^*$ state). However, this explanation does not explain the variation of triplet lifetimes of two orders of magnitude along such retinal-like homologous series.

7.4 **Photoisomerization**

The major driving force for studying the phenomena of photoisomerization in polyenes is that the *cis* to *trans* isomerization of retinal is the crucial initial step in vision. Also, some recent studies (Curley and Fowble 1988) of the photoisomerization of retinoic acid have been reported because of the use of this molecule and its analogues (i.e retinoids) to treat skin disorders (Tsambaos and Orfanos 1981) and because of their possible use as cancer chemopreventive or chemotherapeutic agents (Bertram *et al.* 1987). In addition, there has been much interest in the isomerization of photosynthetic carotenoids. For example, Boucher and Gingras (1984) have described spectroscopic evidence showing that spirilloxanthin bound to the photoreaction centres isolated from *Rhodospirillum rubrum* displays *cis* to *trans* isomerization under conditions that lead to formation of the triplet state of the primary electron donor (blocked electron transfer). On the other, hand Lutz *et al.* (1982) have used time resolved low temperature Raman spectroscopy to show no *cis* to *trans* isomerization of spheroidene and methoxyneurosporene in reaction centres of *Rb. spheroides*. This topic is discussed below (Section 7.7.2).

There have been relatively few studies of C_{40} photoisomerization processes (unlike the many studies of stilbenes, dienes, etc). From the few studies reported, (for example, Land *et al.* 1971; Jensen *et al.* 1982) there is no direct evidence for isomerization occurring via the $1^1B_u^+$ states of carotenoids. Using triplet sensitizers such as chlorophyll *a* and naphthalene, Jensen *et al.* (1982) have shown that all-*trans*- and 15,15′-*cis*-β-carotene undergo isomerization from the triplet state. The overall isomerization efficiency of 10^{-3} mol dm^{-3} all-*trans*-β-carotene was found to be ~13 per cent of that for 15,15′-*cis*-β-carotene when sensitized by chlorophyll *a*. The isomerization efficiency of all-*trans*-β-carotene is dependent upon the concentration as revealed by the concentration-dependent changes in photostationary states (Jensen *et al.* 1982) and by the lower relative isomerization efficiency of 10^{-5} mol dm^{-3} all-*trans*-β-carotene (~6 per cent) reported by H. Hashimoto and Koyama (1988). Furthermore, in a recent study using laser flash spectroscopy with picosecond time resolution, H. Hashimoto *et al.* (1991) have shown no isomerization from the excited singlet state of various β-carotene isomers, but such isomerization was detected via the triplet state. Also, the rate of triplet isomerization was dependent on the starting ground state configuration and was highest for the 15,15′-*cis* central-bent molecule and lowest for the all-*trans* isomer, the latter may occur in antenna complexes (H. Hashimoto and Koyama 1988).

The triplet states of carotenoids have also been studied by the resonance Raman technique used in a time-resolved mode. Time-resolved resonance Raman spectroscopy (TR^3) is a powerful tool to study the molecular structures of electronically excited states such as triplets. For carotenoids, it is usual to use a pump-probe technique in which the pumping laser excites the triplet state of a donor (usually a hydrocarbon) and the subsequent laser probe is in resonance with the $T_1 \rightarrow T_n$ absorption of the carotenoid triplet state. Pulse radiolysis can also be used as the pump (Wilbrandt and Jensen 1983) but, of course, such a method is restricted to non-polar solvents to study excited states (see Chapters 1 and 3). The TR^3 technique has been used to study C_{40} carotenoids, mainly isomers of β-carotene; C_{20} carotenoids, mainly isomers of retinal; and a variety of smaller models such as hexa, hepta-, and octa-triene as models of biological systems (see, for example, Langkilde *et al.* 1987*a*, b).

For the five isomers of β-carotene studied (all-*trans* and 15-*cis*; Wilbrandt and Jensen 1983; all-*trans*, 7-*cis*, 9-*cis*, 13-*cis*, and 15-*cis*: H. Hashimoto and Koyama 1988) it was shown that 15-*cis* and all-*trans* have identical triplet state resonance Raman spectra whereas the remaining mono-*cis* isomers have similar but individually different triplet state spectra. These observations were ascribed to a common triplet species for all-*trans* and 15-*cis* which preferentially reverts to all-*trans* in the ground state. The other mono-*cis* isomers, on the other hand, were assumed to result in different triplet species which either revert to the corresponding ground state *cis* isomer in the case of 7-*cis* and 9-*cis*, or to the all-*trans* ground state in the case of 13-*cis* (H. Hashimoto and Koyama 1988).

The idea that isomeric polyene systems form triplet states which do not undergo equilibration during the lifetime of the triplet states has been substantiated by the results of Langkilde *et al.* (1987*a*, *b*) on the *cis* and *trans* isomers of 2,5-dimethyl-1,3,5-hexatriene and 1,3,5-hexatriene. The methylated derivative has different conformations around the single bonds for the *cis* and *trans* isomers. These differences are retained in the triplet state of the two isomers which give similar but distinct Raman spectra. The *cis* and *trans* isomers of 1,3,5-hexatriene give, on the other hand, identical triplet Raman spectra.

There have been many studies of the photoisomerization of the C_{20} polyenes with particular interest centred on the nature of the excited state precursor of the photoisomerization, i.e. singlet and/or triplet, and also on the effect of excitation wavelength on the isomeric product distribution. There is relatively little agreement in the literature on many aspects of such work. Thus, while Rosenfeld *et al.* (1977) report a substantial variation of the quantum yield of photoisomerization (ϕ_{PI}) of retinal with wavelength, the more recent data of Donahue and Waddell (1984) show little or no variation of ϕ_{PI} for retinal or its *n*-butylamine Schiff base in methanol.

As far as the excited state precursor to retinal photoisomerization is concerned, the excited singlet, a vibrationally excited triplet, and a relaxed triplet have been discussed as possible candidates. In both sensitized and direct excitation, ϕ_{PI} is higher for the various retinal *cis* isomers than for the all-*trans* isomer. In more recent work substantially higher values of triplet as well as singlet isomerization efficiencies have been reported (Jensen *et al.* 1989). These workers studied the photoisomerization of all-*trans*- and 11-*cis*-retinal, and investigated the effect of using sensitizers whose triplet energy levels are above and below the spectroscopic (vertical) triplet energy of retinal (150 kJ mol^{-1}). The technique used was continuous illumination and HPLC to monitor the composition of the photostationary states and the relative photoisomerization quantum yields of the two retinal isomers studied, although laser flash photolysis was used to determine the quenching rate constants for the sensitizer triplet retinal interactions. A number of the results of this recent study are in conflict with earlier data. Thus, Jensen *et al.* find essentially identical efficiencies in sensitizing the isomerization of 11-*cis*-retinal for donor energies above the vertical triplet energy of retinal. These workers also find that the quantum yield of isomerization of the all-*trans* isomer to other isomers is 0.15 ± 0.05 which is markedly higher than previous estimates, but is in good agreement with the value of 0.17 reported by Raubach and Guzzo (1973). The quantum yield of 11-*cis* triplet isomerization was determined as ~ 1.0 by Jensen *et al.* (1989) which is in striking conflict with some previous values using less sophisticated techniques.

For the endothermic sensitizers, the concept of 'non-vertical' energy transfer along some preferred coordinate is invoked to explain the changes in the composition of the mixture of triplet states produced and hence the changes in the composition of the photostationary states. Overall, the observed sensitizer triplet energy dependence of the primary product ratios suggests that the triplet states of the different retinal isomers are different mixtures of triplet excited species. The compositions of these mixtures depend on the starting isomer as well as on the sensitizer triplet energy for energies below the vertical triplet energy of all-*trans* retinal.

For β-carotene isomers there is interest in the concept of isomerization about two double bonds following only one excitation event. The data of Jensen *et al.* (1989) show that 11-*cis*-retinal does undergo one-photon-two-bond isomerization when sensitized by zinc tetraphenylporphyrin.

Finally, in this work, Jensen *et al.* determined the quantum yield of direct isomerization of 11-*cis*-retinal at 254 nm as 0.42. Previously reported values all concern the wavelength range 350–360 nm and vary from 0.12–0.25. Clearly, a wavelength dependence is a possibility and it would be worthwhile to redetermine the direct photoisomerization quantum yields at longer excitation wavelengths to confirm this.

TR^3 has also been applied to various isomers of retinal, see, for example, Hamaguchi *et al.* (1984) and Wilbrandt *et al.* (1985). Most authors have addressed the question as to whether or not different isomers have different or identical relaxed triplet states. Without doubt, the spectrum of the triplet state of the 13-*cis* isomers is quite different from that of the other isomers studied (9-*cis*, 11-*cis*, and all-*trans*) implying that this isomer yields a different relaxed triplet species from the others. Wilbrandt *et al.* (1985), while noting the similarity in positions of the vibrational bands in the transient resonance Raman spectra of the 9-*cis*, 11-*cis*, and all-*trans* isomers, nonetheless suggest that the isomeric retinals form different relaxed species or different mixtures of relaxed species. This conclusion was reached mainly by a consideration of the relative intensities of the transient vibrational bands, and is essentially in agreement with recent HPLC/continuous irradiation studies (Jensen *et al.* 1989).

Finally, in this section it should be noted that another route to photoisomerization of retinol and retinyl acetate arises because of photo-dissociation to a classical carbenium ion followed by isomerization and recombination, that is:

$$\text{all-}trans\text{-ROCOCH}_3 \rightarrow R^+ + CH_3COO^- \rightarrow 11\text{-}cis\text{-ROCOCH}_3$$

7.5 **Radicals**

It was noted above that carotenoid radicals can arise in 'blocked' photosynthetic reaction centres even though they have no role as yet established in normal photosynthesis. Also, radicals arise in the photosynthetic mimetic systems studied by Gust and coworkers (see Section 7.7 below). For the C_{20} retinals, protonation–deprotonation of retinal Schiff bases is intimately associated with the molecular events that take place following excitation of the visual pigment, and various charge transfer theories of the initial steps of vision would be expected to involve retinal radical ions.

Initially pulse radiolysis was used to characterize the radical cations (RC) and radical anions (RA) of a number of carotenoids and retinoids, although, more recently electrochemical methods have been reported (Grant *et al.* 1988). The pulse radiolysis technique has also been used to study the rate of reaction of such radicals with other molecules such as chlorophylls and porphyrins in both homogeneous organic solvents and in micellar systems (see, for example, Lafferty *et al.* 1977; McVie *et al.* 1979; Almgren and Thomas 1980). We have already summarized these early results (Bensasson *et al.* 1983) and Table 7.3 gives both absorption maxima listed in this reference together with more recent data on other carotenoid related molecules.

From our earlier review the following summarizes the major topics:

1. The radicals all absorb to the red side of the ground state absorption bands.
2. For both anions and cations there is a good linear relationship between the radical absorption bands and the number of conjugated double bonds.

Table 7.3 Absorption maxima for carotenoid radical ions

Carotenoid	n[a]	Species	Solvent	λ_{max} (nm)	Ref[b]
Phytoene (*cis/trans*)	3	neutral	hexane	285/285	a
		RC	hexane	470/460	a
		RA	hexane	470/460	a
7,7′-Dihydro-β-carotene	8	neutral	hexane	405	a
		RC	hexane	830	a
		RA	hexane	735	a
		RA	methanol	705	a
Septapreno-β-carotene	9	neutral	hexane	410	a
		RC	hexane	915	a
		RA	hexane	785	a
		RA	methanol	705	a
15,15′-*cis*-β-Carotene	11	neutral	pet. ether	445	a
		RC	hexane	1050	a
		RA	hexane	900	a
		RA	methanol	820	a
Decaprene-β-carotene	15	neutral	pet. ether	495	a
		RC	hexane	1250	a
		RA	hexane	1130	a
		RA	methanol	1010	a
β-Cyclocitral	2	RA	methanol	315	b
		RA	THF	370	b
β-Ionone	3	RC	hexane	375	a
		RC	$C_2H_4Cl_2$	385	b
		RA	hexane	385	a
		RA	alkaline methanol	300	a

Table 7.3 (*cont.*)

Carotenoid	n^a	Species	Solvent	λ_{max} (nm)	Ref[b]
β-Ionone acetaldehyde	4	RC	hexane	470	a
		RA	hexane	460	a
		RA	methanol	350	a
C_{17} aldehyde	5	RC	acetone	515	b
		RA	methanol	395	b
Retinol	5	RC	hexane	600	e
		RA	methanol	370	c
Retinyl acetate	5	RC	hexane	590	d
		RC	acetone	580	e
		RC	$C_2H_4Cl_2$	595	e
		RA	methanol	390	c
Retinal	6	RC	hexane	590	a
		RA	methanol	445	c
Retinoic acid	6	RC	acetone	575	e
		RC	hexane	590	d
		RA	hexane	510	d
		RA	methanol	480	c
Retinal Schiff base	6	RC	hexane	635	e
		RC	acetone	615	e
		RA	methanol	430	c
C_{22} aldehyde	7	RC	acetone	580	e
		RA	methanol	465	b
C_{24} aldehyde	8	RC	acetone	700	b
		RA	methanol	490	b
β-Apo-8′-carotenal (C_{30} aldehyde)	10	RC	hexane	890	a
		RA	hexane	840	a
		RA	methanol	500	a
		RA	THF	725	b
Canthaxanthin	11	neutral	pet. ether	465	a
		RC	hexane	960	a
		RA	hexane	1150	a
Torularhodinaldehyde	14	RC	hexane	1130	a
		RA	hexane	1130	a
		RA	methanol	600	a

[a] *n* is the number of conjugated double bonds.
[b] a, Bensasson *et al.* (1983); b, Bobrowski and Das (1987); c, Raghavan *et al.* (1981); d, Lo *et al.* (1982); e, Bobrowski and Das (1985).

3. Theoretical calculations based on a modified Pariser–Parr–Pople approach give reasonable agreement for the λ_{max} values and also, taking account of relaxation of the radical excited states, predicted correctly the very broad nature of the radical ion absorption bands (e.g. several hundreds of nanometers for dodecapreno-β-carotene).
4. The micellar work showed that the carotenoids are located in the outer mantle of the micelle.
5. Electron transfer reactions involving porphyrins, (see, for example, Lafferty *et al.* 1978; McVie *et al.* 1979) showed fast, near diffusional controlled rate constants for reactions involving carotenoid radical anions ($C^{\cdot -}$) with a range of porphyrins (P).

$$C^{\cdot -} + P \rightarrow C + P^{\cdot -}$$

but very inefficient rates of reaction for carotene radical cations ($C^{\cdot +}$) with the porphyrins, i.e.

$$C^{\cdot +} + P \rightarrow C + P^{\cdot +}$$

except where the porphyrin was chlorophyll *a*, and the possible relevance of this to photosynthesis is worthy of note.

Bobrowski and Das (1987) as well as extending the range of retinal homologues of varying chain length whose radical ions have been characterized, also studied the reactivity of the radical anions to protonation (to yield the alcohol radical) and the reactivity of radical cations to nucleophilic attack. The rate constants for radical anion protonation in alcohols decrease on increasing chain length. Also, the longer lifetime of the anion in CTAB than in Triton-X100 suggests that the ion pairing of the radical anion with the cationic head groups in CTAB results a lower reactivity with water. For the radical cations, Bobrowski and Das (1987) studied the effect of three nucleophiles, water, triethylamine (TEA), and bromide ion (Br^-), and typical data are shown in Table 7.4. As can be seen there is a marked decrease in reactivity upon increasing chain length

Table 7.4 Second order rate constants (k) for the quenching of retinyl radical cations in acetone

Retinal homologues	n[a]	k ($dm^3\ mol^{-1}\ s^{-1}$)		
		$H_2O/10^5$	$TEA/10^8$	$Br^-/10^{10}$
C_{17}	4	1.9	21	5.6
C_{20}	5	0.84	1.8	5.3
C_{22}	6	0.45	0.56	4.1
C_{30}	9	<0.1	0.055	0.012

[a] n is the number of conjugated double bonds.

of the polyenes—this may well be related to a corresponding decrease in the ionization potential of these molecules.

Very recently Bally *et al.* (1992) have stabilized the radical cations of polyenes, via *tert*-butyl capping, and obtained the corresponding absorption spectra. These agree well with the earlier pulse radiolysis data but also show an additional weak absorbance at longer wavelengths than previously reported (Lafferty *et al.* 1977). For β-carotene, for example, this additional band is near 2000 nm.

7.6 Carotenoid–oxygen radical interactions

As noted in Section 7.1 carotenoids including all-*trans* β-carotene may well exhibit anti-cancer properties. Indeed, there is epidemiological evidence that the incidence of cancer may be lower for individuals with high intake of β-carotene (Peto *et al.* 1981). This role of β-carotene may be related to its ability to trap activated oxygen species and Packer *et al.* (1981) have used pulse radiolysis techniques to show that peroxy radicals are quenched by carotenoids. Also, Krinsky and Deneke (1982) have reported that the inclusion of carotenoids in liposomes prevents an iron-catalysed peroxidative attack on lipids, suggesting that carotenoids protect against free radical damage. Burton and Ingold (1984) have used indirect evidence to show that β-carotene interacts with peroxyl radicals ($ROO^{\cdot}$) which may arise from lipid peroxidation. Recently Conn *et al.* (1992) have used pulse radiolysis to study the equilibrium

$$C^{\cdot -} + O_2 \rightleftharpoons C + O_2^{\cdot -}$$

where C is β-carotene and lycopene and $C^{\cdot -}$ the corresponding radical anion. It was reported that β-carotene radical anion efficiently transfers an electron to oxygen but the reverse reaction was not observed, whereas for lycopene a reversible process was established. Also, for β-carotene, slow formation of a new species was observed which was suggested to be due to a β-carotene-superoxide radical complex. Consistent with their interpretation these workers showed that β-carotene readily transfers an electron to lycopene but that the reverse process does not occur.

Pulse radiolysis has also been used to study the association of the radical cations of several retinoids with the parent molecule (Bobrowski and Das 1986). Of course, excimer formation of a wide range of molecules is a well documented phenomena. The singlet excited states of retinoids are too short-lived for bimolecular self-quenching to yield detectable excimers. However, the radical cations live for microseconds or longer and can thus interact with millimolar concentrations of the parent molecule. In the concentration range 10^{-3} – 10^{-2} mol dm^{-3} in oxygen-saturated acetone, Bobrowski and Das report pulse radiolysis generated transients of all-*trans*-retinal, all-*trans*-retinoic acid, and all-*trans*-methyl retinoate that are best understood in terms of association of the radical cations with the parent molecule to give 'dimers'. The equilibrium constants (K) were 220–440 dm^3 mol^{-1} with a substantial increase of K for retinal and retinoic acid in 1,2-dichloroethane. For all-*trans*-retinol and all-*trans*-retinyl acetate, no such 'dimer' formation could be detected. Bobrowski and Das (1986) also note that the formation of such 'dimers' is specially favoured by polyene systems terminating with a heteroatom and in non-polar environments, and speculate that such associative behaviour may well be critical in the role of π-systems as charge mediators along bilayer membranes.

7.7 **Photosynthesis**

It is well established that β-carotene plays two pivotal roles in photosynthesis (see Section 7.1.2). Firstly, β-carotene is an accessory pigment in the antenna system of the photosynthetic apparatus. Secondly, β-carotene protects this apparatus from self-sensitized photo-oxidation by quenching any chlorophyll triplets formed and, in addition, any singlet oxygen produced. Other roles of carotenoids associated with redox reactions have been suggested (e.g. Grant *et al.* 1988). Indeed, Mathis and coworkers (C.C. Schenck *et al.* 1982; Mathis and Rutherford 1984) have reported the formation of carotenoid cation radicals in the blocked PSII reaction centre. Recently this has been unambiguously confirmed by Telfer *et al.* (1991).

7.7.1 *Model systems*

Gust and Moore and co-workers have synthesized a variety of carotenoids linked to a range of porphyrins and pyropheophorbides (see, for example, Gust and Moore 1991). Overall their work can be divided into three directions of study:

(1) the mimicry of the antenna and the photoprotective functions of carotenoids by synthetic dyads which are carotenoporphyrins or carotenopyropheophorbides (A.L. Moore *et al.* 1980, 1982; Bensasson *et al.* 1981; Gust *et al.* 1985; Wasielewski *et al.* 1986);

(2) the mimicry of the photodriven charge separation; and

(3) the mimicry of the transmembrane charge separation.

As far as the role of carotenoids in photosynthesis is concerned the first of these is the most important. In this case the mimicry deals with singlet–singlet and triplet–triplet energy transfer and both the photoprotective and the antenna functions have been successfully mimicked in these models. Such transfer was found to be efficient only when the π-electron systems of the carotenoid and the tetrapyrrole ring system can be arranged in close proximity to each other. Overall, these studies imply that singlet–singlet and triplet–triplet energy transfer occur via an electron exchange mechanism (Gust and Moore 1991).

7.7.2 *Structures of carotenoids in photosynthetic systems*

Probably the most important work in understanding the details of the mechanisms of the photosynthetic reaction centre arises from the X-ray analysis of crystallized photosynthetic bacterial reaction centres (see, for example, Michel 1982; Diesenhofer *et al.* 1984; Diesenhofer and Michel 1988, 1989*a*, *b*). This work has shown that the reaction centres exhibit an approximate twofold symmetry, usually represented as an inverted V, with an A and B branch and a bacteriochlorophyll *a* dimer ('special pair') at the apex. Despite the near symmetry, only the A branch is used in the main electron transfer from the 'special pair', via a bacteriopheophytin, with this initial step possibly facilitated by a monomeric bacteriochlorophyll *a* (Bchl *a*) moiety, and eventually to the quinone system. However, the carotene molecule does not conform to the symmetry and is located only on the (inactive) B branch. It is separated from the 'special pair' by the Bchl *a* monomer of the B branch. Unfortunately, the X-ray data have not yet determined the stereochemistry of the bound carotenoid but the data are consistent with a 15,15′-*cis* structure as suggested by the resonance Raman (Koyama *et al.* 1983) and NMR (Gebhard *et al.* 1991) data. Indeed, all the experimental data currently available suggest that the carotenoids in the reaction centre are in a 15-*cis* configuration with some additional twisting in the conjugated system of the molecule (Koyama *et al.* 1990).

In the light harvesting antenna complexes (LHC) there are, as yet, no equivalent X-ray data. The spectroscopic data available imply an all-*trans* conformation for carotenoids in the LHC of *Rb. spheroides*, *Rs. rubrum*, and *Rps. viridis*. However, there is some debate over the details. Thus the Raman data suggest two different types of conformation for the all-*trans* carotenoids (Koyama 1991) whereas Breton and Nabedryk (1987), using linear dichroism measurements, suggest that only a simple geometric arrangement of the carotenoids occurs.

7.7.3 *Mechanisms*

A major current interest is to discover the precise mechanism by which carotenoids act as light-harvesters and photo-protectors. Questions such as why are 15-*cis* carotenoids selected for reaction centres, and the all-*trans* isomer for the LHC are of interest, and also what is the role of the excited states ($1\,^1B_u$, $2\,^1A_g$, and lowest triplet) are now much discussed (see, for example, Koyama 1991).

The structural organization of the pigments in the reaction centre revealed by the X-ray structure shows that the bacteriochlorophyll monomer in the B branch lies between the 'special pair' and the carotene molecule, ~0.3 nm edge-to-edge from each (Diesenhofer and Michel 1989*a*). This suggests a pathway of triplet energy transfer of the form $(BChl)_2 \rightarrow$ BChl(B branch) $\rightarrow$ 15-*cis* carotene. In solution, 15-*cis*-β-carotene isomerizes to the all-*trans*-β-carotene very efficiently, possibly via a common triplet state (see Section 7.4), and this has led to the proposal that the mechanism for excess energy dissipation by the reaction centre carotenoid includes carotenoid isomerization in the triplet state (Koyama *et al.* 1990). Consistent with this interpretation is the work of Frank and Violette (1989), who showed that the triplet energy transfer to the carotene was inhibited when the BChl (B branch) is dislocated by sodium borohydride treatments. However, the selection of the *cis* isomer may well just be a geometric requirement of the reaction centre complex.

Studies of the mechanistic role of all-*trans*-β-carotene in the LHC have mainly concerned the relative roles of the S_2 ($1^1B_u^+$) and S_1 ($2^1A_g^-$) energy levels of the β-carotene with respect to transfer to the BChl *a*. The S_1 and S_2 lifetimes of about 10 ps and 200 fs, and the corresponding energy levels, suggest that an electron-exchange mechanism *via* S_1 is most important. On the other hand femtosecond studies (Trautman *et al.* 1990) on the B800–850 antenna complex of *Rb. spheroides* 2.4.1 have given energy transfer times similar to that of the S_2 of free spheroidene in solution (~340 fs). Thus, it has also been suggested that singlet energy transfer in the LHC may occur directly from the S_2 state via the dipole mechanism (Shreve *et al.* 1990).

7.8 Concluding remarks

In view of their crucial role in both photosynthesis and vision it is hardly surprising that there continues to be a major interest in the photobiology, photochemistry, and photophysics of a wide range of carotenoids. This is heightened by the epidemiological reports that carotenoids and retinoids might be anticarcinogenic, based on reports that populations with high carotenoid intakes in their diets seem to have a reduced cancer incidence. Mediators of these beneficial effects could be triplet state quenching, singlet oxygen quenching, and radical quenching.

Many of the photo and radiation properties of carotenoid systems have been related to the length of the conjugated system. Thus, plots can be used linking the S_1 and T_1 energy levels to the number of conjugated double bonds (*n*) and also linking the λ_{max} of the radical ions to *n*. In addition, linear correlations can be obtained between the $S_0 \rightarrow S_2$ transition and the triplet–triplet transition. Figure 7.6 shows plots of $^1/S_1$ and $^1/T_1$ energy levels as a function of *n* together with the equations of the lines. Certainly, for these there is a theoretical interpretation of this linearity and the plot for S_1 does give a very reasonable value for the lowest singlet energy level 2^1A_g of β-carotene of 167 kJ mol^{-1}. This corresponds to a $S_0 \rightarrow S_1$ transition around 700 nm.

8

Melanogenesis and melanoma

'Melanin is a pigment of the imagination'

Carroll Williams (Harvard Biologist)

8.1 Introduction

Current interest in melanin is growing, in particular because of the dramatic increase during the last three decades in the incidence of malignant melanoma, possibly through increased exposure to UV radiation (see, for example, MacKie 1983; Bagnara 1988). A rational chemotherapeutic control of melanoma may be possible because of the unique localization of large amounts of the pigment synthesizing enzyme, tyrosinase, in malignant melanocytes.

Human skin colour is largely determined by the size, type, and distribution of cytoplasmic organelles known as melanosomes in which melanin pigments are synthesised and deposited. Melanosomes are the product of specialized glands called melanocytes. Melanocytes are distributed in various sites including the eye, the ear, the central nervous system, the hair, and the skin.

The various stages in the pathway of epidermal melanin pigmentation are illustrated in Fig. 8.1. In the skin, melanocytes project their dendrites into the epidermis where they progressively transfer their melanosomes to keratinocytes. There are several stages in the development of the melanized melanosome including

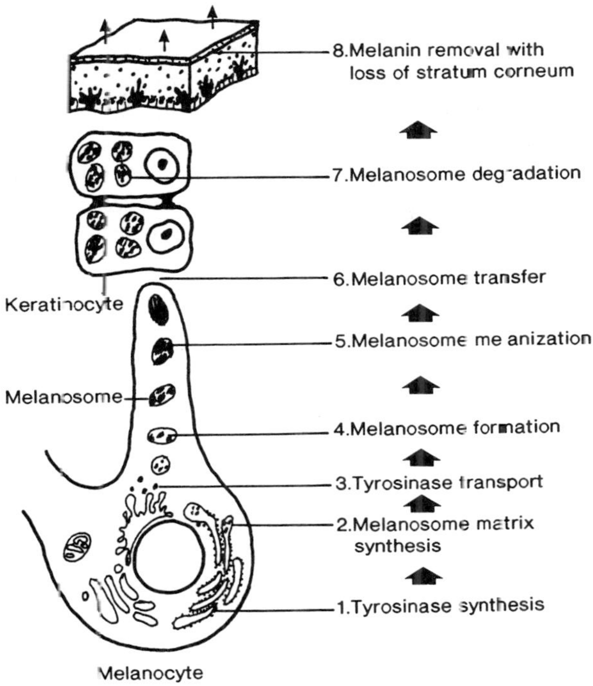

Fig. 8.1 Pathway of epidermal pigmentation (reprinted with permission from Fitzpatrick *et al. Dermatology in general medicine*, 2nd edition, copyright, 1979, McGraw Hill). The arrows represent the direction of movement. For an indication of scale this figure should be compared to Fig. 9.1.

(1) the immature, unmelanized melanosome (premelanosome);

(2) the intermediate form, in which pigment has just begun to accumulate; and

(3) the fully melanised organelle called the melanin granule.

It is the partially melanized melanosome or the melanin granule which is ultimately transferred to the keratinocytes. Degradation of the melanosome occurs within the keratinocyte with disruption of the melanosome and dispersal of melanin particles. This degradation is most marked in the stratum corneum. The whole process of melanosome formation, pigment production, and melanosome transfer is stimulated by the UV component of sunlight, leading to 'delayed tanning'.

The somewhat facetious quotation at the beginning of this chapter is a reflection of the difficulty that has been experienced in defining the chemical structure of melanin (see, for example, Prota 1988). The mammalian pigment is of high molecular weight and intractably insoluble in all solvents. Although the complete structure is unknown, the human epidermal pigment was originally believed to consist of an intimate mixture of two chemically distinct polymers: black or brown *eu*melanin and reddish-brown *phaeo*melanin. Traditionally, these two types of melanin were considered to be chemically distinct as illustrated in Fig. 8.2, essentially an updated version of the classic Raper–Mason scheme for melanogenesis (Raper 1928; H.S. Mason 1948). Both types of melanin are polymers in which all intermediary substances may be co-polymerized. More recent evidence has shown that the two melanins are more similar in that they both contain some sulphur derived from cysteine residues.

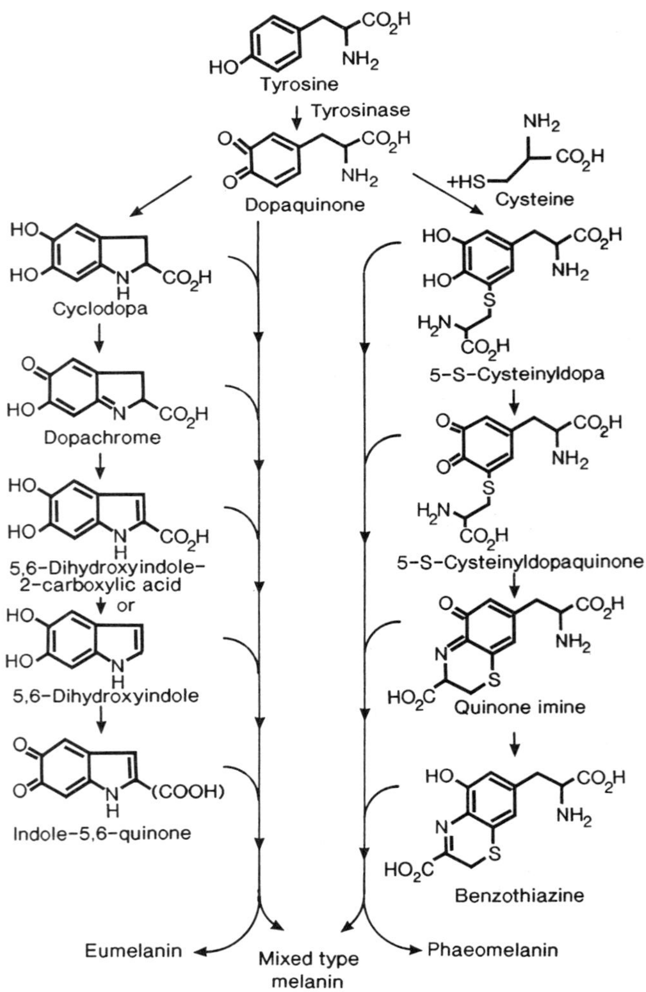

Fig. 8.2 Principal steps in melanin formation (reprinted with permission from Rorsman *et al.* 1983 in *Malignant melanoma, advances of a decade* (ed. R.M. MacKie), S. Karger AG, Basel).

Photochemically the two pigments differ, phaeomelanin being much more photosensitive than eumelanin. Indeed, it has been proposed that the rapid photodegradation of phaeomelanin is responsible for the increased susceptibility of fair-skinned individuals to the deleterious effects of sunlight.

Melanins exhibit many unusual physico-chemical properties for biological polymers. They have significant ion exchange capability (Sarna *et al.* 1976), can undergo reversible redox processes (Gan *et al.* 1974), show persistent electron spin resonance signals of free radical origin (Sealy *et al.* 1980), and are able to scavenge free radical species, for example, those arising from water radiolysis including e^-_{aq}, $OH^\cdot$, and $O_2^{\cdot -}$ (Sarna *et al.* 1986).

Although melanins are thought to be the single most important factor in the protection of human skin from the damaging effects of chronic solar irradiation (Kollias *et al.* 1991), conversely, transformation of the melanocyte following exposure to ultraviolet light can lead to malignant melanoma, often characterized by increased melanin formation. The mechanism whereby malignant melanoma arises is unknown although epidemiological evidence links the development of the disease to sunlight exposure and genetic predisposition. The disease is associated with elevated levels of the phaeomelanin precursor 5-*S*-cysteinyldopa (5-SCD), a natural metabolite of active melanocytes, and there is interest in 5-SCD as a biochemical marker of melanoma metastasis (e.g. Jonsson *et al.* 1980). UV-stimulated melanocytes also produce increased amounts of 5-SCD, which is itself extremely photolabile. Indeed, the photochemical mechanisms operative in UV-induced malignant melanoma may involve the photochemistry of 5-SCD. More recently, efforts have also been made to explore the possibility of using indolic eumelanin precursors and their metabolites as markers for malignant melanoma, 5-hydroxy-6-methoxyindole-2-carboxylic acid being found to be the most useful marker of the indoles tested (Westerhof *et al.* 1987).

As can be seen from Fig. 8.2, the conversion of tyrosine into melanin is proposed to involve a series of oxidation steps, although it is almost certain that this scheme is incomplete. Only some of these intermediates have been characterized. At least two initial steps, the tyrosinase-catalysed oxidation of tyrosine and immediately, in turn, the catecholamine dopa, to the highly reactive dopaquinone, are believed to be under enzymic control. The subsequent reactions leading eventually to melanin polymers may proceed spontaneously. Tyrosinase also oxidizes many different catecholamines to reactive quinones from which other polymeric pigments are derived. Sclerotization in insects also involves the enzymic oxidation of catecholamines.

Much of what is already known about the early chemical stages of melanogenesis has been discovered by enzymatic oxidation studies in which slow enzyme-catalysed steps are rate determining, precluding the build-up of high concentrations of reactive intermediates. Flash photolysis and, in particular, pulse radiolysis have been exploited recently to investigate these reactive intermediates. Such pulsed techniques very rapidly produce, via the semiquinone, high concentrations of quinone and subsequent biologically relevant oxidized species derived from various isolatable intermediates of melanogenesis. This has allowed a further insight into the molecular mechanism of melanogenesis. The isolatable intermediates include dopa, 5-*S*-cysteinyldopa, several dihydroxindoles and metabolites, and the false precursor 4-hydroxyanisole. Such studies provide a detailed kinetic understanding of the chemistry involved.

Some of the unstable intermediates of melanogenesis, in particular the *ortho*-quinones, are likely to be toxic, and malignant melanocytes are known to have damaged, and hence leaky, membranes, allowing the escape of such toxic molecules.

8.2 3,4-Dihydroxyphenylalanines

8.2.1 *Oxidation of 3,4-dihydroxyphenylalanine*

One-electron oxidation of 3,4-dihydroxyphenylalanine (dopa) with ionizing radiation ($N_3^{\cdot}$), see Section 3.1.1.2, using pulse radiolysis (Chedekel *et al.* 1984; Thompson *et al.* 1985) results in the formation of dopasemiquinone (Steenken and Neta 1982; Craw *et al.* 1984*b*), λ_{max}, 305 nm. While this species is probably not on the main path for melanogenesis *in vivo*, its second-order decay almost certainly leads to the formation of dopaquinone, λ_{max} ~ 380 nm, which is a member of the melanogenic pathway. The dopaquinone was found to decay subsequently via a process following overall first-order kinetics into a species with λ_{max} at 480 nm, indicative of dopachrome. The dopaquinone → dopachrome first-order process, which could be followed either via the decay at 380 nm or via the growth at 480 nm, was shown to be strongly base-catalysed, a hundredfold increase in rate being found in going from pH 5.6 to pH 8.6. In the acid region, the appearance of dopachrome also depended upon the azide concentration not being too high (C. Lambert *et al.* 1991).

A similar series of intermediates is obtained on initiating dopasemiquinone formation by 266 nm laser excitation of dopa (Land *et al.* 1986). The quantum yield of hydrated electron formation (9.1 per cent) following flash photolysis of dopa is half the yield of dopasemiquinone (19.6 per cent), implying that dopasemiquinone is formed via two primary photochemical mechanisms: photoionization, giving e^-_{aq}, followed by deprotonation and also photohomolysis, giving $H^{\cdot}$.

The overall first-order kinetics for the transformation dopaquinone → dopachrome may be understood in terms of the scheme shown in Fig. 8.3 in which conversion of dopaquinone ultimately into dopachrome is rate determined, in the pH range studied, by a cyclization of dopaquinone yielding leucodopachrome (cyclodopa), the latter, as soon as it is formed, rapidly (rate constant > $10^9\,dm^3\,mol^{-1}\,s^{-1}$) reacting with unchanged dopaquinone to give dopachrome and reformed dopa. The pH dependence arises because the cyclization, an intramolecular 1,4-Michael addition, is possible only when the amino group is unprotonated; hence the rate of cyclization increases in

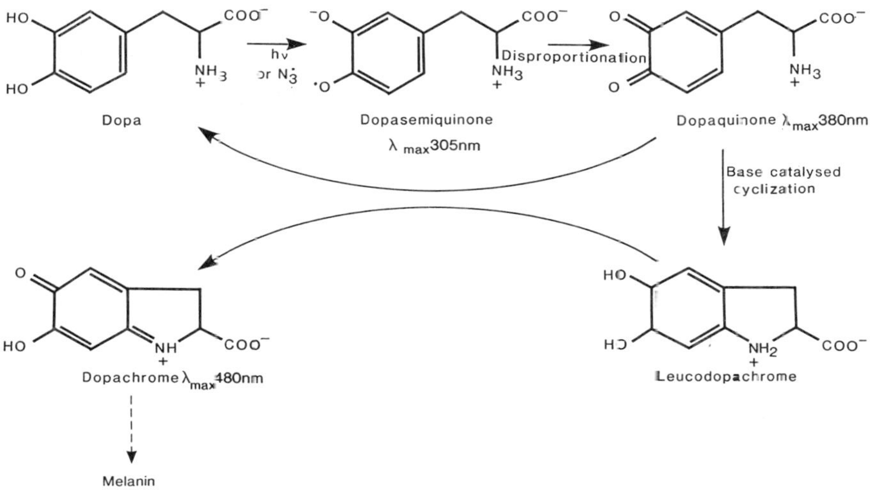

Fig. 8.3 Scheme for oxidation of dopa.

alkaline solution. In support of this mechanism, the corresponding quinone from *N*-acetyldopa cyclizes only extremely slowly, due to the electron-withdrawing effect of the acetyl group which lowers the nucleophilicity of the amino group.

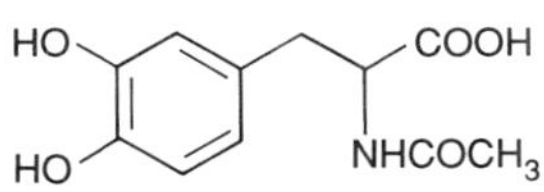

N-Acetyldopa

8.2.2 ***Oxidation of 5-S-cysteinyl-3,4-dihydroxyphenylalanine***

Azide-induced one-electron oxidation of 5-*S*-cysteinyl-3,4-dihydroxyphenylalanine (5-*S*-cysteinyldopa, 5-SCD), the main isomer found in the reaction of dopaquinone with cysteine, results (Thompson *et al.* 1985) in the formation of the corresponding semiquinone which also leads to a series of unstable intermediates, although in this case the later chemistry is different. The decay of the semiquinone does however again follow second-order kinetics and leads to a second transient species with λ_{max} at 380 nm assigned to the corresponding quinone. This second species subsequently decays by first-order kinetics into a third species with two major absorptions, one around 310 nm, the other around 540 nm. Unlike the dopa system, in this case the third species is only transient, the longer wavelength absorption decaying slowly with first-order kinetics $k \sim 3\,s^{-1}$. Similar results are obtained with 2-*S*-cysteinyldopa and 2,5-*S*,*S*′-dicysteinyldopa.

These results can be explained by a different and competing quinone cyclization pathway (Fig. 8.4) compared to that observed during the oxidation of dopa. Cyclization of 5-*S*-cysteinyldopa is accomplished *via* attack of the cysteinyl side-chain amino substituent on the carbonyl at carbon 4 of the phenyl ring, with consequent loss of water. This results in the formation of a cyclic *ortho*-quinone imine intermediate, previously postulated by Prota *et al.* (1970), responsible for the longer wavelength absorption. This *ortho*-quinone imine then rearranges by a hydrogen shift to the more stable phenolic isomer, a benzothiazine, which absorbs only in the UV.

The variation in the modes of cyclization of dopaquinone and 5-*S*-cysteinyldopaquinone is reflected in their different pH dependencies. The cyclization of the cysteinyldopaquinone requires both the initial protonation of the quinone and water elimination, and thus might be expected to be acid catalysed. On the other hand, the reacting species would involve a deprotonated amino group on the cysteinyl side-chain. The observed four fold enhancement in the cysteinyldopaquinone cyclization rate in going from pH 8.0 to pH 5.8, quite different from the equivalent change found with dopaquinone, is a product of these two competing pH dependencies.

Laser flash photolysis of 5-SCD at 256 nm results (Land *et al.* 1986) in a much lower photoionization quantum yield (<0.5 per cent) than obtained with dopa, and the production of an initial transient species whose absorption spectrum was markedly different from the semiquinone spectrum observed after pulse radiolysis-induced azide radical oxidation, and which did not lead to 5-*S*-

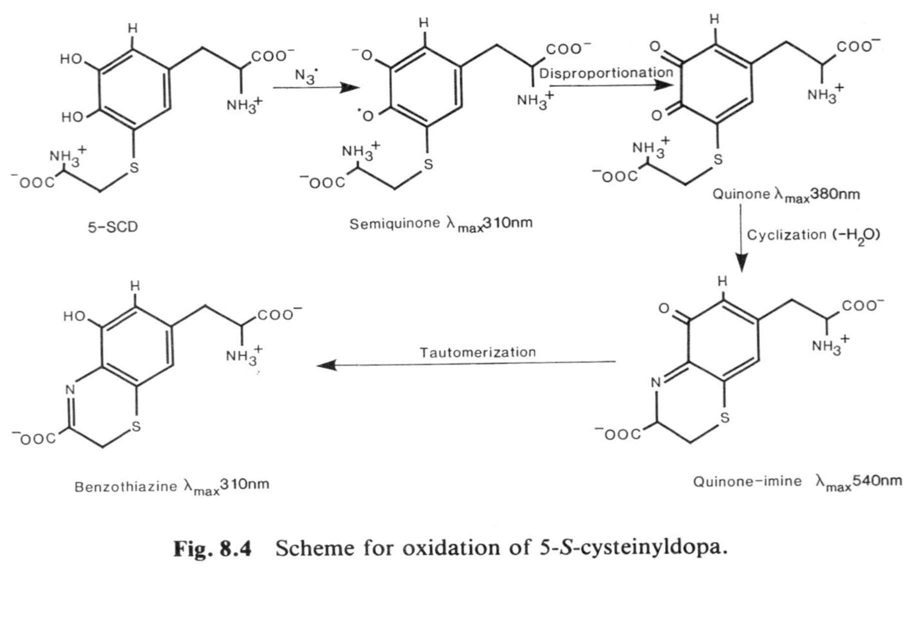

Fig. 8.4 Scheme for oxidation of 5-*S*-cysteinyldopa.

cysteinyldopaquinone nor, subsequently, the quinone imine and benzothiazine. These observations indicate that the primary 5-*S*-cysteinyldopa photochemically produced species is not mainly such a semiquinone formed by photoionization, but rather results from S—C bond homolysis. ESR spin trapping (Koch and Chedekel 1986) is consistent with the following process:

$$(HO)_2C_6H_3SCH_2CH(CO_2^-)NH_3^+ \xrightarrow{h\nu} (HO)_2C_6H_3\dot{S} + \dot{C}H_2CH(CO_2^-)NH_3^+$$

This difference between the photochemistry of dopa and 5-*S*-cysteinyldopa may well be a factor in the increased photosensitivity of fair (phaeomelanic) as compared with dark (eumelanic) skinned individuals.

8.2.3 *Reaction of dopaquinones with thiols*

The interaction of cysteine with melanin precursors is complex since, although cysteine residues are integral parts of some melanin pigments, the presence of cysteine and other thiols can inhibit the formation of melanin *in vitro*. From studies of the enhanced rate of decay of dopaquinone absorption around 380 nm (Chacon *et al.* 1988*a*), cysteine was shown to react with dopaquinone fairly rapidly ($k = 6.5 \times 10^6\ dm^3\ mol^{-1}\ s^{-1}$) presumably to form cysteinyldopa, thus preventing the formation of dopachrome. A number of thiols, including glutathione, cysteamine, and dithiothreitol, were similarly reactive towards dopaquinone. Interestingly, dopasemiquinone was found to be unreactive towards cysteine. So was dopaquinone towards 2-thiouracil, 4-thiouridine, and 8-thioguanosine, which contain sulphur as the C=S tautomer rather than —SH, and penicillamine where the —SH may be unreactive for steric reasons. Dopaquinone was also unreactive towards a number of amino acids and nucleotides including tryptophan, methionine, alanine, histidine, arginine, thymidine, cytidine, uridine, guanosine, and adenosine (Thompson *et al.* 1985).

The rate constant for reaction of 5-*S*-cysteinyldopaquinone with thiols was found to be at least two orders of magnitude lower than that observed with dopaquinone (Thompson *et al.* 1985). Ito and Fujita (1984) oxidized 5-SCD in the presence of cysteine using the Fe^{III}–EDTA system and reported that the reactivity of 5-SCD measured by product analysis was comparable or slightly lower than that of dopa. The pulse radiolysis results (Thompson *et al.* 1985; Chacon *et al.* 1988*a*) would appear to indicate that the reactivity of 5-SCD with cysteine observed by Ito and Fujita may be via a different mechanism than reaction of cysteine with 5-*S*-cysteinyldopaquinone.

8.3 **5,6-Dihydroxyindoles and metabolites**

After dopa, later isolatable intermediates in the series of chemical reactions leading to eumelanin (Fig. 8.2) are 5,6-dihydroxyindole-2-carboxylic acid (DHICA) and its decarboxylation product 5,6-dihydroxyindole (DHI). Current experimental evidence in fact suggests that melanin is a heteropolymer consisting primarily of DHICA- and DHI-derived monomer units (Blois 1978; Ito 1986; A. Palumbo *et al.* 1988). The corresponding indolequinone (Fig. 8.2) oxidation products have been postulated as subsequent intermediates, although they have proved rather elusive until the application of the fast reaction technique of pulse radiolysis.

8.3.1 *Oxidation of 5,6-dihydroxyindoles*

The series of successive transient absorption spectra produced following $N_3^{\cdot}$-induced one-electron oxidation of DHI, DHICA, and the more stable model compound *N*-methyl-5,6-dihydroxyindole (NMDHI) have been recorded (C. Lambert *et al.* 1989).

The initial transient absorption spectra from all three compounds possess absorption bands both at ~310 nm, where O-centred *ortho*-semiquinones have strong absorptions, and at ~500 nm, where N-centred indolyl radicals absorb. Studies of the effect of pH on these absorptions (C. Lambert *et al.* 1989; Al-Kazwini *et al.* 1990*a*), analogies with the assignments of the corresponding radicals from 5-hydroxyindoles by Steenken and Neta (1982) and Jovanovic and Simic (1985*b*), and ESR studies of DHI oxidation in alkaline solution (Sealy *et al.* 1984), led to the conclusion that, in the neutral pH region, the initial radicals obtained by $N_3^{\cdot}$ oxidation of DHI, DHICA, and NMDHI are, respectively, the single, mainly oxygen-centred, radicals:

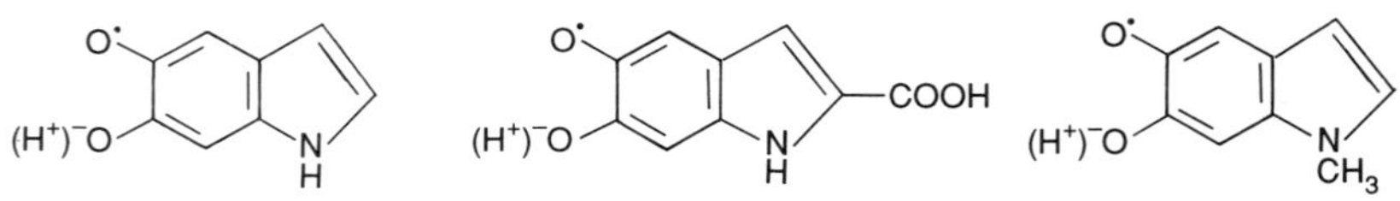

The pK_a values for the hydroxyl groups of the semiquinones derived from DHI and NMDHI are 6.8 ± 0.2 (Al-Kazwini *et al.* 1990).

The decays of all the above radicals resulted in the formation of secondary species most easily monitored around 380–430 nm where they possess weak broad absorption bands or shoulders (C. Lambert *et al.* 1989). In the case of DHI and DHICA, the decays of the secondary species were complex. For DHI, the decay monitored at 400 nm could be fitted to two consecutive first-order processes with rate constants ~150 and ~50 s^{-1}. For DHICA, the corresponding rate constants were ~600 and ~100 s^{-1}. The two pairs of first-order rate constants were independent of the dose and the concentration of parent indole, suggesting that none of these processes involved self reaction or reaction with the parent. For both DHI and DHICA, the decay corresponding to the slower of the first-order decays gave rise to a matching growth of a very broad peak centred around 500–550 nm. In contrast to DHI and DHICA, the decay of the secondary species from the model NMDHI was much slower and mainly took place via a single first-order process, $k \sim 12\,s^{-1}$, which was again dose and parent concentration independent.

A likely mechanism for the second-order decay of the radical derived from DHI is disproportionation giving rise to 5,6-indolequinone, λ_{max} 430 nm:

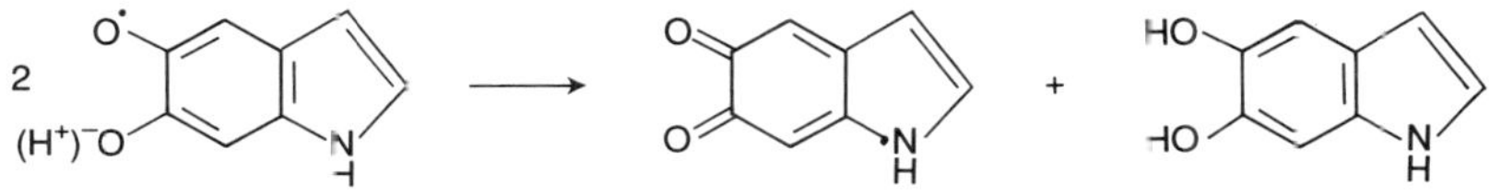

If this is so, then the lack of a variation in the rate of decay of quinone with dose excludes a process of melanochrome formation by self-condensation as postulated by Bu'Lock and Harley-Mason (1951). Melanochrome is another possible intermediate of melanin formation. Other possible products of the bimolecular decay of the radical derived from DHI, or, more likely, products of the decay of 5,6-indolequinone are the tautomeric quinone imine:

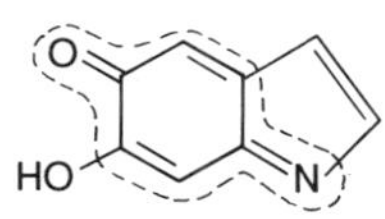

and quinone methide:

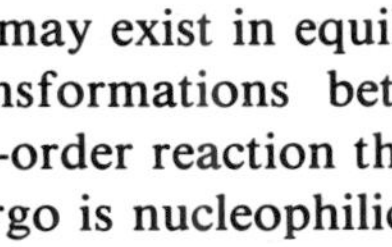

Zwitterionic forms of the above are also conceivable, and quinone, quinone imine, and quinone methide may exist in equilibria with each other.

Besides unimolecular transformations between these different species, another possible pseudo first-order reaction the quinone, quinone imine, and quinone methide could undergo is nucleophilic addition of water. Such addition products could be very shortlived, as soon as they are formed, rapidly reacting with remaining quinone/imine/methide forming dimeric structures. A scheme for melanin polymerization based on this proposed mechanism is given in Fig. 8.5. The stronger nucleophile cysteine is also reactive towards the species with maximum at 430 nm (probably 5,6-indolequinone) ($k \sim 3 \times$

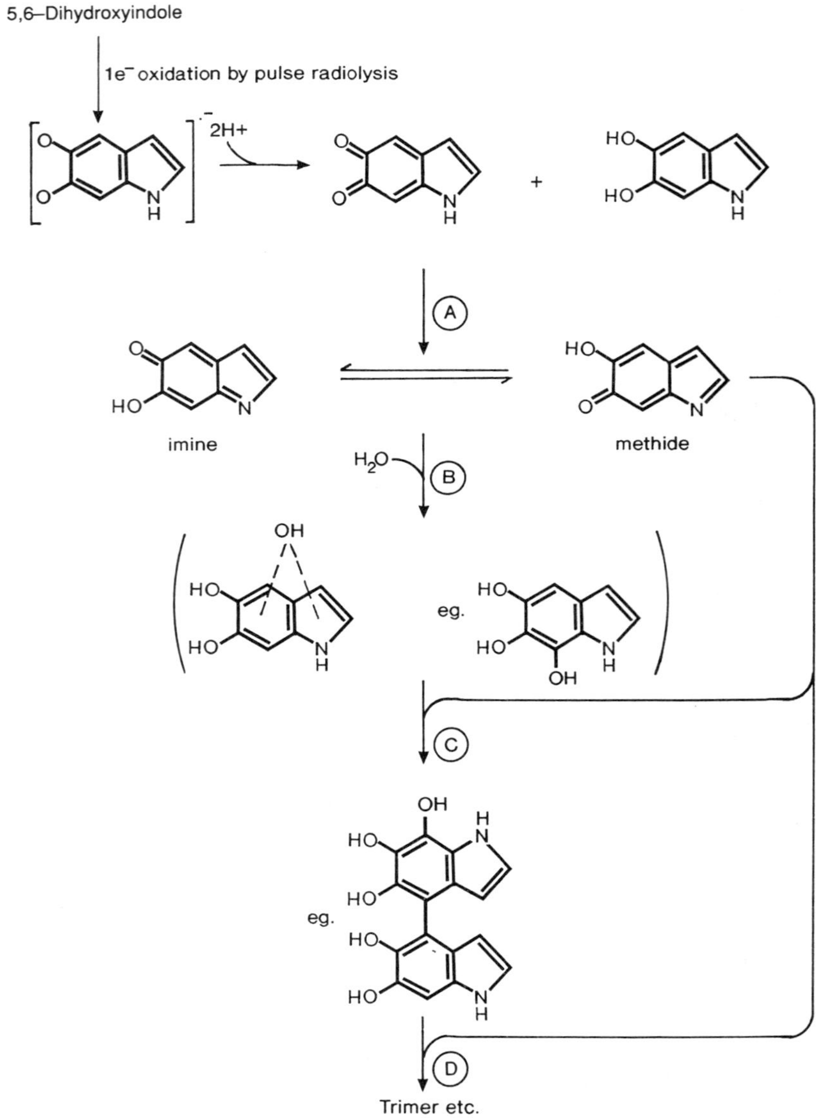

Fig. 8.5 Possible scheme for polymerization in melanogenesis (Lambert *et al.* 1989). Step A may be the faster of the first-order processes observed for the quinone disappearance. Step B involving H_2O addition may be the slower of the first-order processes observed. Step C leads to dimerization via several closely related routes either involving methide addition to C4 of the trihydroxy indole as shown, or a variety of condensation steps not shown. These dimers may contribute to the 550 nm absorption. Step D leads to the trimer and subsequently to the polymer.

$10^6\ dm^3\ mol^{-1}\ s^{-1}$), and this could be related to the formation of sulphur-containing melanins (cf. the dopaquinone reaction with cysteine discussed above). An analogous series of intermediates and reactions can be envisaged from DHICA. In the case of NMDHI the corresponding tautomeric imine and methide are not possible. Here the species with maximum at 420 nm, which is almost certainly due to *N*-methyl-5,6-indolequinone, could also presumably decay via nucleophilic addition of water.

Prota and coworkers (Corradini *et al.* 1986; d'Ischia and Prota 1987; P. Palumbo *et al.* 1987) have recently isolated, in low yield, several types of indole dimer after enzymic, auto-(±metal cations) and photo-oxidation of dihydroxyindoles. These workers suggest that the dimers are formed by coupling of the corresponding phenoxyl radicals. Structures of possible unstable dimers which might isomerize unimolecularly have been suggested (Land 1988). However, the broad peaks in the 380–430 nm region, where it is well established that *ortho*-quinones absorb, obtained in the above pulse radiolysis experiment, are evidence against this radical dimerization hypothesis.

Nevertheless, it would be useful in future to try to provide further evidence to distinguish between the various possible quinone, quinone imine, quinone methide, and dimeric intermediates using a different more structural technique such as time-resolved resonance Raman spectroscopy. It might also be possible to distinguish between the absorption spectra expected of the above tautomers theoretically.

8.3.2 *Mechanism of polymerization of melanin*

The above kinetic data seem to exclude significant melanin polymerization by condensation of semiquinones, reaction of semiquinone with DHI, self-addition of 5,6-indolequinone or tautomers, or reaction of 5,6-indolequinone or tautomers with DHI. These pulse radiolysis data rather suggest that dimerization could occur by reductive addition of a trihydroxyindole to 5,6-indolequinone and/or its isomeric quinone imine or quinone methide. Polymerization of melanin may then occur by stepwise addition of, for example, 5,6-indolequinone methide to reduced oligomeric species as indicated in Fig. 8.5. Of course, large melanoid polymers are not detected in pulse radiolysis experiments due to the discontinuity of production of oxidized species, as distinct from the *in vivo* situation in the biosynthesis of melanin.

8.3.3 *Oxidation of methylated metabolites of 5,6-dihydroxyindoles*

5-Hydroxy-6-methoxyindole (5H6MI) and 6-hydroxy-5-methoxyindole (6H5MI) are close analogues and metabolites of DHI. These methoxylated indoles are found in the urine of normal humans and, in particular, melanoma patients, and are formed by *o*-methylation of 5,6-dihydroxyindoles by the action of catechol-*o*-methyltransferase. These compounds may be usefully compared with dihydroxyindoles on account of the difference in ionizability of O—CH_3 and O—H.

The spectra of the initial radical species resulting from N_3^--induced one-electron oxidation of 5H6MI and 6H5MI are quite different (Al- Kazwini *et al.* 1990*a*; C. Lambert *et al.* 1990*b*). The absorption spectrum of the radical from 6H5MI is very similar to that observed for the corresponding radical from DHI (see above), whereas that from 5H6MI showed two maxima of 340 and 420 nm. These one-electron oxidised radicals have been assigned, respectively, as:

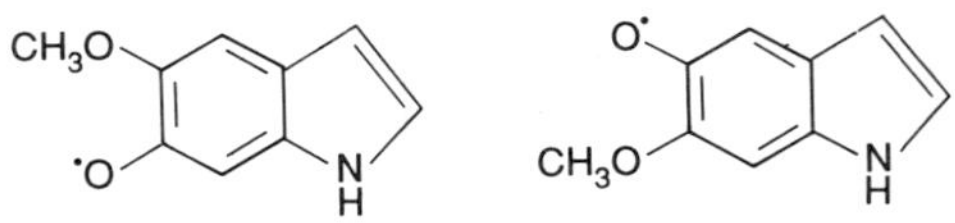

the differences in absorption spectra being explicable in terms of the contrasting structures of the two radicals. The uncharged canonical forms of the radical from 6H5MI allow the free electron to move over both rings of the indole:

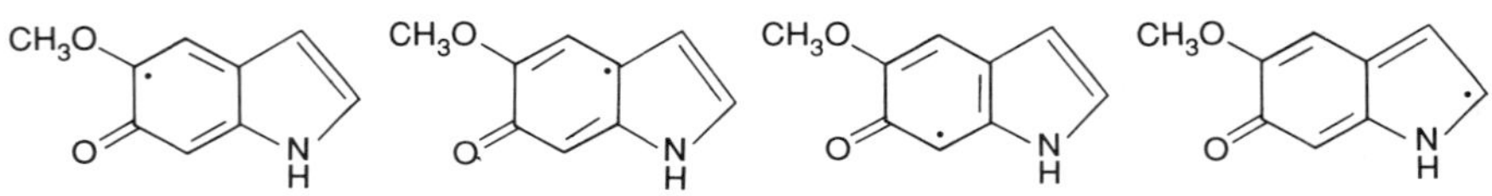

whereas in the case of the radical from 5H6MI only one ring is involved:

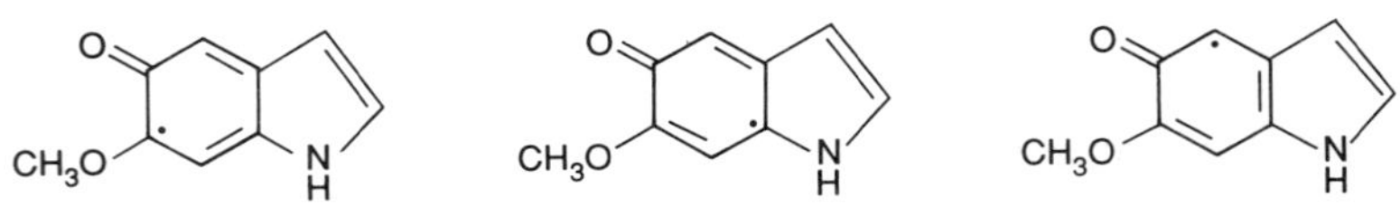

The decay of both radicals followed second-order kinetics and resulted in products which absorbed in the 400–430 nm region (C. Lambert *et al.* 1990*b*). However, the subsequent changes in absorption were quite different for 6H5MI and 5H6MI.

For 6H5MI the subsequent decay monitored around 420 nm fitted two consecutive first-order processes which were both strongly base catalysed. By analogy with the interpretations given to the species derived from DHI and DHICA, the first of these processes was assigned to partial decay via deprotonation of the corresponding quinonoid cation to form an equilibrium mixture of its cation and the corresponding quinone methide:

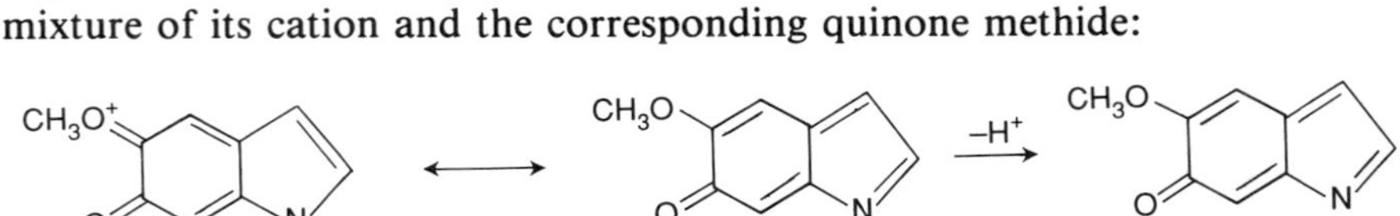

The second process was assigned to reaction of the quinone methide with water yielding hydroxylated products:

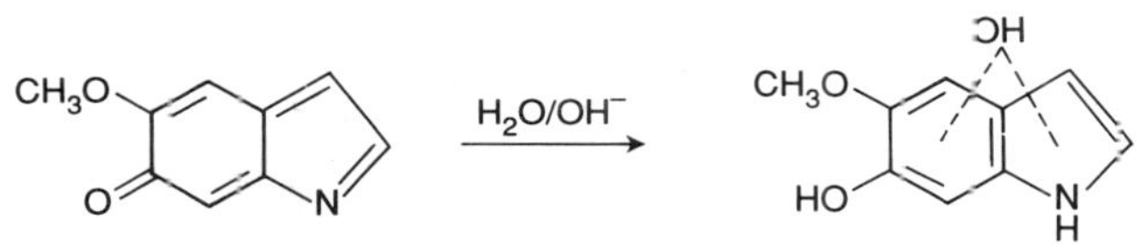

which, like the corresponding species from DHI (Fig. 8.5) may subsequently react with remaining quinonoid cation or quinone methide to give dimeric products, with broad absorption in the 550 nm region detected 0.5 s after the pulse.

For 5H6MI, the decay at 430 nm fitted a single first-order process, which was weakly base catalysed. This process was tentatively attributed to deprotonation of the corresponding quinonoid cation to the corresponding UV-absorbing quinone imine which was stable for at least seconds:

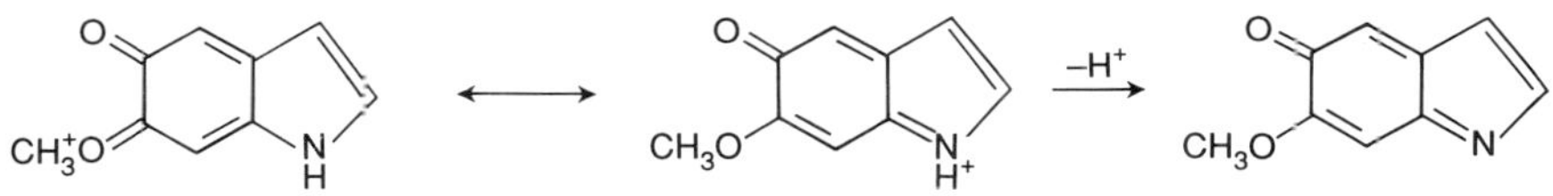

Formation of the quinone methide is not possible in this case.

The contrasting behaviour of the quinonoid cations from 6H5MI and 5H6MI reinforce the probable pivotal role played by a quinone methide (Fig. 8.5) in the melanogenic pathway.

8.4 3,4-Dihydroxymandelic acid

The enzyme tyrosinase is not only responsible for melanogenesis in animals *via* dopa oxidation, but also oxidizes many other catecholamines to reactive quinones from which other polymeric pigments are derived, for example, in the darkening of decaying fruit.

Sclerotization in insects may also involve enzymic generation of quinones. Subsequent tanning mechanisms may then involve nuclear coupling of quinone with cuticular protein to give a protein–catechol adduct, which on oxidation to quinone and further coupling to protein results in crosslinking (Lipke *et al.* 1983). Also, however, evidence has been accumulating for an alternative crosslinking mechanism occurring via quinone methides, involving the β-carbon of the catecholamine side-chain (S.O. Anderson 1979). Such tanning mechanisms may involve nucleophilic addition of protein to the C=C exocyclic double bond of the quinone methide, giving a catecholamine with the protein attached to the side-chain β-carbon. Oxidation of this catechol to another quinone methide and analogous addition to further protein leads to crosslinking.

Sugumaran (1986) has demonstrated that mushroom tyrosinase oxidizes a large number of catechols to the corresponding quinones with the apparent exception of 3,4-dihydroxymandelic acid, which undergoes a novel decarboxylation yielding 3,4-dihydroxybenzaldehyde. This substrate failed to give the usual 400–430 nm *ortho*-quinone peak on interacting with tyrosinase. Since a cuticular enzyme believed to catalyse the formation of quinone methides from 4-alkylcatechols (Sugumaran and Lipke 1983*a*, *b*) also catalyses the oxidative decarboxylation of 3,4-dihydroxymandelic acid (Sugumaran 1986), a quinone methide was suggested as a key intermediate common to both types of enzyme:

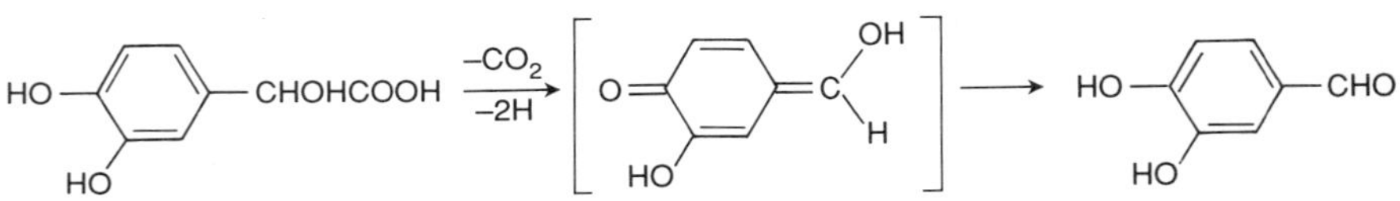

This would therefore represent a new type of activity for tyrosinase.

In view of this result, and the importance of fully understanding the catalytic properties of tyrosinase, pulse radiolysis was used to investigate the mechanism of oxidation of 3,4-dihydroxymandelic acid (Bouheroum *et al.* 1989). The aim was to seek direct evidence for a *short-lived* quinone derived from 3,4-dihydroxymandelic acid, and to establish whether or not this could be an intermediate in the overall decarboxylation mechanism.

Azide radical induced one-electron oxidation of 3,4-dihydroxymandelic acid led to the corresponding typical semiquinone absorption, λ_{max} 310 nm. This decayed over a few milliseconds by disproportionation into a transient absorption with a pronounced shoulder at ~400 nm attributed to 3,4-mandeloquinone:

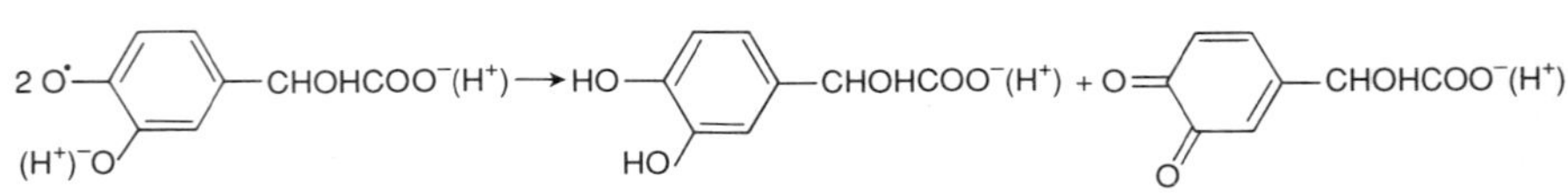

As would be expected of a quinone, the 400 nm species was reactive towards the reductants, cysteine ($k = 1.2 \times 10^6\ dm^3\ mol^{-1}\ s^{-1}$) and ascorbic acid ($k = 8.4 \times 10^5\ dm^3\ mol^{-1}\ s^{-1}$). In the absence of reducing agents, over hundreds of milliseconds, the absorption at 400 nm decayed completely with a concurrent build-up of a strong absorption peaking at 310 nm. The rate of this reaction was independent of pH in the range 6 to 8, and independent of parent concentration in the range 10^{-3} to $10^{-4}\ mol\ dm^{-3}$. The absorption profile of the species with maximum at 310 nm was consistent with it being 3,4-dihydroxybenzaldehyde, as found (Sugumaran 1986) following tyrosinase-catalysed oxidation of 3,4-dihydroxymandelic acid. Consistent with this, the major new component of the irradiated mixture was chromatographically identical to 3,4-dihydroxybenzaldehyde.

The pulse radiolysis results thus clearly show that oxidation of 3,4-dihydroxymandelic acid can lead to a conventional quinone, although in this case it is relatively unstable, decarboxylating to 3,4-dihydroxybenzaldehyde with a half-life of 0.25 s:

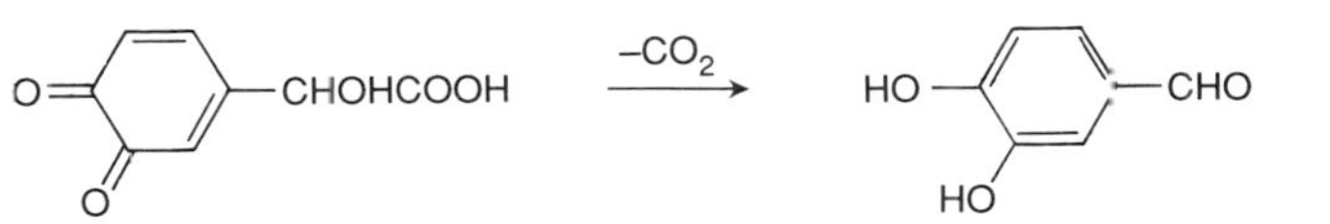

The action of tyrosinase on 3,4-dihydroxymandelic acid is therefore unlikely to represent a new type of activity for this enzyme.

Similar conclusions regarding the intermediacy of a short-lived quinone in the oxidative decarboxylation of 3,4-dihydroxymandelic acid were reached simultaneously by other workers using different techniques including stopped flow and electrochemical oxidation (Cabanes *et al.* 1988; Ortiz *et al.* 1988).

8.5 Therapeutic aspects: 4-hydroxyanisole

One approach towards a rational chemotherapy for malignant melanoma uses the melanogenesis pathway as a means of targeting cytotoxic therapy. Pulse irradiation techniques have been used to provide evidence regarding aspects of the mechanism involved.

Subversion of the chemistry of melanogenesis may be a means of treating malignant melanoma. Thus Riley (1984, 1985) has suggested that the depigmenting agent 4-hydroxyanisole can cause a build-up of high concentrations of toxic intermediates, for example, *ortho*-quinones. As already illustrated for dopaquinone and indolequinones, *ortho*-quinones are very reactive substances, and the generation of such compounds as intermediate metabolites in melanogenesis represents a potential hazard to the well-being of the cells in which the process takes place. A variety of phenolic compounds may be toxic to melanocytes due to the presence in these cells of tyrosinase which converts the non-toxic phenols to potentially damaging quinones. Elevated tyrosinase levels are, in fact, found in the tumour cells and serum of patients with disseminated malignant melanoma.

The phenol 4-hydroxyanisole is one such compound which can be oxidized by tyrosinase to a toxic quinone:

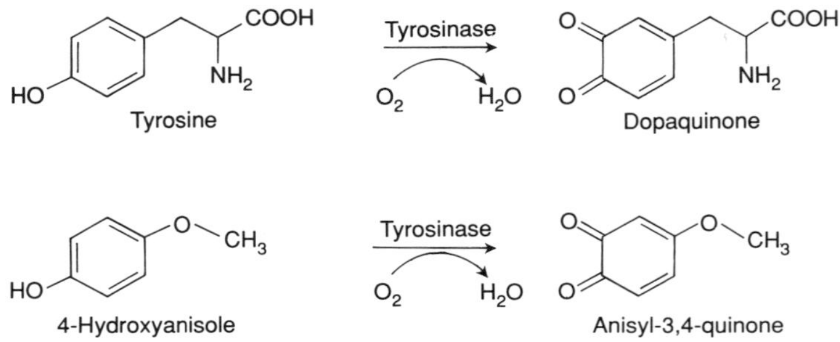

4-Hydroxyanisole is, in fact, a better substrate for tyrosinase than tyrosine itself. *In vitro* experiments have shown 4-hydroxyanisole to be readily oxidized by mushroom tyrosinase to the 3,4-*ortho*-quinone of anisole and the ESR evidence of Nilges *et al.* (1984) shows that, under certain conditions, the corresponding anisyl-3,4-semiquinone radical anion is also formed. It is not known for certain if cytotoxicity is due to the formation of this radical, possibly leading to lipid peroxidation via superoxide formation and metal-catalysed formation of $OH^{\cdot}$, or to further reactions of the quinone, for example, addition reactions with —SH compounds (Takamura *et al.* 1979) or other nucleophiles.

There has been controversy regarding the reactions of anisyl-3,4-semiquinone with oxygen. Thus, Nilges and Swartz (1984) have demonstrated that *in vitro* the semiquinone radical ESR signal intensity is inversely related to the oxygen concentration, and interpret this finding in terms of a cycle of reduction of anisyl-3,4-quinone to the corresponding hydroquinone, with enzymatic reoxidation of the hydroquinone back to the quinone, rather than reaction of the semiquinone with oxygen. Assuming that the semiquinone is formed by reverse dismutation of the anisyl-3,4-quinone and the corresponding hydroquinone, the steady-state concentration of semiquinone will remain low while oxygen is available because the hydroquinone concentration will be small.

However, the influence of tyrosinase on the disappearance of the unstable anisyl-3,4-quinone has been examined by Naish (1985). While the rate of removal of the quinone is significantly diminished in the presence of tyrosinase, this effect cannot be due to reoxidation of the corresponding hydroquinone because it was also observed in a nitrogen atmosphere. Thus the role of oxygen as a regulator is somewhat obscure. The semiquinone of anisyl-3,4-quinone was therefore generated using pulsed radiation techniques in order to provide kinetic evidence regarding its reaction with oxygen, and the reverse reaction of the superoxide anion radical with anisyl-3,4-quinone.

8.5.1 *Oxidation of anisyl-3,4-hydroquinone (3,4-dihydroxyanisole)*

One-electron oxidation of anisyl-3,4-hydroquinone using pulse radiolysis ($N_3^{\cdot}$) (Cooksey *et al.* 1987) or 266 nm laser flash photolysis (Land 1988) led to the anisyl-3,4-semiquinone anion radical. The decay of the semiquinone, formed via ionizing radiation or light, followed second-order kinetics, $2k = 2.5 \times 10^8\ dm^3\ mol^{-1}\ s^{-1}$, due to the disproportionation:

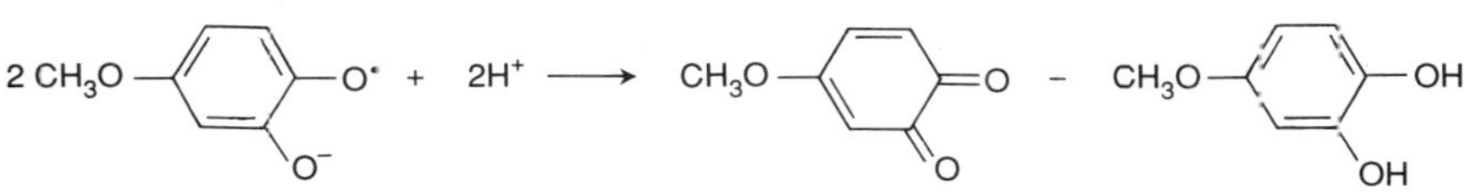

The quinone, monitored at its λ_{max} of 420 nm, was stable for several minutes, although over longer periods it progressively polymerizes.

8.5.2 *Interaction of anisyl-3,4-semiquinone with oxygen and lipid*

The second-order decay of the semiquinone was found to be unaffected by the addition of oxygen (Cooksey *et al.* 1987), consistent with a forward rate constant of $< 10^5\ dm^3\ mol^{-1}\ s^{-1}$ for the reaction:

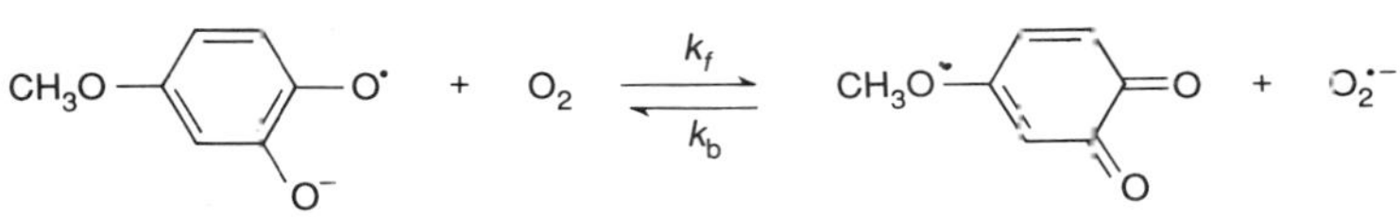

The reverse of this reaction was investigated by studying solutions of anisyl-3,4-quinone in oxygen-saturated formate + buffer. Under such conditions (see Chapter 3), all the primary radicals of water radiolysis are converted into $O_2^{\cdot-}$. Solutions of anisyl-3,4-quinone were obtained by bubbling oxygen through a solution of anisyl-3,4-hydroquinone for short periods of time. Pulse radiolysis of this oxygen-bubbled solution resulted in a transient difference spectrum which, after correction for quinone depletion, was identical to that of the semiquinone. Thus, $O_2^{\cdot-}$ reacts exclusively with the quinone component of the solution ($k = 8.7 \times 10^9\ dm^3\ mol^{-1}\ s^{-1}$) to yield the semiquinone.

The rapid reaction of $O_2^{\cdot-}$ with the quinone is consistent with the one-electron reduction potential of anisyl-3,4-quinone being much more positive than the one-electron reduction potential of oxygen (−155 mV). The estimates of the forward and backward rate constants led (see Chapter 3) to an equilibrium constant limit and hence one-electron potential limit for E_7^1 (anisyl-3, 4-quinone/anisyl-3,4-semiquinone) of > +77 mV. Since 1,4-benzoquinone (E_7^1 (1,4-benzoquinone/1,4-benzosemiquinone) = +99 mV) also had no detectable effect on the decay of anisyl-3,4-semiquinone, it seems likely that the one-electron reduction potential of anisyl-3,4-quinone is much more positive than 0.1 V (versus NHE).

These studies thus support the conclusion that the cytotoxicity mechanism associated with the administration of 4-hydroxyanisole in the treatment of malignant melanoma does not involve the generation of $O_2^{\cdot-}$ from reaction of anisyl-3,4-semiquinone with oxygen, with possible subsequent metal-catalysed production of $OH^{\cdot}$, leading to lipid peroxidation.

An alternate possibility that the semiquinone radical initiates lipid peroxidation by hydrogen abstraction was examined using *trans*-2-butenoic acid as a water-soluble model of unsaturated lipid (Land *et al.* 1990). No such reaction could be detected by pulse radiolysis ($k < 2 \times 10^5\,dm^3\,mol^{-1}\,s^{-1}$). This lack of reactivity implies that in aqueous conditions hydrogen absraction from acyl chains is very inefficient, and so semiquinone-mediated reactions affecting membrane integrity are unlikely to feature prominently among the mechanisms of toxicity of 4-hydroxyanisole oxidation products.

8.5.3 *Addition and redox exchange reactions of anisyl-3,4-quinone*

It seems more likely that the mechanism of the cytotoxic action of 4-hydroxyanisole involves reactions of the derived anisyl-3,4-quinone. *In vivo*, the quinone might be anticipated to undergo nucleophilic addition reactions with —SH and —NH_2 groups, or redox exchanges, e.g. with ascorbate. Table 8.1 lists the rate constants obtained by pulse radiolysis (Land *et al.* 1990) for the thiols: cysteine, glutathione, and dithiothreitol, and with ascorbic acid. No reaction, however, could be detected with the amines: arginine and glutamine, although the former is known to react slowly with anisyl-3,4-quinone, since addition of arginine to a mixture of tyrosinase and 4-hydroxyanisole in PBS buffer leads to an increase in the product quinone decay rate monitored over time scales of minutes to hours (Holden *et al.* 1984).

In view of the proximity of the pK_a values for the thiol equilibria RSH/RS^- to the pH value (7.0) at which the measurements referred to in Table 8.1 were carried out, there is every expectation that the thiol rate constants would be pH sensitive since only RS^- may be reactive.

The rate constants obtained for the reaction of anisyl-3,4-quinone with thiols are consistent with the proposal that one of the major cytotoxic mechanisms could be quinone addition to cellular thiol-containing proteins. There is indeed

Table 8.1 Rate constants for reactions of 4-methoxy-*ortho*-benzoquinone with various substrates[a]

Substrate	Rate constant ($dm^3\,mol^{-1}\,s^{-1}$)
Cysteine	3.5×10^5
Glutathione	3.5×10^5
Dithiothreitol	3.5×10^5
Arginine	<1
Glutamine	<1
Ascorbic acid	1.0×10^4

[a] In addition to substrate, all solutions contained $1–2 \times 10^{-3}\,mol\,dm^{-3}$ 3,4-dihydroxyanisole, $5 \times 10^{-2}\,mol\,dm^{-3}$ NaN_3, and $10^{-1}\,mol\,dm^{-3}$ phosphate buffer pH 7.0, and were flushed with N_2O (Land *et al.* 1990).

evidence that certain crucial protein components of the plasma membrane, such as calcium pumps (Orrenius 1985), are potential targets for reaction with quinones. Redox exchange reactions of the quinone could also lead to antioxidant depletion.

8.6 Summary

As discussed in other chapters, the techniques of flash photolysis and pulse radiolysis have made many significant contributions to knowledge of free radicals and excited states involved in biology and medicine. Less well known are the applications of these techniques to studies of other types of unstable chemicals. A variety of such species, in particular *ortho*-quinones and derived products, arise in the complex processes of melanin formation, and kinetic analysis via pulsed irradiation techniques has provided one of the first direct proofs of the accuracy of the commonly accepted scheme for some of the early chemical steps of melanin formation in mammals (see Fig. 8.2).

Perhaps more importantly, pulse radiolysis has allowed, for the first time, details of the molecular mechanism of the conversion of dihydroxyindole to melanin polymer to be at least partially understood. Nevertheless, further work is required to fully elucidate the role of quinone methides and quinone imines in these processes, and to examine the possible role of 5,6-indolequinone–cysteine reactions in the production of phaeomelanin.

Whilst the semiquinone (and hence quinone) of dopa can be prepared equally well by N_3-induced one-electron oxidation or photoionization, the corresponding semiquinone of 5-*S*-cysteinyldopa can only be prepared by N_3 oxidation. This is because photolysis of 5-*S*-cysteinyldopa does not lead to photoionization, but rather S—C bond homolysis. This difference between the photochemistry of dopa and 5-*S*-cysteinyldopa may well be a factor in the increased photosensitivity of phaeomelanic skin as opposed to its eumelanic counterpart.

The enzyme tyrosinase is responsible for the *in vivo* oxidation of dopa to dopaquinone, the ensuing reactions of which lead to melanin pigments. It has been proposed that this enzyme also possesses a different type of activity as exemplified by the oxidative decarboxylation of 3,4-dihydroxymandelic acid to 3,4-dihydroxybenzaldehyde, which appeared to proceed via a quinone methide rather than a conventional *ortho*-quinone intermediate. Pulse radiolysis has, however, shown this not to be the case, and that the *ortho*-quinone from 3,4-dihydroxymandelic acid simply decays rather quickly, $t_{1/2} \sim 0.25$ s.

Another unstable *ortho*-quinone is anisyl-3,4-quinone formed in the metabolic activation of 4-hydroxyanisole, a possible melanocytotoxic drug under current assessment as a treatment of malignant melanoma. This molecule acts as an analogue substrate for tyrosinase and is converted to anisyl-3,4-quinone from which semiquinones are known to form. Since the semiquinone ESR signal is oxygen dependent, it was thought possible that the toxic action of 4-hydroxyanisole could be the result of active oxygen species generated in the cell from the semiquinone. However, pulse radiolysis studies have shown that anisyl-3,4-semiquinone is unreactive towards oxygen although the reverse reaction of $O_2^{\cdot -}$ with anisyl-3,4-quinone is very rapid. Since anisyl-3,4-semiquinone is also unreactive towards the water-soluble lipid model *trans*-2-butenoic acid, it is unlikely that inhibition of lipid peroxidation by the semiquinone is a major source of cytotoxicity. A more likely toxicity pathway appears to be covalent addition reactions of anisyl-3,4-quinone with cellular nucleophiles, especially thiols, and/or redox exchange reactions of the quinone leading to antioxidant depletion.

4-*n*-Propoxyphenol possesses five-times increased tyrosinase-dependent *in vitro* cytotoxicity towards rat epithelial cells as compared with 4-hydroxyanisole, and the reactivities of the corresponding quinone and semiquinone metabolites are therefore being examined. For the reactions investigated so far, the semiquinone and quinones derived from 4-*n*-propoxyphenol and 4-hydroxyanisole behave almost identically (Cooksey *et al.* 1992). Thus the reasons for the differences in the cytotoxicities *in vitro* of these two compounds must be sought elsewhere.

9

Photodermatology

'I am dark because the sun has scorched me'

Song of Solomon, **1**, 6.

9.1 Introduction

In the first section of this chapter, a description of skin structure and of its optical properties indicates the regions and the depth to which the various wavelengths of light penetrate. Then the biological responses induced by visible light and UV in skin are presented. They have two aspects: firstly relating to the molecular (direct or sensitized) photodamage to the main molecules present in skin (DNA, enzymes, proteins, and lipids); secondly relating to the medical aspects of photosensitization produced by exogenous compounds. Photosensitization can be either harmful, as in the photosensitizing side effects of certain drugs, such as phenothiazines used as tranquillizers and non-steroidal anti-inflammatory drugs, or beneficial, as in the photochemotherapy of psoriasis by furocoumarins.

9.2 Structure of skin

In humans the skin weighs about 2 kg in a 70 kg person (about 3 per cent of the body weight), with a surface area of about 2 m^2. One of the major functions of the skin is to provide a physical barrier which protects the body from the hazards presented by the environment. An important such hazard is sunlight.

Skin is composed of several regions, a schematic of the general structure being given in Fig. 9.1. The outermost part is the epidermis which itself is made up of a series of strata, the tough surface layer being the stratum corneum. This stratum corneum is composed of dead cells and provides protection against water loss and surface abrasion, and attenuates UV radiation before it reaches

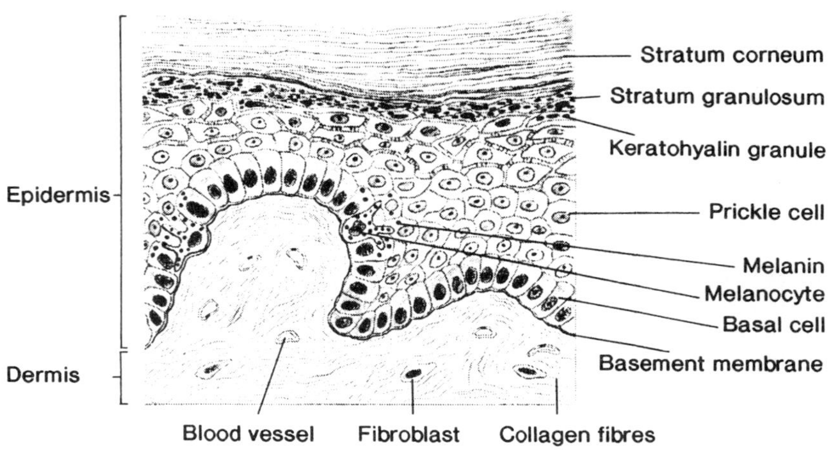

Fig. 9.1 Diagrammatic cross-section of skin (taken from Parrish 1982).

living cells below. The stratum granulosum immediately underneath is so named because of the presence in the cytoplasm of its cells of numerous granules of keratohyalin of varying size. Beneath this granular layer are the prickle cells with their spiny appearance. Epidermal cells in general are known as keratinocytes because they produce the protein keratin. Keratinocytes stem from a single layer of germinative basal cells situated immediately above the basement membrane which separates the epidermis from the dermis. As the basal cells divide, the daughter cells migrate towards the surface. Interspersed amongst the basal cells are melanocytes, the specialized cells which produce the pigment melanin (see Section 8.1) which provides the varying shades of colour seen in human skin. Racial differences in skin colour are mostly due to variations in the productivity of melanocytes.

The epidermis is attached to, and nurtured by, the dermis which is made up of a network of collagen, elastin, and reticular fibres providing a matrix for blood vessels, lymphatics, and nerve fibres. The mean blood flow is much greater than necessary for skin cell nutrition because cutaneous blood flow serves as a heat regulator for the entire organism. Prolonged exposure of the skin to UV makes it possible to irradiate a large proportion of blood cells as they flow through superficial skin vessels.

9.3 Optical properties of skin and solar UV spectrum

Solar UV radiation of wavelengths between 290 and 320 nm (UVB) reaches the earth in relatively small quantities (Fig. 9.2) but is very efficient at causing sunburn of human skin. Radiation of wavelength longer than 320 nm is, on the other hand, relatively inefficient at causing sunburn. Thus, although much greater amounts of UVA are present in the solar spectrum than UVB (Fig. 9.2), the amount of UVA energy required to produce an effect is orders of magnitude higher than for the UVB region.

Regarding the depth to which the various wavelengths of light penetrate Caucasoid skin, it is found that short-wave ultraviolet light (UVC and UVB) is almost completely absorbed by the stratum corneum and the rest of the epidermis. With UVA, visible, and near IR light, on the other hand, considerable portions penetrate the dermis and are absorbed here primarily by the blood vessels (Fig. 9.3).

When light strikes the skin, part is back scattered, part is absorbed in various layers and part is transmitted inward to the successive layers of cells. The term remittance refers to the fraction of incident light which returns from the skin, and transmittance to the fraction which passes through it. Remitted light includes the surface reflectance plus that light which enters the skin and is subsequently back scattered. The surface reflectance component is typically 5 to 7 per cent for perpendicularly incident radiation, is nearly constant for all optical wavelengths and is the same for normal black as for normal white skin.

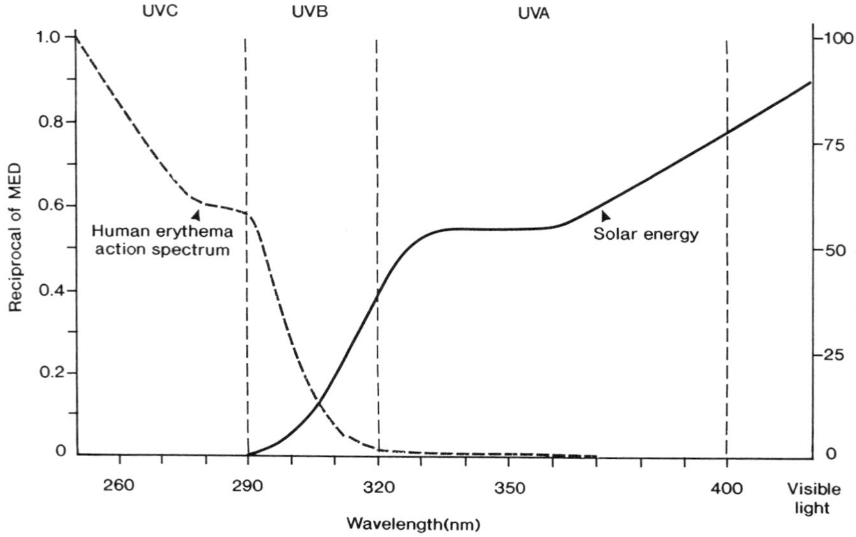

Fig. 9.2 Diagrammatic representation of human erythema action spectrum and terrestrial solar spectrum (MED = minimum erythemal dose = lowest exposure dose that results in erythema) (taken from Parrish 1982).

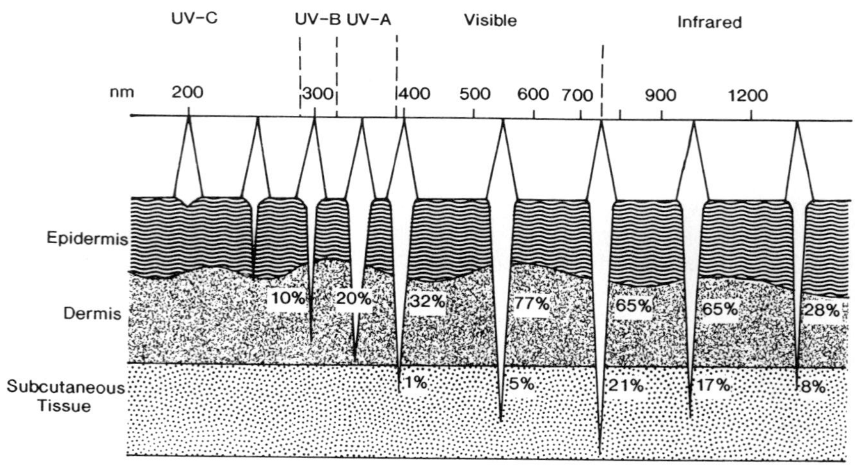

Fig. 9.3 Differential transmission in various layers of skin (adapted from Ippen 1969).

The increased total spectral remittance of fair Caucasian skin as compared with dark Negro skin in the near UV, visible and near IR region results from the smaller amounts of melanin in Caucasians. The remittance of Caucasoid skin in this spectral region is due to light which has traversed the epidermis, has been back-scattered by the dermis, and has returned again through the epidermis.

9.4 Molecular mechanisms of skin damage

The main chromophores involved in the primary processes of skin photo-damage by solar irradiation starting at 290 nm (Fig. 9.2) are:

(1) *in the UV*, the nucleotides and the aromatic amino acids, tryptophan and tyrosine which have respectively, at 290 nm, molar absorption coefficients of 1500, 4000, and 82 $dm^3 mol^{-1} cm^{-1}$, but do not absorb above 320 nm, and NADH absorbing up to 400 nm;

(2) *in the visible*, the porphyrins, bilirubin, flavins, quinones, and melanin.

9.4.1 *Photodamage to DNA*

In the case of the nucleotides, components of nucleic acids absorbing exclusively in the UVB (290–320 nm), the main significant photoproducts are considered to be the (5–5, 6–6) cyclobutane pyrimidine dimers (CPD) and the (6–4) pyrimidine–pyrimidone dimers (Cleaver *et al.* 1988; Mitchell 1988). The (6–4) photoproducts are more important in *E. coli* and the cyclobutane dimers more important in mammalian cells (Brash 1988). Since the discovery of both these forms of DNA photodamage, they have been considered to be the major photobiological lesions responsible for solar carcinogenesis.

However, their importance in biological inactivation, mutagenesis, and carcinogenesis has probably been overemphasized. Using published data for UV transmission of human skin and solar spectral irradiance, Tyrrell and Pidoux (1987) have estimated the relative biological effectiveness of the UVB and UVA regions of sunlight for cytotoxicity at the basal layer of the epidermis, and shown that UVA erythema can account for up to 60–80 per cent of the cytotoxic effectiveness of the combined UVA and UVB regions. There is little evidence that a DNA target is involved in these cytotoxic effects since DNA does not absorb UVA.

9.4.2 *Photodamage to amino acids and proteins*

Increasingly, new data are pointing to proteins as the main target:

1. Proteins *in vivo* are in contact with many other molecules including DNA and a number of photochemical heteroadducts to the DNA, leading to

DNA–protein crosslinks (DPC), can be formed in cells (K.C. Smith 1976*a*, *b*). *In vitro*, it has been shown that 15 amino acids can photoreact with polynucleotides (Shetlar *et al.* 1985). In epidermal DNA, DPC have also been found (Kornhauser 1976). The action spectrum for formation of the DPC in human teratocarcinoma cells is similar to that of the DNA single-strand breaks up to about 320 nm, and has a second peak maximum three orders of magnitude lower at about 405 nm (J.G. Peak *et al.* 1985; M. J. Peak *et al.* 1985), suggesting the role of porphyrins.

2. The action spectra of other types of UV cell damage are also slightly red shifted when compared to the DNA absorption spectrum (Coohill 1984), suggesting the possible role of proteins. An example of this role is given by Lebedeva *et al.* (1978) for UV-induced isochromatid breaks.
3. Tyrrell and Pidoux (1988) have shown that the presence of endogenous glutathione, a free radical scavenger, in human skin fibroblasts induces resistance to solar ultra violet radiation. The mechanism of this photoprotection indicates that UV acts via radicals and that non-radical DNA damage is not the major lesion. Aromatic amino acids are the likely candidates for the photoproduction of these radicals.
4. Zamansky and Chou (1988) have demonstrated that, in normal diploid human skin fibroblasts, UV light disrupts the cytoskeletal microtubules and microfilaments, suggesting tryptophan as a major chromophore.

Flash photolysis studies of tryptophan and tyrosine in aqueous solution have shown that the primary act is a photoionization producing hydrated electrons and, respectively, tryptophyl and tyrosyl radicals (Grossweiner *et al.* 1963; Baugher and Grossweiner 1978). This photoionization and these radicals are also observed by flash photolysis of proteins and enzymes in aqueous solution (see Chapter 6). These radicals can be responsible for DNA–protein crosslinks, protein–protein crosslinks in collagen and elastin, and oxidation of sulphydryl groups leading to disulphide crosslinks and loss of skin elasticity. They can produce inactivation of protective enzymes, such as superoxide dismutases, catalase, and glutathione peroxidase which defend the cells against the action of superoxide anion and hydrogen peroxide. These radicals can also damage other types of enzymes such as the photolyases which are able to repair the DNA CPD. Photolyases, also called 'photoreactivating enzymes', are able to catalyze the light-dependent (300–600 nm) opening of the cyclobutane ring of pyrimidine dimers in DNA, formed by UVB (290–320 nm). They are found in prokaryotic and eukaryotic organisms (Sutherland 1981). In the case of *E. coli*, the enzyme has been identified as a flavoprotein, and has been studied by flash photolysis (Heelis and Sancar 1986). The tryptophyl and tyrosyl radicals can also cause lipid peroxidation by a chain reaction leading to a loss of the membrane integrity.

9.4.3 *Endogenous photodynamic effects*

Cellular damage in the near UV region is strongly dependent upon the presence of oxygen, indicating the importance of 'photodynamic effects' (Webb and Brown 1979; J.G. Peak *et al.* 1983). According to photochemical investigations, two main types of pathways exist for 'photodynamic action', they usually occur via the triplet states of the photosensitizers, in the presence of oxygen (G.O. Schenck and Koch 1960; Gollnick 1968):

In *Type I* reactions, interaction between the substrate, and the electronically excited (usually a triplet state) sensitizer, leads to electron or hydrogen atom transfer and gives rise to pairs of reactive radicals of both substrate and photosensitizer. Subsequent reactions vary, but, commonly, molecular oxygen is involved to form oxidized products. With a reducing photosensitizer PH and an oxidizing substrate A the reactions would be:

$$\left.\begin{array}{l} {}^3PH + A \rightarrow PH^{\cdot +} + A^{\cdot -} \\ {}^3PH + A \rightarrow P^{\cdot} + AH^{\cdot} \end{array}\right\} + O_2 \rightarrow \text{oxidized products}$$

With an oxidizing photosensitiser P and a reducing substrate RH the reactions would be:

$$\left.\begin{array}{l} {}^3P + RH \rightarrow P^{\cdot -} + RH^{\cdot +} \\ {}^3P + RH \rightarrow PH^{\cdot} + R^{\cdot} \end{array}\right\} + O_2 \rightarrow \text{oxidized products}$$

Reactions of oxygen with the reduced forms of the photosensitizer, or with the substrate, can produce $O_2^{\cdot -}$ and its conjugate acid $HO_2^{\cdot}$, which can react further. Oxidized forms of the sensitizer, or of the substrate, can react with oxygen to give peroxyradicals, and can initiate a free radical chain autoxidation.

Additionally, a direct electron transfer from the excited sensitizer to oxygen can occur:

$${}^{1,3}P + O_2 \rightarrow P^{\cdot +} + O_2^{\cdot -}$$

This can be considered as being also included in Type I reactions.

Type II reactions occur via energy transfer (see also Sections 4.2.1.1 and 12.3). They involve transfer of excitation energy from the triplet excited photosensitizer to ground state triplet oxygen, forming singlet excited oxygen:

$${}^3P + O_2({}^3\Sigma_g^-) \rightarrow P + O_2({}^1\Delta_g)$$

Candidate photodynamic sensitizers for UV irradiation are tryptophan and its photo-oxidized product *N*-formylkynurenine (Walrant and Santus 1974*a*, *b*), melanin precursors (Land *et al.* 1986), nicotinamide adenine dinucleotide (reduced form, NADH) and nicotinamide adenine dinucleotide phosphate (reduced form, NADPH) (Cunningham *et al.* 1985*a*), and, for visible light, porphyrins, flavins, and quinones (Jagger 1983).

Tryptophan and related molecules are probably among the most efficient photosensitizers. Tryptophan, *N*-formylkynurenine, and the tryptophan-rich enzyme, lyzosyme, are able to photoproduce H_2O_2, which is lethal to cells (McCormick *et al.* 1976), probably via photoionization leading to $O_2^{\cdot -}$. The quantum yield for the monophotonic photoionization of trytophan, 7.5×10^{-2}, has been measured by flash photolysis (Bazin *et al.* 1983). The photoionization of NADH, also studied by flash photolysis, is exclusively detected after biphotonic excitation (Lindqvist *et al.* 1985). However, J.G. Peak *et al.* (1984) have described photosensitized DNA breakage by continuous 334 nm excitation of NADH. The photosensitized production of the superoxide anion from NADH and NADPH has been observed with a rather low quantum yield in the range of 10^{-7}–10^{-8} for 290 nm excitation and of 10^{-9} at 334, 365, and 405 nm (Cunningham *et al.* 1985*a,b*). Czochralska *et al.* (1984) have also observed hydrogen peroxide, a dismutation product of the superoxide anion, produced in NADH and NADPH aqueous solutions by near-UV irradiation.

Quinones so far do not appear to be implicated in photosensitization or in direct photoeffects with cellular consequences. In contrast to benzoquinone and duroquinone, the biological quinone ubiquinone has a very low ϕ_T of 0.04 (Bensasson *et al.* 1972).

The photoproduction of melanin, the skin pigment which is a biopolymer of quinone and hydroquinone monomer units, involves the formation of catecholic precursors, such as dopa and 5-*S*-cysteinyldopa, which can be efficient photosensitizers. These melanin precursors, via photoionization and photohomolysis, give solvated electrons and alanyl, phenoxyl, and thiyl radicals (Land *et al.* 1986), leading to skin damage including lipid peroxidation and DNA damage (Miranda *et al.* 1984; Koch and Chedekel 1986), (for more details see Chapter 8).

In the presence of oxygen, flavins (for example; lumiflavin, riboflavin (vitamin B_2), and flavin mono nucleotide) are efficient photosensitizers able to oxidize amino acids, proteins, DNA and nucleotides, and lipids (Heelis 1982) via a Type II or a Type I mechanism, depending upon the substrate and the oxygen concentration (Heelis *et al.* 1978).

Most porphyrins in living systems contain a paramagnetic iron atom, such as the haem of haemoglobin (ferric protoporphyrin IX), and are not photosensitizers due to very short triplet lifetimes. Free base porphyrins might be responsible for the peak at 405 nm of the action spectrum observed for DNA–protein crosslinks in human cells (Peak *et al.* 1985*a,b*). Natural skin photosensitivity by endogenous free base uro-, copro-or proto-porphyrins occurs in certain porphyrias, hereditary human diseases due to an excess of these pigments in the blood up to 100 times their normal concentration (see Section 9.8). Singlet oxygen is regarded as being involved in porphyrin photosensitization as a number of photo-oxidation end products have been identified as specific to an attack by this activated form of oxygen. Examples include the formation of the specific cholesterol allylic hydroperoxide (Lamola *et al.* 1973; J.P.Thomas *et al.* 1987) and of methionine sulphoxide (Jori *et al.* 1969). Porphyrins also undergo weak monophotonic photoionization (Bonnett *et al.* 1982).

9.4.4 *Photoinduced lipid peroxidation*

Skin lipid peroxidation can be induced by UV (Meffert *et al.* 1976). The target molecules are the polyunsaturated fatty acids (PUFA), essential components of cell membranes. They can be attacked by the radicals R· photoproduced via the reactions described above (Sections 9.4.2–3) and/or by singlet oxygen. The PUFA radical thus formed reacts with oxygen to yield $PUFAO_2^{\cdot}$:

$$R^{\cdot} + PUFA \rightarrow RH + PUFA^{\cdot}$$

$$PUFA^{\cdot} + O_2 \rightarrow PUFAOO^{\cdot}$$

This peroxy radical propagates the damage by abstracting a hydrogen atom from another PUFA:

$$PUFAOO^{\cdot} + PUFA \rightarrow PUFAOOH + \dot{P}UFA$$

Such chain reactions have been observed in UV-irradiated skin, particularly at the level of the lysosomal membranes, by B.E. Johnson and Daniels (1969). *In vivo*, PUFAOOH hydroperoxides are unstable as they react with biological metal complexes to produce alkoxy and peroxy radicals contributing to an enhancement of lipid peroxidation:

$$PUFAOOH + M^{n+} \rightarrow PUFAO^{\cdot} + M^{(n+1)+} + OH^-$$

$$PUFAOOH + M^{(n+1)+} \rightarrow PUFAOO^{\cdot} + M^{n+} + H^+$$

One of the products of PUFA peroxidation is a fluorescent brownish pigment, lipofuscin, a granulized cellular material which accumulates in lysosome vacuoles.

The cytotoxic consequences of oxidizing lipid radicals are numerous, including loss of fluidity (Hicks and Gebicki 1978) and oxidation of proteins and nucleic acids (Borg and Schaich 1984; Summerfield and Tappel 1984; Ueda *et al.* 1985; Akasaka 1986) leading to cataract formation (Babizaev 1985), mutagenicity (Pietronigro *et al.* 1977), and cancer (McBrien and Slater 1982).

9.4.5 *Photoageing and photocarcinogenesis*

Repeated photodamage of the DNA and proteins of the skin will lead to a number of histopathological changes called photoageing or dermatoheliosis, and can also lead to cancer by cumulative effects. Sunlight has been called 'the most ubiquitous carcinogen of all' (Forbes *et al.* 1977). Skin cancer was first related to sunlight around 1910 by the dermatologist William Dubreuilh who linked the high incidence of skin cancer among Bordeaux vineyard workers to overexposure to the sun (Garfield 1979).

UVB (290–320 nm) induces direct DNA photodamage, and has been considered in the past as the main cause of skin photoageing and carcinogenesis. This was based on the 10^3–10^4 times greater UVB efficacy than that of UVA at inducing acute skin effects and photocarcinogenesis (C. A. Cole *et al.* 1986). Since the human use of tanning equipment delivering large doses of UVA over short times, UVA hazards have been extensively studied, especially in the skin. Recent results have shown that human skin tanning, whether with UVB or UVA, results in a carcinogenic risk of a similar order of magnitude (Roza *et al.* 1989).

In addition to direct photodamage of DNA and protein, another process leading to photoageing and photocarcinogenesis is skin lipid peroxidation induced by UV, as described in Section 9.4.4. The importance of this process is emphasized by the observation (Black 1987*a,b*) that dietary lipids and lipophilic antioxidants can modify the photocarcinogenic response of skin in hairless mice.

Photodamage is increased in the case of skin disorders manifest by decreased or absent melanin, e.g. vitiligo or albinism, or by inadequate DNA repair as in *Xeroderma Pigmentosum*, and can lead to an enhanced risk of skin cancer.

9.4.6 *Exogeneous photosensitization, phototoxicity, and photoallergy*

In these processes, normally harmless UVA and visible doses of radiation are absorbed by 'foreign' molecules present at non-damaging concentrations, which trigger inflammatory reactions. These inflammatory tissue reactions (e.g. erythema, oedema) are included in the broad clinical category of 'phototoxicity'. The exogenous photosensitizers can be plant products (including furocoumarins present in figs), perfumes (including 5-methoxypsoralen present in 'Eau de Cologne'), dyestuffs (including thiazines, anthraquinones) or drugs (including tranquillizers such as phenothiazines or anti-inflammatory compounds such as piroxicam). Some examples of phototoxic drugs are described below in more detail. They are usually photodynamic sensitizers acting via Type I or Type II reactions. In the case of furocoumarins (psoralens), another type of reaction occurs, which does not require the presence of oxygen. It is a concerted reaction between the double bond of a furocoumarin and the double bond of a pyrimidine base, leading to the formation of a cyclobutane ring (see Section 9.7).

In the less common situation where the immune system is activated by photosensitization, clinicians use the term photoallergy which is characterized by a histopathology similar to that of contact allergic dermatitis.

The screening of potential phototoxic drugs can be performed by the photohaemolysis method, which uses red blood cell UV irradiation in the presence of the drug (Hetherington and Johnson 1984). The phototoxic damage is determined via a quantitative spectrophotometric measurement of the concentration of haemoglobin released by the haemolysis. However, an inert drug can be transformed *in vivo* into a phototoxic molecule, thus *in vitro* studies are insufficient and must always be complemented by metabolic and pharmacokinetic studies.

It is very important to emphasize that drugs, which cause photosensitized oxidative damage to the skin, will very likely cause similar damage to the eye (J.E. Roberts 1984), mainly to the retina which is irrigated by blood and also to the lens where the oxygen pressure is lower but can reach 72 mm Hg (Kwan *et al.* 1971).

9.5 **Inflammatory responses: erythema, tanning, and thickening**

The clinical inflammatory responses of skin to UV exposure sufficient to produce observable changes are:

1. *Erythema*, often called sunburn, which is the most visible and best-known component of UV inflammation. UV erythema (redness) is caused by vasodilation and consequent increased blood content in the dermis.
2. *Tanning*, which consists of two types: (i) immediate tanning (IT), otherwise known as immediate pigment darkening (IPD) or immediate pigmentation; and (ii) delayed tanning (DT), otherwise known as delayed pigmentation or true melanogenesis. IT, which can be induced by wavelengths from 320 to 500 nm, may begin immediately and fade within seconds to minutes, or may persist longer with very high doses or after prolonged exposure. DT results in the familiar tan induced by repeated sun exposure in outdoor workers and sunbathers. The stimulus for induction of DT is not known, but is possibly related to UV injury to melanocytes or keratinocytes (see Chapter 8).
3. *Thickening*. After UV injury there is an initial depression in many aspects of cellular activity, a subsequent increase towards normal, and then above-normal cell replication during repair. The increase in activity persists leading to increased skin thickness until, in the absence of further irradiation, there is a gradual reduction to normal replication and thickness.

9.6 **Drugs inducing photosensitizing side effects**

9.6.1 ***Phenothiazine psychiatric drugs***

Many psychiatric drugs are derivatives of phenothiazine, chlorpromazine (CLP) having been the first major psychiatric agent. Long-term, high dose chlorpromazine therapy leads to a number of side effects which include photosensitization of the skin and eye tissues. Such phototoxic and photoallergic side effects have been attributed to the formation of the radical cation ($CLP^{\cdot +}$) produced by ionization of an excited state of CLP. It has also been

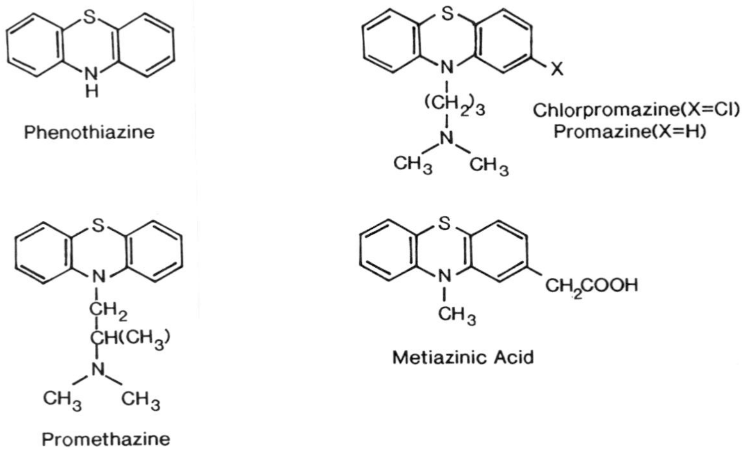

Fig. 9.4 Structures of phenothiazine and derivatives.

suggested that the radical cations may be important intermediates in the biochemical action of these drugs. The structure of phenothiazine and several derivatives that have been studied by pulsed irradiation methods are given in Fig. 9.4.

The triplet state of chlorpromazine (λ_{max} 460 nm) has been characterized in a number of solvents including ethanol (Iwaoka and Kondo 1974), isopropanol (A.K. Davies *et al.* 1976), glycerol, dioxane, acetonitrile (Iwaoka and Kondo 1977), and water (Navaratnam *et al.* 1978). Evidence for the formation of promazine and chlorpromazine triplet states in isopropanol was provided by energy transfer experiments with β-carotene, both drugs sensitizing intense absorption at 530 nm due to the β-carotene triplet (A.K. Davies *et al.* 1976).

The quantum yield for production of triplet chlorpromazine decreases considerably with increase in solvent polarity, the yield falling sixfold in passing from isopropanol to water where its yield has been established as $(8.0 \pm 1.0) \times 10^{-2}$ for 347 nm excitation (Navaratnam *et al.* 1978).

Oxygen efficiently quenches the triplet state of chlorpromazine and this reaction has been suggested to be a factor contributing towards the phototoxicity of this drug (A.K. Davies *et al.* 1976). Since from indirect measurements there does not appear to be agreement as to whether or not $^1O_2$ is produced in this reaction (Piette *et al.* 1986), direct measurements of any 1270 nm $^1O_2$ phosphorescence following laser flash excitation of chlorpromazine would be useful.

The initial products observed after flash photolysis of CLP include $CLP^{\cdot +}$ and e^-_{aq} as well as ^3CLP (Navaratnam *et al.* 1978). Photoproduction of $CLP^{\cdot +}$, λ_{max} 525 nm, results from a stepwise biphotonic photoionization process (Buettner *et al.* 1989, Hall *et al.* 1991):

$$CLP \xrightarrow{h\nu_1} {}^1CLP^*$$

$${}^1CLP^* \longrightarrow {}^3CLP$$

$${}^3CLP \xrightarrow{h\nu_2} CLP^{\cdot +} + e^-_{aq}$$

The sensitivity of phenothiazines towards chemical oxidation is well known and their radical cations may be prepared by pulse radiolysis in aqueous solution under oxidative conditions (see A.K. Davies *et al.* 1979, for example). Absolute rate constants have been measured for the reaction of metiazinic acid (Bahnermann *et al.* 1981), chlorpromazine, and promazine (Bahnermann *et al.* 1983*a*) with a variety of electrophilic radicals including $OH^{\cdot}$, $Br_2^{\cdot -}$, $(SCN)_2^{\cdot -}$, $I_2^{\cdot -}$, $N_3^{\cdot}$, Tl^{2+}, $Tl(OH)^{+}$, $(t\text{-}C_4H_9SSC_4H_9)^{\cdot +}$, $(CH_3SSCH_3)^{\cdot +}$, and lipoate$^{\cdot +}$, the values lying in the range 10^8 to 10^{10} dm^3 mol^{-1} s^{-1}. Whereas most of these oxidizing radicals react predominantly, if not exclusively, via simple electron transfer, the reaction between $OH^{\cdot}$ and phenothiazines is more complex (A.K. Davies *et al.* 1979; Bahnermann *et al.* 1983*a*). Here, as well as reacting partially via direct one electron transfer, $OH^{\cdot}$ can add at various positions in the aromatic system as well as abstracting hydrogen atoms. Subsequently, some further radical cation formation occurs via acid-catalysed elimination of OH^- from the $OH^{\cdot}$ adducts. Quantitative measurements at pH 3.5 of an equilibrium between $I_2^{\cdot -}$, promethazine, its reduced cation, and I^- are consistent with a standard one-electron reduction potential of +0.865 V for promethazine (Bahnermann *et al.* 1983b). The highly oxidative nature of phenothiazine radical cations is also illustrated by the ability of the cations of promethazine and metiazinic acid to oxidize ascorbic acid and α-tocopherol (Bahnermann *et al.* 1983*b*). In 1 mol dm^{-3} HCl, however, the direction of electron transfer in the ascorbate reaction is reversed (Pelizzetti *et al.* 1979).

Hydrated electrons generated by photoionization of promazine may attach themselves to either the sulphur atom or its aromatic systems with equal probability (A.K. Davies *et al.* 1979). Addition to the sulphur atom probably leads to immediate cleavage of a sulphur–carbon bond whereas addition to the aromatic system probably produces an anion radical which can undergo an electron transfer reaction with a further promazine molecule to produce the promazine radical cation. With chlorpromazine, e^-_{aq} probably reacts with the aromatic system to eliminate Cl^-.

The stepwise biphotonic photoionization of chlorpromazine, and probably promazine as well (Buettner *et al.* 1989), suggests that photoionization is not an important mechanism for the observed photosensitizing actions of these drugs. The drug triplet states and/or derived $^1O_2$ are more likely to be the source of such phototoxic and photoallergic side effects.

9.6.2 *Non-steroidal anti-inflammatory drugs*

A number of non-steroidal anti-inflammatory drugs (NSAID) of widely differing structure have gained prominence recently as useful therapeutic agents in the management of rheumatoid- and osteo-arthritis, free of the long-term side-effects associated with corticosteroids. However, several NSAID have been implicated in adverse photosensitivity reactions. Three examples: piroxicam, benoxaprofen, and naproxen, will be considered here although pulsed irradiation studies have also been carried out on others, for example, azapropazone (Jones *et al.* 1985, 1988).

9.6.2.1 *Piroxicam*

Shortly after piroxicam (see Fig. 9.5 for structure) became widely available for clinical use, it became apparent that the agent caused photosensitization in a small percentage of patients. Attempts to demonstrate phototoxicity to piroxicam in humans, laboratory animals, and *in vitro* cell assays were nevertheless unsuccessful. However, at high concentrations, a metabolite of piroxicam (2-methyl-4-oxo-2*H*-1,2-benzothiazine-1, 1-dioxide, often referred to as Compound 1, Fig. 9.5), was found to be phototoxic in animal studies and *in vitro* assays (Kochevar *et al.* 1986). This led to the hypothesis that piroxicam photosensitivity was due to a metabolite of piroxicam preferentially formed or accumulated in the skin of affected patients.

Based on their absorption spectra, Compound 1 did not appear to be a likely photosensitizer compared to piroxicam. The longest wavelength absorption maximum for Compound 1 at pH 7.4 is at 330 nm with a low ε of 1.81×10^2 $dm^3 mol^{-1} cm^{-1}$, whereas that for piroxicam is at 355 nm with a much higher ε of $1.45 \times 10^4 dm^3 mol^{-1} cm^{-1}$.

Pulse radiolysis of piroxicam and Compound 1 alone in benzene (Bensasson and Land, unpublished results) gave rise to transient absorptions which decayed by wavelength independent first-order kinetics, $k = 5.7 \times 10^4$ and

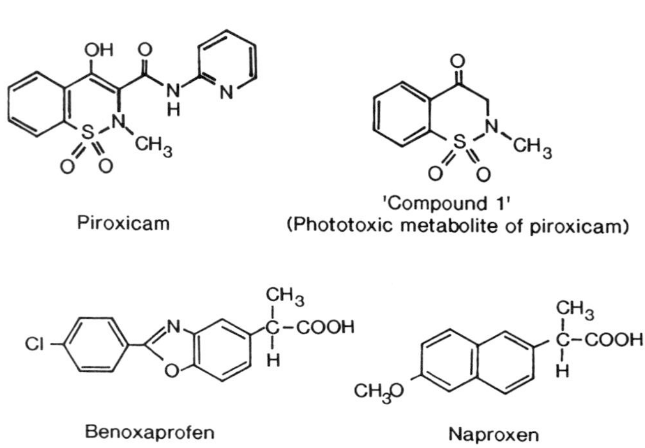

Fig. 9.5 Structures of non-steroidal anti-inflammatory drugs and related compounds.

$1.8 \times 10^5\,s^{-1}$, respectively, indicative of triplet states. For piroxicam, the absorption was broad with a single peak at 430 nm whereas that from Compound 1 had a main peak at 660 nm with a smaller one at 430 nm. Confirmation that these absorptions were due to triplet states was obtained by showing that both species sensitized β-carotene triplet. Similar results for piroxicam have been obtained by laser flash photolysis (R.S. Becker *et al.* 1990).

Pulse radiolysis of biphenyl in the presence of low concentrations of piroxicam led to the quenching of biphenyl triplet, λ_{max} 360 nm, and a corresponding grow-in of the 430 nm absorption due to piroxicam triplet. Such biphenyl to piroxicam triplet transfer is consistent with the triplet level of piroxicam being below that of biphenyl (E_T = 273 kJ mol^{-1}). No such transfer from biphenyl triplet could be seen with Compound I suggesting that its triplet level lies above that of biphenyl. The ε_T of both piroxicam and Compound 1 triplet were estimated using the energy-transfer technique with biphenyl triplet as donor for piroxicam, and anthracene as acceptor for Compound 1, values of 11 900 dm^3 mol^{-1} cm^{-1} at 430 nm and 2 130 dm^3 mol^{-1} cm^{-1} at 660 nm, respectively, being obtained. Using these absorption coefficients, estimates of ϕ_T for 355 nm excitation were obtained *via* laser flash photolysis and the comparative technique with anthracene as standard, values of 0.006 and 0.82 being found for piroxicam and Compound 1. This large difference in triplet yields was also reflected in the corresponding $^1O_2$ quantum yields, ϕ_Δ, determined through time-resolved measurements of the $^1O_2$ emission at 1270 nm (Western *et al.* 1987). The ϕ_Δ was 0.19 and 0.35 for Compound 1 and $<2 \times 10^{-3}$ and $<10^{-2}$ for piroxicam in D_2O and C_6H_6, respectively.

In view of the wide differences in the ϕ_T and ϕ_Δ values of Compound 1 and piroxicam, it appears likely that a Type II photodynamic mechanism would explain the phototoxicity of Compound I compared to the non-phototoxicity of piroxicam.

9.6.2.2 *Benoxaprofen*

The adverse photosensitivity reports on benoxaprofen (see Fig. 9.5 for structure) led to its withdrawal from clinical use in 1982. The photosensitivity appears to be phototoxic in nature, no cases of photoallergy having been reported (Kligman and Kaidbey 1982).

Benoxaprofen in aqueous solution exhibits a lowest energy absorption band centred at 305 nm (ε = 25 000 dm^3 mol^{-1} cm^{-1}). Change of solvent polarity did not produce any significant shift in λ_{max}, indicating that the lowest excited singlet state is probably $\pi\pi^*$ (Navaratnam *et al.* 1985). The fluorescence quantum yield in ethanol is 0.15 and the phosphorescence spectrum implies a high singlet–triplet splitting (67 kJ mol^{-1}), indicative of the lowest excited triplet level also being $\pi\pi^*$ in nature.

The transient difference spectrum observed after 266 nm excitation of benoxaprofen in deaerated aqueous solution shows a maximum at 410 nm with a shoulder at 350 nm (Navaratnam *et al.* 1985). The 410 nm maximum is somewhat solvent dependent (R.S. Becker *et al.* 1990). In water, oxygen quenched the 410 nm absorption but not the 350 nm shoulder implying the presence of at least two species. The 410 nm peak was assigned to the triplet state, the rate constant for its reaction with oxygen being 1.2×10^9 dm^3 mol^{-1} s^{-1}. The peak at 410 nm could be sensitized by xanthone triplet excited at 355 nm, consistent with the 410 nm peak being due to benoxaprofen triplet state. In agreement with this, subtraction of the spectrum measured in oxygen-saturated solution from the initial spectrum measured under nitrogen indeed gave the same spectrum confirming its triplet nature. An ε^T value of 17 500 dm^3 mol^{-1} cm^{-1} was obtained for benoxaprofen at 410 nm *via* the energy transfer technique. From this value, a ϕ_T of 0.19 for 266 nm excitation of benoxaprofen was obtained by the comparative method using as standard the photoionization of $K_3Fe(CN)_6$ for which a quantum yield of 0.52 for e^-_{aq} formation has been measured (Mialocq *et al.* 1980). Oxygen uptake measurements in the added presence of histidine suggested a singlet oxygen quantum yield via the benzoxaprofen triplet of 0.18. Photolysis of benoxaprofen also results in decarboxylation but this must occur via the singlet rather than the triplet excited state since the decarboxylation yield, also 0.18, is unaffected by oxygen.

Since the lysis of erythrocytes has been shown to occur in both anaerobic and aerobic benoxaprofen solutions (Webster *et al.* 1982; Sik *et al.* 1983), it would appear that more than one mechanism occurs. Decarboxylation (via singlet excited benaxoprofen) and singlet oxygen production (via triplet excited benoxaprofen) may be the appropriate initiators of phototoxicity for these respective conditions. No evidence was obtained for photoionization playing a role in either case.

9.6.2.3 *Naproxen*

The chromophore of naproxen is the naphthalene system (see Fig. 9.5 for structure). Like benoxaprofen, it is propionic acid-derived, but has much less extended conjugation than benoxaprofen, with no heteroatoms, so has a substantially lower absorption coefficient. It has a fluorescence quantum yield of 0.40, significantly higher than that of benaxoprofen, (D.E. Moore and Chappuis 1988).

The transient absorption obtained after 266 nm laser flash photolysis of naproxen exhibits peaks at 440 nm assigned to the triplet state, and at 700 nm, clearly indicating the presence of the hydrated electron (D.E. Moore and Chappuis 1988). Oxygen saturation of the solution resulted in the rapid decay, within 0.4 and 0.1 μs, of the 440 and 700 nm signals, respectively. Spikes (1977) showed that the $^1O_2$ quantum yield of naproxen was ~0.2 and hence rather similar to that of benoxaprofen. Naproxen also undergoes decarboxylation but in this case it was proposed to occur via CO_2 loss from the radical cation formed via photoionization.

Since both naproxen and benoxaprofen lead to decarboxylation and to similar yields of $^1O_2$, the major factor accounting for the decreased phototoxicity of naproxen as compared with benoxaprofen could be the 10 times lower ground state molar absorption coefficient of the former.

9.6.3 *Anti-bacterial drugs*

Nalidixic acid (NA) (1-ethyl-1,4-dihydro-7-methyl-4-oxo-1,8-naphthyridine-3-carboxylic acid) is an example of a phototoxic anti-bacterial agent. Laser flash photolysis studies by Dayhaw-Baker and Truscott (1988), and by Vermeersch *et al.* (1991), have shown that the NA triplet is an oxidant for tryptophan and tyrosine and that the molecular mechanism of the phototoxicity may occur via Type I and Type II photosensitizations, involving singlet oxygen ($\phi_\Delta = 0.2$), the radical dianion of NA, and the superoxide anion $O_2^{\cdot -}$.

9.7 Furocoumarins and PUVA

A well established form of photochemotherapy is that used in the treatment of psoriasis, and also mycosis fungoides (a rare form of cancer), vitiligo, atopic eczema, and polymorphous light eruption. The technique is called PUVA

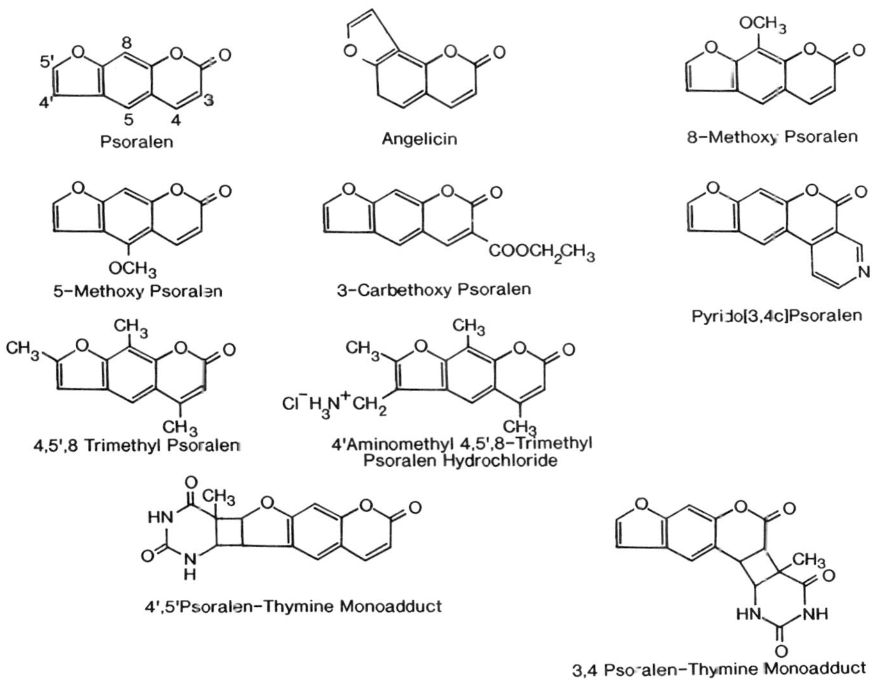

Fig. 9.6 Structures of some important furocoumarins.

therapy, (Psoralen + UVA), the drug being applied topically or orally. To be effective, both the psoralen derivative, usually 8-methoxypsoralen (8-MOP), but sometimes 5-methoxypsoralen (5-MOP), and the UVA radiation, must reach the basal layer of the epidermis. The technique has also recently been adapted to treat extracorporeally the blood of leukaemic cutaneous T-cell lymphoma patients (Edelson *et al.* 1987), and patients suffering from rheumatoid arthritis (Malawista *et al.* 1991).

Psoralens belong to the group of heterocyclic compounds known as furocoumarins. The structures of a selection of members of this group, important in this context, are given in Fig. 9.6.

9.7.1 *Photocycloadditions*

Diseases such as psoriasis involve the loss of control of cellular proliferation and DNA synthesis. PUVA therapy is thought to be mediated, at least in part, by two successive photoadditions of a single psoralen molecule to pyrimidines in the two strands of duplex DNA, resulting in crosslinks which inhibit DNA replication. Monofunctional lesions in DNA caused by psoralens may also be important and are easier to repair than crosslinks. The interaction of photo-excited drug with DNA is facilitated by the ability of psoralens to form intercalation complexes with DNA in the absence of light.

The photoreactive sites in 8-MOP are the 3,4 and 4′,5′ double bonds. There

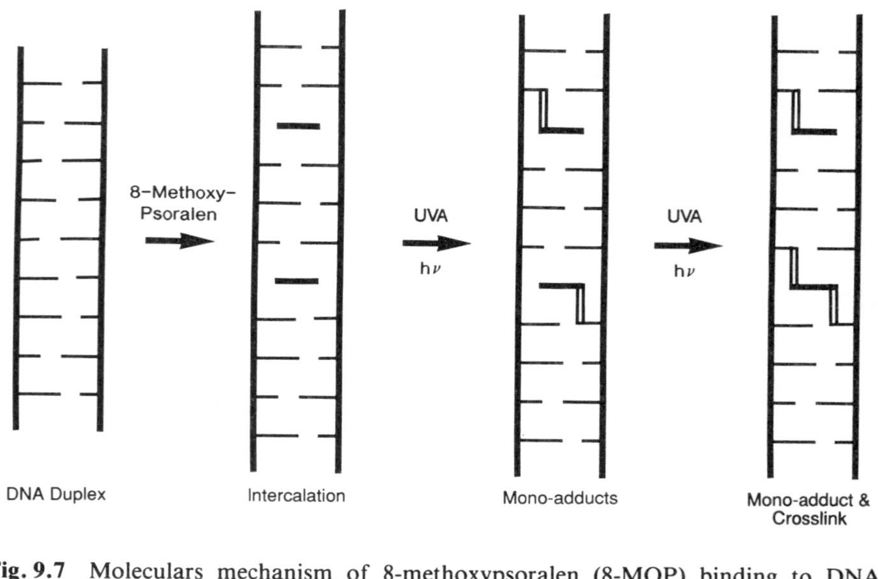

Fig. 9.7 Moleculars mechanism of 8-methoxypsoralen (8-MOP) binding to DNA to form single strand adducts and interstrand crosslinks (adapted from Cole and Sinden 1975).

are therefore two types of monoadduct, one resulting from the 8-MOP 3,4 pyran double bond reacting with a pyrimidine base 5,6 double bond, forming a cyclobutane ring, and the other resulting from the 8-MOP 4′,5′ furan double bond adding onto a pyrimidine base 5,6 double bond, forming another cyclobutane ring.

In order to form a crosslink, one quantum of light must first be absorbed by the 8-MOP to form a monoadduct. The 8-MOP thus becomes associated with a pyrimidine on one strand of duplex DNA. The second quantum of UVA must then be absorbed by the monoadduct causing the other reactive double bond to become connected with a pyrimidine on the opposite strand of the duplex DNA, forming a biadduct. These processes are shown schematically in Fig. 9.7. Thymine is usually chosen as the model for the pyrimidine base in DNA with which the 8-MOP bonds.

One difference between the two types of monoadducts is that the 3,4 pyran

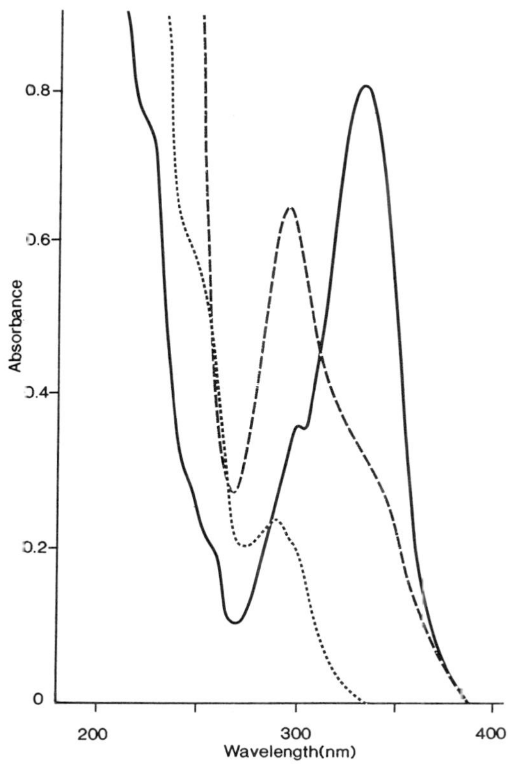

Fig. 9.8 Ground state absorption spectra of psoralen, ---; 3,4-psoralen–thymine monoadduct, ... (taken from Musajo *et al.* 1967*b*); and 4′5′-psoralen–thymine monoadduct, —— (taken from Musajo *et al.* 1967*a*).

adduct does not absorb UVA radiation whereas the 4′5′ furan adduct does (Fig. 9.8). Thus, regarding the mechanism of crosslink formation, after the first quantum of UVA has been absorbed by the psoralen itself to form a monoadduct, the second quantum has to be absorbed by a 4′,5′ monoadduct, rather than a 3,4 monoadduct, to form a crosslink.

Information on the configuration of the 4′,5′ 8-MOP–thymine monoadduct has been obtained by X-ray crystallographic studies (Land *et al.* 1982; Peckler *et al.* 1982). A sample of the adduct, formed on UV irradiation of 8-MOP in the presence of calf-thymus DNA, then isolated by acid hydrolysis (Kanne *et al.* 1982), was shown to consist mainly of a mixture of the two *cis-syn* enantiomers (mirror image isomers) (Peckler *et al.* 1982). Here the adduct conformation was determined by the constraints imposed by the DNA helix on the geometry of the non-covalent intercalation complex formed by 8-MOP and DNA prior to irradiation. Somewhat surprisingly, UV irradiation of 8-MOP and thymine in an ice–methanol matrix in the presence of the triplet photosensitizer benzophenone yielded the same mixture of *cis-syn* enantiomers (Land *et al.* 1982),

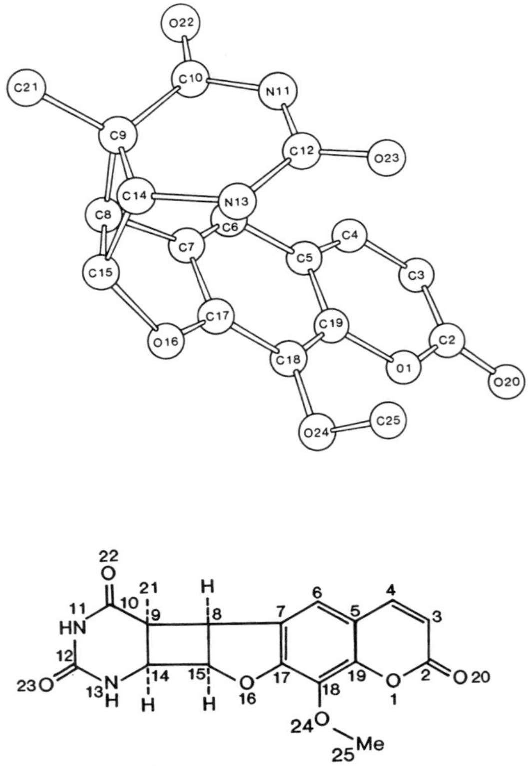

Fig. 9.9 Molecular configuration of *cis-syn* 8-methoxypsoralen–4′,5′-thymine mono-adduct obtained by X-ray crystallography (taken from Land *et al.* 1982).

suggesting that 8-MOP and thymine have an inherent tendency to complex together in this particular way. The molecular configuration of one of these *cis-syn* enantiomers is given in Fig. 9.9.

The X-ray results lead to several conclusions regarding the interactions of the adduct with DNA:

1. Since a mixture of enantiomers was found, this suggests that 8-MOP can form monoadducts at A–T (A = adenine, T = thymine) as well as T–A sequences in double helical DNA.
2. The three unique molecules in the crystal unit cell have 8-MOP-thymine moiety interplanar angles between 44° and 53°, implying a limited range of flexibility depending on the environment.
3. Model building studies on the basis of the X-ray results with the monoadduct, suggest that for thymine-8-MOP-thymine diadducts the configuration at the pyrone ring will also be *cis-syn*.

Thus, by assuming similar angles of 44–53° between the 8-MOP and thymine moieties at the pyrone ring, one can predict that a large kink of up to 70° in the DNA will occur where the crosslink is formed (Fig. 9.10). As it has also been

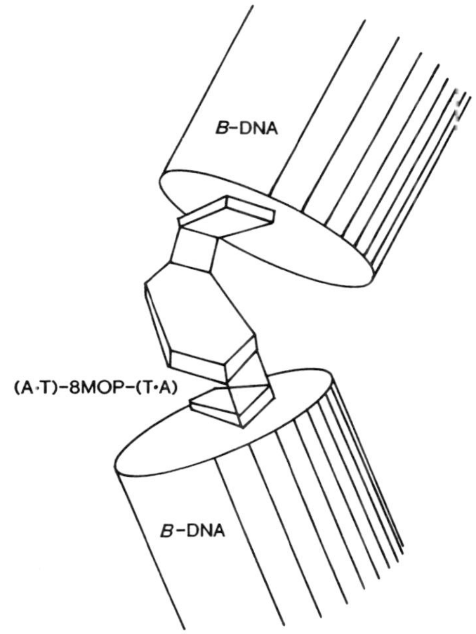

Fig. 9.10 Schematic drawing of double-helical *B*-DNA linked by a thymine–8-MOP–thymine crosslink. The T–A base-pairs on both sides of the psoralen moiety are shown as rectangular plates (taken from Peckler *et al.* 1982).

shown by Weisehahn and Hearst (1978), from superhelical density studies, that psoralen crosslinking causes DNA unwinding, it may be concluded that psoralen–DNA crosslinking may only occur in those regions of duplex DNA that are flexible and are not constrained by proteins or the three-dimensional structure.

Pulsed irradiation methods have been extensively employed to investigate properties of the excited, in particular triplet, states of furocoumarins and derivatives including 4′,5′ thymine monoadducts. Such studies are aimed at further understanding the photobiological properties of furocoumarins and developing more efficient photochemotherapeutic agents than 8-MOP (and 5-MOP). Furthermore, the long-term use of 8-MOP and 5-MOP has been suspected to lead to carcinogenesis (R.S. Stern *et al.* 1979).

9.7.2 *Ground state absorption spectra*

The ground state absorption spectra of most furocoumarins are similar to that of psoralen (Fig. 9.8). They show an absorption band of $\pi\pi^*$ character in the near UV around 300 nm with an ε between 10 000 and 15 000 $dm^3\,mol^{-1}\,cm^{-1}$. Another band often appears as a shoulder around 330 nm, with an ε of 7000 $dm^3\,mol^{-1}\,cm^{-1}$ for psoralen. The lowest singlet excited state of furocoumarins is assigned to a $^1\pi\pi^*$ state slightly below the $^1n\pi^*$ state of the carbonyl (Song and Tapley 1979).

9.7.3 *Triplet photophysical properties*

Early theoretical studies (Song *et al.* 1971; Mantulin and Song 1973; Harrigan *et al.* 1976) predicted that the reactive state is the triplet rather than the singlet. This prediction is consistent with the findings of Bevilacqua and Bordin (1973) that oxygen and paramagnetic ions have a quenching effect on the rate of photoreaction between psoralen and thymine.

9.7.3.1 *Absorption spectra*

The lowest triplet state has been assigned to a $^3\pi\pi^*$ transition with a strong localization of the triplet excitation in the 3,4 C=C of the pyrone moiety (Mantulin and Song 1973). Figure 9.11 includes the triplet–triplet absorption spectrum of psoralen obtained on flash photolysis of psoralen in benzene (Bensasson *et al.* 1978*a*). Similar transient absorption spectra are obtained from psoralen on pulse radiolysis in benzene or flash photolysis in water. Confirmation that the absorption was due to psoralen triplet was obtained by showing that this transient species sensitized β-carotene triplet.

In the case of 8-MOP, whose triplet absorption was first detected by Poppe and Grossweiner (1975), a marked blue shift in the triplet absorption maximum was found in going from benzene to water as solvent (Fig. 9.11) (Bensasson *et al.* 1978*a*). A large change was also noted in the triplet spec-

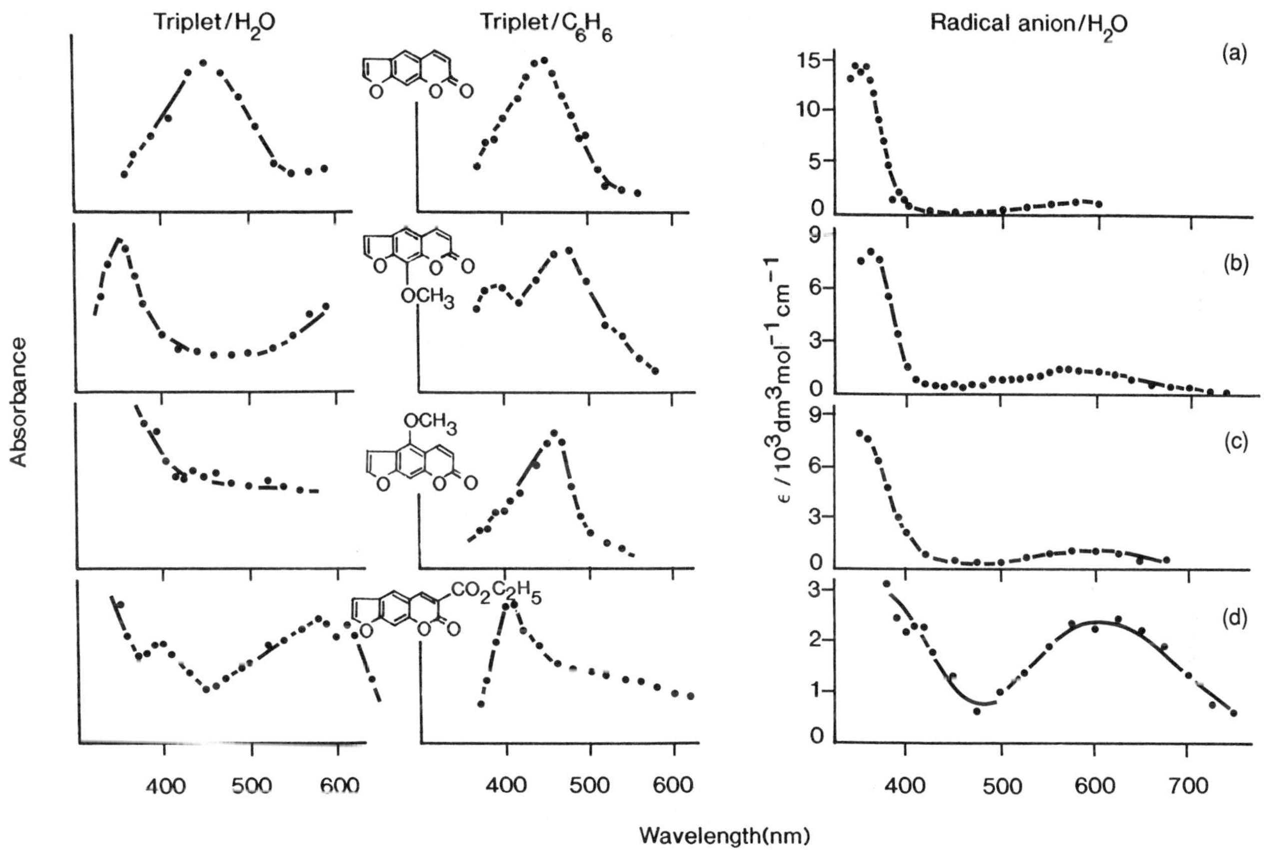

Fig. 9.11 Triplet absorptions of (a) psoralen, (b) 8-methoxypsoralen, (c) 5-methoxypsoralen, and (d) 3-carbethoxypsoralen in water and benzene, and corresponding radical anion absorptions in water. (Reprinted with permission from *Photochemistry and Photobiology*, **39**, Bensasson, R. V., Chalvet, O., Land, E. J. and Ronfard-Haret, J. C., Triplet, radical anion and radical cation spectra of furocoumarins, Copyright (1984) Pergamon Press Ltd.)

Table 9.1 Molar absorption coefficients, ε_T, of furocoumarin triplets

Furocoumarin	Solvent[a]	Wavelength (nm)	ε_T ($dm^3\ mol^{-1}\ cm^{-1}$)	Reference[b]
Psoralen	B	450	8 100	a
	M	440	30 100	b
8-Methoxypsoralen	B	480	10 000	a
8-Methoxypsoralen-4′5′-thymine monoadduct	M	510	6 000	c
5-Methoxypsoralen	B	450	10 200	a
	E	440	9 500	d
	D	440	11 600	e
3-Carbethoxypsoralen	B	410	11 500	f
	W	580	8 300	f
	E	450	5 460	g
4,5′,8-Trimethylpsoralen	M	470	33 000	h
	B	450	7 750	i
4′ Aminomethyl-4,5′,8-trimethylpsoralen	M	460	24 200	j
5,8-Dimethoxypsoralen	B	450	3 800	i
3-Carbethoxypseudopsoralen	B	510	16 500	i
Pseudopsoralen	B	470	15 600	i
5’-Aceto-8-methylpsoralen	B	430	20 900	k
5’-Aceto-4,8-dimethylpsoralen	B	430	21 300	k
Pyrido[3, 4-*c*] psoralen	E	470	19 000	l
7-Methylpyrido [3, 4-*c*]-psoralen	E	450	28 000	l
4,7-Dimethylallopsoralen	B	440	9 500	m
4,7,4′-Trimethylallopsoralen	B	470	7 140	m
4,7,5′-Trimethylallopsoralen	B	450	9 060	m
4′5′ Dihydropsoralen	B	500	15 700	n
4′,5′ Psoralen–thymine monoadduct	M	500	8 500	o
Allopsoralen	B	450	6 200	k
3-Acetoallopsoralen	B	450	13 300	k
Isopseudopsoralen	B	480	7 560	k

3-Acetoisopseudopsoralen	B	510	9 890	k
Pseudoisopsoralen	B	450	11 700	k
3-Acetopseudoisopsoralen	B	450	18 800	k
Angelicin	B	450	4 700	a
	B	430	5 500	k
	M	450	8 000	p
3-Acetoangelicin	B	430	13 200	k
4-Methylangelicin	M	410	11 800	p
5-Methylangelicin	M	420	12 200	p
5′-Methylangelicin	M	410	12 700	p
4′-Methylangelicin	M	410	12 700	g
	E	410	13 800	g
4′,4-Dimethylangelicin	M	410	9 000	p
	B	450	5 610	k
4,5′-Dimethylangelicin	M	410	9 900	q
	B	450	5 510	k
4′5-Dimethylangelicin	M	410	9 000	g
	E	410	10 700	g
6,4′-Dimethylangelicin	B	450	6 480	k
6,4-Dimethylangelicin	B	450	8 750	k
6,4,4′-Trimethylangelicin	B	450	6 300	k
4′-Aminomethyl-4,5′-dimethylangelicin	M	410	12 000	q
4′-Hydroxymethyl-4,5′-dimethylangelicin	M	450	9 500	p
4′-Methoxymethyl-4,5′-dimethylangelicin	M	410	9 600	p
4′-Acetoxymethyl-4,5′-dimethylangelicin	M	450	7 500	p
4′-*N*,*N*-dimethylamino-ethoxymenthyl-4,5′-dimethylangelicin	M	450	12 200	q

[a]B, benzene; M, methanol; E, ethanol; W, water; D, dioxane.

[b] a, Bensasson *et al.* (1978*a*); b, Sa e Melo *et al.* (1979); c, Land (unpublished); d, Craw *et al.* (1983); e, Sa e Melo *et al.* (1988); f, Ronfard-Haret *et al.* (1982); g, Averbeck *et al.* (1984); h, Beaumont *et al.* (1979*b*); i, Knox *et al.* (1986); j, Salet *et al.* (1980); k, Knox *et al.* (1988); l, Ronfard-Haret *et al.* (1987); m, Vedaldi *et al*, (1987); n, Land and Truscott (1979); o, Bensasson *et al.* (1980); p, Craw *et al.* (1984*a*); q, Guiotto *et al.* (1981).

trum of 3-carbethoxypsoralen (3-CPs) (Ronfard-Haret *et al.* 1982) and 5-methoxypsoralen (Bensasson *et al.* 1978*a*; Craw *et al.* 1983) in going from benzene to water. Since the spectra of 3-CPs and 8-MOP triplets in water are similar to that of the corresponding radical anions (see Fig. 9.11) it may be that these triplets in water have some charge-transfer state character, consistent with exciplexes of the type: $^3(FC^- . H_2O^+)$ (Bensasson *et al.* 1984*b*). These psoralens are examples of molecules with a triplet absorption spectrum strongly modified when going from an H-bonding solvent such as water to a non-H-bonding solvent such as benzene.

A weak transition found in the red region on laser flash photolysis of 8-MOP in water was shown by Sloper *et al.* (1979), using N_2O as quencher, to be due to hydrated electrons resulting from biphotonic photoionization occurring at high laser intensities. Triplet 8-MOP production was shown to be monophotonic. A transient spectrum obtained on 355 nm laser flash photolysis of the 4′,5′ 8-MOP–thymine monoadduct (Land, unpublished results) also shows a band at 700 nm indicative of e^-_{aq} in addition to a triplet band at 510 nm. Again the laser intensity dependence was consistent with biphotonic production of e^-_{aq} and monophotonic production of the triplet.

Table 9.1 lists the triplet molar absorption coefficients for a number of furocoumarins. The absorption coefficients were mainly obtained by the energy transfer comparative technique (see Section 3.1.2.1) using as standard biphenyl triplet in benzene (pulse radiolysis) or biphenyl, retinol or β-carotene triplet in methanol or ethanol (flash photolysis).

9.7.3.2 *Lifetimes*

Observed furocoumarin triplet lifetimes, which are in the microsecond range, depend on the concentration of furocoumarin used. Two competing reactions occur:

$$^3FC \xrightarrow{k_I} FC \quad \text{Reaction I = Intrinsic relaxation to ground state}$$

$$^3FC + FC \xrightarrow{k_{II}} \text{Products} \quad \text{Reaction II = Ground state quenching}$$

Table 9.2 gives the self-quenching and intrinsic rate constants that have been measured for various furocoumarins. The first-order decay rate constants of many other furocoumarin triplets have been measured at single parent concentrations (e.g. Guiotto *et al.* 1981; Averbeck *et al.* 1984; Craw *et al.* 1984a; Ronfard-Haret *et al.* 1987), those of the extra carbonyl containing 5′-aceto-8-methylpsoralen, 3-acetoallopsoralen, 3-acetoisopseudopsoralen and 3-acetopseudoisopsoralen being particularly long-lived, $\tau \sim 40$–$60\ \mu s$, under the conditions chosen (Knox *et al.* 1988).

9.7.3.3 *Intersystem crossing yields*

Estimates of the values of ϕ_T for various furocoumarins in benzene, water, methanol, ethanol, and carbon tetrachloride, obtained using the comparative

Table 9.2 Triplet decay rate constants of furocoumarins

Furocoumarin	Solvent[a]	Triplet + Ground state rate constant ($dm^3\ mol^{-1}\ s^{-1}$)	Intrinsic first-order decay rate constant (s^{-1})	Reference[b]
Psoralen	W	2.3×10^9	1.4×10^5	a
8-Methoxypsoralen	W	3.0×10^9	2.0×10^5	b
	W	4×10^9	4.2×10^5	c
	M	1.7×10^9	2.3×10^5	b
	B	1.9×10^8	9.0×10^5	a
4,5′,8-Trimethylpsoralen	M	3.2×10^9	1.4×10^5	b
4′-Aminomethyl-4,5′,8-trimethylpsoralen	W	10^9	10^4	d
3-Carbethoxypsoralen	W	1.5×10^9	0.6×10^5	e
	SDS/W	$< 10^8$	0.3×10^5	e
	E	1.1×10^9	1.8×10^5	e
	B	1.1×10^9	0.5×10^5	e
Angelicin	W	3.6×10^9	1.7×10^5	a
5-Geranoxypsoralen	E	3.0×10^9	1.6×10^6	f

[a] W, water; M, methanol; E, ethanol; B, benzene; SDS/W, sodium dodecyl sulphate in water.
[b] a, Bensasson *et al.* (1978*a*); b, Beaumont *et al.* (1979*a*); c, Craw *et al.* (1983); d, Salet *et al.* (1980); e, Ronfard-Haret *et al.* (1982); f, Morlière *et al.* (1991).

method described in Section 3.1.2.3, are collected in Table 9.3. It may be noted that there is often a change in intersystem crossing quantum yield in going from the non-polar solvent benzene to the polar solvent water. This suggests that the actual medium in which the furocoumarin is contained in skin would be important with respect to photoreactions occurring via the triplet.

Furocoumarins possess a lowest energy $n\pi^*$ state that is close in energy to the lowest $\pi\pi^*$ state. This phenomenon, termed the 'proximity effect' (Lai *et al.* 1982), has been found to lead to very efficient radiationless decay from the lower of the two excited states (Lim 1977).

For the furocoumarins not substituted by an extra carbonyl, the proximity of the lowest $n\pi^*$ and $\pi\pi^*$ states may be the dominant factor and radiationless $S_1 \rightarrow S_0$ internal conversion would then be expected to predominate, resulting in the low ϕ_T values generally observed. If the same principle can be extended to the T_1 and S_0 levels, it is also possible to understand the reason for the short intrinsic lifetime of the triplet states (see Section 9.7.3.2), as again there will be efficient radiationless transitions back to the ground state.

For the furocoumarins substituted with an extra carbonyl, there is an extra $n\pi^*$ state which lowers the S_1 energy gap from the ground state, typically

Table 9.3 Singlet triplet intersystem crossing quantum yield ϕ_T, $^1O_2$ quantum yield (ϕ_Δ) and fraction of triplets which, when quenched by O_2, produce $^1O_2$ (S_Δ) of furocoumarins

Furocoumarin	Solvent[a]	Excitation wavelength (nm)	ϕ_T	ϕ_Δ	S_Δ	Reference[b]
Psoralen	B	355	0.034			a
	B	355		0.012	0.34	b
	W	355	0.12			c
	D	337		0.04		d
	E	265	0.06			c
	E	355	0.06			c
	M	337		0.05		d
	C	—		0.0055		e
8-Methoxypsoralen	B	355	0.011			a
	B	355	0.0044	0.40		b
	W	355	0.058			c
	E	265	0.03			c
	E	355	0.04			c
	M	347	0.012			c,f
	C	—		0.002		e
	M	337		0.009		d
	D	337		0.02		d
8-Methyoxypsoralen-4′,5′-thymine monoadduct	M	355	0.34			g
	W	355	0.18			g
5-Methoxypsoralen	B	355	0.067			a
	B	355		0.021	0.32	b
	W	355	<0.01			a
3-Carbethoxypsoralen	B	355	0.30			h
	B	355		0.30	1.0	b
	W	355	>0.35			h
	E	355	0.44			h
4,5′,8-Trimethylpsoralen	M	347	0.093			f
	B	355	0.21	0.084	0.41	b
4′ Aminomethyl-4,5′8-trimethylpsoralen	W	355	0.2			i
5,8-Dimethoxypsoralen	B	355	0.039	0.0040	0.10	b
3-Carbethoxypseudopsoralen	B	355	0.51	0.49	0.96	b
Pseudopsoralen	B	355	0.045	0.026	0.57	b
5′-Aceto-8-methylpsoralen	B	355	0.28	0.11	0.39	j
5′-Aceto-4,8-dimethylpsoralen	B	355	0.52	0.38	0.72	j
Pyrido[3,4-*c*]psoralen	E	355	0.04			k
	B	355	0.02	0.008	0.40	k
7-Methylpyrido[3,4-*c*]-psoralen	E	355	0.02			k
	B	355	<0.01	0.003	>0.3	k
4,7-Dimethylallopsoralen	B	355	0.13	0.064	0.49	l
4,7,4-Trimethylallopsoralen	B	355	0.28	0.073	0.26	l

4,7,5′-Trimethylallopsoralen	B	355	0.18	0.12	0.68	l
4′,5′-Dihydropsoralen	W	355	0.068			m
Psoralen–4′,5′-thymine monoadduct	W	355	0.076			n
Allopsoralen	B	355	0.17	0.07	0.41	j
3-Acetoallopsoralen	B	355	0.68	0.55	0.81	j
Isopseudopsoralen	B	355	0.15	0.10	0.67	j
3-Acetoisopseudopsoralen	B	355	0.65	0.64	0.99	j
Pseudoisopsoralen	B	355	0.03	0.01	0.33	j
3-Acetopseudoisopsoralen	B	355	0.56	0.55	0.99	j
Angelicin	B	355	0.009			a
	B	355	0.03	0.01	0.33	j
	C	—		0.0026		e
	M	337		0.03		d
	D	337		0.02		d
	W	355	0.33			a
	M	265	0.031			o
3-Acetoangelicin	B	355	0.54	0.40	0.73	j
4-Methylangelicin	M	265	0.12			o
5-Methylangelicin	M	265	0.08			o
5′-Methylangelicin	M	265	0.08			o
4′-Methylangelicin	M	265	0.08			p
4′,4-Dimethylangelicin	M	265	0.11			p
	B	355	0.11	0.02	0.18	j
4,5′-Dimethylangelicin	M	265	0.063			q
	B	355	0.08	0.02	0.25	j
4′,5-Dimethylangelicin	M	265	0.11			o
6,4′-Dimethylangelicin	B	355	0.07	0.01	0.14	j
6,4-Dimethylangelicin	B	355	0.04	0.01	0.25	j
6,4,4′-Trimethylangelicin	B	355	0.13	0.02	0.16	j
4′-Aminomethyl-4,5′ dimethylangelicin	M	265	0.028			q
4′-Hydroxymethyl-4, 5′dimethylangelicin	M	265	0.11			o
4′-Methoxymethyl-4,5′-dimethylangelicin	M	265	0.11			o
4′-Acetoxymethyl-4,5′-dimethylangelicin	M	265	0.13			o
4′-*N*,*N*-dimethylamino-ethoxymethyl-4,5′-dimethylangelicin	M	265	0.059			q

[a]B, benzene; W, water; D, deuterated water (D_2O); E, ethanol; M, methanol; C, carbon tetrachloride.

[b]a, Bensasson *et al.* (1978*a*); b, Knox *et al.* (1986); c, Sa e Melo *et al.* (1979); d, Krasnovsky *et al.* (1986); e, Krasnovsky *et al.* (1983); f, Beaumont *et al.* (1979*b*); g, Land (unpublished); h, Ronfard-Haret *et al.* (1982); i, Salet *et al.* (1980); j, Knox *et al.* (1988); k, Ronfard-Haret *et al.* (1987); l, Vedaldi *et al.*. (1987); m, Land and Truscott (1979); n, Bensasson *et al.* (1980); o, Craw *et al.* (1984*a*); p, Averbeck *et al.* (1984); q, Guiotto *et al.* (1981).

by about 40 kJ mol^{-1}. This may reduce the proximity effect in the singlet manifold, allowing for an increase in $S_1 \rightarrow T_1$ intersystem crossing and hence a larger ϕ_T as observed (see Table 9.3). Again, for the T_1 and S_0 levels, the extra nπ* state reduces the effect of the radiationless transitions and hence the triplet lifetime is considerably longer, as noted in Section 9.7.3.2.

9.7.4 *Triplet reactions*

9.7.4.1 *Energy transfer: singlet oxygen formation*

A number of triplet–triplet energy transfer rate constants involving furocoumarin triplets have been determined as a byproduct of triplet absorption coefficient determinations by the energy transfer method (see, for example, Bensasson *et al.* 1978*a*; Land and Truscott 1979; Sa e Melo *et al.* 1979). They tend to be close to the diffusion-controlled limit and are consistent with the triplet energy levels that have been measured using phosphorescence techniques by Mantulin and Song (1973).

Perhaps the most important energy transfer reaction undertaken by furocoumarins is with oxygen. Poppe and Grossweiner (1975) showed early on that 8-MOP triplet is quenched by oxygen, rate constant $\sim 1 \times 10^9$ dm^3 mol^{-1} s^{-1}. Since then, the rate constants for the oxygen quenching of many furocoumarin triplets have been shown to lie around 1–3 $\times 10^9$ dm^3 mol^{-1} s^{-1} (see, for example, Ronfard-Haret *et al.* 1982; Craw *et al.* 1984*a*).

Ronfard-Haret *et al.* (1982) compared survival curves after UVA irradiation of the yeast *Saccharomyces cerevisiae* in the presence of 8-MOP and 3-CPs, with and without oxygen. In the case of 3-CPs, the yeast cells were much less sensitive to UVA under hypoxic conditions than under oxygenated conditions, whereas with 8-MOP there was very little difference between the survival curves with or without oxygen. Similar results to those found with 8-MOP have more recently been obtained with two pyridopsoralens (Ronfard-Haret *et al.* 1987). All these results are consistent with 8-MOP and the pyridopsoralen cell killing not being mainly due to Type II photodynamic action involving singlet oxygen ($^1O_2$), but rather to furocoumarin photoaddition to DNA. With 3-CPs on the other hand, with its 10$\times$ higher triplet yield, singlet oxygen may play a more significant role. The pyridopsoralens have also been shown to have exceptionally high triplet levels which allow the additional process of triplet-energy transfer to thymine leading to thymine dimer formation (Costalat *et al.* 1990).

Quantitative data on the $^1O_2$ producing efficiencies (ϕ_Δ) of a wide variety of furocoumarins have been obtained by direct observation of the weak 1270 nm emission of $^1O_2$ detected after laser flash photolysis, mainly in oxygen-saturated benzene solution. The ϕ_Δ values thus obtained are included in Table 9.3, together, where available, with the corresponding S_Δ values, S_Δ being the fraction of triplets which, when quenched by O_2, lead to $^1O_2$.

It can be seen that few of the S_Δ values are close to unity, and it has been found that neither the electron transfer process,

$$^3FC + O_2 \rightarrow FC^{\cdot +} + O_2^{\cdot -}$$

nor furocoumarin excited singlet state quenching by oxygen, are important regarding the ϕ_Δ and S_Δ values obtained (Knox *et al.* 1986).

A possible reason for the S_Δ values being higher for the carbonyl-substituted furocoumarins than for the non-substituted molecules is linked to the energy of the furocoumarin triplet with respect to the energy of singlet oxygen. In any energy transfer reaction the energy of the donor (D) must be higher than that of the acceptor. However, this excess energy also has to be lost:

$$^3D + O_2 \rightarrow D + {}^1O_2 + \Delta E$$

where ΔE is the excess energy. If the triplet energy of the donor molecule is much greater than that of the acceptor, then energy transfer is not efficient as there is a large excess of energy to be lost (L.K. Patterson *et al.* 1970). If, however, the energy of the donor is only slightly higher than that of the acceptor molecule, then the intersystem crossing is efficient as there is very little excess energy to be lost.

Because of the extra $n\pi^*$ state in the carbonyl-substituted furocoumarins, they are likely to have lower triplet energies than the non-substituted furocoumarins, and hence less energy to lose, so that energy transfer to produce $^1O_2$ will be more efficient, i.e. S_Δ will be nearer unity for those furocoumarins with relatively low triplet energies.

An alternative proposal, to account for the range of S_Δ values obtained, is that the relatively high S_Δ values (with aceto-substituted molecules) arise when the lowest triplet is mainly $\pi\pi^*$ in character, while the relatively low S_Δ values arise when the lowest triplet is predominantly $n\pi^*$ in character (Redmond and Braslavsky 1988).

The relative yields of singlet oxygen from some furocoumarins in D_2O in the presence and absence of DNA have also been measured using the time-resolved-luminescence technique (Beaumont *et al.* 1985). For both psoralen and 4′-aminoethyl-4,5′,8-trimethylpsoralen (AMT), increasing DNA concentrations led to a decrease in the yield of singlet oxygen, there being a linear relation between the yield of singlet oxygen and the observed triplet yield determined by laser flash photolysis. It thus seems unlikely that singlet oxygen production occurs from these furocoumarins when they are complexed to DNA.

9.7.4.2 *Nucleic acid bases and DNA*

Of particular relevance to the possible involvement of triplet states in monoadduct and crosslink formation are the reactivities of furocoumarin and monoadduct triplet states with nucleic acid bases. Some of the rate constants that have been obtained in water are collected in Table 9.4.

Energetic considerations exclude the possibility that any of these reactions correspond to triplet–triplet energy transfer, since the nucleic acid bases all have higher triplet energies than the furocoumarin derivatives concerned. It can be

Table 9.4 Rate constants for furocoumarin triplet quenching by nucleic acid bases and amino acids in water ($dm^3\ mol^{-1}\ s^{-1}$)

Quencher	Psoralen[a]	Angelicin[a]	8-Methoxypsoralen–thymine-4′,5′ mono-adduct[b]
Thymine	7.5×10^8	1.1×10^9	6.4×10^7
Uracil	1.1×10^8	2.1×10^8	5.6×10^6
Cytosine	1.7×10^7	2.6×10^7	8×10^5
Adenosine	4.1×10^7	4.7×10^7	$<10^6$
Guanosine	7.3×10^8	7.2×10^8	2.2×10^7
Tryptophan	3.1×10^9	2.1×10^9	1.8×10^9
Tyrosine	2.1×10^8	1.6×10^8	1.2×10^7
Histidine	3.3×10^7	3.7×10^7	1.5×10^6
Phenylalanine	$\sim 3 \times 10^6$	$\sim 5 \times 10^6$	$<10^6$
Glycine	$<10^5$	$<10^5$	$<10^5$
Cysteine	$<10^5$	$<10^5$	6×10^6

[a] Bensasson *et al.* (1978*a*).
[b] Land (unpublished).

seen that wide variations in rate constant occur depending upon the furocoumarin derivative and the nucleic acid base quencher. The similarity between psoralen and angelicin triplets is perhaps surprising in view of the differences in their skin photosensitizing properties. This might support the view that the photosensitizing differences are due to geometric variations which allow psoralen, but much less so angelicin, to form crosslinks after intercalation in DNA. 8-MOP triplet in methanol did not react with any of the nucleic acid bases tested (Beaumont *et al.* 1979*b*). In view of the extensive use of 8-MOP in photochemotherapy, this may mean that the triplet excited state of this psoralen itself is not involved in photosensitization. It could be that the 4′,5′ mono-adduct is formed via the singlet excited state after absorption of the first photon. Absorption of the second photon by the intercalated mono-adduct could then lead to crosslink formation via the corresponding triplet state. Consistent with this, the 8-MOP–thymine 4′,5′ mono-adduct has a much higher ϕ_T than 8-MOP and its triplet is reactive towards thymine.

Laser flash photolysis experiments have been carried out with psoralen and 8-MOP complexed with DNA (Beaumont *et al.* 1980). It was found that whereas the amount of psoralen triplet observed decreased as DNA was added, no corresponding reduction in triplet state absorption occurred with 8-MOP. Thus different excited state mechanisms may pertain to different psoralen derivatives. The progressively reduced amounts of psoralen triplet obtained on adding increasing amounts of DNA fit quite well the residual concentrations of free psoralen remaining. It thus appears that triplet psoralen is not observed when the psoralen is complexed with DNA.

With AMT, which forms much stronger complexes with DNA (Isaacs *et al.*

1977), the yield of triplet psoralen derivative detected falls much faster, and for low [DNA] addition no observable triplet was found (Salet *et al.* 1980). The finding (Beaumont *et al.* 1983) that the fluorescence lifetime of AMT is reduced from 2.0 ns to 0.3 ns on binding to DNA has been interpreted in terms of a singlet excited state mechanism for photoinduced mono-adduct formation between AMT and DNA.

Additional evidence concerning the mechanism of mono-and bi-adduct formation with AMT was gained by Johnston *et al.* (1977, 1981) and Johnston and Hearst (1981) who carried out some novel experiments involving the direct measurement of DNA crosslink formation after one or two successive 15 ns 347 nm laser pulses. They showed that a small amount of crosslinking occurs after a single laser flash, but, as the time between the two flashes was progressively increased, there was a rise in the number of crosslinks produced which followed first-order kinetics and a relaxation time of 1.3 μs. Since this relaxation time correlated with reported relaxation times for unstacking of base pairs at the end of helical regions of DNA, they postulated that the 1.3 μs time delay was required for some conformational change in DNA to occur before a newly formed mono-adduct can absorb a second photon to give a crosslink. It was suggested that a subpopulation of intercalated AMT molecules is capable of forming crosslinks within the lifetime of a single laser pulse.

9.7.4.3 *Amino acids*

Since it is possible that furocoumarin photosensitization might occur at the level of proteins as well as DNA, the reactions of several furocoumarin triplets with various amino acids have been investigated. The rate constants obtained for psoralen, angelicin, and the 8-MOP–4′,5′-thymine monoadduct are included in Table 9.4. As with nucleic acid base quenching, the possibility that these reactions with amino acids are due to energy transfer may be excluded on energetic grounds.

Tryptophan may be seen to be the most reactive amino acid and, in general, weak long-lived transient intermediates were found to remain after the furocoumarin triplet had been quenched by tryptophan. These long-lived intermediates may provide an insight into the main type of quenching occurring with tryptophan, and possibly pyrimidines and purines as well, since they possess their most intense absorption bands in the region of 360 nm, where the electron adducts of the furocoumarin derivatives have their main peaks (see Fig. 9.11). Such radical anions could be formed via a charge-transfer reaction of the type:

$$^{3}FC + TrpH \rightarrow FC^{\cdot -} + Trp^{\cdot} + H^{+}$$

The ~360 nm peak would be expected to dominate a composite spectrum containing equal quantities of $Trp^{\cdot}$ and $FC^{\cdot -}$. Analogous results were obtained with the somewhat less reactive quencher tyrosine.

Since nucleic acid bases are known to possess electron donor properties (Hush and Cheung 1975), it seems possible that pyrimidines likewise quench furocoumarin triplets via a charge-transfer mechanism, at least in part.

In a similar type of reaction, the tyrosine analogue dopa efficiently quenches ^{3}8-MOP (rate constant ~2×10^9 dm^3 mol^{-1} s^{-1}) to form 8-MOP$^{\cdot -}$ and the dopasemiquinone radical (Craw *et al.* 1984*b*). Since, as described in Chapter 8, dopasemiquinone can lead to the series of chemical intermediates which ultimately form melanin pigments, such a reaction of ^{3}8-MOP might play some part in the increased pigmentation caused by PUVA therapy.

9.8 Diseases associated with defective tetrapyrrole metabolism

9.8.1 *Bilirubinaemia*

The use of phototherapy as a treatment for neonatal hyperbilirubinaemia was first proposed by Cremer *et al.* (1958) and has now become one of the most widely used treatments for jaundice in the first few days of life. In fact, skin yellowing (or jaundice) was known as a sign of disease to both Galien and Hippocrates, and the North American Indians are thought to have been aware of the beneficial effects of the Sun in reducing the yellow colour of a baby exposed to such light (see, for example, L. Stern 1984).

Naturally occurring bilirubin is insoluble in water but soluble in lipids. In the liver it is glucuronidated (or conjugated) to render it soluble so that it may be excreted. The unconjugated bilirubin is bound to albumin in plasma, and this is related to its transport to the liver and the avoidance of damage due to free bilirubin. The binding of bilirubin by human serum albumin (HSA) is dependent upon pH and the presence of other species, including some drugs, capable of displacing bilirubin from its protein site. Thus, a large amount of free bilirubin can arise despite the presence of apparently sufficient albumin in the system, and the binding capacity of the HSA is an important factor in considering treatment of the infant.

The molecular mechanisms associated with the phototherapy of jaundice have been debated for more than 20 years with the major contribution of laser flash photolysis and pulse radiolysis being made about 10 years ago.

At least three processes have been claimed to be involved in the photo clearance of bilirubin during phototherapy:

(1) photobleaching via oxidative reactions;

(2) photoinduced *cis–trans* isomerization; and

(3) photocyclization.

Laser flash photolysis of bilirubin has shown that ϕ_T is very low with typical values being ≤ 0.005 in chloroform and ~ 0.01 in benzene (Sloper and Truscott 1980). However, this technique has also shown a quantum yield for *cis–trans* isomerization of ~0.1–0.2 (Sloper and Truscott 1982), a similar value being obtained by a fluorescence technique (Lamola *et al.* 1982). The fluorescence quantum yield of bilirubin is very low and dependent on the environment, with a value of less than 0.001 in typical low viscosity solvents, although a much higher value arises on binding to albumin or on cooling. No phosphorescence has been reported for bilirubin even in a glass at 77 K. Picosecond excitation has determined the singlet lifetime as near 20 ps for bilirubin bound to HSA (Green *et al.* 1981; Tran and Beddard 1981).

Clearly these results are consistent with only a very minor contribution to bilirubin photoclearance arising via photo-oxidation. *Z* to *E* configurational isomerisation (see Fig. 9.12) is the most quantum efficient photochemical reaction of bilirubin. Since bilirubin contains two non-identical exocyclic double bonds there are three additional isomeric forms to the native 4*Z*, 15*Z* form. These are the 4*Z*, 15*E*; 4*E*, 15*Z*; and 4*E*,15*E* isomers. There have been many studies of these isomerization processes for bilirubin both bound to HSA and free in solution. Often the work is based on fitting ϕ values to the observed ratios of isomers, and typical studies are those of Pratesi and coworkers (Fusi *et al.* 1985). These and other results are reviewed by Ennever (1988). As can be seen from the ϕ values given in Fig. 9.12 there is a marked regioselectivity and the ratio of the 4*Z*,15*E* isomer to the 4*E*,15*Z* is $\sim$ 100.

Perhaps surprisingly, despite the relatively high quantum efficiency of the 4*Z*,15*E* production, it is not now believed to be the most important route for pigment elimination. This may be due to re-isomerization back to the 4*Z*,15*Z* (natural) bilirubin in the bile. Instead, it now appears that the formation of the photocyclization product, known as lumirubin (see Fig. 9.12) formed via the 4*E*,15*Z* isomer, is the principal mechanism for bilirubin excretion. An interesting and related observation is that green fluorescence lamps are more effective in the therapy than the usually used blue or white lamps. This may be related to a reduced yield of the 4*Z*,15*E* isomer which is not a precursor of lumirubin and hence possibly to an increased amount of available 4*Z*,15*Z* form. Thus Greenberg *et al.* (1987) have shown that ϕ for lumirubin formation is 0.0015 at 450 nm and 0.0032 at 510 nm.

There are still many unanswered questions associated with the molecular mechanisms of bilirubin phototherapy. Many of these are related to the role of regioselectivity and stereoselectivity effects of HSA. These, in turn, may well be related to exciton coupling between the two pyrromethenone chromophores (see Agati and Fusi 1990). The time may be ripe for a re-investigation of these processes via pulsed laser studies, possibly using time-resolved resonance Raman monitoring.

9.8.2 *Porphyria*

Skin photosensitivity associated with the various forms of porphyric disease (rather rare, mostly hereditary) is due to the presence of excess porphyrin in the skin (see Fig. 9.13 for structures). Photosensitivity due to porphyrins was first reported by Hausmann (1911) and Meyer-Betz (1913). Furthermore, it has been

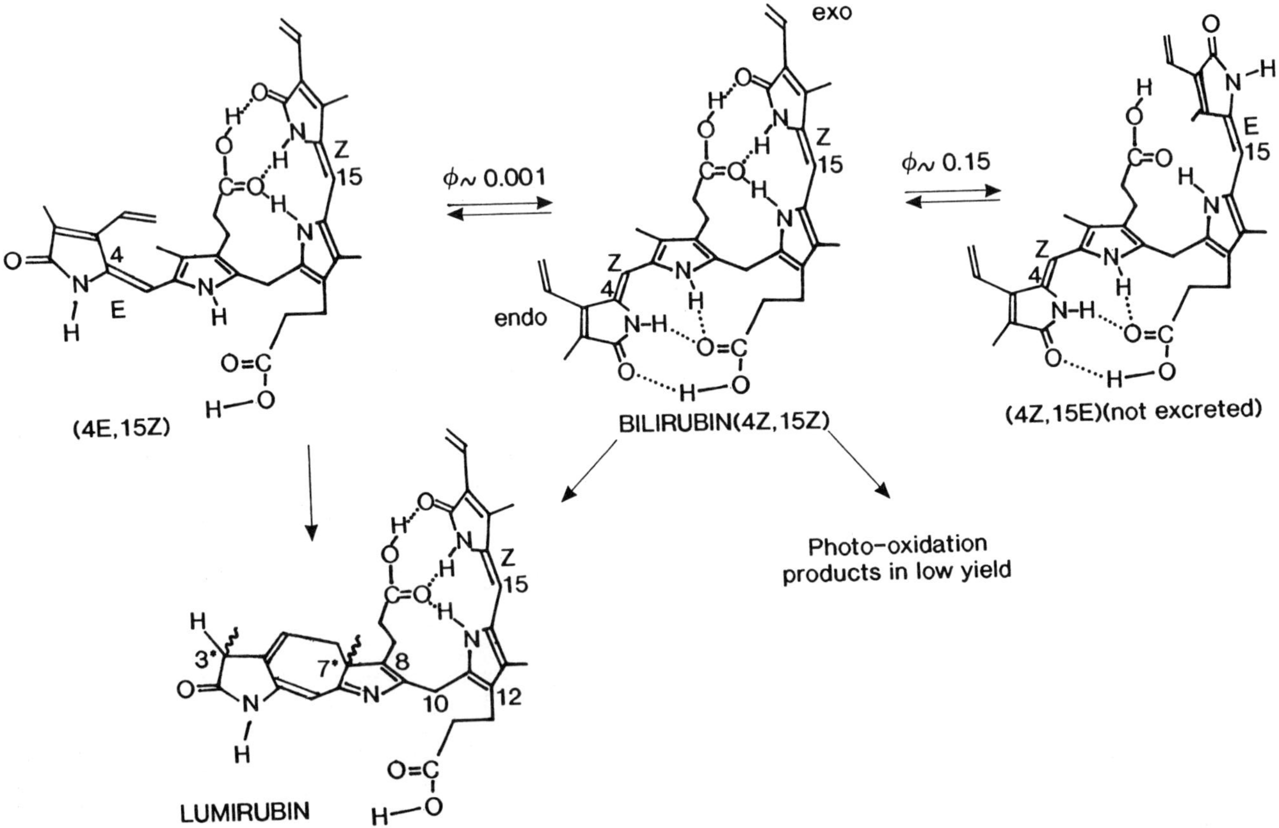

Fig. 9.12 Photochemical mechanism of bilirubin phototherapy.

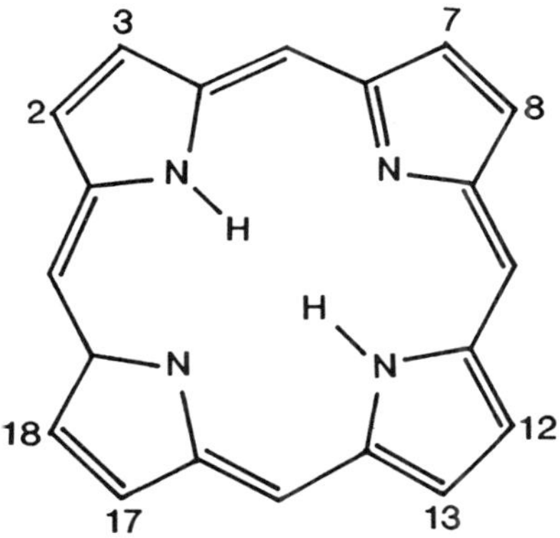

Porphyrin	2	3	7	8	12	13	17	18
Haematoporphyrin	Me	E	Me	E	Me	P^{me}	P^{me}	Me
Protoporphyrin	Me	V	Me	V	Me	P^{me}	P^{me}	Me
Coproporphyrin I	Me	P^{me}	Me	P^{me}	Me	P^{me}	Me	P^{me}
Coproporphyrin III	Me	P^{me}	Me	P^{me}	Me	P^{me}	P^{me}	Me
Uroporphyrin I	A^{me}	P^{me}	A^{me}	P^{me}	A^{me}	P^{me}	A^{me}	P^{me}
Uroporphyrin III	A^{me}	P^{me}	A^{me}	P^{me}	A^{me}	P^{me}	P^{me}	A^{me}

The isomers which occur in normal haem biosynthesis are coproporphyrin III and uroporphyrin III, however, coproporphyrin I and uroporphyrin I accumulate as a result of porphyric diseases.

Side chain abbreviations: Me=Methyl, E=$-CH_2OHCH_3$, V=Vinyl, P^{me}=$-CH_2CH_2COOCH_3$, and A^{me}=$-CH_2COOCH_3$

Fig. 9.13 Structures of porphyrins (as esters).

suggested that the natural photosensitivity associated with the porphyric diseases may be related to the various Werewolf legends (Milgrom 1984). The porphyric disease itself is classified according to the principal site of abnormal porphyrin biosynthesis, and six types of porphyria have been identified (M. R. Moore and Brodie 1985).

The porphyrins which occur in the porphyrias arise by a disfunction of haem biosynthesis and vary in structure (and properties) depending on which particular step in the biosynthetic pathway is blocked. The defects diminish haem production, and since haem regulates its own synthesis by inhibiting the synthesis of the first enzyme in the pathway, the accumulation of porphyrin intermediates is aggravated by the decreased formation of haem. Thus, in erythropoietic protoporphyria (EPP), an excess of lipid soluble protoporphyrin arises, while in congenital erythropoietic porphyria (Gunther's disease) there are high levels of water soluble uroporphyrin. As discussed in Chapter 12, a related skin photosensitivity arises as a side effect of the photodynamic therapy (PDT) of cancer. In this case it is haematoporphyrin related species which accumulate in the skin, and can lead to severe skin photosensitivity of the treated cancer patient for many weeks or even months.

The ground state absorption spectra of such (aetio free base) porphyrins show a strong band centred near 400 nm called the Soret band, together with four much weaker (~1/10 intensity) bands in the visible region called the Q bands. In fact, the number and intensity of the Q bands is reduced on protonation or metallation of the imino (—N=) nitrogen and the pyrrole (—NH—) nitrogens of the porphyrin (tetrapyrrole). The spectra of all types of porphyrins have been the subject of several reviews both from an experimental and theoretical view point (see Pottier and Truscott 1986, and references therein).

The fluorescence quantum yields tend to be near 0.1 and the fluorescence lifetimes (τ_f) rather long (15–25 ns). For haematoporphyrin there has been some interest in the aggregated form and for such (rather ill-defined) material τ_f is usually much shorter.

The most important photophysical parameters associated with the skin photosensitivity caused by porphyrins, which are monomeric *in vivo*, are their triplet state properties and the subsequent production of singlet oxygen by exchange energy transfer (see Chapter 4). Therapeutic protection against this skin photosensitivity may proceed via energy transfer reactions associated with β-carotene (see Chapter 7). The triplet–triplet absorption spectra of several porphyrins have been reported and all show a rather broad band with a maximum at ~440 nm and with ε_T values in the region of $1\text{–}3 \times 10^4$ dm^3 mol^{-1} cm^{-1} (Bonnett *et al.* 1980).

The inherent triplet lifetimes of porphyrins in typical solvents are very long and are controlled simply by the concentration of adventious impurities (e.g. trace oxygen). However, in microheterogeneous environments, e.g. some liposomes, there is a reduction of triplet lifetime, presumably due to interaction of the triplet and phosphatidyl choline. Also, such studies on porphyrins in liposomes, cell ghosts, very low density lipoprotein (VLDL), and HSA imply more than one porphyrin environment (C. Lambert *et al.* 1985; Truscott 1986).

Triplet quantum yields, ϕ_T, for porphyrins are high and typically lie in the range 0.8 ± 0.1, so that $\phi_T + \phi_F \simeq 1$. Of particular interest for skin photosensitivity is the quantum yield of singlet oxygen production (ϕ_Δ) with values for protoporphyrin and haematoporphyrin (as the dimethyl esters) in benzene of 0.57 and 0.53, respectively (see, for example, Truscott 1988).

An alleviation of the skin photosensitivity from EPP (but not the other porphyrias) can be achieved using high dosages of β-carotene. Certainly, as noted in Chapter 7, carotenoids can protect against skin damage both by quenching the porphyrin triplet and any singlet oxygen which is formed. In some ways this is analogous to the protective role of carotenoids in photosynthesis. The fact that other forms of porphyria are not alleviated by carotene treatment may simply be related to the differing solubilities involved, i.e. β-carotene is lipid soluble whereas uroporphyrin is water soluble.

9.9 Summary and conclusions

Light transmission through the skin varies as a function of wavelength, the epidermis being more protected by the physical passive absorption of the stratum corneum from the UVB than from the UVA. UVA can account for 60–80 per cent of the toxic effects of the combined solar UVA and UVB on epidermal keratinocytes, when UV transmission of human skin and spectral solar irradiance data are taken into account (Tyrrell and Pidoux 1987).

Skin damage can be caused by light-induced damage to DNA, proteins, and enzymes, sometimes leading to lipid peroxidation. Table 9.5 lists some of the main endogenous photosensitizing molecules present in skin and the initial radicals formed therefrom on light absorption. Particularly important amongst these endogenous sensitizers are the porphyrins present in excess in patients suffering from the various porphyric diseases. Further details of the nature and reactions of the radicals and excited states of some of these molecules are provided in Chapters 5, 6, and 12.

Photobiologists are now increasingly interested in the UVA region where the DNA is not the chromophore because:

(1) a tryptophan photoproduct *N*-formyl kynurenine absorbing up to 400 nm, catechol, and catecholamine melanin precursors, absorbing up to 350 nm,

Table 9.5 Endogenous photosensitizers and their light-induced intermediates causing photodynamic effects

Photosensitizer	Reactive intermediates
Tryptophan (TrpH)	^{3}TrpH, Trp$^\cdot$, e^-_{aq}, $O_2^{\cdot -}$, $^1O_2$
Tyrosine (TyrOH)	^{3}TyrOH, TyrO$^\cdot$, e^-_{aq}, $O_2^{\cdot -}$, $^1O_2$
N-Formylkynurenine (NFK)	^{3}FK, FK$^{\cdot -}$, FKH$^\cdot$, $^1O_2$
NADH	NAD$^\cdot$, e^-_{aq}, $O_2^{\cdot -}$
Riboflavin (Rf)	^{3}Rf, RfH$^\cdot$, $O_2^{\cdot -}$, $^1O_2$
Porphyrins (P)	^{3}P, P$^{\cdot -}$, P$^{\cdot +}$, e^-_{aq} $^1O_2$
Dopa	Dopasemiquinone, e^-_{aq}, $O_2^{\cdot -}$

may photosensitize the formation of the cell-lethal product, hydrogen peroxide (McCormick *et al.* 1976; Graham *et al.* 1978; J. McCord 1979);

(2) singlet oxygen has been shown to be involved in the photoinactivation of cultured human skin fibroblasts (Tyrrell and Pidoux 1989); the source of this activated oxygen species has not yet been clearly identified; and

(3) antioxidants are able to diminish the UV-mediated responses in the skin of animals and inhibit photocarcinogenesis.

UV light has deleterious effects on cell architecture, cytoskeleton and membranes being the targets. Lipid peroxidation, initiated via free radicals and/or singlet oxygen and occurring by a chain reaction, is a process which proceeds even in the dark but is accelerated by light and is linked to photocarcinogenesis. Although the carcinogenesis risk by UVB is at least 10^3–10^4 times greater than that of UVA, tanning, whether with UVB or UVA, results in similar carcinogenic risks.

A wide variety of exogeneous skin photosensitizers are known, some of the more important ones being listed in Table 9.6. Many of these are drugs and the Office of Science and Technology in the US Department of Health and Human Services has recently provided a listing of medications and other agents that may increase sensitivity to light (Levine 1990 and references therein). Besides the psychiatric and anti-inflammatory drugs which exhibit photosensitizing side effects, perhaps the most important group of exogeneous photosensitisers are the furocoumarins.

It seems clear from the data obtained by pulsed irradiation methods that

Table 9.6 Exogenous photosensitizers (taken from B.E. Johnson 1988)

Drugs Antibacterials: tetracyclines, sulphonamides, nalidixic acid
Tranquillizers: phenothiazines
Antidepressants: protryptiline
Antiarrhythmics or antihypertensives: amiodarone, methyldopa, propanolol
Anti-inflammatory non-steroidals: benoxaprofen, naproxen, azapropazone, piroxicam
Anti-fungal: griseofulvin
Topical bacteriostat or antifugal: halogenated salicylanilides, biothionol, fentichlor
Plant materials Furocoumarins, alpha-terthienyl, and related polyacetylenes
Dyestuffs Thiazines: methylene blue, toluidine blue
Xanthenes: fluorescein, eosin, rose bengal, erythrosin
Anthraquinones: anthraquinone, benzanthrone
Pitch, coal tar, and similar
Polycyclic hydrocarbon products Anthracene, acridine, 1,2-benzanthracene
Perfumes and cosmetics 5-Methoxypsoralen, 6-methyl coumarin
Sunscreens and printing inks Anisyl *o*-dimethyl-aminobenzoic acid, *p*-aminobenzoic acid
Tattoos Cadmium sulphide

several mechanisms operate in the photosensitizing action of furocoumarins. Both Type I and Type II processes probably play a role in certain circumstances. However, the most important mechanism is via oxygen-independent photocycloaddition reactions in duplex DNA. Such lesions are likely to be the main origin of the antiproliferative effects of furocoumarins exploited in photochemotherapy. Lipid–furocoumarin adducts have also been isolated (Specht *et al.* 1988), indicating that the photobiological effects of furocoumarins are due to the combined effects of several different molecular events occurring in a variety of cellular components.

Photosensitization was used thousands of years ago by the Egyptians, Chinese, and Indians to cure a range of disorders such as psoriasis, vitiligo, and rickets. Today, a number of skin defects are treated via phototherapy as well as those treated by photochemotherapy. These, as well as hyperbilirubinaemia, psoriasis, and vitiligo discussed above, include port wine stain and pruritus, although little is known of the molecular mechanisms involved in either case. An additional cosmetic use of laser light is in the removal of unwanted tattoos.

10

Cancer: radiotherapy

'In radiotherapy a partial response is a complete failure'

Lester J. Peters (1985)

10.1 The chemical basis of radiobiology and radiotherapy

Both normal and malignant cells are sensitive to high energy radiation. Although a long-term effect of high energy irradiation can be cancer induction, such radiation can also be used to treat cancer. Cancer is characterized by a progressive accumulation of malignant cells and the object of radiotherapy is to sterilize cancer cells so that they lose their capacity for unlimited proliferation. One of the important aims of radiation chemical and radiation biological studies is to provide information which may lead to the optimization of radiotherapy.

10.1.1 *Time scale of high energy radiation action*

Damage to any biological material exposed to high energy radiation stems from the molecular events initiated in the complex mixtures of chemicals which make up such biological systems. These events span a time scale of at least 25 orders of magnitude, and may be very roughly divided into three main phases:

1. The physical stage, 10^{-18}–10^{-12} s, during which the energy of the incident radiation is absorbed (mainly by water—see Table 1.3) and chemically reactive species are formed via ionization and excitation.
2. The chemical stage, 10^{-12} – 1s, during which the highly reactive excited states and free radicals formed in the primary events react with each other and with the cellular constituents.
3. The biological stage, from seconds up to decades, during which the modifications in biochemical and cellular processes resulting from the radiation induced chemical changes take effect.

High time resolution pulsed radiation studies have made major contributions to knowledge of the physical and, in particular, the chemical stage.

10.1.2 *Effects of different types of high energy radiation*

All the types of radiation that are used in radiotherapy, consisting of fast charged particles (e.g. α, β/electrons), neutral particles (e.g. neutrons—fast or slow), and electromagnetic radiation (e.g. γ-rays and X-rays), lead to both ionization and excitation. In polar media, such as water, which makes up ~80 per cent of the mass of mammalian cells, any excited states formed are very short-lived and dissociative, and so the major observable initial process is ionization giving predominantly reducing e^-_{aq} and oxidizing $OH^{\cdot}$ (see Chapter 2). In less polar media relatively long-lived excited states can be formed. Such environments may exist in the residual ~20 per cent of organic matter which make up the rest of the mass of mammalian cells, for example within lipids.

Although hydrated electrons and $OH^{\cdot}$ radicals are the principal active species formed on irradiation of water, some other molecular products are formed whose relative amounts vary with the different types of radiation. These molecular products are hydrogen and hydrogen peroxide formed via the reactions:

$$2e^-_{aq} + 2H_2O \longrightarrow H_2 + 2OH^-$$

$$2OH^{\cdot} \longrightarrow H_2O_2$$

and tend to be formed most readily when the ion pairs generated by the radiation are formed close together. With low LET radiation, e.g. hard X-rays, γ-rays, and fast electrons (*L*inear *E*nergy *T*ransfer being a term used to define the local rate of energy dissipation—see Chapter 1), the ion pairs formed are well spaced so the yields of e^-_{aq} and $OH^{\cdot}$ are high and the yields of H_2O_2 and H_2 are low. With high LET radiation on the other hand, e.g. α particles, the ions are formed very close together. Such α particles have a bigger mass than electrons so they travel much more slowly, producing a much higher ionization density in the track. This enhances molecular H_2 and H_2O_2 and thereby reduces the yield of e^-_{aq} and $OH^\cdot$. The density of ionization of the tracks of neutrons lies somewhere in between those for fast electrons and α particles. The primary radicals are much more reactive than the molecular products, so the hydrated electrons and $OH^\cdot$ radicals are the most important precursors of radiobiological and hence radiotherapeutic effects.

10.1.3 *Indirect action, direct action, and target theory*

Attack on biological targets mediated by the species from water radiolysis is often referred to as indirect action. The number of solute molecules interacting with the free radicals produced in the solvent increases, with increasing solute concentration, to a plateau value corresponding to all the radicals formed undergoing reaction. At very high solute concentrations (> 1 per cent) significant direct action on the solute also occurs.

Exactly the same considerations apply to the mammalian cell, if one considers it as being a fairly dilute, although complex and inhomogeneous, aqueous solution. Important contributions to the effect of radiation on cells are due to indirect action.

At one time it was considered that all radiation effects on cells were due to direct action, via the energy absorbed in some of the ~20 per cent organic constituents of cells. Target theory is based on the postulate that certain biological changes follow from direct ionization or excitation of particular regions or targets. For a cell to be inactivated, a target, or number of targets, within the cell must be inactivated, and each target is inactivated by a single direct hit. An obvious candidate for the target is the DNA in chromosomes in the nucleus.

Although the general notion of direct and indirect effects has been useful, conceptually the idea is somewhat imprecise and it is difficult to verify or refute mechanisms based on these as distinct classes.

10.1.4 *Effects of radiation on individual cell components*

Pulse radiolysis studies have contributed much information on the interaction of e^-_{aq} and $OH^\cdot$ with individual cell components in dilute aqueous solution. Flash photolysis studies of dilute solutions of such components using ionizing UV excitation are also sometimes good models for the direct interaction of high energy radiation with cell components. Such irradiation studies of DNA bases and DNA itself, the important *in vivo* cell target, are described in Chapter 5, and amino acids, peptides and proteins in Chapter 6.

10.1.5 *Radioprotection*

The phenomenon of radioprotection would be useful to radiotherapy if one could protect normal tissue inevitably irradiated when treating tumours. It might be expected that the blood supply would be poorer in tumours, leading to some selectivity of radioprotector take-up by normal tissue. Radioprotection can arise from both endogenous and exogenous materials. Unfortunately, exogenous drugs tried so far have been disappointing in the clinical situation due to their toxicities.

It is convenient to classify mechanisms of radiation protection into those involving fast initial radiation chemical processes and those in which the protection is a result of changes in cell biochemistry or physiology. In the first case, the protective compounds are believed to act by direct participation in the free radical reactions associated with damage to vital molecules. In the biochemical mechanisms, protection is assumed to be a result of the action of the protector compound, either on the metabolism of the cell, or on the biochemical processes that affect the response of the cell to the initial chemical damage after the fast radiation chemical reactions have been completed.

The most efficient radioprotective agents are organic mercaptans, RSH, in particular the amino thiols, cysteamine and cysteine, which contain labile hydrogen atoms. Less efficient radioprotectors are certain other amino acids and alcohols.

The most popular model to explain chemical radioprotection by RSH compounds postulates that a generalized macromolecular free radical ($\dot{\wedge\!\wedge}$) formed in a cellular target, either by direct or indirect action, is restored or repaired by hydrogen atom transfer from the labile SH group of the protector:

$$\dot{\wedge\!\wedge} + \mathrm{RSH} \rightarrow \overset{\mathrm{H}}{\wedge\!\wedge} + \mathrm{RS}^{\cdot}$$

This process arises because the S—H bond energy in thiols is only 382 ± 6 kJ mol^{-1}, while the C—H bond energy is somewhat higher.

Repair actions of this type have indeed been observed by pulse radiolysis. Thus, from studies of the growth of the readily observable $\mathrm{RSSR}^{\cdot-}$ absorption around 400 nm, formed via the equilibrium:

$$\mathrm{RS}^{\cdot} + \mathrm{RS}^{-} \rightleftharpoons \mathrm{RSSR}^{\cdot-}$$

G.E. Adams *et al.* (1967*a*, 1968*b*, 1969) showed that cysteamine will transfer an $\mathrm{H}^{\cdot}$ atom to a number of model organic radicals with rate constants between 0.18 and $4.2 \times 10^{8}\ \mathrm{dm}^{3}\ \mathrm{mol}^{-1}\ \mathrm{s}^{-1}$, and repair polyethylene oxide radicals with rate constants decreasing with increasing polymer size. Similar H atom transfers from glutathione, dithiothreitol, and cysteine have been reported (Redpath 1973; Baker *et al.* 1982; Willson 1983).

Thiyl radicals, considered in the past to be rather unreactive, were thought to, in turn, disappear mainly via a diffusion controlled recombination to the disulphide:

$$\mathrm{RS}^{\cdot} + \mathrm{RS}^{\cdot} \rightarrow \mathrm{RSSR}$$

Since the steady state concentration of $\mathrm{RS}^{\cdot}$ *in vivo* is low, other pathways are more likely to occur and it has now been demonstrated that the thiyl radical can in fact be an oxidant (Forni and Willson 1986) and can react with C—H bonds (Elliot *et al.* 1984; Akhlaq *et al.* 1987; Schöneich *et al.* 1989*a*, *b*).

Although only ~20 per cent of the radiolytically produced OH· radicals reacting with DNA do so with the sugar moiety, the radicals so produced are very important because they lead to critical damage such as DNA strand breaks and base release. Accordingly it has been shown that glutathione can transfer an H· atom to carbohydrate radicals including radicals derived from the DNA sugar deoxyribose (Baker *et al.* 1982; Tamba and Quintiliani 1984), the extent of repair being dependent upon the particular carbohydrate. The H·-atom transfer from dithiothreitol to OH·-damaged deoxyribose has been found to exhibit both fast and slow components, the latter proceeding with a rate constant of 2.4×10^7 dm^3 mol^{-1} s^{-1} (Held *et al.* 1985). Hydrogen donation to OH·-damaged DNA bases by cysteine has been reported to have rate constants of $<10^6$ dm^3 mol^{-1} s^{-1} (Nucifora *et al.* 1972), whereas nucleosides, nucleotides, and DNA radicals themselves react with cysteine with rate constants of around 10^8 dm^3 mol^{-1} s^{-1} (Greenstock and Dunlop 1975), consistent with predominant reaction with the sugar component. Several thiols and thiolates interact with OH adducts of 2′-deoxyguanosine and 2′-deoxyguanosine-5-monophosphate which have oxidizing properties (O'Neill 1983). This reaction occurs via an electron transfer process with rate constants at pH 7.4–7.6 of $<2 \times 10^6$ to 7.8×10^7 dm^3 mol^{-1} s^{-1}, depending on the particular thiol. The rate constant can be increased by more than an order of magnitude at higher pH because RS^- is more reactive than RSH in an electron transfer reaction.

10.1.6 *Oxygen effect*

In very general terms, in the absence of a protective agent, the macromolecular free radical $\wedge\dot{M}\wedge$ is a precursor of molecular damage expressed later at the cellular level. In the presence of oxygen, the radical can be converted into a peroxyl radical:

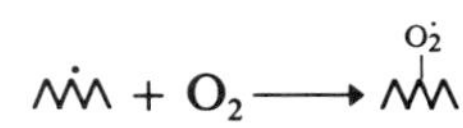

Indeed, competition between endogenous hydrogen donors such as RSH and oxygen may be a determinant of the extent of damage.

The above reaction is, in fact, the radiation chemical interpretation of the oxygen effect. If $\wedge\dot{M}\wedge$ represents an important target molecule, for example, DNA, converted to a free radical, this DNA free radical can in turn be converted into a peroxyl radical which appears to be irrepairable. This is termed 'fixing of the damage' (for further discussion of peroxyl radicals, see Chapter 4).

Oxygen effects are universal phenomena in radiobiology. The sensitization effects on cells irradiated in the presence of oxygen, compared with cells irradiated in anoxia, is evident throughout cellular radiobiology, both *in vitro* and *in vivo*. It is generally true that the presence of oxygen is sensitizing for the lethal effects of radiation and indeed for most biological manifestations of damage, such as the induction of division delay, gene mutations or chromosome aberrations.

The sensitizing effect of oxygen is different for different types of high energy radiation. Thus, the ratio of doses to produce a given effect without and with oxygen present (*O*xygen *E*nhancement *R*atio, OER) is about three with low LET radiation, such as X- or γ-rays, but much less for high LET radiation, such as α particles, where the OER is usually around 1.2–1.5. This decreased oxygen effect with densely ionizing radiation has been ascribed, in part, to free radicals formed in the dense regions of the track reacting differently from those in less dense regions. As described in Section 10.1.2, for water itself more molecular products are formed by radical–radical reactions with high LET radiation than with low LET radiation. As the LET increases, intra-track radical–radical reactions tend to predominate over radical–substrate reactions outside the track. Since 'oxygen enhancement' involves a radical–substrate reaction, then clearly oxygen enhancement ratios would be expected to decrease with increasing LET. In fact, it is a general feature of high LET radiation that the radiation chemical reactions thus induced are far more difficult to influence than those produced by low LET radiation.

The relative radioresistance of hypoxic cells present in some human tumours has been considered to be an important cause of failure in radiotherapy. In most normal tissues, the blood supply maintains an intracellular oxygen concentration of $\sim 3 \times 10^{-5}$ mol dm^{-3}. In tumours, disorganized cell proliferation outstrips the growth of blood supply, leaving some cells with so little oxygen that they are relatively radioresistant. Radiation doses sufficient to kill all fully oxygenated tumour cells fail to sterilize these radioresistant cells, which then ultimately lead to the regrowth of tumours.

One approach towards overcoming the hypoxia problem has been to carry out radiotherapy treatment in high pressure oxygen chambers. However, the results of clinical trials using hyperbaric oxygen have been disappointing—probably because the high rate of oxygen metabolism restricts the distance which oxygen can diffuse out from capillaries, all hypoxic cells not being reached. Another approach, where in fact pulse radiolysis studies have had considerable impact, has been the use of chemical radiosensitizers.

10.2 **Radiosensitization**

For some time there has been wide interest in the development of chemical radiosensitizers which can selectively increase the radiosensitivity of hypoxic cells without affecting the radiation response of well-oxygenated normal tissue. The structures of some of the more important compounds that have been tested are given in Fig. 10.1. It will be noticed that most are nitro compounds,

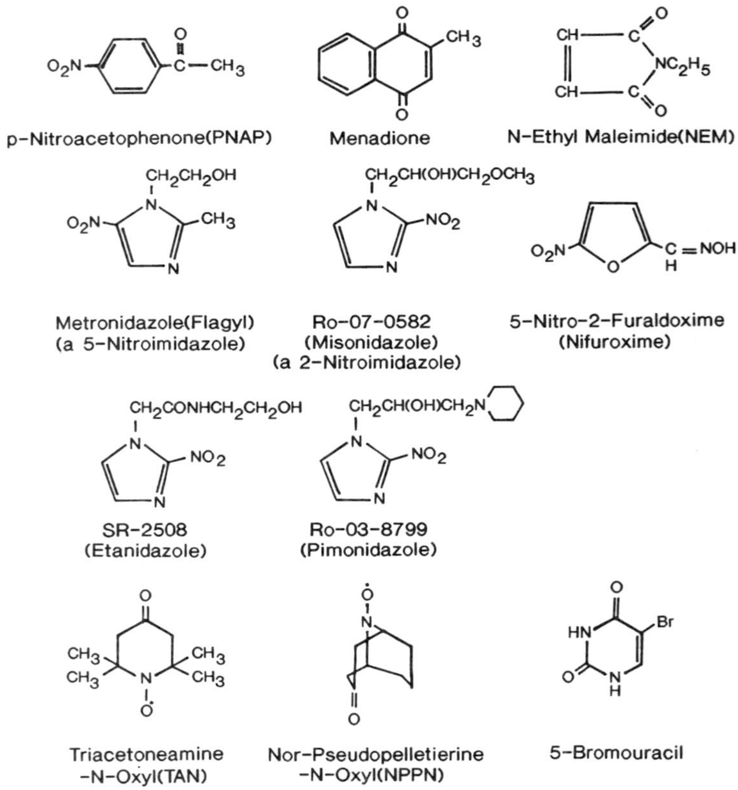

Fig. 10.1 Structures of radiosensitizers.

the nitro group having a strong electron affinity. Compounds often used as radiosensitizers are the nitroimidazoles, in particular, metronidazole and misonidazole. So far, neurotoxic side effects have severely limited their use in the clinic, although Phase III trials are currently in progress for etanidazole, a less lipophilic analogue of misonidazole, both alone and in combination with another more electron affinic analogue, pimonidazole.

10.2.1 *Electron affinic radiosensitizers*

As noted previously (Bensasson *et al.* 1983), it was suggested some time ago (G.E. Adams and Dewey 1963) that efficiencies of radiosensitization are related to the electron affinities of the sensitizers. Molecules of high electron affinity, acting as electron carriers and electron transfer agents, may be able to enhance radiobiological damage by increasing the effective radius of hydrated electrons produced by the radiation.

10.2.1.1 *Cascade electron transfer*

The ease with which electron transfer processes can take place in simple aqueous solution can readily be demonstrated by pulse radiolysis (G. E. Adams *et al.* 1967*b*, 1968*c*; Adams and Willson 1973). On irradiation of a multisolute system, e_{aq}^- reacts initially with the solute for which the product of the reaction rate constant and concentration is the greatest. In a system containing different solutes A, B, C, and D, each of which reacts rapidly with e_{aq}^- and which are present in concentrations such that $[A] \gg [B] \gg [C] \gg [D]$, the electron adduct $A^{\cdot -}$ will be formed initially. Subsequently, electron transfer processes may occur, with $A^{\cdot -}$ reacting with B to give $B^{\cdot -}$ and so on, until eventually $D^{\cdot -}$ is formed (Fig. 10.2). As an illustration, an aqueous solution (pH 12.5) containing acetone and isopropanol (both 0.3 mol dm^{-3}), acetophenone (3×10^{-3} mol dm^{-3}), benzophenone (10^{-4} mol dm^{-3}), and *p*-nitroacetophenone (5×10^{-6} mol dm^{-3}) was pulse-irradiated (G.E. Adams and Willson 1973) and the transient spectra recorded (Fig. 10.3). During the pulse e_{aq}^- reacts with acetone, and $H^\cdot$ and $OH^\cdot$ react with isopropanol, each to give $(CH_3)_2CO^{\cdot -}$. The radical undergoes electron transfer to acetophenone, as shown by the presence, 2 μs after the pulse, of the spectrum of the acetophenone radical

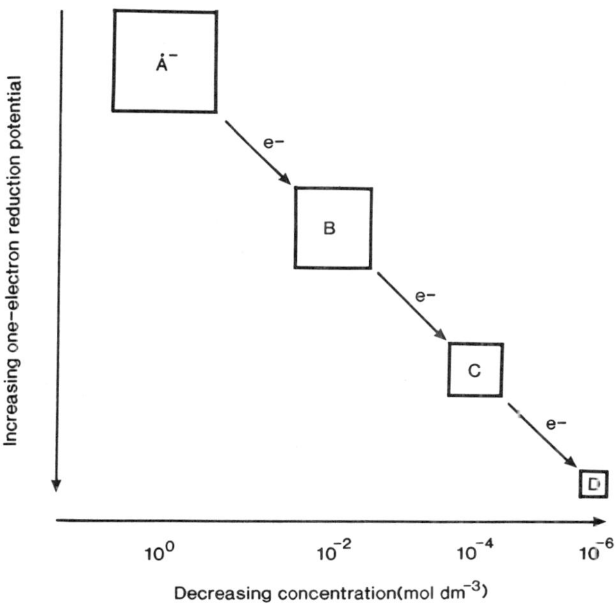

Fig. 10.2 Electron transfer from compounds of low reduction potential, but in high concentration, to compounds of high potential in low concentration. (Reprinted with permission from Willson in *Biochemical mechanisms of liver injury* (ed. Slater) pp. 123–234. Copyright 1978 by Academic Press Inc. (London) Ltd.)

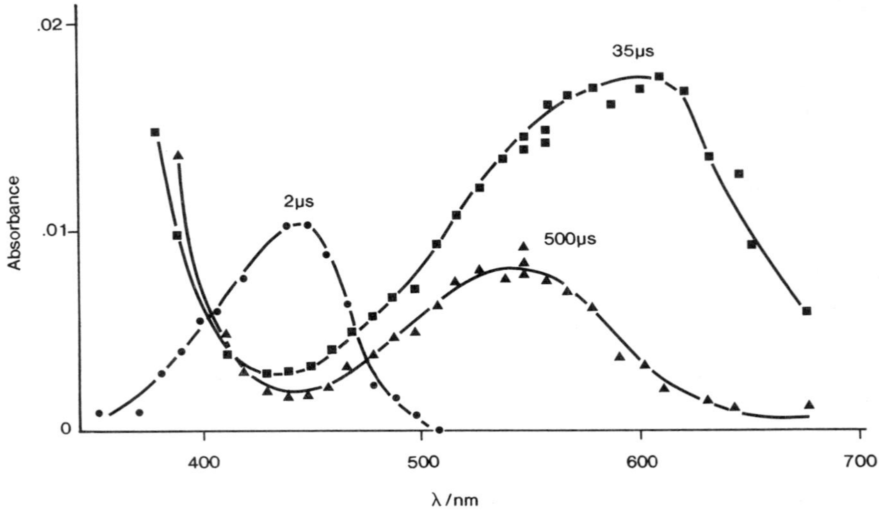

Fig. 10.3 Cascade electron transfer. Transient absorption spectra after 2, 35, and 500 μs on pulse radiolysis of aqueous solutions (pH 12.5) containing isopropanol (0.3 mol dm^{-3}), acetone (0.3 mol dm^{-3}), acetophenone (3×10^{-3} mol dm^{-3}), benzophenone (10^{-4} mol dm^{-3}), and *p*-nitroacetophenone (5×10^{-6} mol dm^{-3}) (taken from Adams and Willson 1973).

anion (λ_{max} 445 nm). This decays, and after 35 μs is replaced by that of benzophenone (λ_{max} 610 nm). In turn this spectrum decays and about 0.5 ms later is replaced with the spectrum of the PNAP radical anion (λ_{max} 545 nm). These spectral changes are due to chain or cascade electron transfer through the multicomponent system:

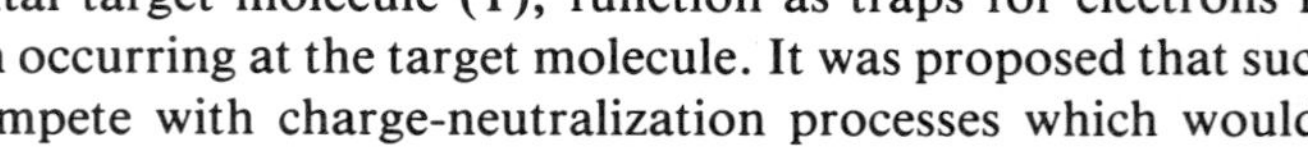

10.2.1.2 *Electron affinic mechanism*

The electron affinic mechanism of radiosensitization (G. E. Adams and Cooke 1969) postulates that electron affinic sensitizers bound to, or in close proximity with, a vital target molecule (T), function as traps for electrons liberated in ionization occurring at the target molecule. It was proposed that such reactions would compete with charge-neutralization processes which would otherwise tend to protect against irreversible damage to the molecule:

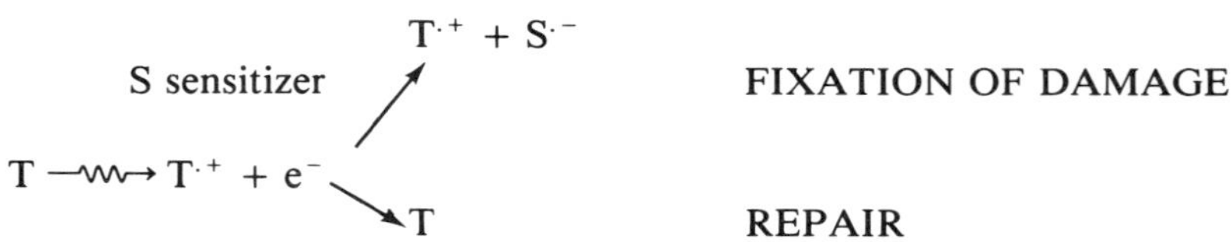

Table 10.1 Rate constants for electron transfer from the thymine anion to various radiosensitizers

Radiosensitizer	Rate constant/10^9 ($dm^3\ mol^{-1}\ s^{-1}$)	Reference[a]
P-Nitroacetophenone (pH 7)	4.8	a
Benzophenone (pH 12)	3.8	a
Menadione (pH 7)	4.0	a
Nifuroxime	6.4	b
Metronidazole	3.1	c
Misonidazole	3.1	c
4-Nitroimidazole	4.0	d

[a]a, G.E. Adams *et al.* (1972*c*); b, Greenstock and Dunlop (1973); c, Whillans *et al.* (1975); d, Greenstock *et al.* (1976).

($T^{\cdot +}$ and electrons may also be formed via indirect action).

It may be noted that, in general terms, radiosensitizers are oxidants whereas radioprotectors are reductants.

Much evidence suggests that DNA is the principal radiobiological target within a cell, and pulse radiolysis studies have shown that pyrimidines and purines are extremely reactive towards hydrated electrons and hydroxyl radicals (see Chapter 5). The rate of electron transfer from nucleic acid base electron adducts to various electron affinic compounds is almost diffusion-controlled. For example, Table 10.1 lists the rate constants for electron transfer from the thymine radical anion to a number of radiosensitizers, all the rate constants lying in the range $3.1–6.4 \times 10^9\ dm^3\ mol^{-1}\ s^{-1}$. The corresponding constants obtained with other pyrimidine and purine bases and their nucleosides and nucleotides lie in a similar range (G.E. Adams *et al.* 1972*c*; Whillans and Adams 1975). Rate constants for the oxidation of the products of $OH^{\cdot}$ reaction with nucleic acid bases by radiosensitizers are much slower. In this case the oxidation transfer efficiency is low, usually less than 10 per cent, indicating that a competing reaction occurs, often presumed to be adduct formation.

One of the earliest radiosensitizers tested was *N*-ethyl maleimide (NEM) (Fig. 10.1) (Bridges 1960). Since NEM was known to be a sulphydryl poison in some industrial processes, it was originally argued that if the natural radiosensitivity of cells is partly conditioned by sulphydryl compounds RSH occurring naturally in the cell as radioprotectors (see Section 10.1.5), removal of these compounds by chemical reaction might increase the radiosensitivity:

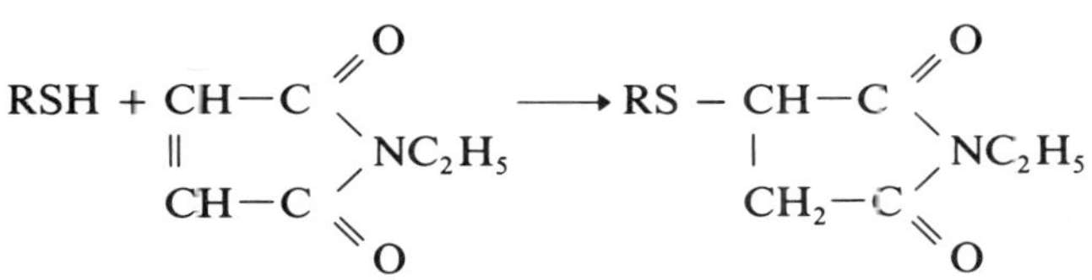

Although the sensitizing effect of NEM has been demonstrated in many bacterial cell systems, it is now believed that sulphydryl binding is not the major cause of sensitization by this compound (Bridges 1961). Thus bacteria incubated with NEM and then washed free of the sensitizer are not sensitized to subsequent irradiation. Furthermore, rapid mixing studies (see Section 10.2.3) by G.E. Adams *et al.* (1968*a*) showed that the time scale of sensitization of anoxic bacteria by NEM is much shorter than a reasonable estimate of the time required for binding of NEM with intracellular sulphydryl groups. NEM was subsequently considered to sensitize by fixing damage rather like oxygen, competing with intracellular sulphydryl compounds which otherwise might repair. The general hypothesis that sensitizers, by acting as electron transport agents, increase the efficiency of damage caused by reaction of hydrated electrons within the cell, was originally proposed to explain the sensitizing effect of NEM. Consistent with this, the thymine radical anion has been shown to transfer electrons efficiently to NEM with a rate constant of 8.0×10^8 dm^3 mol^{-1} s^{-1} (Greenstock *et al.* 1970).

10.2.1.3 *One-electron reduction potential-sensitization efficiency*

One-electron reduction potentials can be considered as measures of electron affinities and such redox properties of a large number of potentially useful hypoxic cell sensitizers have been determined. The principle of the pulse radiolysis method of determining reduction potentials was described in Chapter 3. Table 10.2 records the one-electron reduction potentials thus obtained for several representative potential radiosensitizers. A comprehensive list of such potentials has been compiled by Wardman (1989).

To achieve a given level of cell survival, a lower radiation dose is required in the presence of a sensitizer than in its absence. This radiation dose modification factor (enhancement ratio, ER) increases with the sensitizer concentration. Sensitization efficiency is defined as the concentration of the drug required to

Table 10.2 One-electron reduction potentials (E_7) of sensitizers at pH7

Sensitizer	E_7 (mV) (versus NHE)	Reference[a]
Metronidazole	−486	a
Misonidazole	−389	a
Etanidazole	−388	b
4-Nitroacetophenone	−355	c
Pimonidazole	−346	d
Nifuroxime	−253	c
Menadione	−240	e
Oxygen	−155	e

[a]a, Wardman and Clarke (1976); b, J.M. Brown and Workman (1980); c, Meisel and Neta (1975); d, Smithen *et al.* (1980); e, Meisel and Czapski (1975).

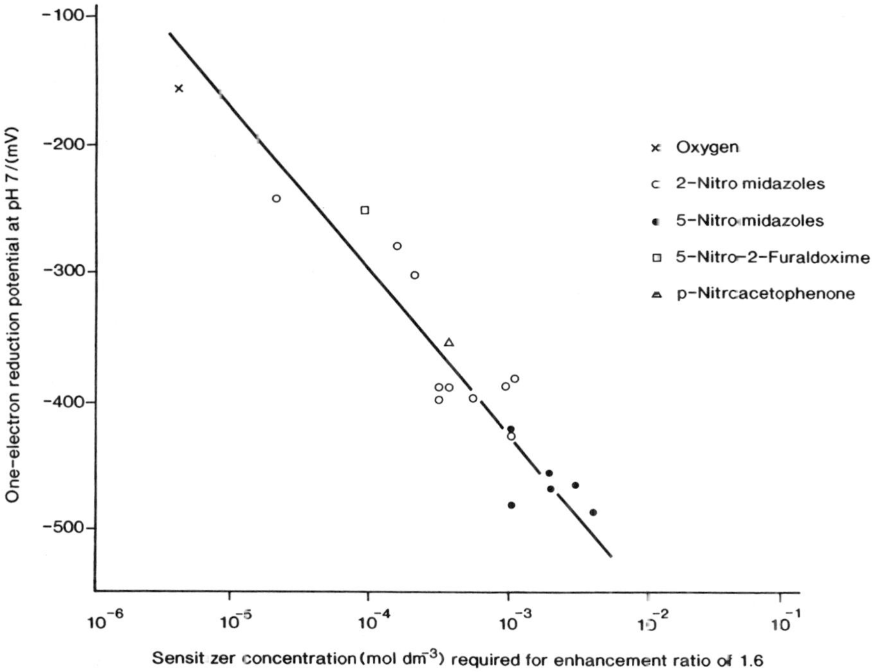

Fig. 10.4 Dependence of radiosensitization efficiency of nitro compounds with hypoxic Chinese hamster V79-379A cells *in vitro* with one-electron reduction potential (taken from Adams *et al.* 1976; Adams and Wardman 1977).

achieve a given ER. Figure 10.4 shows a plot of sensitization efficiencies of a range of hypoxic mammalian cell radiosensitizers as a function of their one-electron reduction potentials, (G.E. Adams *et al.* 1976; G.E. Adams and Wardman 1977). The one-electron reduction potential data used in this classic plot represent perhaps the most important contribution of pulsed irradiation methods to this field. The efficiencies were measured as the concentration of sensitizer in the growth medium required to give an ER of 1.6 in hypoxic Chinese hamster cells. The plot shows that as the one-electron reduction potential increases, so the concentration of sensitizer required to achieve an ER of 1.6 decreases. Oxygen, itself the most efficient radiosensitizer, appears to fall on the same straight line plot. A similar trend relating sensitization efficiency with reduction potential was also found for anoxic bacterial spores (Simic and Powers 1974).

Such 'electron-affinity relationships' can be described by the general equation:

$$-\log C = b_0 + b_1 E$$

where C is the concentration of sensitizer required to achieve a constant radiosensitization response, E is the one-electron reduction potential of the radiosensitizer and b_0 and b_1 are constants (G.E. Adams *et al.* 1979*a*). A typical value of $b_1 = 9.54\ V^{-1}$, derived (Wardman 1982) from the behaviour of 56 neutral nitro aryl compounds in mammalian cells (G.E. Adams *et al.* 1976, 1979*a*), indicates that a reduction potential increase of 0.1 V corresponds to an increased radiosensitization efficiency of around a 10-fold lower concentration to achieve the same response.

These, and subsequent studies (G.E. Adams *et al.* 1979*b*), show that electron affinity is the dominant property affecting the ability of a compound to act as a hypoxic cell radiosensitizer, although radiosensitization efficiencies greater than would be predicted on the basis of reduction potentials can be achieved with nitroimidazoles substituted with groups carrying basic functions (G.E. Adams *et al.* 1980). The importance of differing hydrophilicity, ionization, and proton-accepting properties in determining extracellular/intercellular sensitizer concentration gradients has also been stressed (Wardman 1984, 1987).

To date, most pulsed irradiation studies in the field of radiosensitization have dealt with dilute fluid solutions. Diffuse reflectance studies of dry DNA doped with various amounts of misonidazole (O'Neill *et al.* 1989*a*,*b*, 1990) indicate that the spectral characteristics of directly formed DNA-derived transient species, probably radicals, are modified by the radiosensitizer. The change in the yield of DNA derived species was found at one misonidazole molecule per ~50–100 base pairs, consistent with energy/charge trapping by misonidazole. Such processes can also be followed using 'in pulse' luminescence spectroscopy (Al-Kazwini *et al.* 1990*b*).

10.2.1.4 *Excited state reactions*

Although excited states are not generally thought to play a major role in the radiolysis of DNA, it is possible that some directly formed electronically excited DNA bases could transfer electrons to electron affinic drugs. Support for this is provided by the flash photolysis studies of Kemp *et al.* (1985, 1987*a*) which showed that the triplet states of thymine, thymidine, and uracil are quenched by a range of electron affinic molecules, including metronidazole and misonidazole, at rates which accord with the predictions of the Rehm–Weller relation (see Chapter 2) signifying an electron-transfer mechanism. An alternative pathway could be for electronic energy to be transferred from DNA to the drug, yielding excited drug which could abstract electrons or H atoms from ground state DNA. To help assess such a possibility, Kemp *et al.* (1987*b*) measured the triplet levels of metronidazole and misonidazole via energy transfer, both being 244 kJ mol^{-1}. These E_T levels, in conjunction with the one-electron reduction potentials of the ground states, led via the equation E_O (^{3}Metro/Metro$^{\cdot -}$) = E_O (Metro/Metro$^{\cdot -}$) + E_T to estimates of 2.04 and 2.14 V for the one-electron reduction potentials of triplet metronidazole and misonidazole, respectively (Kemp and Martins 1981). The reduction potentials of these triplet nitroheterocyclics are thus of the same order as that of OH·, and so strongly oxidizing properties would be anticipated.

10.2.2 *Other types of radiosensitizer*

It was noted above that most molecules considered as possible radiosensitizers are nitro-compounds. Somewhat related to these are the stable organic nitroxyl free radicals which are derivatives of nitric oxide (NO). These were investigated because of the similarity between oxygen and nitric oxide, and Fig. 10.1 includes the structures of two such radicals, triacetoneamine-*N*-oxyl (TAN) and nor-pseudopelletierine-*N*-oxyl (NPPN). Another class of radiosensitizers are the halogenated pyrimidines, for example, 5-bromouracil and 5-bromodeoxyuridine (5-BUdR).

For the nitroxyl free radicals it is suggested that the mechanism of sensitization resembles the fixation process proposed for oxygen in that these radicals form adducts with free radical centres in DNA. To illustrate this it has been shown that TAN reacts rapidly with the electron adduct of thymine (Willson 1971), and also with the OH· adducts such as thymine–OH· (Emmerson and Willson 1968), although such adduct formation may not be a simple bimolecular process (P.B. Roberts and Fielden 1971). The NO group of NPPN is less sterically blocked than is TAN and this results in a somewhat greater efficiency for this compound as a radiosensitizer (see, for example, Emmerson *et al.* 1971; P.B. Roberts and Fielden 1971).

For the halogenated pyrimidines it should be noted that since the bromine atom is about the same size as the methyl group, 5-bromouracil can replace thymine in DNA. G.E. Adams (1967) has suggested that the enhanced sensitivity of this replacement is due to the high electronegativity of the bromine atom in the incorporated base.

10.2.3 *Rapid mixing studies*

Consistent with the suggestion that the mechanism of action of radiosensitizers involves fast free radical processes, are the novel fast-mixing and irradiation studies of Michael *et al.* (1973, 1978, 1979). For example, using 'bursts' of oxygen at varying times before and after the electron pulse and studying bacterial (*Serratia marcescens*) survival time allowed these workers to monitor the kinetics of two types of oxygen-dependent damage and to distinguish somewhat different modes of damage fixation with different radiosensitizers. Analogous kinetic studies of the reactions of oxygen and sulphydryls with irradiated transforming DNA from *Bacillus subtilis* (Held *et al.* 1984), and of the repair of radical precursors of DNA single- and double-strand breaks in Chinese hamster V79 cells (Michael *et al.* 1986), have also been carried out.

10.3 Summary

As mammalian cells consist of ~80 per cent water, the radiation chemistry of water is very important with respect to the mechanisms involved in radiotherapy. Different types of high energy radiation produce varying relative amounts of radical and molecular products. With low LET radiation (e.g. hard X-rays, γ-rays, and fast electrons), ions formed in the track are well spaced and few molecular products are formed. With high LET radiation (e.g. α particles) high ionization densities are formed in the track favouring the molecular products.

Radiation protection mechanisms may involve fast initial radiation chemical processes or changes in cell biochemistry or physiology. Protectors may thus interfere with free radical reactions associated with damage to vital molecules, or modify the metabolism of cells affected by these initial chemical reactions. One model to explain radioprotection by sulphydryl compounds postulates that a radical formed in a cellular target, either by direct or indirect action, is restored by H atom transfer from the labile SH group.

In the presence of oxygen this target radical can be converted into a peroxyl radical which cannot be repaired, this being the radiation chemical interpretation of the oxygen effect. Oxygen effects are universal phenomena in radiobiology. Oxygen sensitizes lethal effects and the induction of division delay, gene mutations, and chromosome aberrations. The sensitization effect is different for different types of high energy radiation, the oxygen enhancement ratio (OER) being ~3 for low LET radiation and generally 1.2–1.5 for high LET radiation, the variation being due to the track differences mentioned earlier. The feature that high LET radiation-induced reactions are more difficult to influence than low LET radiation-induced reactions applies not only to oxygen effects but also to radioprotection.

The relative radioresistance of hypoxic (i.e. low oxygen content) cells present in some human tumours is considered to be an important cause of failure of radiotherapy. An attempt to overcome this problem has been the development over many years of a group of mainly nitro compounds which selectively increase the radiosensitivity of hypoxic cells without affecting the ionizing radiation response of well-oxygenated normal tissue. Such electron affinic sensitizers are believed to function as traps for electrons liberated by ionization of the target molecule, these reactions competing with charge neutralization processes which would tend to protect against irreversible damage. DNA is the principal radiobiological target, and electron adducts of DNA bases efficiently transfer electrons to many such sensitizers. A correlation has been demonstrated between radiosensitization efficiency and the sensitizer one-electron reduction potentials measured by pulse radiolysis. Oxygen, itself the most effective radiosensitizer, appears to fit the same correlation.

The fast mixing and irradiation technique developed to study chemical processes involved in radiosensitization lasting up to 10^{-2} s after cellular energy deposition appears to confirm that sensitizers act primarily by fast free radical processes.

It is important to note that, so far, toxic side effects have limited the use of radiosensitizers and radioprotectors to clinical trials.

11

Cancer: chemotherapy

Diseases desperate grown,
By desperate appliances are reliev'd
Or not at all

William Shakespeare
Hamlet iii 9

11.1 Introduction

The first compound tested for antitumour activity by the National Cancer Institute of the USA was the simple quinone, 2-methyl-*p*-benzoquinone. Since 1955 over 1500 potential drugs have been screened for antitumour activity, the majority of which are quinones.

The number of cancer chemotherapeutic agents that have been studied by flash photolysis and pulse radiolysis methods is increasing rapidly. In this chapter, we consider a group of quinone antitumour agents (anthracyclines and mitomycin C) where such studies have contributed towards an understanding of their mode of action or side effects. Pulsed radiation techniques have also recently begun to be used to study other non-quinone antitumour chemotherapeutic agents including the metal-chelator bleomycin (Goldstein and Czapski 1987) 3-amino-1,2,4-benzotriazine-1,4-dioxide (SR 4233) (Laderoute *et al.* 1988), and several ellipticines (Sekaki *et al.* 1989). Studies relating to a chemotherapeutic approach towards malignant melanoma, based on exploiting the enhanced pigment producing systems of melanotic melanomas to generate toxic agents exclusively within such melanocytes, are described in Chapter 8.

11.2 Anthracyclines

The anthracycline glycosides are the largest class of clinically used antitumour quinones, several hundred anthracycline analogues having been prepared or extracted. Adriamycin (doxorubicin) and daunomycin (daunorubicin) (see Fig. 11.1) are by far the most widely used and it is estimated that more than two million cancer patients have received adriamycin. Despite their central position in the chemotherapeutic control of cancer, they remain antibiotics of last resort and exhibit toxicity to the host organism as well as the neoplasm.

A characteristic feature of the quinone moiety is its ability to undergo reversible reduction and form semiquinone radicals. Although there is evidence that the interaction of the anthracyclines with DNA and membranes may comprise the core of the antitumour effects, with neither event requiring redox chemistry, semiquinone radicals may lead to cytotoxicity and, in particular, cardiotoxicity (Abdella and Fisher 1985; Powis 1987, 1989). Adriamycin, like other anthracyclines, is known to augment the flow of electrons from NADPH to molecular oxygen. In this reaction, an intermediate semiquinone radical is formed by interaction with mammalian microsomes (Sato *et al.* 1977; Bachur *et al.* 1978).

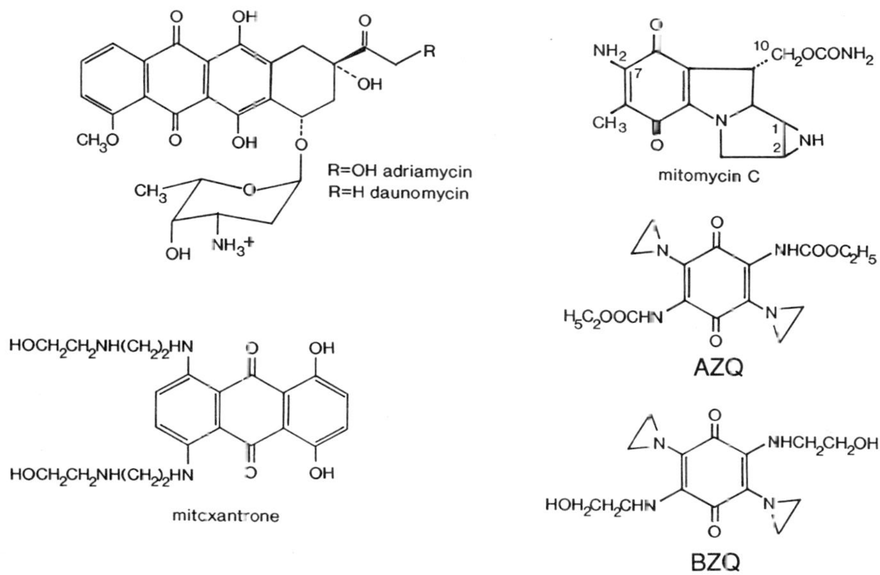

Fig. 11.1 Structures of some quinone antitumour agents.

From ESR studies it has been shown that the semiquinone radicals may serve to shuttle the electrons to oxygen, a reaction catalysed by NADPH–cytochrome P-450 reductase (Bachur *et al.* 1979). The semiquinone radicals could be responsible for toxicity, and be site-specific, binding to DNA. They could either react directly, or they could generate other radicals such as $O_2^{\cdot-}$ or $OH^{\cdot}$, which could react with the proximal DNA, producing the strand scission reported for adriamycin (Neidle and Waring 1983).

Figure 11.2 shows a general scheme that has been proposed for enzymatic reductive activation of the anthracycline and subsequent generation of $O_2^{\cdot-}$, H_2O_2, and $OH^{\cdot}$. This figure also shows that deglycosylation of the hydroquinone can also occur leading to an unstable quinone methide which tautomerizes to the 7-deoxyaglycone. Pulse radiolysis methods have contributed significantly towards our understanding of the properties of anthracycline

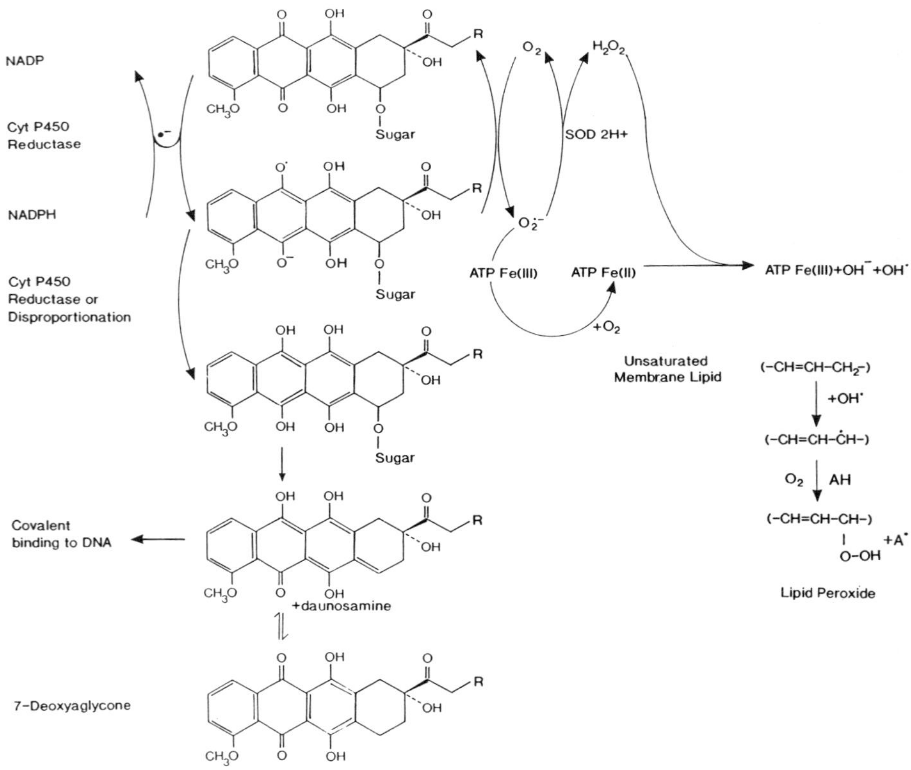

Fig. 11.2 Enzymatic reductive activation of anthracyline leading to deglycosylation and formation of the aglycone, and concomitant interaction with molecular oxygen leading to active oxygen species and lipid peroxidation. (Reprinted with permission from *Adv. Free Radical Biology and Medicine*, **1**, Lown, J. W., Molecular mechanisms of action of anticancer agents involving free radical intermediates. Copyright (1985) Pergamon Press Ltd.)

semiquinones and also the reactive aglycone tautomers obtained on reduction under anaerobic conditions. They provide a firm physico-chemical basis on which theories concerning the action of the agents can rest.

Although anthracyclines absorb in the visible, flash photolysis measurements tend to suggest that it is unlikely that these compounds would be useful as photodynamic sensitizers (see Chapter 12) on account of their relatively poor abilities to photosensitize singlet oxygen formation (Andreoni *et al.* 1989). However, the photosensitized degradation of DNA by daunomycin has been observed (P. J. Gray *et al.* 1982) and also the photogeneration of $O_2^{\cdot -}$ (Carmichael *et al.* 1983). Indeed, some examples of new classes of anticancer chemotherapeutic agents, including anthrapyrazoles and anthracenediones (cf. mitoxantrone; Fig. 11.1), are strongly phototoxic to cancer cells *in vitro*, and have been proposed as alternatives to the cardiotoxic clinical anthracyclines (Reszka *et al.* 1990). These new agents are more resistant to enzymatic reduction (see Section 11.4) while retaining the structural characteristics necessary for DNA binding.

11.2.1 *The model naphthazarin*

It is useful to study the ways in which simple quinones form radicals in order to gain insight into the way more complex quinones might produce their toxic effects. Naphthazarin provides a simple model of the internally hydrogen-bonded redox active quinonoid part of adriamycin and daunomycin.

In aqueous solution naphthazarin can exist in the fully protonated form, represented as NH_2, or in the forms NH^- or N^{2-} depending on the pH:

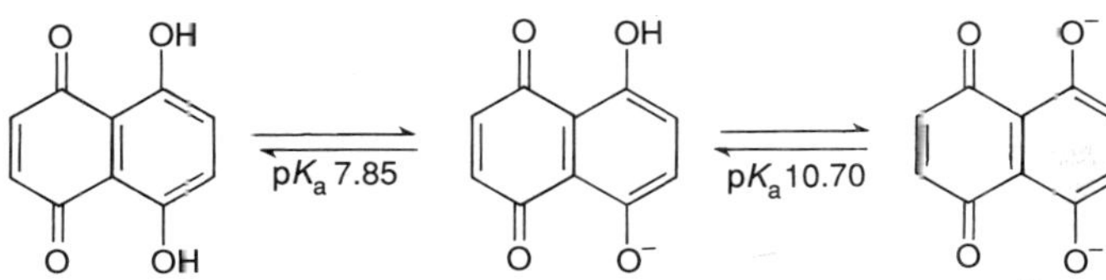

NH_2 NH^- N^{2-}

11.2.1.1 *Addition of a single electron*

Pulse radiolytic one-electron reduction of dilute aqueous solutions of naphthazarin containing high concentrations of formate or formic acid (Land *et al.* 1983*a*) leads to various protonated forms of the semiquinone:

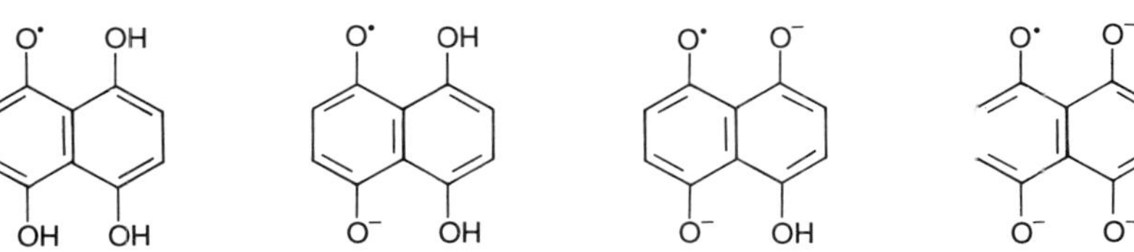

$NH_3^{\cdot}$ $NH_2^{\cdot -}$ $NH^{\cdot 2-}$ $N^{\cdot 3-}$

Measurement of changes in semiquinone spectra with pH allowed three of the four forms of the semiquinone to be distinguished with pK_a values of 2.7 and $>$13.8. The charge on the semiquinone at pH 4.65 was found to be -1 by determination of the effects of ionic strength on the rate of mutual reaction of the radicals. The radical present at this pH is thus assigned to $NH_2^{\cdot -}$, the pK_a of 2.7 being attributed to $NH_3^{\cdot} \rightleftharpoons NH_2^{\cdot -} + H^+$. This p$K_a$ value is 1.4 units less than the pK_a of the protonated form of 1,4-naphthasemiquinone, probably because the anion $NH_2^{\cdot -}$ is stabilized by intramolecular hydrogen bonding involving the 5- and 8-hydroxy groups:

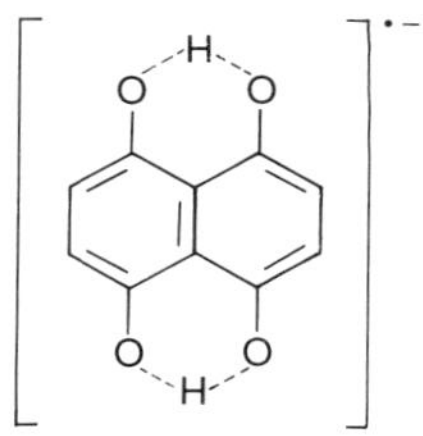

Because of this high stabilization, the pK_a > 13.8 is attributed to the dissociation of $NH_2^{\cdot -}$.

The one-electron reduction potential of naphthazarin at pH 5.5 was obtained by setting up an equilibrium between naphthazarin and its semi-reduced form, and the known standard 2,6 dimethyl-1,4-benzoquinone and its semi-reduced form. An $E^1_{5.5}$ of -104 mV was found for $NH_2/NH_2^{\cdot -}$.

Such equilibrium measurements were repeated for a range of pH values using various redox standards including oxygen, riboflavin, and a viologen, yielding the experimental values shown in Fig. 11.3(a). Inspection of this figure shows that the one-electron reduction potential of naphthazarin is almost independent of pH in the range 5–7. This shows that no proton is involved in the reduction process and so provides further evidence that the radical exists as $NH_2^{\cdot -}$ in this region.

An expression can be developed to relate the reduction potential at any one pH to the reduction potential at any other provided the pK_a values of the parent and semiquinone are known. The solid line of Fig. 11.3(a) was calculated from this expression on the basis of $E_{5.5} = -104$ mV (versus NHE), and it can be seen that there is good agreement between the calculated and experimental values.

11.2.1.2 *Second one-electron step*

A number of examples are known where semiquinones are present at appreciable concentrations in solutions containing the quinone and the corresponding hydroquinone. While studying the one-electron reduction of naphthazarin, it was found (Land *et al.* 1983b) that although the semiquinone dispropor-

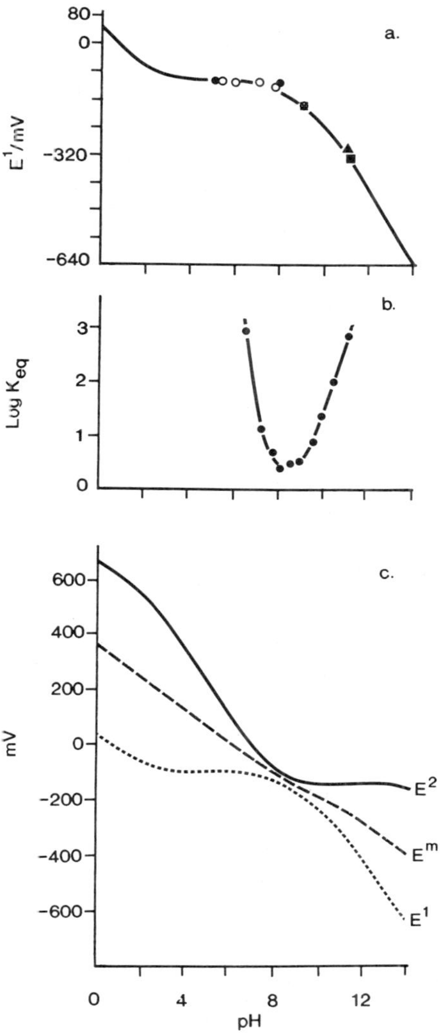

Fig. 11.3 Variation with pH of (a) the one-electron reduction potential of naphthazarin (the standards used were: ○, 2,6-dimethyl-1,4-benzoquinone; ●, oxygen; X, duroquinone; ■, riboflavin; and ▲, hydroxyethylviologen); (b) log K_{eq} for naphthazarin; (c) first one-electron reduction potential (E^1, . . .), second one-electron reduction potential (E^2, —), and two-electron reduction potential of naphthazarin (E^m, -----) (taken from Land *et al.* 1983*a*, *b*).

tionates completely to quinone + hydroquinone at extremes of pH, it is intrinsically rather stable in weak alkali. The reaction:

$$2 \text{ semiquinone} \rightleftharpoons \text{quinone} + \text{hydroquinone}$$

was studied over the pH range 1–14. Within the approximate range 6–11, the semiquinone produced by the pulse did not disappear completely but became reduced to a concentration which remained constant for hundreds of milliseconds at least. The concentrations of semiquinone radicals found at equilibrium were used to calculate the corresponding stability constants (K_{eq}) shown in Fig. 11.3(b). Such stability constants are related to the one-electron reduction potentials for the quinone and the semiquinone (at 25°C) via the expression:

$$E(\text{quinone/semiquinone}) - E(\text{semiquinone/hydroquinone}) = -59 \log K_{eq}$$

The two one-electron reduction potentials are related in turn to the two-electron reduction potential of the quinone, *E*(quinone/hydroquinone), by the expression:

$$E(\text{quinone/semiquinone}) + E(\text{semiquinone/hydroquinone}) = 2E(\text{quinone/hydroquinone})$$

The second one-electron and the two-electron reduction potentials thus calculated from the naphthazarin first one-electron reduction potential given in Fig. 11.3(a), and the stability constants given in Fig. 11.3(b), are shown in Fig. 11.3(c).

11.2.1.3 *Removal of a single electron*

As a dihydroxyquinone, naphthazarin can form semiquinones by one-electron oxidation as well as one-electron reduction. Indeed, the one-electron oxidation of adriamycin has been implicated in one ESR study in aqueous solution at physiological pH (Pietronigo *et al.* 1979).

The reactions of the oxidizing radicals $OH^{\cdot}$, $O^{\cdot -}$, $N_3^{\cdot}$, $Br_2^{\cdot -}$, $SO_4^{\cdot -}$, $\dot{C}H_2CHO$, and $\dot{C}HCHO^-$ with naphthazarin have been studied by pulse radiolysis (Mukherjee *et al.* 1988). Although all these oxidants give some semioxidized naphthazarin, other species such as adducts may also be formed. The radicals $SO_4^{\cdot -}$, $O^{\cdot -}$, and $N_3^{\cdot}$ in strong alkali appear to lead predominantly to semioxidized naphthazarin which, above pH 5, exists in the anionic form:

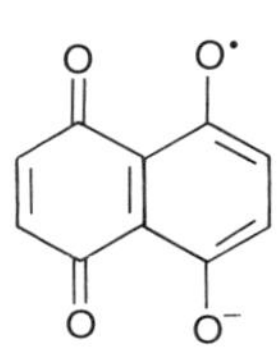

The decay of semioxidized naphthazarin, like semireduced naphthazarin, was strongly pH dependent, although in the former case a satisfactory explanation for the varying kinetics observed is not yet available.

11.2.2 *Adriamycin and daunomycin*

In aqueous solution adriamycin can, like naphthazarin, exist in several states of protonation, and Sturgeon and Schulman (1977) have estimated the

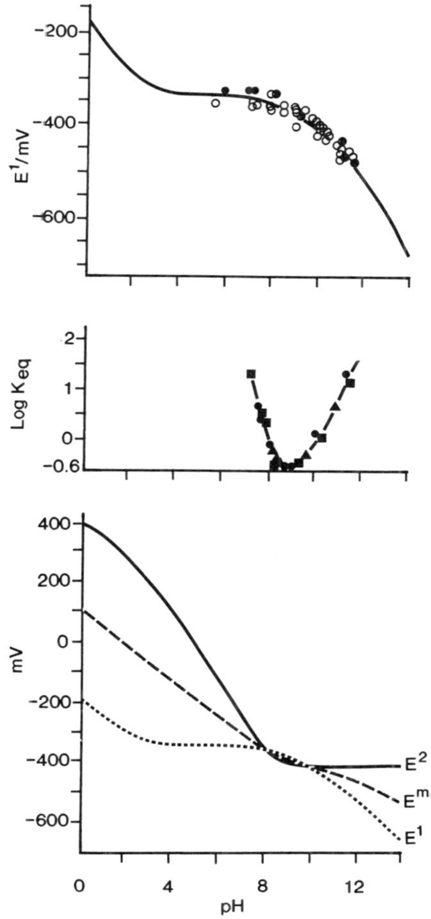

Fig. 11.4 Variation with pH of (a) the one-electron reduction potentials of adriamycin (●) and daunomycin (○); (b) log K_{eq} for adriamycin (●), daunomycin (■), and for both adriamycin *and* daunomycin (▲); (c) reduction potentials of adriamycin and daunomycin: first one-electron reduction potential (E^1, . . .), second one-electron reduction potential (E^2, —), two electron reduction potential (E^m, -----) (taken from Mukherjee *et al.* 1989).

following pK_a values.

$$^{+}HQH_2 \rightleftharpoons H^+ + QH_2 \qquad pK_a\ 8.22$$

$$^{+}HQH_2 \rightleftharpoons H^+ + {}^{+}HQH^- \qquad pK_a\ 9.01$$

$$^{+}HQH^- \rightleftharpoons H^+ + QH^- \qquad pK_a\ 9.36$$

$$QH_2 \rightleftharpoons H^+ + QH^- \qquad pK_a\ 10.1$$

$$QH^- \rightleftharpoons H^+ + Q^{2-} \qquad pK_a\ 13.2$$

Fully protonated adriamycin is represented as $^{+}HQH_2$ where the first H refers to that on the amino sugar and the other two to those on the hydroquinone.

11.2.2.1 *Addition of a single electron*

Pulse radiolysis of aqueous solutions of adriamycin under reducing conditions leads to semiquinone radicals in various states of protonation (Svingen and Powis 1981, Houée-Levin *et al.* 1985; Land *et al.* 1985). In the acid region, the variation with pH of the resultant absorption changes are consistent with a semiquinone pK_a of 2.9. The corresponding pK_a value found with daunomycin (Mukherjee *et al.* 1989) was the same within experimental error.

The one-electron reduction potential of adriamycin and daunomycin was estimated as a function of pH (Land *et al.* 1983*c*; Mukherjee *et al.* 1989) in exactly the same manner as employed with naphthazarin. The results obtained, which were again identical for adriamycin and daunomycin, are shown in Fig. 11.4(a). The near independence of pH of the one-electron reduction potential in the region around pH 7 shows that the reduction process in this region involves neither the uptake nor the liberation of a proton. The amino group in the parent molecule is protonated at such pH values while both hydroxy groups on the hydroquinone are undissociated. Since the reduction around pH 7 must take place at the quinone part of the molecule, the reduction process in this region is:

$$^{+}HQH_2 + e^- \rightarrow {}^{+}HQH_2^{\cdot -}$$

Consequently, the pK_a of the adriamycin semiquinone at 2.9 must be due to the equilibrium:

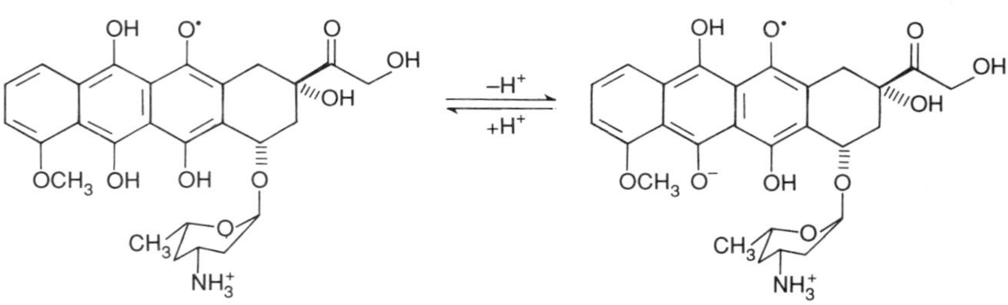

The best fit of the calculated and experimental reduction potentials against pH shown in Fig. 11.4(a) were based on a pK_a of 9.2 for the NH_3^+ group of the semiquinone sugar moiety, and a value of $E_7^1 = -341$ mV for both adriamycin and daunomycin. Several other closely related anthracyclines including 4-epiadriamycin, 4-deoxyadriamycin, 4-demethoxydaunorubicin, and carminomycin have closely similar E_7^1 values (Hoey *et al.* 1988*a*).

Adriamycin semiquinone reacts with oxygen with a rate constant of 3.5×10^8 dm^3 mol^{-1} s^{-1} (Land *et al.* 1983*c*, 1985). The reduction potential of −341 mV can be combined with the reduction potential of oxygen (−155 mV) to calculate the equilibrium constant, *K*, for the reaction:

$$O_2^{\cdot -} + \text{adriamycin} \rightleftharpoons O_2 + \text{adriamycin semiquinone}.$$

The value at pH 7 is 7.04×10^{-4} and so this equilibrium is well over to the left. The *K* value remains the same down to about pH 5 where protonation of $O_2^{\cdot -}$ begins to be significant. On the alkaline side, where the adriamycin semiquinone is in the negatively charged form instead of the zwitterionic form and the parent deprotonates, the equilibrium would be even more to the left. Similar considerations apply to the corresponding semiquinones of daunomycin.

11.2.2.2 *Second one-electron step*

In strongly acid and alkaline solutions in the absence of oxygen, the adriamycin and daunomycin semiquinone absorptions decay in a second-order manner attributed to disproportionation of the semiquinone to yield parent and hydroquinone (Land *et al.* 1985; Mukherjee *et al.* 1989). Within the pH range ~5 to 11, the semiquinone absorption at first decreased and then, within 1 or 2 ms, attained a value which remained constant for at least 10–20 ms. This behaviour, analogous to naphthazarin, is due to the equilibrium:

$$2 \text{ semiquinone} \rightleftharpoons \text{quinone} + \text{hydroquinone}$$

although with adriamycin and daunomycin the equilibrium appears to be much more in favour of the semiquinone. Similar results at pH 7 were obtained for daunomycin by Houée-Levin *et al.* (1985).

Fig. 11.4(b) shows the variation in the equilibrium constant, K_{eq}, with pH for solutions containing adriamycin or daunomycin. As for naphthazarin, these stability constants were used in conjunction with the one-electron reduction potentials given in Fig. 11.4(a) to calculate the second one-electron and two-electron reduction potentials of adriamycin and daunomycin illustrated in Fig. 11.4(c).

11.2.2.3 *Deglycosylation*

In the case of naphthazarin, the above equilibrium could be maintained indefinitely in the absence of oxygen, as confirmed by ESR measurements of the radicals in the same solutions (Dodd and Mukherjee 1984; Mukherjee *et al.* 1987). In the case of adriamycin (Lea *et al.* 1989) and daunomycin (Houée-Levin *et al.* 1986) on the other hand, further changes in absorption occur over hundreds of milliseconds and longer. Mainly on the basis of earlier studies involving the chemical reduction of daunomycin with the radical dimer bi(3,5,5-trimethyl-2-oxomorpholin-3-yl) in methanol (Kleyer and Koch 1983), these slow changes lasting seconds have been attributed to deglycosylation of the hydroquinone to form the quinone methide tautomer. However, since the semiquinone is in equilibrium with hydroquinone and parent, loss of the sugar from semiquinone and hydroquinone are indistinguishable. The quinone methide decays over minutes to 7-deoxyadriamycinone. Analogous reactions occur with daunomycin.

To test the fit of this mechanism, shown in Fig. 11.5, to the pulse radiolytic data (Lea *et al.* 1989), the expected concentrations of the various species were calculated as a function of time by numerical integration of the corresponding rate equations. Optimum agreement between the changes in concentrations of the relevant species calculated in this way and the experimental results could be obtained if the rate constant for the deglycosylation reaction is $1.1\ s^{-1}$ and the rate constant for the quinone methide to 7-deoxyadriamycinone reaction is $1.5 \times 10^{-2}\ s^{-1}$, all at pH 7. In contrast to the findings of Houée-Levin *et al.* (1986), Lea *et al.* (1989) found no evidence for any additional intermediates involved in the process hydroquinone to quinone methide tautomer.

It is interesting to note that similar aglycones have been found as metabolites in the plasma of cancer patients treated with anthracyclines (Cummings *et al.* 1987*a*, *b*).

11.3 **Mitomycin C**

The aziridine mitomycin C (Fig. 11.1) is believed to function by inhibiting DNA replication, this being brought about by covalent binding to cellular DNA followed by crosslinking between the complementary strands. It is generally accepted that mitomycin C must first of all be activated by reduction before binding can occur via aziridine ring opening.

It was originally believed that the bioactivation consisted of a two-electron reduction of mitomycin C to the hydroquinone, resulting in the molecular rearrangements required to produce reactive centres at carbons 1 and 10. Another more widely held view was that the drug is activated by a one-electron pathway in which the semiquinone radical undergoes the changes intramolecularly. Radiolytic methods have been used (Hoey *et al.* 1988*b*; Machtalere *et al.* 1988) to help account for the number of different products produced when mitomycin C is reduced by one-electron enzyme systems, and to decide whether ring opening occurs at the semiquinone or hydroquinone stage.

Both enzymic (using hypoxanthine/xanthine oxidase) and radiolytic (using $CO_2^{\cdot -}$) reduction of mitomycin C was shown by HPLC to give rise to the same three products, the *cis* and *trans* forms of 2,7-deamino-1-hydroxymitosene (B, Fig. 11.6) and 2,7-diaminomitosene (C, Fig. 11.6), in the same proportion (Hoey *et al.* 1988*b*).

Pulse radiolysis of mitomycin C in the presence of excess formate gave rise

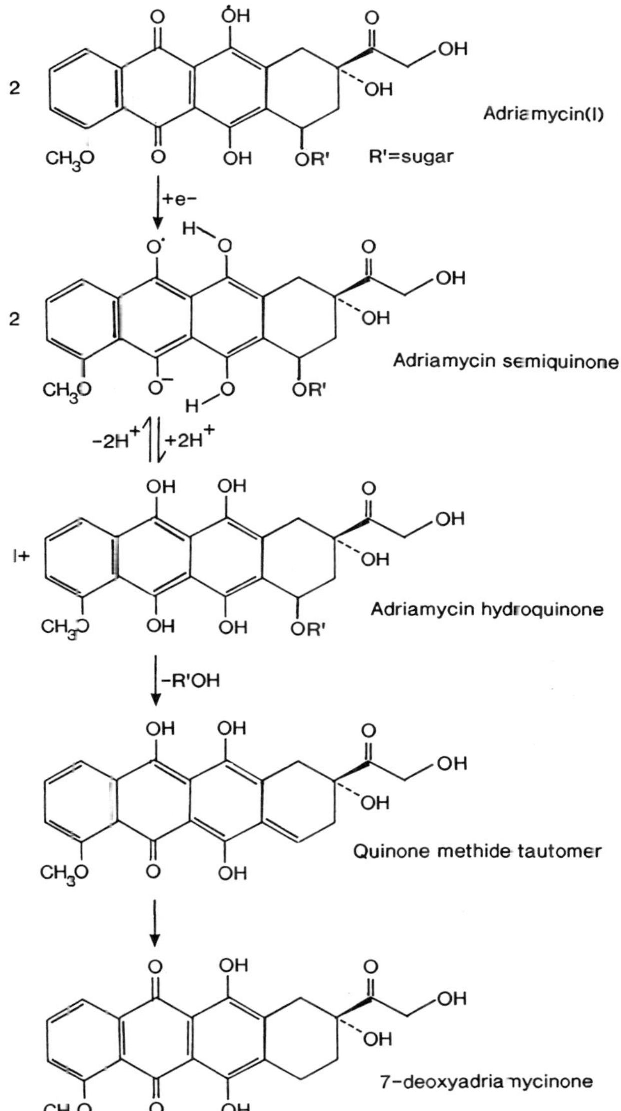

Fig. 11.5 Scheme for reductive deglycosylation of adriamycin (taken from Lea *et al.* 1989).

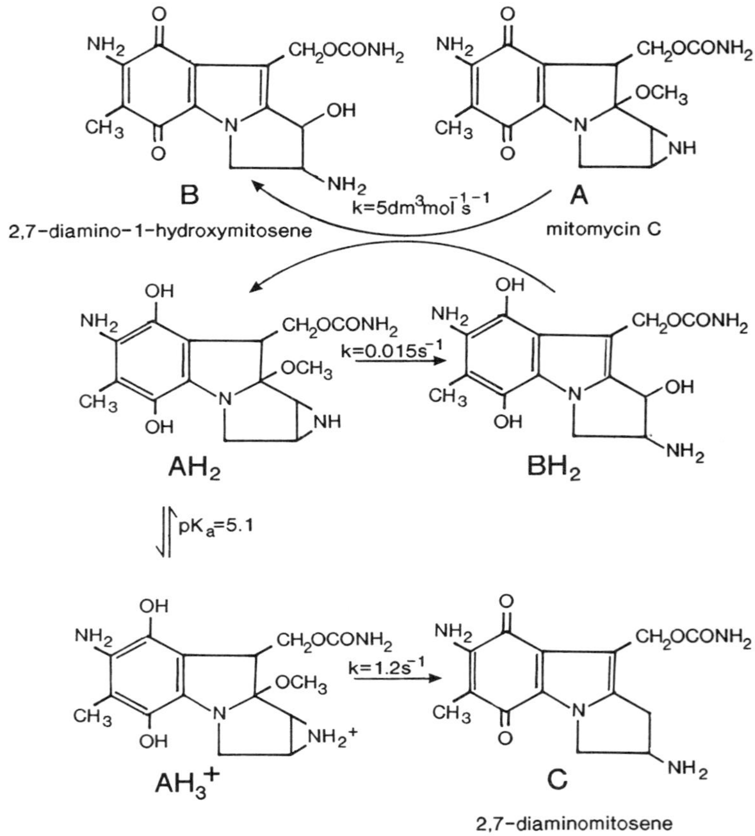

Fig. 11.6 Scheme for reactions taking place after reduction of mitomycin C (taken from Hoey *et al.* 1988*b*).

to the semiquinone which then decayed exclusively and completely by disproportionation into the hydroquinone and parent quinone (Hoey *et al.* 1988*b*). The rate constant for this disproportionation at pH 7.0 was $2k = 5.1 \times 10^7$ $dm^3\ mol^{-1}\ s^{-1}$, independent of dose and mitomycin C concentration. This suggests that aziridine ring opening must occur after the hydroquinone has been formed.

Admission of oxygen within a second of formation of the hydroquinone at pH 7.0 caused the absorption spectrum to revert to that of the unirradiated solution, again showing that the semiquinone did not undergo rearrangement before dismutation.

Further evidence against the ring opening reaction occurring at the semiquinone stage was obtained by analysing the yields of products formed from a dose of radiation delivered from an accelerator (which favours dismutation of the semiquinone), and a low dose-rate ^{137}Cs source (which would prolong the lifetime of the semiquinone). At pH 7.0, the yields of mitosene products were the same in the two cases, consistent with their being formed from the hydroquinone.

The concentrations of the mitosenes formed and their relative amounts were strongly dependent upon pH. Higher yields of products were obtained in alkaline solution.

The hydroquinone of mitomycin C formed on pulse radiolysis was observed to undergo further reactions over periods of tens of seconds, the spectral changes being consistent with the formation of mitosene quinones as found after enzymic reduction. The complete mechanism was determined using pulse radiolysis over time scales between 0.1 s and 1 min at various pH values, the scheme which fits the results is shown in Fig. 11.6. The essence of this scheme is that below pH 7.0 a protonated form of the hydroquinone (AH_3^+) undergoes an intramolecular rearrangement to produce 2,7-diaminomitosene (C), while at higher pH a conversion of the deprotonated form (AH_2) to BH_2 predominates. The BH_2 then takes part in a chain reaction involving unchanged mitomycin C:

$$AH_2 \rightarrow BH_2$$

$$BH_2 + A \rightarrow B + AH_2$$

The cytotoxicity of mitomycin C is affected by the availability of oxygen and the pH of the environment. Since the mitomycin C semiquinone reacts rapidly with oxygen ($k = 2.2 \times 10^8$ dm^3 mol^{-1} s^{-1}; Butler *et al.* 1985) to form $O_2^{\cdot-}$, it is possible that some killing of aerobic cells by mitomycin C could occur via $O_2^{\cdot-}$. The semiquinone and hydroquinone of mitomycin C react rapidly with oxygen, whereas the reductive processes are quite slow. Activation (reduction) of mitomycin C may therefore only occur when redox coupling of the drug and oxygen has significantly reduced the oxygen tension.

Crosslinking of DNA in EMT6 tumour cells by mitomycin C is known to be strongly pH dependent increasing from pH 7.5 to 5.7 (K.A. Kennedy *et al.* 1985). The pulse radiolysis results suggest that the crosslinking precursor could be the protonated hydroquinone, and that competition could exist between such beneficial adduct formation and the fruitless production of mitosenes.

11.4 Correlation between one-electron reduction potential and ease of enzymatic reduction

It is generally accepted that there is no specific quinone-reductase involved in the activation of antitumour quinones. It is believed, however, that quinones can be reduced by a variety of enzyme systems including dehydrogenases,

Table 11.1 One-electron reduction potentials and rates of reduction by xanthine oxidase/hypoxanthine for various quinone antitumour agents (adapted from Butler *et al.* 1987)

Compound	E^1_7 (mV) (versus NHE)	nmol.min.$^{-1}$unit of enzyme^{-1}
AZQ	−70	172
Mitomycin C	−310[a]	89
Adriamycin	−341	37
BZQ	−376	32

[a]Butler and Hoey (1986).

NADPH cytochrome P-450 reductase, DT diaphorase, and xanthine oxidase. There are no specific quinone-binding sites on these enzymes and the quinones merely divert reducing equivalents from the normal pathways.

Table 11.1 demonstrates a correlation between the values of the one-electron reduction potentials of several quinone antitumour agents, measured by pulse radiolysis, and the rates at which these quinones accept electrons from the xanthine/hypoxanthine system (Butler *et al.* 1987). A similar dependence was reported earlier for the abilities of quinones to be reduced by NADPH cytochrome P-450 reductase (Svingen and Powis 1981).

The relationship between E^1_7 and enzymatic reduction rate is not simple, but it is expected that at some point the reduction potential of the quinone couple must become so negative that the enzyme system cannot reduce them even at a very slow rate. Mitoxantrone (Fig. 11.1), generally considered to be less cardiotoxic than adriamycin ($E^1_7 = -341$ mV), comes into this category ($E^1_7 = -527$ mV; Butler and Hoey 1987). The dramatic difference in one-electron reduction potentials of AZQ and BZQ (Fig. 11.1) (Butler *et al.* 1987), reflected in the difference in the cytotoxic mechanisms associated with these diaziridinyl drugs (Butler *et al.* 1990), may be due to the greater electron-donating ability of the amine in the $-NHCH_2CH_2OH$ group of BZQ towards the quinone moiety, compared with that of the amide in the $-NHCOOC_2H_5$ group of AZQ. The cytotoxicities of eight analogues of AZQ have also been recently shown to correlate with the ease of reduction of the compounds reflected by their one-electron reduction potentials (Dzielendziak *et al.* 1990).

11.5 Summary

The success of adriamycin and daunomycin in antitumour activity has shown that DNA intercalators can be effective 'broad band' anticancer agents (see, for example, Waring 1981, 1991; Pullman and Jortner 1990). However, the lack of understanding of the molecular mechanisms of the cytotoxic action is a drawback with respect to the design of more effective drugs of this class. It is now believed that an important aspect of the mechanism of the intercalating drugs is the effects on DNA, interactive proteins, and, in particular, the disruption of one specific enzyme, namely topoisomerase (Liu 1989). The process of intercalation itself may not be responsible for damage to topoisomerase but some other property of the drug is important which allows it to interact in some way with the enzyme once intercalated in DNA.

It is well known that suitably sized flat ring structures are potential intercalating molecules (see Chapter 9, the furocoumarins as further examples). The factors which control intercalation, apart from geometric considerations, are presumably stacking and charge transfer interactions, the latter involving the aromatic moiety of the drug and the DNA bases. The pulse radiolysis studies described above have mainly concerned the redox chemistry of the quinones, while this may not be directly related to the enzyme poisoning properties, useful information may arise on the nature of the charge transfer interaction and also on the role of the semiquinones and their redox cycling (production of $O_2^{\cdot -}$, H_2O_2, and $OH^{\cdot}$) in cytotoxicity and, in particular, cardiotoxicity. Much of the redox data obtained on adriamycin and daunomycin has been better understood due to the detailed earlier studies of the model quinone, naphthazarin. Thus, the one- and two-electron reduction potentials for the quinones and semiquinones have been derived and the effects of pH on these parameters quantified.

For mitomycin C, pulse radiolysis has been useful in several respects and, for example, has clearly shown that drug activation, via ring opening, must occur after the corresponding hydroquinone has been formed.

It is well known that the development of multidrug resistance by tumour cells is a serious clinical problem and it is noteworthy that some of this class of drugs (e.g. mitoxantrone) may show only partial cross-resistance to the other intercalating agents (see, for example, Faulds *et al.* 1991). An interesting and alternative approach may be the enhancement of antitumour drug cytotoxicity via laser photoactivation. Thus, Andreoni *et al.* (1991) have recently shown that laser photoactivation can be effectively achieved using at least some of the anthracycline derivatives, with the rather subtle structural differences within this class of drugs being of marked importance with respect to the degree of light-activation.

There is still much to be learned about the mechanisms of the chemotherapy of the anthracyclines, and pulse radiolysis can be expected to contribute to the overall pool of physico-chemical knowledge on which such mechanisms must be based.

12

Cancer: photodynamic therapy (PDT)

'Se non e vero e ben trovato'

Giordano Bruno

12.1 Introduction

It is about one hundred years since the *in vivo* photosensitivity associated with porphyrins was established (see, for example, Meyer-Betz 1913). The relatively new, photochemotherapeutic cancer treatment known as photodynamic therapy (PDT) is closely related to such a photosensitivity (photodynamic) effect (see Chapter 9). This cancer treatment has experienced a huge upsurge in research activity in the last decade or so, substantially initiated by the pioneering work of Dougherty (see, for example, Dougherty *et al.* 1978).

PDT can be considered as a specialized or targeted form of chemotherapy, in which the drug itself is harmless, but is activated where and when required by exposure to light. Thus PDT is a localized form of therapy and the unwanted side effects of chemotherapy may be avoided. Alternatively, PDT may be considered as an extension of radiotherapy, where, because of the prior sensitization of the tissue, cytotoxicity is induced at much lower radiation energies than with conventional radiotherapy. In fact, however, PDT is a totally unique

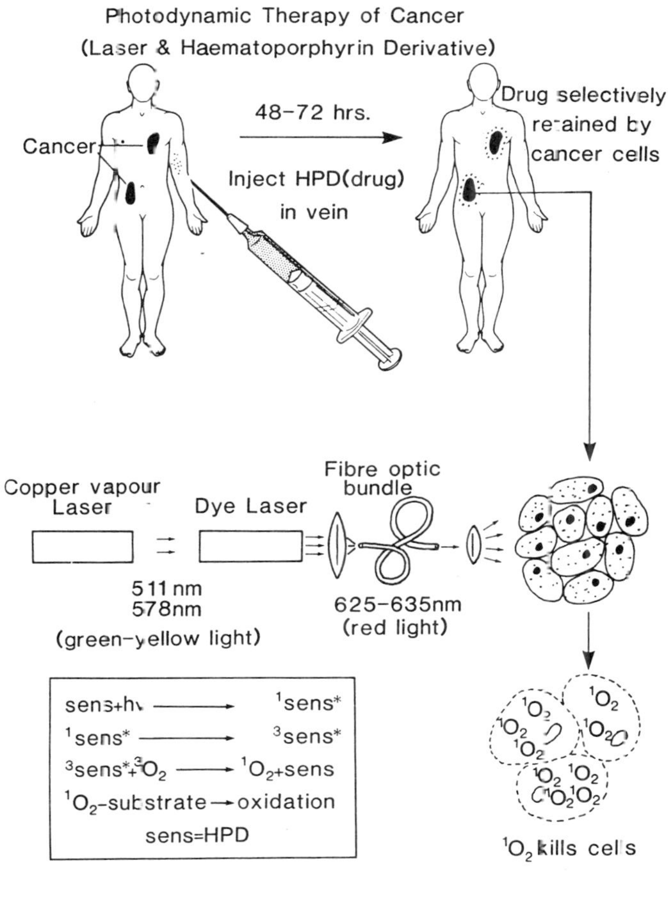

Fig. 12.1 Principle of the photodynamic therapy of cancer.

approach, both in the mechanisms by which cellular damage is caused and in the remarkable way in which any consequential normal tissue damage heals.

Currently PDT (Fig. 12.1) involves red (often laser) irradiation of a tumour following the administering of a light absorbing drug. The drug currently used is a complex mixture of haematoporphyrin and porphyrin related material, see Figs 12.2 and 12.3 for typical structures (C—C linked porphyrins are also present), and the absorption of light leads to the production of cytotoxic species. These porphyrin mixtures are known as haematoporphyrin derivative (HPD) or, as a somewhat purified form of HPD, by a variety of names including dihaematoporphyrin ether/ester (DHE) (see below) and the commercial product, Photofrin II (now called Photofrin^R or polyporphyrin ester). Usually the light irradiation system involves a tunable dye or gold vapour laser in the

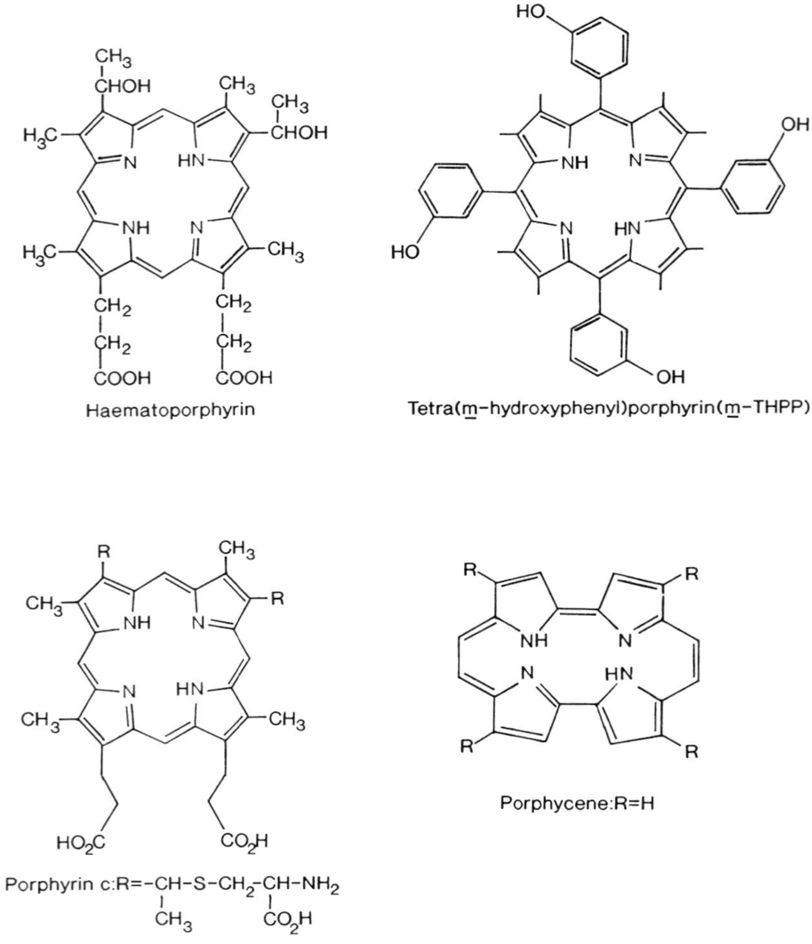

Fig. 12.2 Structures of some porphyrins and related compounds.

630 nm spectral region see (Table 1.1). The involvement of endogenous oxygen is necessary for an effective photosensitization. In fact, the clamping of blood flow in the legs of mice before and during illumination prevents tumour response due to HPD + light (Dougherty *et al* 1981).

Much current interest involves the search for alternative (pure) drugs with enhanced and red-shifted absorbance characteristics to take advantage of increased light penetration at longer wavelengths (approximately doubling between 600 and 800 nm, see Fig. 9.3 and, for example, B.C. Wilson 1989). The structures of typical examples of such 'second generation' photosensitizers are given in Figs 12.2, 12.4, and 12.5. PDT is also being considered for other

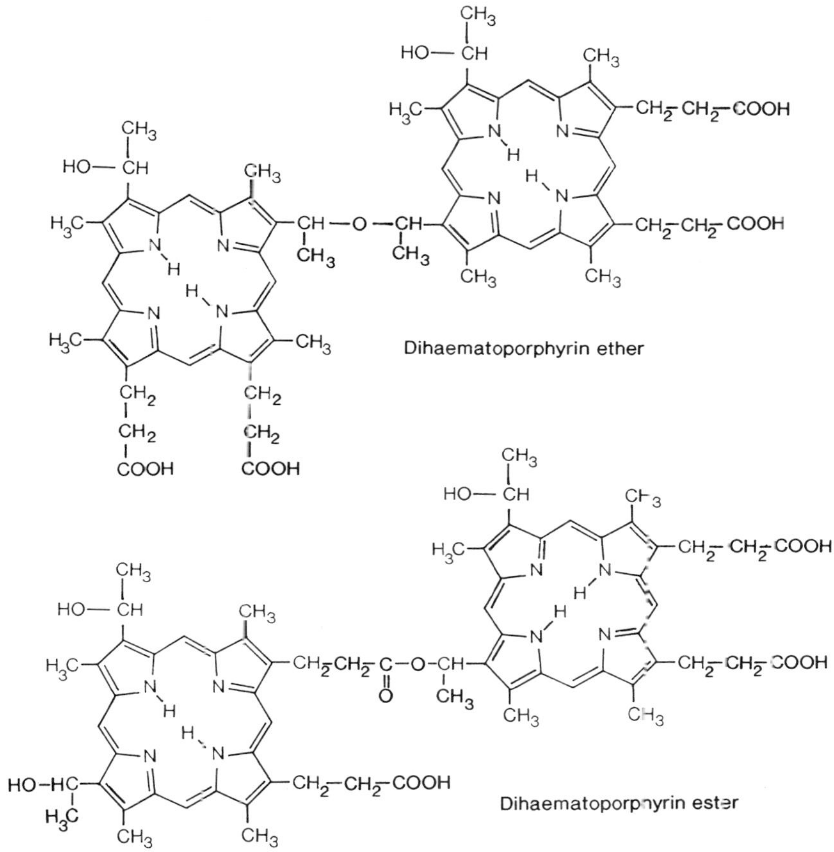

Fig. 12.3 Structures of some linked porphyrins.

medical uses such as the treatment of psoriasis, atheromatous plaques, and viral and microbial infections (see, for example, Spikes and Jori 1987).

Porphyrins are cyclic tetrapyrroles (see Fig. 12.2) and these molecules are pivotal to many aspects of life. The iron complex of protoporphyrin IX (PP) is haem and defects in haem biosynthesis lead to the group of diseases called porphyrias (see Section 9.8). Several porphyrias lead to skin photosensitivity and such a 'natural' effect is due to an excess of porphyrins such as PP in the skin. This porphyrin induced photosensitivity is closely related to that which arises as an unwanted side effect in PDT. Indeed, the structural relationships between haematoporphyrin (HP) and its dehydration products, PP and monohydroxyethylvinyldeuteroporphyrin (HVD) suggest that these molecules all arise in varying amounts in HPD and/Photofrin^{R}. In discussing the

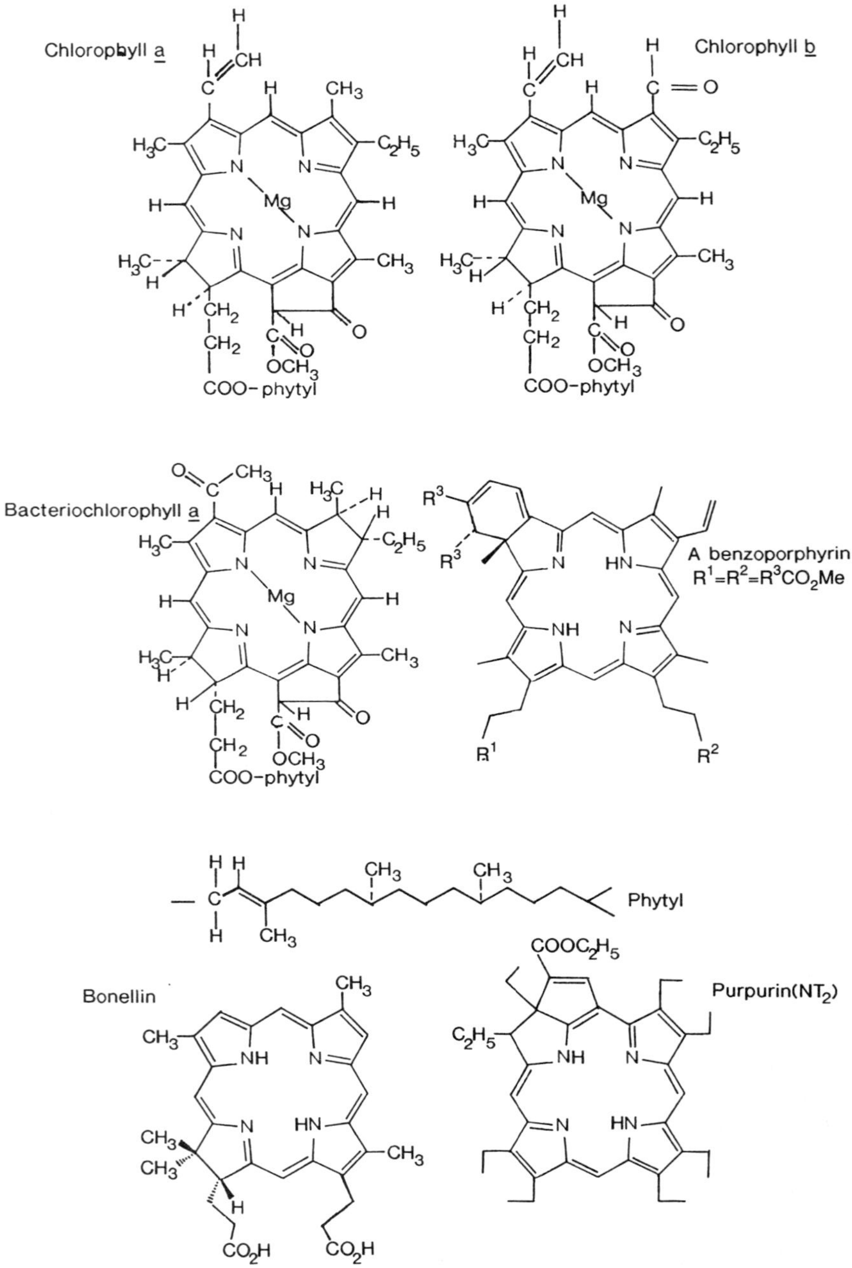

Fig. 12.4 Structures of some chlorins, a bacteriochlorin, a purpurin, and a benzoporphyrin.

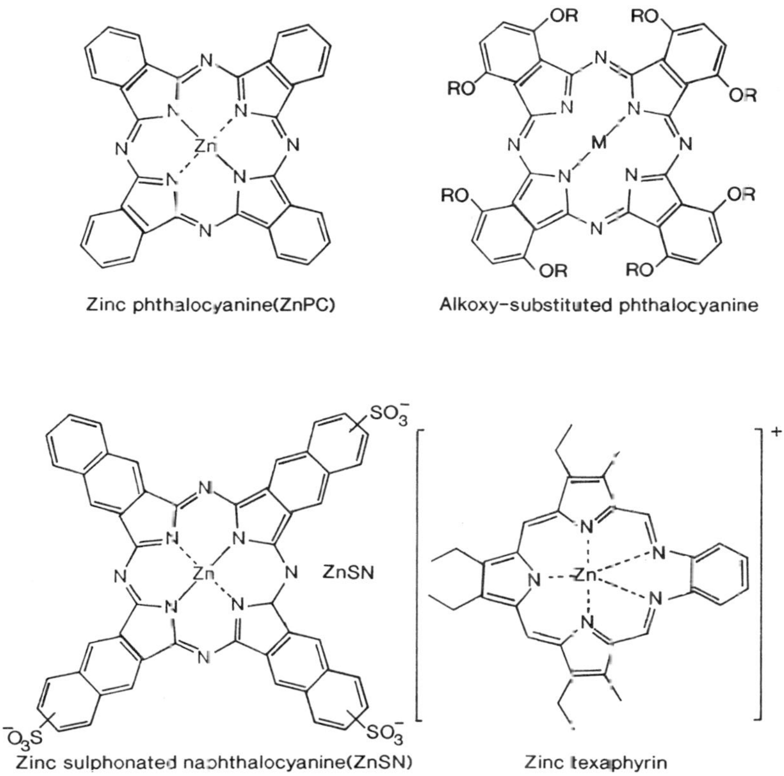

Fig. 12.5 Structures of some phthalocyanines, a naphthalocyanine, and a texaphyrin.

photochemical aspects of PDT it is critical to remember that all forms of HPD are not only a complex mixture (discussed below), but also that aggregation of porphyrins will arise particularly in aqueous based solvents.

12.2 **Preparation and nature of HPD and related mixtures**

HPD was first described by Lipson and Baldes (1960) and is obtained by the action of a mixture of sulphuric and acetic acids on HP. The product of this treatment is mainly haematoporphyrin diacetate (Bonnett *et al.* 1981) and this is treated with sodium hydroxide to obtain HPD. The neutralization treatment leads to hydrolysis to give HP, PP, and HVD and a complex mixture of condensation products (dimers and higher oligomers of HP and possibly of PP and HVD). The nature of the linkage of the porphyrins in the oligomeric fraction has led to much debate particularly as the tumour localizing activity is associated with this dimeric/oligomeric fraction. Amongst the bonding (between HP moieties) discussed are esters and ethers as shown in Fig. 12.3, although a C—C linkage has also been suggested (Bonnett *et al.* 1981; Byrne *et al.* 1990). Currently, it is believed that freshly prepared HPD is mainly composed of HP moieties which are ester linked but that on standing at room temperature these spontaneously convert to about a 50/50 mixture of ester and ether links. The mechanism of this conversion probably involves unreacted acetate groups which undergo a reaction with free secondary alcohol groups to form ethers, while the less stable esters are gradually hydrolysed.

Pandey and Dougherty (1988, 1989) have synthesized HP 'dimers' linked unambiguously by ether and ester links, respectively. Comparison with Photofrin^R suggested that the ester linked dimer was not present in Protofrin^R, but it is possible that higher molecular weight fractions of Photofrin^R contain porphyrins joined by both ester and ether links.

HPD is typically 55 per cent dimer/oligomer plus 45 per cent monomeric porphyrins. DHE is prepared from HPD via gel filtration chromatography and consists of typically 85 per cent dimer/oligomer fraction plus 15 per cent of monomeric porphyrins. Since the oligomeric (or polymeric) fraction is the component which is localized and retained by tumours, alternative names such as PPE (polyporphyrin ester) have been suggested instead of DHE.

As noted above, aggregation will occur in aqueous solvent systems, HPD and DHE being typically used at pH 7.2–7.4 in sterile saline solutions. In organic environments, including proteins and lipids, the porphyrins will be at least partially monomeric while the oligomeric fractions may well change conformation and be disaggregated (i.e. one polymeric chain from another), but the polymer itself will remain mainly intact. It is to be expected that *in vivo* the ester links will be more susceptible to hydrolysis than other forms of bonding.

With the continuing expansion of clinical applications of PDT and related basic biological research, there is an increasing need for the accurate determination of porphyrin sensitizers in cells, body tissues, and fluids. Such work, including the use of radio-labelled porphyrins, has been discussed by Brown and Vernon (1990).

12.3 **Primary molecular photo-processes involved in PDT**

Excitation of porphyrins (P) and related photosensitizers with red light leads to the production of the first excited singlet state (^1P). This state is responsible for two competitive processes—fluorescence back to the ground state, P, and intersystem crossing to yield the lowest triplet state (^3P). While the red fluorescence emission is of some interest as a means of locating a malignancy, it is the relatively long-lived triplet state that undergoes exchange electron energy transfer (see Chapter 2) with oxygen to yield singlet oxygen, and it is this form of activated oxygen which is believed to initiate tumour destruction and also to cause the unwanted side effect of persistent skin photosensitivity (Chapter 9). Overall these processes can be written as:

$$P + h\nu(\text{red}) \rightarrow {}^1P$$

$${}^1P \rightarrow P + \text{fluorescence (red)}$$

$${}^1P \rightarrow {}^3P$$

$${}^3P + O_2 \rightarrow P + O_2({}^1\Delta_g)$$

$$O_2({}^1\Delta_g) + \text{biological substrate} \rightarrow \text{oxidation products}$$

However, while the involvement of singlet oxygen in PDT is generally accepted, this has not been unambiguously proven. Alternative suggestions, for example, based on spin-trapping/ESR techniques have proposed a role for $OH^{\cdot}$. Furthermore it should be noted that singlet oxygen can be converted to $O_2^{\cdot-}$ in the presence of reducing agents such as thiols (see, for example, Buettner and Need 1985; Buettner and Hall 1987).

12.3.1 *Absorbance spectra*

Porphyrin ground state absorbance spectra (Fig. 12.6) are characterized by a strong band near 400 nm, known as the Soret or B band ($\varepsilon \sim 10^5$ dm^3 mol^{-1} cm^{-1}) together with weaker (~1/10) bands in the visible which are called the Q bands. The longest wavelength band for HP at physiological pH is near 625 nm with a similar absorption for HPD (630 nm). This is the band excited during PDT and the molar absorption coefficient for this HP band is $\sim 4 \times 10^3$ dm^3 mol^{-1} cm^{-1}.

The number of Q bands depends on the protonation state of the pyrrole nitrogens and also on metal chelation at the centre of the porphyrin ring. Thus the four Q bands are due to two electronic transitions Q_x and Q_y in which the transition dipole lies along the x and y axes and to the 0–0 and 0–1 vibrational bands of each of Q_x and Q_y. When the pyrrole nitrogens are protonated or with metal substitution the Q_x and Q_y transitions are degenerate and the four banded spectrum becomes a two banded spectrum, see Fig. 12.6 for the tin(IV) PP. Reduction of one double bond of the macrocyclic porphyrin ring leads to a chlorin, and further reduction to a bacteriochlorin. These conversions lead to an increase in the energy difference between the Q_x and Q_y transitions and to a red shift and intensification of the longest wavelength absorption band. These spectral features make chlorins (and the closely related purpurins and benzoporphyrin derivatives) and bacteriochlorins potentially useful drugs for PDT. These spectral features also explain why blood is red and grass is green! Figures 12.6 and 12.7 compare the spectrum of HP with a typical purpurin, benzoporphyrin, and bacteriochlorophyll *a*. In addition, water soluble derivatives of bacteriochlorophyll *a* have similar spectra (Beems *et al.* 1987). Overall, the characteristics of cyclic tetrapyrrole absorbance spectra have been reviewed

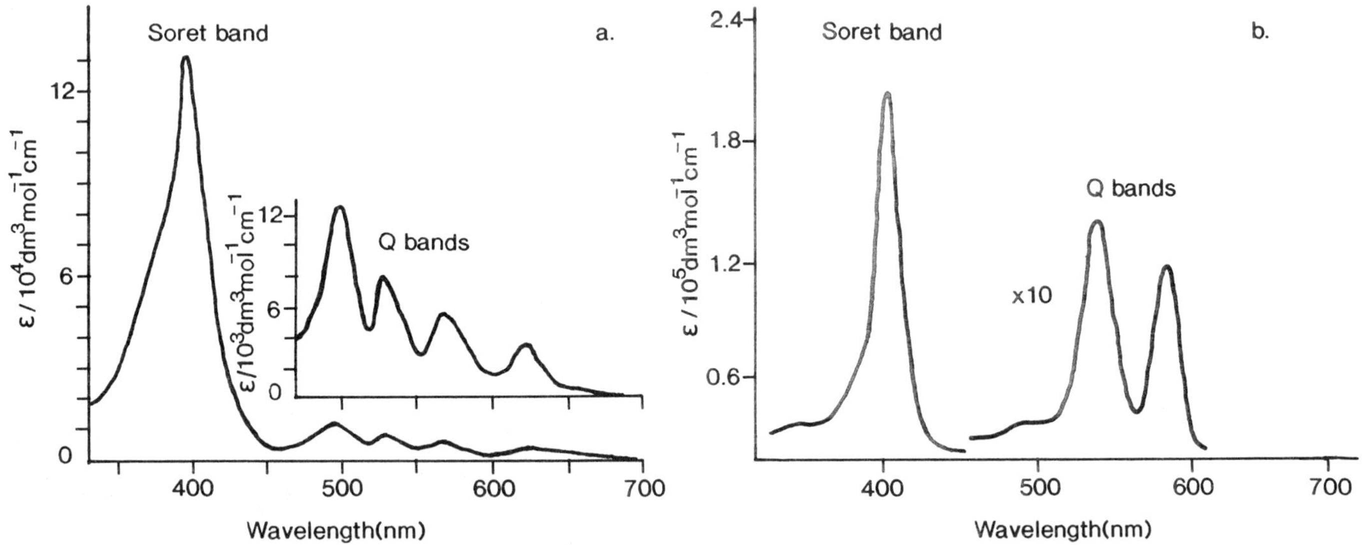

Fig. 12.6 Typical ground state absorption spectra for porphyrins: (a) haematoporphyrin as a monomer in a typical organic solvent; (b) tin protoporphyrin in methanol.

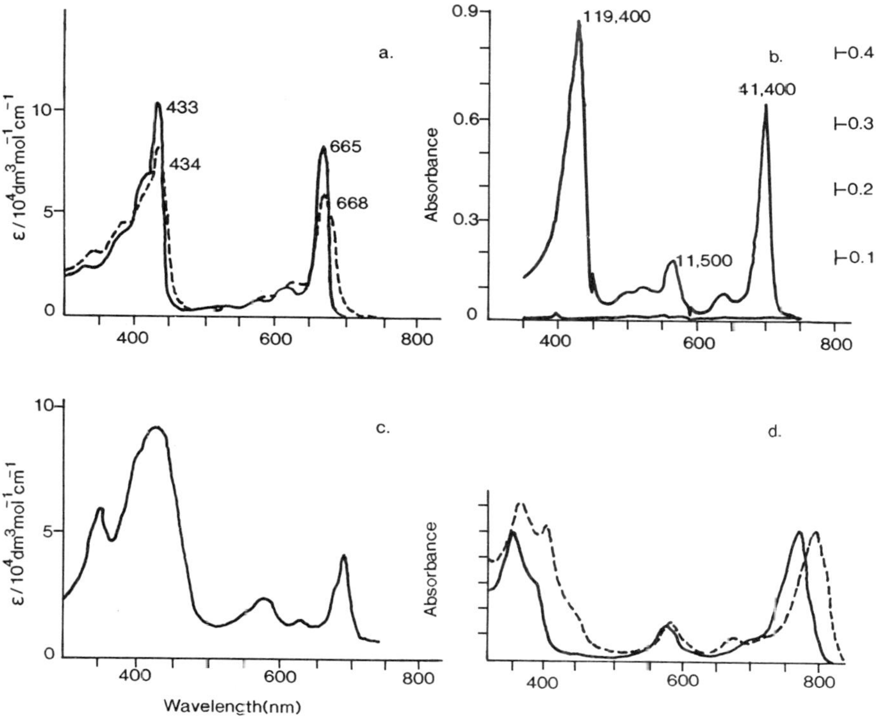

Fig. 12.7 Typical ground state absorption spectra for a chlorin, purpurin, benzoporphyrin, and bacteriochlorin. (a) Chlorophyll *a* CCl_4; —, monomer; -----, dimer. Molar absorption coefficients are given per mole of monomer in the dimer. (Reprinted with permission from Sauer, Lindsay-Smith, and Schultz, *J. Am. Chem. Soc.*, **88**, 2681–8, copyright 1966 American Chemical Society. (b) Purpurin in benzene (taken from Borland *et al.* 1988). (c) A typical benzoporphyrin dimethyl ester in methanol. (d) Bacteriochlorophyll *a* (—) and bacteriochlorophyll *b* (-----) in acetone (taken from Eimhjellen *et al.* 1963).

both from an experimental and theoretical standpoint (see, for example, Corwin *et al.* 1968; K.M. Smith 1975; Gouterman 1978).

As well as the tetrapyrrole related system there has been much interest in the use of phthalocyanines (octasubstituted tetra-aza porphyrins) in PDT (see Fig. 12.5). Of course, phthalocyanines (PC) have other, wide-ranging applications, varying from use as dyes and pigments to passive Q-switches in laser cavities and as possible light harvesters for solar energy conversion.

Unsubstituted phthalocyanines are extremely insoluble but can sometimes be 'solubilized' in liposomal preparations or in rather unusual solvents such as 1-chloronaphthalene. Sulphonation or carboxylation of the benzene rings of a phthalocyanine leads to solubility in aqueous solutions. Figure 12.8 gives the ground state absorbance spectra of the tetrasulphonated forms of metal-free phthalocyanine (PCS_4) and of $ZnPCS_4$ in water. The spectrum of sulphonated AlPC is very similar to that of $ZnPCS_4$. As for the porphyrins, the changes in the spectrum on metallation are related to the symmetry of the molecule and the corresponding degeneracy of the transition dipoles.

Both sulphonated (e.g. aluminium phthalocyanine as the disulphonate, $AlPCS_2$) and non-sulphonated (e.g. zinc phthalocyanine, ZnPC, in liposomes) are being considered for PDT.

Closely related to the phthalocyanines are the naphthalocyanines (NC) in which the four benzene rings of phthalocyanine are replaced by naphthalene rings. This further shifts the red absorption from about 670–700 nm to near 770 nm which is quite similar to that of bacteriochlorins.

In summary, naphthalocyanines and bacteriochlorins typically exhibit molar absorption coefficients *c.* 10^5 $dm^{3}mol^{-1}cm^{-1}$ at near 770 nm and phthalocyanines at 670–700 nm. Chlorins, as well as purpurins and benzoporphyrins, typically show molar absorption coefficients *c.* 10^4 at 660–770 nm while porphyrins have molar absorption coefficients *c.* 10^3 $dm^3 mol^{-1} cm^{-1}$ at near 625 nm. Several of the specific wavelength maxima which could be used in PDT are given in Table 12.3.

12.3.2 *Fluorescence*

12.3.2.1 *Steady-state fluorescence*

As noted above the fluorescence emission of HP has been used as a diagnostic method for tumour detection for some time (Rasmussen-Taxdal *et al.* 1955). Such emission at neutral pH consists of bands at 615 and 675 nm with an intensity ratio of ~2:1 (for a review of the effect of pH on fluorescence spectra, see Pottier and Truscott 1986). Vever-Bizet *et al.* (1985) have measured the pK_a values of porphyrins, including HP, in a range of environments via fluorescence titration methods. Of particular interest was that incorporation of PP into egg phosphatidyl choline small unilamellar vesicles (SUV) shifted one of the pK_a values associated with the carboxylic groups to the physiological region. As noted by these authors this could be of relevance to the uptake and retention of porphyrins in tumours which are known to be somewhat more acidic than normal tissue. The fluorescence properties of porphyrins in microheterogeneous environments in general are quite complex and probably reflect differing 'solubilization' sites (Beltramini *et al.* 1987).

There have been many reports of HPD fluorescence from a wide range of tumours (see, for example, Gomer 1984; Moan 1984; Profio 1984). The identification of the fluorescing molecules and the determination of their concentrations are not easy due to HPD being a mixture, and to more than one ionic form arising in the pH range studied.

More recently, Kinoshita *et al.* (1988) have compared the fluorescence spec-

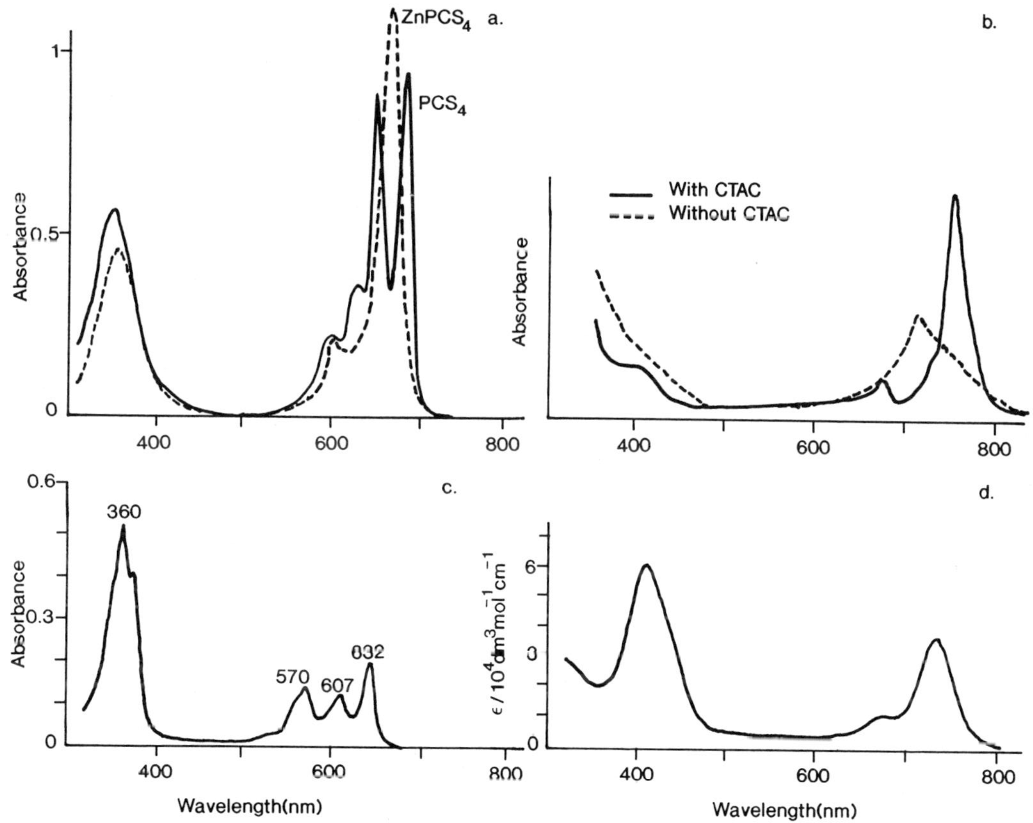

Fig. 12.8 Typical ground state absorption spectra of phthalocyanines, naphthalocyanines, and other red-enhanced photosensitizers: (a) Metal free and zinc phthalocyanine; (b) zinc sulphonated naphthalocyanine; (c) porphycene in DPPC-SUVs; (d) texaphyrin in methanol.

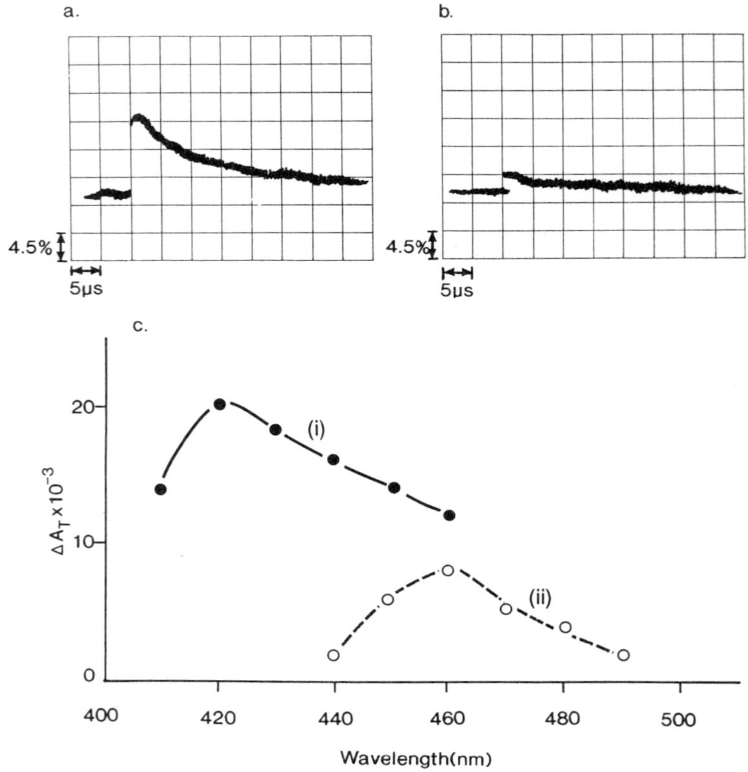

Fig. 12.9 Transient species obtained from HPD in BHK 21 cell suspensions. (a) HPD bound to cells in aqueous buffer (pH 7.4). (b) Supernatent from (a) added to blank cells. (c) Difference spectra from 10 μg ml^{-1} HPD in aqueous buffer (i), in BHK 21 cells (ii) (taken from Truscott *et al.* 1988).

tra (and lifetimes) of HP and HPD in leukaemic cells and normal lymphocytes with those in solution at a range of porphyrin concentrations. The results imply that HP is much more concentrated in the cells than in the medium. Furthermore, it was reported that HP was rapidly changed to a photo-product during irradiation. Clearly, this result is of considerable significance at least to the quantitative study of HP in cells and suggests that future studies should only use very weak light sources. Photodegradation of Photofrin^{R} in tumour cells has also been reported by Moan and Kessel (1988), and for phthalocyanines (McCubbin and Phillips 1986) and for some purpurins (Kessel 1989; Macpherson *et al.* 1990).

The fluorescence quantum yields (ϕ_F) of many monomeric porphyrins are ~0.1, although a value of 0.03 has been reported for porphyrin *c* (the

Table 12.1 Fluorescence yields of typical porphyrins in methanol

Porphyrin	Environment	ϕ_F
TPP	Neutral	0.13
	Acidic	0.095
	Basic	0.13
m-THPP	Neutral	0.12
	Acidic	0.015
	Basic	0.03

porphyrin chromophore present in cytochrome *c*) (Ghiggino *et al.* 1988). If aggregation arises, the ϕ_F is substantially reduced. Differing ionic forms of porphyrins can also lead to substantial effects on ϕ_F and this is illustrated for TPP and a typical hydroxy derivative in Table 12.1 (Bonnett *et al.* 1988). As can be seen the hydroxy substituent has little effect upon the ϕ_F of TPP in neutral methanol. However, protonation of the central nitrogen atoms, to form the porphyrin dication, causes a substantial reduction in ϕ_F. This effect is probably due to a buckling of the porphyrin ring on protonation, with the meso substituents then leading to severe distortion in the dication structure and consequent reduction in ϕ_F.

For many of the systems being considered as alternatives to HPD for PDT, the ϕ_F values are rather similar to the porphyrins. However, the phthalocyanines are usually strongly fluorescent with, for example, ϕ_F for PCS_4, $AlPCS_4$, and $ZnPCS_4$ reported as 0.62, 0.58, and 0.32, respectively (Harriman and Richoux 1980). Of course, as with the porphyrins, the ϕ_F value of phthalocyanines is much reduced when complexed to heavy or paramagnetic metals, a well-known commercial blue pigment with virtually zero fluorescence being CuPC.

12.3.2.2 *Fluorescence lifetime*

In general, fluorescence lifetimes of (τ_f) of monomeric porphyrins are rather long, Chantrell *et al.* (1977) reported a value of 23 ns for PP DME in benzene. There have been many studies of the τ_f values of porphyrins such as HP. Thus, for example, Andreoni and Cubeddu (1984) have reported that where HP is totally monomeric, such as in detergents above the critical micelle concentration, and also in methanol solution, a single τ_f of ~14 ns is observed. In aqueous solutions, two τ_f values arise with τ_f ~14 ns, assigned to the monomer, and τ_f ~4 ns, assigned to an aggregated form of HP. More recently, Kinoshita *et al.* (1988) have assigned τ_f as ~15 ns for the monomer but only ~0.3 ns for the aggregate which is in agreement with a value given for HPD by Yamashita *et al.* (1984). Kinoshita *et al.* (1988) interpret species with τ_f ~2.5 ns as due to both photo-products associated with the use of intense light sources and/or trace impurities in solution.

Picosecond studies have concerned HPD in tumour cells (Yamashita *et al.* 1984) and HPD and Photofrin^R in mammary carcinoma mitochondria (Hanzlik *et al.* 1989). The major conclusion of Yamashita *et al.* is that an aggregated (oligomeric) component of HPD with τ_f ~100 ps and with a red shifted maximum ~600 nm selectively accumulates in the tumour cells to an increasing degree with increased incubation time. Hanzlik *et al.* compared HPD and Photofrin^R τ_f values *in vitro* (water) and in subcellular organelles prepared from tumours borne on treated animals, and showed that cancer cell mitochondria accumulate the most hydrophobic components which exhibit very short τ_f values (~20 ps).

Using time-gated fluorescence spectroscopy Cubeddu *et al.* (1989) studied the tumour localizing fraction obtained from HPD via reverse phase chromatography. Three fluorescence lifetimes (700 ps, 3 ns and 15 ns) were obtained in different wavelength regions in micellar systems. Below the critical micelle concentration, the results imply an energy transfer mechanism with a conversion from molecules emitting at 626 nm to those fluorescing at 660–670 nm, such energy transfer being precluded at higher detergent concentrations due to unfolding and 'insulation' of the polymeric molecules. These workers also suggest that their preliminary measurements with cultured cells show evidence of such energy transfer processes.

There have been fluorescence lifetime studies on some of the other molecules being considered for PDT. Typically, for those porphyrins which exhibit aggregation phenomena a long (monomer) and a short (dimer/aggregation) τ_f is observed, whereas for porphyrins, such as chlorin e_6, which do not aggregate, only a single τ_f (4.3 ns and 5.3 ns for chlorin e_6 in phosphate buffer and ethanol, respectively) is observed (Roeder and Wabnitz 1987). As might be expected chlorin e_6 has a rather similar τ_f to that of chlorophyll *a* and chlorophyll *b*, with values reported for the chlorophylls ranging from 1–6 ns (Butler and Norris 1963; Holten *et al.* 1976; Huppert *et al.* 1976; Connolly *et al.* 1982; Jabben *et al.* 1986).

Purpurins are being actively pursued as alternatives to HPD and these molecules are also chlorin-type systems (see Fig. 12.4). Kessel (1989) has reported τ_f values of 2.0 and 0.65 ns for NT2 and a tin derivative ($SnNT2H_2$), rather similar to the values for other chlorins. Macpherson *et al.* (1990) using synchrotron radiation excitation also report similar τ_f values. Since NT2 is known to readily photo-oxidize, they also noted the τ_f value of the methyl acetal photo-oxidation product from NT2 (τ_f = 4.2 ns), such photo-oxidation may well be important if this molecule is to be used in PDT. The τ_f values of a number of benzoporphyrins have also recently been reported to be in the region of 5 ns (Sternberg *et al.* 1992).

There have been several measurements of τ_f values of phthalocyanines. Typically values, between 3 and 6 ns are reported depending on the (diamagnetic) central metal. Sulphonated phthalocyanines exhibit τ_f values essentially invariant with the degree of sulphonation in both methanol and buffer solutions (τ_f ~5 ns) (see, for example, Rousseau *et al.* 1990; Ambroz *et al.* 1991). Of course, phthalocyanines are well known to aggregate and it is generally accepted that fluorescence can only be detected from the monomeric state (Harriman and Richoux 1980).

There have been relatively few studies of the related naphthalocyanines (NC) but Firey and Rodgers (1987) report a τ_f for SiNC of 2.85 ns in benzene and McCubbin and Phillips (1986) a value of 2.5 ns for aluminium sulphonated NC. That is within the range reported for phthalocyanines.

12.3.3 *Triplet state*

This is the precursor of the cytotoxic species singlet oxygen so it is not surprising that there have been many studies of the triplet state of porphyrins and other molecules being considered for PDT. Some of the earliest uses of conventional flash photolysis (using discharge lamps rather than lasers as excitation sources) established many important features of porphyrin and chlorin triplet absorption spectra. More recently there have been studies of the triplet yield (ϕ_T), triplet minus singlet molar absorption coefficient ($\varepsilon_T - \varepsilon_G$), absorption spectra, lifetimes, and rates of reaction with many other species. The driving force for these extensive studies is not just interest in PDT but also in photosynthesis and solar energy conversion in general (Darwent *et al.* 1982). In some cases data are available for a number of homogeneous and microheterogeneous environments including cell model systems, but data on photosensitizers actually accumulated by tumour cells are limited at this time. Table 12.2 lists typical data for porphyrins related to PDT. As can be seen where the porphyrins are not aggregated the ϕ_T values are quite high ($\phi_T = 0.7 \pm 0.2$) and in many cases $\phi_T + \phi_F$ is close to unity. Corresponding data for chlorins, phthalocyanines, and some other systems are given in Tables 12.3, 12.4, and 12.5.

There has been interest in studying photosensitizers not only in simple solutions and detergents but also in other microheterogeneous environments such as bound to protein, lipoprotein, and to erythrocyte ghosts (see, for example, Reddi *et al.* 1983; Redmond *et al.* 1984; C. Lambert *et al.* 1985; Firey and Rodgers 1988). In some cases more than one triplet lifetime can arise and this can be either due to monomer–dimer equilibria or to more than one site existing for the photosensitizer in the macromolecular system.

A study of the binding of HP, HPD and Photofrin[R] to human serum albumin (HSA) in aqueous solution using the laser flash photolysis technique has been reported by Redmond *et al.* (1984). They found that the detected triplet absorption increases substantially for HP with increasing values of the ratio [HSA]/[HP] to reach a maximum at a 1:1 ratio. All HP molecules are then in the monomeric state with HSA exhibiting one site for binding as has been determined elsewhere by other methods. It is of interest to note that singlet oxygen can readily be detected from the HSA . . . HP complex although it cannot be detected in cells (see below). On the other hand Photofrin[R] shows virtually no triplet (or singlet oxygen) formation either in the presence or absence of HSA, while HPD itself shows intermediate behaviour. Thus *in vivo* HP is likely to be bound entirely as the monomer, while HPD will have only some monomeric species bound to HSA. For Photofrin^R on the other hand it may be that virtually all the porphyrin material is bound as a polymeric or aggregated species.

More recently there has been interest in extending studies of triplet states (and singlet oxygen, see below) to the cellular environment. Thus, Craw *et al.* (1984*c*, Truscott *et al.* (1988), and Firey *et al.* (1988*b*) have reported the detection of the triplet states of HP, HPD, and phthalocyanines in cellular environments. Typical results for triplet transients in cells are those shown in Fig. 12.9 following HPD incubation with transformed baby hamster kidney cells. As can be seen there are significant differences in the spectrum of the triplet state compared to aqueous environments. Kinetic studies show more than just the triplet state and imply that the radical cation may also be formed.

Recently Takemura *et al.* (1989) have indicated the importance of triplet lifetime of sensitizers with respect to photodynamic action and have reported triplet lifetimes of metallo-porphyrins and Photofrin^R when absorbed on filter paper as a model of the cellular environment.

12.3.4 *Main cytotoxic species*

With current photosensitizers, $^1O_2$, resulting from energy transfer from the sensitizer triplet state, is thought to be the cytotoxic agent in photodynamic therapy. However, the biological damage might be due to other reactive oxygen species such as, superoxide anions and hydroxyl radicals, ultimately formed from $^1O_2$. For example, $^1O_2$ can be reduced by NADH and other agents, yielding superoxide anions (G. Peters and Rodgers 1981), which would generate in turn highly reactive hydroxyl radicals via the metal catalysed Fenton–Haber–Weiss reactions (see Section 4.2.3.1). There have been many studies of the relative yields of singlet oxygen via a range of chemical trapping techniques (for a review, see Jori and Spikes, 1984). More recently, singlet oxygen has been detected by its direct near infra-red phosphorescence at ~1270 nm with a few results reported via the thermal lensing technique. Under most conditions the ratio ϕ_T/ϕ_Δ, usually called S_Δ, is a measure of those triplets which give singlet oxygen on reaction with oxygen (see Section 4.2.1.1). Thus for a sensitizer to be an efficient producer of singlet oxygen both ϕ_T and ϕ_Δ should be as close as possible to unity.

Tables 12.2–12.5 report the quantum yields of singlet oxygen, ϕ_Δ, for a range of sensitizers and environments. Generally, ϕ_Δ values are similar or somewhat lower for the chlorins, bacteriochlorins, and phthalocyanines than the porphyrins such as HP. Nevertheless, such values are sufficiently high to render all these molecules extremely good photosensitizers, particularly taking into account their high molar absorption coefficients in the red spectral region. It is relatively easy to observe the production of singlet oxygen in homogeneous

Table 12.2 Typical values of ϕ_T and ϕ_Δ for porphyrins

Porphyrin	Solvent	ϕ_T	ϕ_Δ	Reference[a]
HP	H_2O	0.63	0.32	a
HP	MeOD	0.71	0.53	b
HP	MeOH/H_2O	0.83	—	a
HP	MeOH/D_2O	0.83	0.65	c
HP	SDS/H_2O	0.87	—	c
HP	SDS/D_2O	—	0.27	c
HP	CTAB/H_2O	0.70	—	c
HP	CTAB/D_2O	—	0.22	c
HP	TX100/H_2O	0.78	—	c
HP DME	Benzene	0.72	0.50	d
DHE	CTAB/D_2O	—	0.29	b
$TPPS_4$	H_2O	—	0.63	e
m-THPP	MeOH	0.69	0.57	f
SnPP	MeOH	0.83	0.70	g

[a] a, Rodgers (1985); b, Keir *et al.* (1987); c, Reddi and Jori (1988); d, Keene *et al.* (1986); e, Kimel *et al.* (1989); f, Bonnett *et al.* (1988); g, Land *et al.* (1988).

Table 12.3 Typical values of ϕ_T and ϕ_Δ for chlorins, bacteriochlorins, and purpurins

Sensitizer	λ(nm)[a]	Solvent	ϕ_T	ϕ_Δ	Ref.[b]
Chlorophyll *a*	670	Ether	0.64	—	a
Chlorophyll *b*	650	Ether	0.38	—	a
Chlorophyll *a*	670	Benzene	—	0.37	b
Chlorophyll *b*	650	Benzene	—	0.44	b
Pheophytin *a*	668	CCl_4	—	0.8	c
Pheophytin *a*	668	Ether	—	0.75	c
Benzoporphyrin BPD	688	MeOH	—	0.47	d
Purpurin NT2	700	MeOH	0.82	0.68	e
Bonellin	640	EtOH	0.85	—	f
Mono-L-aspartyl-chlorin e_6	654	H_2O	—	0.47	g
Di-L-aspartyl-chlorin e_6	654	H_2O	—	0.20	g
Bchl *a*	773	Benzene	0.32	0.32	h
Bpheo *a*	760	Benzene	0.73	0.46	h

[a] Longest wavelength absorption band.
[b] a, Bowers and Porter (1967); b, Stirling and Truscott, unpublished; c, Krasnovski *et al.* (1988); d, Sternberg *et al.* (1992); e, Macpherson *et al.* (1990); f, Mathews *et al.* (1980); g, Kimel *et al.* (1989); h, Borland *et al.* (1987).

Table 12.4 Typical values of ϕ_T and ϕ_Δ for phthalocyanines (PC) and naphthalocyanines (NC)

Sensitizer	λ(nm)[a]	Solvent	ϕ_T	ϕ_Δ	Reference[b]
AlPC	678	MeOD	0.42	0.30	a
ZnPC	672	MeOD	0.56	0.36	a
AlPC	675	H_2O	—	0.085	b
CuPC	678	Chloronaphthalene	1.0	~0	c
PC	698	Chloronaphthalene	0.14	—	c
Octa-*n*-butoxy PC	739	Benzene	0.1	—	d
Zn-octa-*n*-butoxy PC	739	Benzene	0.46	0.45	d
SiNC	(800)	Benzene	0.39	0.35	e

[a] Longest wavelength absorption band.
[b] b, Keir *et al.* (1987); b, Kimel *et al.* (1989); c, McVie *et al.* (1978); d, Ford *et al.* (1989); e, Firey and Rodgers (1987).

Table 12.5 Typical values of ϕ_T and ϕ_Δ for other sensitizers for PDT

Sensitizer	λ(nm)[a]	Solvent	ϕ_T	ϕ_Δ	Reference[b]
Porphycene	633	Toluene	0.42	0.30	a
Tetra-*n*-propyl-porphycene	637	Toluene	—	0.35	a
Metallo-texaphyrins	745	MeOH	0.9	0.68	b

[a] Longest wavelength absorption band.
[b] a, Aramendia *et al.* (1986); b, Harriman *et al.* (1989).

environments such as benzene, alcohol or aqueous solutions, and in microheterogeneous cell models such as micelles, lipoproteins, and erythrocyte cell ghosts. However, the kinetics in the microheterogeneous environments are complex (Rodgers 1985, Langlois *et al.* 1986; Valduga *et al.* 1988).

To date there has been no generally accepted observation of singlet oxygen from sensitizers incorporated in cells and tissues. Such an observation has been reported by J. Parker (1987), but other groups have failed to confirm this (Rodgers 1988; B.C. Wilson 1989). The lack of $^1O_2$ luminescence has been ascribed to its short lifetime in a cellular environment. An estimation of this lifetime *in vivo* of less than 1 μs has been made by Moan *et al.* (1979) and of less than 0.5 μs by Patterson *et al.* (1990), these short lifetimes are below the technical performance of the present infra-red detectors.

However, Baker and Kanofsky (1991) have detected the 1270 nm emission of singlet oxygen from leukaemia cells incubated with Photofrin[R]. Most of this emission was assigned to singlet oxygen generated near the surface of the cells which then diffused into the D_2O buffer.

Generally, the hypothesis of singlet oxygen involvement in photosensitization by porphyrins, phthalocyanines, and methylene blue in biological environments is supported by the following experimental observations:

1. Singlet oxygen quenchers such as DABCO (1,4-diazabicyclo[2,2,2] octane) and β-carotene protect cells, whereas D_2O, as solvent, enhances the cell killing (Weishaupt *et al.* 1976).
2. The absence of $O_2^{\cdot -}$ production by irradiation of HPD or of methylene blue has been reported by Gibson *et al.* (1984). These authors have used three different detection systems for $O_2^{\cdot -}$ namely oxidation of epinephrin to adrenochrome, reduction of cytochrome *c*, and reduction of nitro blue tetrazolium. Moreover, they observed that HPD and red light, could inhibit mitochondrial cytochrome oxidase activity but that the O_2^{-} generating systems (xanthine–xanthine oxidase and irradiation of riboflavin) could not.
3. Singlet oxygen is the ultimately reactive species in *Salmonella typhimirium* DNA damage induced by illumination of methylene blue by visible light (Epe *et al.* 1989). DNA modifications induced by methylene blue photosensitization are recognized and incised by specific repair endonucleases. They are similar to those induced by a chemical source of singlet oxygen (thermal decomposition of a 1,4-etheno-2,3-benzodioxin) and are different from those induced by radicals (Epe *et al.* 1989; Müller *et al.* 1990).
4. The predominant role of singlet oxygen in the photohaemolysis of red blood cells sensitized by phthalocyanine sulphonates was shown via a definite deuterium isotope effect (Sonoda *et al.* 1987).
5. The specific product of $^1O_2$ attack on cholesterol, 3β-hydroxy-5α-cholest-6-ene-5-hydroperoxide, was identified in the photohaemolysed red blood cells of patients possessing the genetic disorder erythropoietic protoporphyria (Lamola *et al.* 1973). This specific hydroperoxide was also observed in erythrocyte membranes photosensitized by HPD, while relatively insignificant amounts of free radical-derived hydroperoxides were detected (J.P. Thomas *et al.* 1987).
6. The effects of various scavengers (tryptophan, cysteamine, catalase, superoxide dismutase . . .) of activated forms of oxygen on survival of EMT6 and CHO cells following photodynamic treatment by porphyrins or γ irradiation were compared. They affirmed the predominant role of $^1O_2$ in the photodynamic cell killing (Henderson and Miller 1986).

12.3.5 *Sensitizer photofading*

There is considerable interest in the photofading of the sensitizers being considered for PDT. This arises both as a route to reduce the duration of the unwanted skin photosensitivity side effects (see, for example, McCubbin and Phillips 1986; W.G. Roberts *et al.* 1989) and as a way of increasing light penetration during therapy. While several mechanisms can lead to dye photo-induced fading a major route is direct reaction with singlet oxygen and the rate constant for the interaction between sensitizer and singlet oxygen is some measure of this. Certainly, for example, the bacteriochlorins exhibit both a rapid tendency to photobleach and a high rate constant, k_Δ, for interaction with singlet oxygen. In a given environment it seems that the value of k_Δ varies with the type of sensitizer and the presence of metal coordination. Typically, values of 10^9, 10^8, and 10^7 $dm^3 mol^{-1} s^{-1}$ are observed for bacteriochlorins, chlorins, and purpurins; porphyrins; and phthalocyanines, respectively (Borland *et al.* 1987, 1988). Metallo-coordinated photosensitizers show a reduced k_Δ as compared with metal free systems. For purpurins, the zinc complex reacts nearly 10 times faster than the metal free system. A similar, though less marked, trend can be observed by comparing the magnesium containing Bchl *a* and Bpheo *a*.

12.4 **Molecular targets of photosensitization**

A review of the photosensitizing effects of porphyrins and chlorins has been given by Jori and Spikes (1984). The porphyrin photosensitization, *in vitro* and *in vivo*, of a wide range of biological molecules has been reported including:

(1) lipids and steroids, components of biological membranes,

(2) proteins, and

(3) nucleic acids.

The biological membranes (Valenzeno 1987, Girotti 1990) and in particular mitochondrial membranes (Salet and Moreno 1990) are considered as the critical targets for cell killing by photosensitization. HP photosensitized oxidation of cholesterol, a determinant of eukaryotic membrane mechanical stability, gives its 5-α-hydroperoxide, this reaction being specific to a singlet oxygen mechanism. A similar 'ene' type reaction is observed in the HP photosensitized oxidation of unsaturated fatty acids (Chacon *et al.* 1988*b*). Lipid peroxidation leads to structural and functional damage of cell membranes and of their surroundings including crosslinking for proteins with aminolipids, inactivation of intrinsic enzymes, and increased permeability (Girotti 1990).

Photo-oxidation of histidine, tryptophan, tyrosine, methionine, and cysteine residues can produce reactions with amino groups and intramolecular crosslinking in proteins (Dubbelman *et al.* 1978; Verweij and van Steveninck 1982).

Although PDT acts mainly via membrane damage, cancer induction by photodynamic processes have been reported via photosensitized damage of DNA (Moan *et al.* 1990). In cells, DNA–protein crosslinks and DNA single-strand breaks have been observed, all of them via the singlet oxygen pathway. At equitoxic doses, X-rays induce about 80 per cent more single-strand breaks in DNA of human cells than photosensitization with HP (Moan *et al.* 1990).

12.5 Mechanisms of tumour necrosis

This chapter has been mainly concerned with the primary photophysical/photochemical processes following light absorption by the porphyrins in HPD and also the alternative molecules being considered as 'second generation' photosensitizers for PDT. In this section, however, we briefly comment on the mechanisms of the drug delivery and tumour necrosis subsequent to the photochemical events.

Sensitizer transport in the body and subsequent specific localization depend on its hydrophobicity, aggregation state, and specific ionic state. The main carriers in the bloodstream are lipoproteins and serum albumin (see review by Mazière *et al.* 1991). Among reasons for the drug accumulation and retention in a tumour are

(1) the relatively large interstitial space and the greater permeability of tumour vessels to sensitizer loaded macromolecules, with the reduced lymphatic drainage also increasing the sensitizer concentration in the interstitial space;
(2) the high levels of low density lipoprotein (LDL) receptors in proliferating cells associated with cholesterol accumulation (Kessel *et al.* 1987); and
(3) the lower pH often found in the interstitial fluid of neoplastic tumours.

In general, porphyrins and other photosensitizers often show a moderate selectivity for retention by neoplastic tissues which can sometimes be considerably increased by covalent binding to serum albumin (Hünerbein *et al.* 1991).

The primary mechanism of tumour destruction by PDT was originally thought to be the selective accumulation of HPD in neoplastic cells and their subsequent direct killing via photoactivation of the intracellular photosensitizer (one or more components of HPD). However, several observations show that the effectiveness of PDT may not be related to directly destroying tumour cells (see, for example, Chang and Dougherty 1978; Bugelski *et al.* 1981; Moan *et al.* 1981; Henderson *et al.* 1984, 1985; Star *et al.* 1986). Recent data (see Zhou 1989, for numerous references) indicate that initial vascular damage and subsequent tumour cell anoxia may well be an alternative indirect mechanism leading to tumour necrosis. Certainly the mechanism of PDT may well involve the killing of tumour cells as mainly a secondary step following damage to the microvasculature and perturbation of the tumour microcirculation. Thus the vascular endothelium may well be the main target of tumour photosensitization by HPD (Henderson and Dougherty 1992).

12.6 Future directions and new photosensitizers for PDT

The drugs used in clinical applications of PDT are HPD and a purified version, Photofrin^R. The phase III trials carried out by a large number of hospitals throughout the world have started and are under the main supervision of

American Cyanamid Company, Lederle Laboratories, and QLT Phototherapeutics of Vancouver. Dougherty *et al.* (1990) have reviewed the present status for PDT treatment of different cancers including endobronchial, oesophageal, bladder, and ocular tumours.

The future of PDT will be improved by finding new drugs:

(1) with light absorption more to the red, in the spectral region where most tissues transmit optimally (750–900 nm, see B.C. Wilson 1989);

(2) with better tumour-localizing properties;

(3) which are single pure substances; and

(4) with little or no skin photosensitizing side effects.

The search for new effective photosensitizers often uses cell models, and photosensitized cell killing (Jori and Spikes 1984) has been extended to other sensitizers such as the purpurins (Morgan *et al.* 1988) and phthalocyanines (Ben-Hur *et al.* 1987; Brasseur *et al.* 1987; Tralau *et al.* 1987). Other dyes currently being evaluated in preclinical assays include chlorins, bacteriochlorins, and benzoporphyrins (see Figs 12.4 and 12.5 for typical structures). They have large molar absorbance coefficients in the 680–800 nm range, more to the red than HPD, and allow the injection of significantly lower drug concentrations. The most promising drugs, ready for clinical trials, are phthalocyanines (Barr *et al.* 1990; Van Lier 1990; Jori 1992), benzoporphyrins (Richter *et al.* 1987, 1990), and hydroxy-substituted chlorins (K.R. Adams *et al.* 1992).

A particularly interesting development for PDT is the use of 5-aminolaevulinic acid (ALA). This molecule is the precursor of PP in haem biosynthesis. Since the conversion of PP into haem is rather slow, certain cells have the capacity to synthesize and accumulate PP when exposed to ALA. The topical application of ALA to patients with skin cancer is becoming an established treatment (J.C. Kennedy *et al.* 1990). One of the major advantages of this treatment is that no persistent skin photosensitization arises because ALA penetration through abnormal keratin is favoured over normal keratin. Certainly, this seems a most promising development of PDT since, in the clinical trial reported by Kennedy, the response rate for basal cell carcinomers following a single treatment showed a 90 per cent complete response for the first 80 lesions treated.

An interesting direction of research has been the tumour-targeted photosentizers with porphyrins and chlorins covalently coupled to monoclonal antibodies directed against antigens present at the surface of neoplastic cells, or the binding of the sensitizers to low density lipoproteins. These strategies for tumour targeting have been reviewed by Jori and Reddi (1990).

Compared to the other cancer therapies, PDT has advantages such as:

(1) the irradiation wavelengths are specifically absorbed by the photosensitizers;

(2) the irradiation is directed locally to the tumour; and

(3) there is a somewhat larger retention of the sensitizer in the tumour tissue than in the adjacent healthy tissues.

Moreover, for equitoxic doses, DNA damage induced by PDT is small when compared to that induced by other cancer therapies (chemotherapeutic drugs and ionizing radiation) and thus the induction of a secondary cancer from PDT should have a low probability.

As pointed out by Pratesi (1987) the development of low-cost, compact, and efficient red diode lasers such as the cw diode laser with 1 W at ~780 nm and the development of new photosensitizers such as phthalocyanines and benzoporphyrins absorbing in the far-red within the spectral range of emission of these new lasers, will further stimulate interest in PDT which suffers presently from the high cost of the laser sources used for Photofrin^{R}.

12.7 Concluding remarks

In recent years, there has been a substantial increase in research on the photochemistry of HPD, on unravelling the nature of its components, and on developing alternative sensitizers with enhanced and red-shifted absorption bands corresponding to increased light penetration of tissue. While much of the basic photochemistry and photophysics in 'simple' homogeneous environments is now established, there are still problems in interpreting the data in cell models such as liposomes. Future research, even using the sophisticated techniques associated with, for example, singlet oxygen detection via its near infra-red luminescence, single photon counting, and triplet studies via laser flash photolysis needs to concern environments nearer to the physiological situation. In applying these and other techniques it is vital to remember that studies of complex mixtures such as HPD and possibly sulphonated phthalocyanines may give information very difficult to interpret. On the other hand there could be a clinical advantage in using mixtures since different components could localize in different cell sites (see, for example, Peng *et al.* 1991).

Possibly one of the most important developments of PDT to date is the topical use of ALA. Not only does this avoid the unwanted side effect of skin photosensitivity after therapy, but during irradiation the PP probably converts to corresponding chlorins so that increased light penetration at ~670 nm is also achieved (Charlesworth and Truscott 1993; König *et al.* 1993).

Finally, while there have been many claims of the efficacy of PDT from a wide range of research groups for over 10 years, it is still not an accepted treatment. One reason for this is that PDT is a topic that requires expertise in a range of specialist subjects and many communication gaps can arise between the clinical and non-clinical world. As stated by Bown (1990) the clinical use of PDT requires a new way of thinking about human disease and it will take time, effort, and high quality data to convince established practitioners of its efficacy.

References

Abdella, B. R. J. and Fisher, J. (1985). A chemical perspective on the anthracycline antitumour antibiotics. *Environmental Health Perspectives*, **64**, 3–18.

Abramovitch, S. and Rabani, J. (1976). Pulse radiolytic investigations of peroxy radicals in aqueous solutions of acetate and glycine. *J. Phys. Chem.*, **80**, 1562–5.

Adams, G. E. (1967). The general application of pulse radiolysis to current problems in radiobiology. *Current Topics Radiat. Res.*, **3**, 35–93.

Adams, G. E. and Cooke, M. S. (1969). Electron-affinic sensitization. I. A structural basis for chemical radiosensitizers in bacteria. *Int. J. Radiat. Biol.*, **15**, 457–71.

Adams, G. E. and Dewey, D. L. (1963). Hydrated electrons and radiobiological sensitisation. *Biochem. Biophys. Res. Commun.*, **12**, 473–7.

Adams, G. E. and Wardman, P. (1977). Free radicals in biology: the pulse radiolysis approach. In *Free radicals in biology*, vol. 3 (ed. W. A. Pryor), pp. 53–95. Academic Press, New York.

Adams, G. E. and Willson, R. L. (1973). Ketyl radicals in aqueous solution: pulse radiolysis study. *J. Chem. Soc., Faraday Trans. I*, **69**, 719–29.

Adams, G. E., Michael, B. D., and Land E. J. (1966). Pulse radiolysis investigation of unimolecular elimination reactions of free radicals produced by OH addition to hydroxy-benzenes. *Nature*, **211**, 293–4.

Adams, G. E., McNaughton, G. S., and Michael, B. D. (1967*a*). The pulse radiolysis of sulphur compounds. Part 1. Cysteamine and cystamine. In *The Chemistry of ionization and excitation* (ed. G. R. A. Johnson and G. Scholes), pp. 281–93. Taylor and Francis, London.

Adams, G. E., Michael, B. D., and Richards, J. T. (1967*b*). Three stage electron transfer in aqueous and alcoholic solutions. *Nature*, **215**, 1248–50.

Adams, G. E., Cooke, M. S., and Michael, B. D. (1968*a*). Rapid mixing in radiobiology. *Nature*, **219**, 1368–9.

Adams, G. E., McNaughton, G. S., and Michael, B. D. (1968*b*). Pulse radiolysis of sulphur compounds. Part 2. Free radical 'repair' by hydrogen transfer from sulphydryl compounds. *Trans. Faraday Soc.*, **64**, 902–10.

Adams, G. E., Michael, B. D., and Willson, R. L. (1968*c*). Electron transfer studies by pulse radiolysis. *Adv. Chem. Ser.*, **81**, 289–308.

Adams, G. E., Armstrong, R. C., Charlesby, A., Michael, B. D., and Willson, R. L. (1969). Pulse radiolysis of sulphur compounds. Part 3. Repair by hydrogen transfer of a macromolecule irradiated in aqueous solution. *Trans. Faraday Soc.*, **65**, 732–42.

Adams G. E., Aldrich, J. E., Bisby R. H., Cundall R. B., Redpath J. L., and Willson R. L. (1972*a*). Selective free radical reactions with proteins and enzymes: reaction of inorganic radical anions with amino acids. *Radiat. Res.*, **49**, 278–89.

Adams, G. E., Redpath, J. L., Bisby, R. H., and Cundall, R. B. (1972*b*). The use of free radical probes in the study of mechanisms of enzyme inactivation. *Isr. J. Chem.*, **10**, 1079–93.

Adams, G. E., Greenstock, C. L., Van-Hemmen, J. J., and Willson, R. L. (1972*c*). Radical oxidation mechanisms in cellular radiosensitisation: Electron transfer in the pulse radiolysis of aqueous nucleotide solutions. *Radiat. Res.*, **49**, 85–95.

Adams, G. E., Flockhart, I. R., Smithen, C. E., Stratford, I. J., Wardman, P., and Watts, M. E. (1976). Electron affinic sensitization. VII. A correlation between structures, one-electron reduction potentials, and efficiencies of nitroimidazoles as hypoxic-cell radiosensitizers. *Radiat. Res.*, **67**, 9–20.

Adams, G.E., Clarke, E.D., Flockhart, I.R., Jacobs, R.S., Sehmi, D.S., Stratford, I.J., Wardman, P., Watts, M.E., Parrick, J., Wallace, R.G., and Smithen, C.E. (1979*a*). Structure–activity relationships in the development of hypoxic cell radiosensitizers. I. Sensitization efficiency. *Int. J. Radiat. Biol.*, **35**, 133–50.

Adams, G.E., Clarke, E.D., Gray, P., Jacobs, R.S., Stratford, I.J., Wardman, P., Watts, M.E., Parrick, J., Wallace, R.G., and Smithen, C.E. (1979*b*). Structure–activity relationships in the development of hypoxic cell radiosensitizers. II. Cytotoxicity and therapeutic ratio. *Int. J. Radiat. Biol.*, **35**, 151–60.

Adams, G.E., Ahmed, I., Clarke, E.D., O'Neill, P., Parrick, J., Stratford, I.J., Wallace, R.G., Wardman, P., and Watts, M.E. (1980). Structure–activity relationships in the development of hypoxic cell radiosensitizers. III. Effects of basic substituents in nitroimidazole sidechains. *Int. J. Radiat. Biol.*, **38**, 613–26.

Adams, K.R., Berenbaum, M.C., Bonnett, R., Nizhnik, A.N., Salgado, A., and Valles, M.A. (1992). Second generation tumour photosensitizers: the synthesis and biological activity of octaalkyl chlorins and bacteriochlorins with graded amphiphilic character. *J. Chem. Soc., Perkin Trans. I*, 1465–70.

Adelman, R., Saul, R.L., and Ames, B.N. (1988). Oxidative damage to DNA: relation to species, metabolic rate and life span. *Proc. Natl. Acad. Sci. USA.*, **85**, 2706–8.

Agati, G. and Fusi, F. (1990). Recent advances in bilirubin photophysics *J. Photochem. Photobiol., B. Biol.*, **7**, 1–14.

Agmon, N. and Levine, R.D. (1977). Energy, entropy and the reaction coordinate: Thermodynamic-like relations in chemical kinetics. *Chem. Phys. Lett.*, **52**, 197–201.

Akasaka, S. (1986). Inactivation of transforming activity of plasmid DNA by lipid peroxidation. *Biochim. Biophys. Acta*, **867**, 201–8.

Akhlaq, M.S. and von Sonntag, C. (1987). Intermolecular H.- abstraction of thiyl radicals from thiols and the intramolecular complexing of the thiyl radical with the thiol group in 1, 4-dithiothreitol. A pulse radiolysis study. *Z. Naturforsch.*, **42c**, 134–40.

Akhlaq, M.S., Schuchmann, H.P., and v. Sonntag, C. (1987). The reverse of the repair reaction of thiols: H-abstraction at carbon by thiyl radicals. *Int. J. Radiat. Biol.*, **51**, 91–102.

Aldrich, J.E., Foldvary, P., Hunt, J.W., Taylor, W.B., and Wolff R.K. (1972). A general technique for the spectroscopy of transient absorbing species. *Rev. Sci. Instrum.*, **43**, 991–5.

Alfassi, Z.B. and Schuler, R.H. (1985). Reaction of azide radicals with aromatic compounds. Azide as a selective oxidant. *J. Phys. Chem.*, **89**, 3359–63.

Algar, B.E. and Stevens, B. (1970). The photoperoxidation of unsaturated organic molecules. VI. The inhibited reaction. *J. Phys. Chem.*, **34**, 3029–34.

Al-Kazwini, A.T., O'Neill, P., Adams, G.E., Cundall, R.B., Jacquet, B., Lang, G., and Junino, A. (1990*a*). One-electron oxidation of methoxylated and hydroxylated indoles by N_3: 1. Characterization of the primary indolic radicals, *J. Phys. Chem.*, **94**, 6666–70.

Al-Kazwini, A.T., O'Neill, P., Adams, G.E., and Fielden, E.M. (1990*b*). Radiation-induced energy migration within solid DNA: The role of misonidazole as an electron trap. *Radiat. Res.*, **121**, 149–53.

Allen, A.O. (1976). *Yields of free ions formed in liquids by radiation*. NSRDS- NBS-57.

Allison, S., Austin, R., and Hogan, M. (1989). Binding and twisting dynamics of short linear DNAs. Analysis of the triplet anisotropy decay of a 209 base pair fragment by Brownian simulation. *J. Chem. Phys.*, **90**, 3843–54.

Almgren, M. (1991). Kinetics of excited state processes in micellar media. In *Kinetics and catalysis in microheterogeneous systems* (ed. M. Gratzel and K. Kalyanasundaram), pp. 63–113. Marcel Dekker, New York.

Almgren, M. and Thomas, J.K. (1980). Interfacial electron transfer involving radical ions of carotene and diphenylhexatriene in micelles and vesicles. *Photochem. Photobiol.*, **31**, 329–35.

Amand, B. and Bensasson, R. (1974). Contribution à l'étude du spectre d'absorption du premier état singulet ($S_n \leftarrow S_1$) du naphtalene en solution. *C.R. Acad. Sci. Paris*, **278**, 981–3.

Amand, B. and Bensasson, R.V. (1975). Determination of triplet quantum yields by laser flash absorption spectroscopy. *Chem. Phys. Lett.*, **34**, 44–8.

Ambroz, M., Beeby, A., MacRobert, A.J., Simpson, M.S.C., Svensen, R.K., and Phillips, D. (1991). Preparative, analytical and fluorescence spectroscopic studies of sulphonated aluminium phthalocyanine photosensitizers. *J. Photochem. Photobiol., B: Biol.*, **9**, 87–95.

Ames, B.N. (1989). Endogenous oxidative DNA damage, aging, and cancer. *Free Rad. Res. Commun.*, **7**, 121–8.

Amouyal, E. and Bensasson, R.V. (1976). Duroquinone triplet reduction in cyclohexane, ethanol and water, and by durohydroquinone *J. Chem. Soc., Faraday Trans. I*, **72**, 1274–87.

Amouyal, E. and Bensasson, R.V. (1977). Interaction of duroquinone lowest triplet with amines. *J. Chem. Soc., Faraday Trans. I*, **73**, 1561–8.

Anderson, P.O., Gillbro, T., Ferguson, L., and Cogdell, R.J. (1991). Absorption spectral shifts of carotenoids related to medium polarizability. *Photochem. Photobiol.*, **54**, 353–60.

Anderson, P.O., Gillbro, T., Asato, A.E., and Liu, R.S.H. (1992). Dual singlet state emission in a series of mini-carotenes. *J. Lumin.*, **51**, 11–20.

Anderson, S.O. (1979). Biochemistry of insect cuticle. *Ann. Rev. Entomol.*, **24**, 29–61.

Andreoni, A. and Cubeddu, R. (1984). Photophysical properties of photofrin II in different solvents. *Chem. Phys. Lett.*, **108**, 141–4.

Andreoni, A., Land, E.J., Malatesta, V., McLean, A.J., and Truscott, T.G. (1989). Triplet state characteristics and singlet oxygen generation properties of anthracyclines. *Biochim. Biophys. Acta*, **990**, 190–7.

Andreoni A., Colasanti, A., Malatesta, V., Riccio, P., and Roberti, G. (1991). Enhancement of antitumour drug cytotoxicity via laser photoactivation. *Photochem. Photobiol.*, **53**, 797–805.

Andrieux, C.P., Hapiot, P., and Saveant, J.-M. (1990). Fast kinetics by means of direct and indirect electrochemical techniques. *Chem. Rev.*, **90**, 723–38.

Aramendia, P.F., Redmond, R.W., Nonell, S., Schuster, W., Braslavsky, S.E., Schaffner, K., and Vogel, E. (1986). The photophysical properties of porphycenes: Potential photo-dynamic agents. *Photochem. Photobiol.*, **44**, 555–9.

Arce, R. and Pacheco, M. (1986). Intermediates and quantum yields in the photolysis of guanine and its derivatives in neutral glasses at 77 K. *J. Photochem.*, **34**, 89–104.

Arce, R. and Rodriquez, G. (1985). The photochemistry and quantum yields in the photolysis of guanine and its derivatives in neutral glasses at 77 K. *J. Photochem.*, **28**, 111–30.

Arce, R. and Rodriquez, G. (1986). An electron paramagnetic resonance study of the excited triplet state of adenine and some of its derivatives in aqueous glasses at 77 K. *J. Photochem.*, **33**, 89–97.

Arce, R., Jimenez, L.A., Riviera, V., and Torres, C. (1980). Intermediates in the room temperature flash photolysis and low temperature photolysis of purine solutions. *Photochem. Photobiol.*, **32**, 91–5.

Arnold, J.S., Browne, R.J., and Ogryzlo, E.A. (1965). The red emission bands of molecular oxygen. *Photochem. Photobiol.*, **4**, 963–9.

Asher, S. and Sauer, K. (1976). Resonance Raman spectroscopy of Mn III etioporphyrin I at the π-π^* and charge transfer bands: The use of charge transfer bands to monitor the complexation state of metalloporphyrins. *J. Chem. Phys.*, **64**, 4115–25.

Asmus, K.-D., and Janata, E. (1982*a*). Conductivity monitoring techniques. In *The study of fast processes and transient species by electron pulse radiolysis* (ed. J.H. Baxendale and F. Busi), pp. 91–113. D. Reidel, New York.

Asmus, K.-D., and Janata, E. (1982*b*). Polarography monitoring techniques. In *The study of fast processes and transient species by electron pulse radiolysis.* (ed. J.H. Baxendale and F. Busi), pp. 115–28. D. Reidel, New York.

Asmus, K.-D., Bahneman, D., Mönig, J., Searle, A., and Willson, R.L. (1979). Electrophilic free radicals and phenothiazines: a pulse radiolysis study. In *Radiation biology and chemistry, research developments* (ed. H.E. Edwards, S. Navaratnam, B.J. Parsons, and G.O. Phillips), pp. 39–47. Elsevier, Amsterdam.

Asmus, K.-D., Gobl, M., Hiller, K.-O., Mahling, S., and Monig, J. (1985). S∴N and S∴O Three-electron-bonded radicals and radical cations in aqueous solution. *J. Chem. Soc., Perkin Trans. II*, 641–6.

Aubry, J.-M. (1985). Search for singlet oxygen in the decomposition of hydrogen peroxide by mineral compounds in aqueous solutions. *J. Am. Chem. Soc.*, **107**, 5844–9.

Aubry, J.M., Rigaudy, J., Ferradini, C., and Pucheault, J. (1981). A search for singlet oxygen in the disproportionation of superoxide anion. *J. Am. Chem. Soc.*, **103**, 4965–6.

Auger, P. (1925). Sur l'effet photoélectrique composé. *Journal de Physique*, **6**, 205–8.

Auger, P. (1926). Sur le rendement de la fluorescence dans le domaine des rayons X. *C. R. Acad. Sci. Paris*, **182**, 773–5.

Averbeck, D., Dubertret, L., Craw, M., Truscott, T.G., Dall'Acqua, F. Rodighiero, P., Vedaldi, D., and Land, E.J. (1984). Photophysical, photochemical and photobiological studies of 4'-methylangelicins, potential agents for photochemotherapy. *Il Farmaco*, **39**, 57–69.

Babizaev, M.A. (1985). Accumulation of lipid peroxidation products in the human lens during cataract maturation. *Vopr. Med. Khim.* **31**, 100–4.

Bachur, N.R., Gordon, S.L., and Gee, M.V. (1978). A general mechanism for microsomal activation of quinone anticancer agents to free radicals. *Cancer Res.*, **38**, 1745–50.

Bachur, N.R., Gordon, S.L., Gee, M.V., and Kon, H. (1979). NADPH cytochrome P-450 reductase activation of quinone anticancer agents to free radicals. *Proc. Natl. Acad. Sci.*, **76**, 954–7.

Backstrom, H.L.J. and Sandros, K. (1958). The quenching of the long-lived fluorecence of biacetyl in solution. *Acta Chem. Scand.*, **12**, 828–32.

Bagnara, J.T. (ed.) (1988). *Advances in pigment cell research. Progress in clinical and biological research*, Vol. 256. Alan R. Liss, New York.

Bahnermann, D., Asmus, K.-D., and Willson, R.L. (1981). Free radical reactions of the phenothiazine, metiazinic acid. *J. Chem. Soc., Perkin Trans. II*, 890–5.

Bahnermann, D., Asmus, K.-D., and Willson, R.L. (1983*c*). Free radical induced one-electron oxidation of the phenothiazines chlorpromazine and promethazine. *J. Chem. Soc., Perkin Trans. II.*, 1661–8.

Bahnermann, D., Asmus, K.-D., and Willson, R.L. (1983*b*). Phenothiazine radical-cations: Electron transfer equilibria with iodide ions and the determination of one-electron redox potentials by pulse radiolysis. *J. Chem. Soc., Perkin Trans. II*, 1669–73.

Baker, A. and Kanofsky, J.R. (1991). Direct observation of singlet oxygen phosphorescence at 1270 nm from L1210 leukaemia cells exposed to polyporphyrin and light. *Arch. Biochem. Biophys.*, **286**, 70–5.

Baker, M.Z., Badiello, R., Tamba, M., Quintiliani, M., and Gorin, G. (1982). Pulse radiolytic study of hydrogen transfer from glutathione to organic radicals. *Int. J. Radiat. Biol.*, **41**, 595–602.

Ballini, J.P., Daniels, M., and Vigny, P. (1988). In *Light in biology*, Vol. 1 (ed. R.H. Douglas, J. Moan, and F. Dall'Acqua), pp. 31–8. Plenum Press, New York.

Bally, T., Roth, K., Tang, W., Schrock, R.R., Knoll, K., and Park, L.Y. (1992). Stable polarons in polyacetylene oligomers: optical spectra of long polyene radical cations. *J. Am. Chem. Soc.*, **114**, 2440–6.

Balzani, V. and Scandola, F. (1981). Photochemical electron transfer reactions in homogeneous solutions. In *Photochemical conversion and storage of solar energy* (ed. J.S. Connolly), pp. 97–129. Academic Press, New York.

Balzani, V. and Scandola, F. (1991). *Supramolecular photochemistry*. Ellis Horwood series in Physical Chemistry (ed. T.J. Kemp). Ellis Horwood, New York.

Balzani, V., Bolletta, F., Gandolfi, M.T., and Mestri, M. (1978). Bimolecular electron transfer reactions of the excited states of transition metal compounds. *Top. Curr. Chem.*, **75**, 1–64.

Balzani, V., Bolletta, F., Scandola, F., and Ballardini, R. (1979). Excited state electron transfer reactions of transition metal complexes. *Pure Appl. Chem.*, **51**, 229–311.

Bansal, K.M. and Fessenden, R.W. (1976). Pulse radiolysis studies of the oxidation of phenols by $SO_4^{\cdot-}$ and $Br_2^{\cdot-}$ in aqueous solutions. *Radiat. Res.*, **67**, 1–8.

Barbara, P.F. and Jarzeba, W. (1990). Ultrafast photochemical charge transfer and excited state solvation. In *Advances in photochemistry*, Vol. 15 (ed. D.H. Volman, G.S. Hammond, and K. Gollnick), pp. 1–68. J. Wiley, New York.

Barker, G.C. and Bozlan, J.A. (1974). The influence of homogeneous hydrogen abstraction on the nitrous oxide photocurrent. Part 1, Underlying theory. *Electroanal Chem. Interfac. Electrochem.*, **49**, 227–38.

Barr, H., Tralau, C.J., Boulos, P.B., MacRobert, A.J., Lewis, M.R., Phillips, D., and Bown, S.G. (1990). Selective necrosis in experimental colon cancer using photodynamic therapy with phthalocyanine photosensitisation. *Gastroenterology*, **98**, 1532–7.

Baugher, J.F. and Grossweiner, L.I. (1978). Photolysis mechanism of aqueous tyrosine and tyrosyl peptides. *Photochem. Photobiol.*, **28**, 175–84.

Baxendale, J.H. and Khan, A.A. (1969). The pulse radiolysis of *p*-nitrosodimethylalanine in aqueous solution. *Int J. Radiat. Phys. Chem.*, **1**, 11–24.

Bayliss, N.S. and McRae, E.G. (1954). Solvent effects in organic spectra: dipole forces and the Franck–Condon principle. *J. Phys. Chem.*, **58**, 1002–6.

Bazin, M. and Ebbesen, T.W. (1983). Distortions in laser flash photolysis absorption measurements. The overlap problem. *Photochem. Photobiol.*, **37**, 675–8.

Bazin, M., Patterson, L.K., and Santus, R. (1983). Direct observation of monophotonic photoionization in tryptophan excited by 300-nm radiation. A laser photolysis study. *J. Phys. Chem.*, **87**, 189–90.

Beaumont, P.C., Parsons, B.J., Navaratnam, S., Phillips, G.O., and Allen, J.C. (1979*a*). Laser flash photolysis and fluorescence studies of the interaction of the excited states of furocoumarins with DNA in solution. In *Radiation biology and chemistry; Research developments* (ed. H.E. Edwards, S. Navaratnam, B.J.

Parsons, and G.O. Phillips), pp. 441–51. Elsevier, Amsterdam.

Beaumont, P.C., Parsons, B.J., Phillips. G.O., and Allen, J.C. (1979*b*). A laser flash photolysis study of the reactivities of the triplet states of 8-methoxypsoralen and 4, 5', 8-trimethylpsoralen with nucleic acid bases in solution. *Biochim. Biophys. Acta*, **562**, 214–21.

Beaumont, P.C., Parsons, B.J., Navaratnam, S., Phillips, G.O., and Allen, J.C. (1980). The reactivities of furocoumarin excited states with DNA in solution. A laser flash photolysis and fluorescence study. *Biochim. Biophys. Acta*, **608**, 259–65.

Beaumont, P.C., Parsons, B.J., Navaratnam, S., and Phillips, G.O. (1983). A laser flash photolysis and fluorescence study of aminomethyltrimethylpsoralen in the presence and absence of DNA. *Photobiochem. Photobiophys.*, **5**, 359–64.

Beaumont, P.C., Rodgers, M.A.J., Parsons, B.J., and Phillips, G.O. (1985). Singlet oxygen production by some furocoumarin derivatives in the presence of DNA: Time resolved luminescence measurements. *Photochem. Photobiol.*, **42**, 605–8.

Bechtold, R., Kuehn, C., Lepre, C., and Isied, S.S. (1986). Directional electron transfer in ruthenium-modified horse heart cytochrome *c*. *Nature*, **322**, 286–8.

Beck, G. (1979). Transient conductivity measurements with subnanosecond time resolution. *Rev. Sci. Instr.*, **50**, 1147–50.

Beck, G., Kiwi, J., Lindenau, D., and Schnabel, W. (1974). On the kinetics of polymer degradation in solution-1. Laser flash photolysis and pulse radiolysis studies using the light scattering detection method. *European Polymer J.*, **10**, 1069–75.

Becker, M.M. and Wang, J.C. (1984). Use of light for footprinting DNA *in vivo*. *Nature*, **309**, 682–7.

Becker, R.S. (1969). *Theory and interpretation of fluorescence and phosphorescence*. Wiley Interscience, New York.

Becker, R.S. (1988). The visual process: photophysics and photoisomerisation of model visual pigments and the primary reaction. *Photochem. Photobiol.*, **48**, 369–99.

Becker, R.S. and Kogan, G. (1980). Photophysical properties of nucleic acid components- 1. The pyrimidines: thymine, uracil, *N*, *N*-dimethyl derivatives and thymidine. *Photochem. Photobiol.*, **31**, 5–13.

Becker, R.S., Hug, G., Das, P.K., Schaffer, A.M., Takemura, T., Yamamoto, N., and Waddell, W. (1976). Visual pigments 4. Comprehensive consideration of the spectroscopy and photochemistry of model visual pigments. *J. Phys. Chem.*, **80**, 2265–73.

Becker, R.S., Bensasson, R.V., Lafferty, J., Truscott, T.G., and Land, E.J. (1978). Triplet excited states of carbonyl-containing polyenes. *J. Chem. Soc., Faraday Trans. II*, **74**, 2246–55.

Becker, R.S., Chakravorti, S., and Yoon, M. (1990). Photochemical and photophysical properties of piroxican and benoxaprofen in various solvents. *Photochem, Photobiol.*, **51**, 151–4.

Beckman, J.S. (1990). Ischaemic injury mediator. *Nature*, **345**, 27–8.

Beckwith, A.L.J., Bowry, V.W., and Ingold, K.U. (1992). Kinetics of nitroxide radical trapping. 1. Solvent effects. *J. Am. Chem. Soc.*, **114**, 4983–92.

Becquerel, H. (1896). Notes à L'Académie des Sciences 24 Février, 2, 9, & 30 Mars, 18 Mai, Paris.

Beeby, A., Parker, A.W., Simpson, M.S.C., and Phillips, D. (1992). The effect of solvent deuteration on the photophysics of sulphonated aluminium phthalocyanine. *J. Photochem. Photobiol. B Biol.*, **16**, 73–81.

Beems, E.M., Dubbelman, T.M.A.R., Lugtenburg, J., Van Best, J.A., Smeets, F.M.A., and Boegheim, J.P.J. (1987). Photosensitizing properties of bacteriochlorophyllin *a* and bacteriochlorin *a*. Two derivatives of bacteriochlorophyll *a*. *Photochem. Photobiol.*, **46**, 639–43.

Beer, A. (1852). Bestimmung der Absorption des rothen Lichts in farbigen Flüssigkeiten. *Ann. Physik.*, **86**, 76–88.

Behar, D., Czapski, G., Rabani, J., Dorfman, L. M., and Schwarz, H. A. (1970). The acid dissociation constant and decay kinetics of the perhydroxyl radical. *J. Phys. Chem.*, **74**, 3209–13.

Beltramini, M., Firey, P. A., Ricchelli, F., Rodgers, M. A. J., and Jori, G. (1987). Steady-state and time-resolved spectroscopic studies on the haematoporphyrin-lipoprotein complex. *Biochemistry*, **26**, 6852–8.

Ben-Hur, E., Green, M., Prager, A., Kol, R., and Rosenthal, I. (1987). Phthalocyanine photosensitization of mammalian cells: Biochemical and ultra structural effects. *Photochem. Photobiol.*, **46**, 651–6.

Ben-Ishai, R., Green, M., Graff, E., Elad, D., Steinmaus, H., and Solomon, J. (1973). Photoalkylation of purines in DNA. *Photochem. Photobiol.*, **17**, 155–7.

Bensasson, R. and Land, E. J. (1971). Triplet–triplet extinction coefficients via energy transfer. *Trans. Faraday Soc.* **67**, 1904–15.

Bensasson, R. and Land, E. J. (1973). Optical and kinetic properties of semireduced plastoquinone and ubiquinone: electron acceptors in photosynthesis. *Biochim. Biophys. Acta*, **325**, 175–81.

Bensasson, R. and Land, E. J. (1978*a*). Physical properties of excited states: A general method for measuring triplet–triplet extinction coefficients, singlet–triplet intersystem crossing efficiencies and related parameters. *Photochem. Photobiol. Rev.*, **3**, 163–91.

Bensasson, R. and Land, E. J. (1978*b*). Intersystem crossing efficiencies of retinal isomers in different solvents measured by laser flash absorption spectrophotometry. *Nouv. J. Chim.*, **2**, 503–7.

Bensasson, R., Bonneau, R., Faure, J., and Joussot-Dubien, J. (1970). Utilisation d'un laser déclenché au néodyme en spectroscopie par éclair. Application à l'étude du naphtalene et du benzène. In Transitions non radiatives dans les molécules, XXème Réunion de la Société de Chimie Physique, 27–30 Mai 1969. *J. Chim. Phys., Sup.*, 133–6.

Bensasson, R., Chachaty, C., Land, E. J., and Salet, C. (1972). Nanosecond irradiation studies of biological molecules-1 Coenzyme Q6 (Ubiquinone-30). *Photochem. Photobiol.*, **16**, 27–37.

Bensasson, R., Land, E. J., and Truscott, T. G. (1975). Wavelength dependence of the intersystem crossing efficiency for retinal isomers. *Photochem. Photobiol.*, **21**, 419–21.

Bensasson, R. V., Land, E. J., and Maudinas, B. (1976). Triplet states of carotenoids from photosynthetic bacteria studied by nanosecond ultraviolet and electron pulse irradiation. *Photochem. Photobiol.*, **23**, 189–93.

Bensasson, R., Dawe, E. A., Long, D. A., and Land, E. J. (1977). Singlet → triplet intersystem crossing quantum yields of photosynthetic and related polyenes. *J. Chem . Soc., Faraday Trans. I*, **73**, 1319–25.

Bensasson, R. V., Land, E. J., and Salet, C. (1978*a*). Triplet excited state of furocoumarins: reaction with nucleic acid bases and amino acids. *Photochem. Photobiol.*, **27**, 273–80.

Bensasson, R. V., Goldschmidt, C. R., Land, E. J., and Truscott, T. G. (1978*b*). Laser intensity and the comparative method for determination of triplet quantum yields. *Photochem. Photobiol.*, **28**, 277–81.

Bensasson, R. V., Salet, C., Land, E. J., and Rushton, F. A. P., (1980). Triplet excited state of the 4',5' photoadduct of psoralen and thymine. *Photochem. Photobiol.*, **31**, 129–33.

Bensasson, R. V., Land, E. J., Moore, A. L., Crouch, R. L., Dirks, G., Moore, T. A., and Gust D. (1981). Mimicry of antenna and photoprotective carotenoid functions by a synthetic carotenoporphyrin. *Nature,* **290**, 239-332.

Bensasson, R. V., Land, E. J., and Truscott, T. G. (1983). *Pulse radiolysis and flash photolysis: Contributions to the chemistry of biology and medicine*. Pergamon, Oxford.

Bensasson, R. V., Land, E. J., Liu, R. S. H, Lo, K. K. N., and Truscott, T. G. (1984*a*). Triplet states of isomers of the C_{15}-aldehyde and the C_{18}-ketone: lower homologues of retinal. *Photochem. Photobiol.*, **39**, 263–5.

Bensasson, R. V., Chalvet, O., Land, E. J., and Ronfard-Haret, J. C. (1984*b*). Triplet, radical anion and radical cation spectra of furocoumarins. *Photochem. Photobiol.*, **39**, 287–91.

Bent, D. V. and Hayon, E. (1975*a*). Excited state chemistry of aromatic amino acids and related peptides. I Tyrosine. *J. Am. Chem. Soc.*, **97**, 2599–606.

Bent, D. V. and Hayon, E. (1975*b*). Excited state chemistry of aromatic amino acids and related peptides. II Phenylalanine. *J. Am. Chem. Soc.*, **97**, 2606–12.

Bent, D. V. and Hayon, E. (1975*c*). Excited state chemistry of aromatic amino acids and related peptides. II. Tryptophan. *J. Am. Chem. Soc.*, **97**, 2612–19.

Berg, H. and Reissmann, P. (1970). Photo-polarographie XXIII Mitt. Uber die ursachen von photo-reststromen *J. Electroanal. Chem. Interfacial Electrochem.*, **24**, 427–34.

Bernhard, W. A. (1981). Solid-state radiation chemistry of DNA: The bases. *Adv. Radiat. Biol.*, **9**, 199–280.

Bernier, M., Kusama, V., Borgers, M., Ver Donck, L., Valdes-Aguilera, O., Neckers, D. C., and Hearse, D. J. (1991). Pharmacological studies of arrhythmias induced by Rose-Bengal photoactivation. *Free Rad. Biol. Med.*, **10**, 287–96.

Bernstein, M., Rothberg L. J., and Peters, K. S. (1982). Time-resolved photoacoustic spectroscopy in the picosecond regime. *Chem. Phys. Lett.*, **91**, 315–8.

Bert, P. (1871). Recherches expérimentales sur l'influence que les changements dans la pression atmosphérique exercent sur les phénomènes de la vie. *C. R. Acad. Sci.*, **73**, 213–6 and 503–7.

Bert, P. (1878). *Barometric pressure: Researches in experimental physiology* (trans. M. A. Hitchcock and F. A. Hitchcock). College Book Co., Columbus, Ohio, 1943.

Bertram, J. S., Kolonel, L. K., and Meyskens, Jr, F. L. (1987). Rationale and strategies for chemoprevention of cancer in humans. *Cancer Res.*, **47**, 3012–31.

Bertrand, P. (1991). Application of electron transfer theories to biological systems. In *Long-range electron transfer in biology* (ed. G. Palmer), pp. 1–47. Springer-Verlag, Berlin.

Bevilacqua, R. and Bordin, F. (1973). Photo-C_4-cycloaddition of psoralen and pyrimidine bases: Effect of oxygen and paramagnetic ions. *Photochem. Photobiol.*, **17**, 191–4.

Bielski, B. H. J. (1978). Reevaluation of the spectral and kinetic properties of $HO_2^{\cdot}$ and $O_2^{-\cdot}$ free radicals *Photochem. Photobiol.*, **28**, 645–9.

Bielski, B. H. J. (1990). Generation of iron(IV) and iron(V) complexes in aqueous solutions. In *Methods in enzymology*, Vol. 186, Oxygen radicals in biological systems, Part B, Oxygen radicals and antioxidants, pp. 108–13. Academic Press, San Diego.

Bielski, B. H. J. and Arudi, R. L. (1983). Preparation and stabilisation of aqueous ethanolic superoxide solutions. *Anal. Biochem.*, **133**, 170–8.

Bielski, B. H. J. and Cabelli, D. E. (1991). Highlights of current research involving superoxide and perhydroxyl radicals in aqueous solutions. *Int. J. Radiat. Biol.*, **59**, 291–319.

Bielski, B.H.J., Arudi, R.L., and Sutherland, M.W. (1983). A study of the reactivity of HO_2/O_2^- ion with unsaturated fatty acids. *J. Biol. Chem.*, **258**, 4759–61.
Bielski, B.H.J., Cabelli, D.E., and Arudi, R.L. (1985). Reactivity of $HO_2^{\cdot}/O_2^{-\cdot}$ radicals in aqueous solution. *J. Phys. Chem. Ref. Data*, **14**, 1041–100.
Birge, R.R. (1981). Photophysics of light transduction in rhodopsin and bacteriorhodopsin. *Ann. Rev. Biophys. Bioeng.*, **10**, 315–54.
Birge, R.R. (1986). Two photon spectroscopy of protein-bound chromophores. *Acc. Chem. Res.*, **19**, 138–46.
Birge, R.R. and Pierce, B.M. (1979). A theoretical analysis of the two-photon properties of linear polyenes and the visual chromophores. *J. Chem. Phys.*, **70**, 165–78.
Birks, J.B. (1975). Photophysics of aromatic molecules–A postcript. In *Organic molecular photophysics*, Vol. 2 (ed. J.B. Birks), pp. 409–597. Wiley, London.
Black, H.S. (1987*a*). Photocarcinogenesis and diet. *Fed. Proc.*, **46**, 1901–5.
Black, H.S. (1987*b*). Potential involvement of free radical reactions in ultraviolet light-mediated cutaneous damage. *Photochem. Photobiol.*, **46**, 213–21.
Blanchetière, M. (1913). Oxydation et luminescence. *C.R. Acad. Sci.*, **157**, 118–21.
Blazek, E.R., Peak, J.G., and Peak, M.J. (1989). Singlet oxygen induces frank breaks as well as alkali- and piperidine-labile sites in supercoiled plasmid DNA. *Photochem. Photobiol.*, **49**, 607–13.
Blois, M.S. (1978). The melanins: Their synthesis and structure. *Photochem. Photobiol. Rev.*, **3**, 115–34.
Boag, J.W. and Steel, R.E. (1960). Studies of the kinetics of radiation induced chemical reactions in liquids by means of absorption spectroscopy. *British Empire Cancer Campaign Report*, **38**, Part II, 251–2.
Bobrowski, K. and Das P.K. (1982). Transient phenomena in the pulse radiolysis of retinyl polyenes. 2. Protonation kinetics. *J. Am. Chem. Soc.*, **104**, 1704–9.
Bobrowski, K. and Das, P.K. (1985). Transient phenomena in the pulse radiolysis of retinyl polyenes. 4. Environmental effects on absorption maximum of retinal radical anion. *J. Phys. Chem.*, **89**, 5733–8.
Bobrowski, K. and Das, P.K. (1986). Transient phenomena in the pulse radiolysis of retinyl polyenes. 5. Association of radical cations with parent molecules. *J. Phys. Chem.*, **90**, 927–31.
Bobrowski, K. and Das, P.K. (1987). Transient phenomena in the pulse radiolysis of retinyl polyenes. 6. Radical ions of retinal homologues. *J. Phys. Chem.*, **91**, 1210–5.
Bobrowski, K. and Holcman, J. (1987). Formation of three-electron bonds in one-electron oxidized methionine dipeptides: a pulse radiolysis study. *Int. J. Radiat. Biol.*, **52**, 139–44.
Bobrowski, K. and Lubis, R. (1986). Intermolecular charge transfer involving tryptophan, tyrosine and three-electron-bonded intermediates derived from methionine. *Int. J. Radiat. Biol.*, **50**, 1039–50.
Bobrowski, K., Wierzchowski, K.L., Holcman, J., and Ciurak, M. (1990). Intramolecular electron transfer in peptides containing methionine, tryptophan and tyrosine: a pulse radiolysis study. *Int. J. Radiat. Biol.*, **57**, 919–32.
Bondarev S.L., Dvornikov, S.S., and Bachilo, S.M. (1988). Fluorescence of β-carotene at 77 and 4.2 K. *Opt. Spectrosc. (USSR)*, **64**, 268–70.
Bondarev, S.L., Bachilo, S.M., Dvornikov, S.S., and Tikhomirov, S.A. (1989). $S_2 \rightarrow S_0$ fluorescence and transient $S_n \leftarrow S_1$ absorption of all-*trans*-β-carotene in solid and liquid solutions. *J. Photochem. Photobiol. A: Chem.*, **46**, 315–22.
Bonneau, R., Faure, J., Joussot-Dubien, J., Lindqvist, L., and Barthelemy, C. (1968).

Spectroscopie par éclair à excitation par laser déclenché: application à l'étude du naphtalène. *C.R. Acad. Sci. Ser. B*, **267**, 412–15.

Bonneau, R., Joussot-Dubien, J., and Bensasson, R. (1969). Giant laser pulse photolysis of benzene and mesitylene. *Chem. Phys. Lett.*, **3**, 353–6.

Bonneau, R., Carmichael, I., and Hug, G.L. (1991). Molar absorption coefficients of transient species in solution. *Pure Appl. Chem.*, **63**, 289–99.

Bonnett, R., Charalambides, A.A., Land, E.J., Sinclair R.S., Tait, D., and Trauscott, T.G. (1980). Triplet states of porphyrin esters *J. Chem. Soc., Faraday Trans. 1*, **76**, 852–9.

Bonnett, R., Ridge, R.J., Scourides, P.A., and Berenbaum, M.C. (1981). On the nature of haematoporphyrin derivative. *J. Chem. Soc., Perkin Trans. I*, 3135–40.

Bonnett, R., Ridge, R.J., Land, E.J., Sinclair, R.S., Tait, D., and Truscott, T.G. (1982). Pulsed irradiation of water-soluble porphysins. *J. Chem. Soc., Faraday Trans. I*, **78**, 127–36.

Bonnett, R., McGarvey, D.J., Harriman, A., Land, E.J., Truscott, T.G., and Winfield, U.-J. (1988). Photophysical properties of *meso*-tetraphenylporphyrin and some *meso*-tetra (hydroxyphenyl) porphyrins. *Photochem. Photobiol.*, **48**, 271–6.

Boone, C.W., Kelloff, G.J., and Malone, W.E. (1990). Identification of candidate cancer chemopreventive agents and their evaluation in animal models and human clinical trials: a review. *Cancer Res.*, **50**, 2–9.

Borg, D.C., and Schaich, K.M. (1984). Cytotoxicity from coupled redox cycling of autoxidizing xenobiotics and metals. *Isr. J. Chem.*, **24**, 38–53.

Borg, D.C., Schaich, K.M., Elmore, J.J., and Bell, J.A. (1978). Cytotoxic reactions of free radical species of oxygen. *Photochem. Photobiol.*, **28**, 887–907.

Borland, C.F., McGarvey, D.J., Truscott, T.G., Cogdell, R.J., and Land, E.J. (1987). Photophysical studies of bacteriochlorophyll *a* and bacteriopheophytin *a*—Singlet oxygen generation. *J. Photochem. Photobiol., B: Biol.*, **1**, 93–101.

Borland, C.F., McGarvey, D.J., Morgan, A.R., and Truscott, T.G. (1988). Laser flash photolysis of purpurins: Novel potential photosensitisers of interest in photodynamic therapy. *J. Photochem. Photobiol., B: Biol.*, **2**, 427–34.

Borland, C.F., Cogdell, R.J., Land, E.J., and Truscott, T.G. (1989). Bacteriochlorophyll a triplet state and its interactions with bacterial carotenoids and oxygen. *J. Photochem. Photobiol., B: Biol.*, **3**, 237–45.

Bors, W., Lengfelder, E., Saran, M., Fuchs, C., and Michel, C. (1976). Reactions of oxygen radical species with methional: a pulse radiolysis study. *Biochem. Biophys. Res. Comm.*, **70**, 81–7.

Bors, W., Heller, W., Michel, C., and Saran, M. (1990). Flavonoids as antioxidants: determination of radical scavenging efficiencies. In *Methods in enzymology*, Vol. 186, Oxygen radicals in biological systems, Part B, Oxygen radicals and antioxidants (ed. L. Packer and A.N. Glazer), pp. 343–55. Academic Press, San Diego.

Bothe, E., Görner, H., Opitz, J., Schulte-Frohlinde, D., Siddiqi, A., and Wala, M. (1990). Single- and double-strand break formation in double-stranded DNA upon nanosecond laser-induced photoionization. *Photochem. Photobiol.*, **52**, 949–59.

Boucher, F. and Gingras, G. (1984). Spectral evidence for photo-induced isomerization of carotenoids in bacterial reaction center. *Photochem. Photobiol.*, **40**, 277–81.

Bouguer, P. (1729). *Essai d'optique sur le graduation de la lumière*. Le Havre.

Bouguer, P. (1760). Traité d'optique sur la graduation de la lumière. Paris (Opus posthum.) (trans. W.E.K. Middleton) (1961). University of Toronto Press, Toronto.

Bouheroum, M., Bruce, J.M., and Land, E.J. (1989). The mechanism of the oxidative decarboxylation of 3,4-dihydroxymandelic acid: Identification of 3,4-mandeloquinone as a key intermediate. *Biochim. Biophys. Acta*, **998**, 57–62.

Boulnois, J.-L. (1986). Photophysical processes in recent medical laser developments: a review. *Lasers Med. Sci.*, **1**, 47–66.

Boulnois, J.L. (1989). Photophysical processes in laser-tissue interactions. In *Primer on laser angioplasty* (ed. R. Ginsburg and J.C. White). Futura Publishing Inc., Mount Kisco, N.Y.

Bowers, P.G. and Porter, G. (1967). Quantum yields of triplet formation in solutions of chlorophyll. *Proc. Roy. Soc.*, **A296**, 435–41.

Bown, S.G. (1990). Photodynamic therapy to scientists and clinicians—one world or two? *J. Photochem. Photobiol., B: Biol.*, **6**, 1–12.

Bowry, V.W. and Ingold, K.U. (1992). Kinetics of nitroxide radical trapping. 2. Structural effects. *J. Am. Chem. Soc.*, **114**, 4992–6.

Bradley, D.J. (1982). Generation and measurement of picosecond pulses, *J. Phys. Chem.*, **82**, 2259–68.

Brady, J.E. and Carr, P.W. (1985). An analysis of dielectric models of solvatochromism. *J. Phys. Chem.*, **89**, 5759–66.

Brash, D.E. (1988). UV mutagenic photoproducts in *Escherichia coli* and human cells: a molecular genetics perspective on human skin cancer. *Photochem. Photobiol.*, **48**, 59–66.

Braslavsky, S.E. (1985). Time resolved photoacoustic and photothermal methods. Application to substances of biological interest. In *Primary processes in biology and medicine* (ed. R.V. Bensasson, G. Jori, E.J. Land, and T.G. Truscott), NATO ASI A85, pp. 147–58. Plenum, New York.

Braslavsky, S.E. (1987). Time-resolved photoacoustic and photothermal methods. Applications to phytochrome and tetrapyrroles. *Photobiochem. Photobiophys. Suppl.* 83–91.

Braslavsky, S.E. and Heihoff, K. (1989). Photothermal methods. In *Handbook of organic photochemistry* (ed. J.C. Scaiano). CRC Press, Boca Raton.

Brasseur, N., Ali, H., Langlois, R., and van Lier, J.E. (1987). Biological activities of phthalocyanines—VII. Photoinactivation of V-79 Chinese Hamster Cells by selectively sulphonated gallium phthalocyanines. *Photochem. Photobiol.*, **46**, 739–44.

Breton, J. and Nabedryk, E. (1987). Pigment and protein organization in reaction center and antenna complexes. In *The light reactions* (ed. J. Barber), pp. 159–95. Elsevier, Amsterdam.

Bridges, B.A. (1960). Sensitization of *Escherichia coli* to gamma-radiation by *N*-ethylmaleimide. *Nature*, **188**, 415.

Bridges, B.A. (1961). The effect of *N*-ethylmaleimide on the radiation sensitivity of bacteria. *J. Gen. Microbiol.*, **24**, 467–72.

de Broglie, L. (1924). Thesis, Paris. A tentative theory of light quanta. *Philos. Mag.*, **47**, 446–58.

Bromberg, A. and Foote, C.S. (1989). Solvent shift of singlet oxygen emission wavelength. *J. Phys. Chem.*, **93**, 3968–9.

Bronskill, M.J. and Hunt, J.W. (1968). A pulse-radiolysis system for the observation of short-lived transients. *J. Phys. Chem.*, **72**, 3762–6.

Bronskill, M.J., Taylor, W.B., Wolff R.K., and Hunt, J.W. (1970). Design and performance of a pulse radiolysis system capable of picosecond time resolution. *Rev. Sci. Instrum.*, **41**, 333–40.

Brown, J.M. and Workman, P. (1980). Partition coefficient as a guide to the development of radiosensitisers which are less toxic than misonidazole. *Radiat. Res.*, **82**,

171–90 (O'Neill, P. and Ho, S., personal communication quoted in above).

Brown, S.B. and Vernon, D.I. (1990). The quantitative determination of porphyrins in tissues and body fluids: Applications in studies of photodynamic therapy. In *Photodynamic therapy of neoplastic disease*, Vol. 1 (ed. D. Kessel), pp. 109–28. CRC Press, Boca Raton.

Brune, W.H. (1992). Stalking the elusive atmospheric hydroxyl radical. *Science*, **256**, 1154–6.

Brunschwig, B.S., Ehrenson, S., and Sutin, N. (1987). Solvent reorganization in optical and thermal electron-transfer processes: solvatochromism and intramolecular electron-transfer barriers in spherical molecules. *J. Phys. Chem.*, **91**, 4714–23.

Buettner, G.R. (1990). On the reaction of superoxide with DMPO/$^{\bullet}$OOH. *Free Rad. Res. Commun.*, **10**, 11–5.

Buettner, G.R. and Hall, R.D. (1987). Superoxide, hydrogen peroxide and singlet oxygen in haematoporphyrin derivative-cysteine, -NADH and -light systems. *Biochim. Biophys. Acta*, **923**, 501–7.

Buettner, G.R. and Need, M.J. (1985). Hydrogen peroxide and hydroxyl free radical production by haematoporphyrin derivative, ascorbate and light. *Cancer Lett.*, **25**, 297–304.

Buettner, G.R., Hall, R.D., Chignall, C.F., and Motten, A.G. (1989). The stepwise biphotonic photoionization of chlorpromazine as seen by laser flash photolysis. *Photochem. Photobiol.*, **49**, 249–56.

Bugelski, P.F., Porter, C.W., and Dougherty, T.J. (1981). Autoradiographic distribution of haematoporphyrin derivative in normal and tumour tissue of the mouse. *Cancer Res.*, **41**, 4606–12.

Bu'Lock, J.D. and Harley-Mason, J. (1951). Melanin and its precursors. Part II, Model experiments on the reactions between quinones and indoles, and consideration of a possible structure for the melanin polymer. *J. Chem. Soc.*, 703–12.

Burton, G.W. and Ingold, K.U. (1984). β-Carotene: an unusual type of lipid antioxidant. *Science*, **224**, 569–75.

Butler, J., and Hoey, B.M. (1986). The apparent inhibition of superoxide dismutase activity by quinones. *J. Free Rad. Biol. Med.*, **2**, 77–81.

Butler, J. and Hoey, B.M. (1987). Are reduced quinones necessarily involved in the antitumour activity of quinone drugs? *Br. J. Cancer*, **55**, Suppl VIII, 53–9.

Butler, J., Land, E.J., Swallow, A.J. and Prutz, W. (1984). The azide radical and its reaction with tryptophan and tyrosine. *Radiat. Phys. Chem.*, **23**, 265–70.

Butler, J., Hoey, B.M., and Swallow, A.J. (1985). Reaction of the semiquinone free radicals of anti-tumour agents with oxygen and iron complexes. *FEBS Lett.*, **182**, 95–8.

Butler, J., Land, E.J., Prutz, W.A., and Swallow, A.J. (1986). Reversibility of charge transfer between tryptophan and tyrosine. *J. Chem. Soc., Chem. Commun.*, 348–9.

Butler, J., Hoey, B.M., and Lea, J.S. (1987). The reduction of anti-tumour diaziridinyl benzoquinones. *Biochim. Biophys. Acta*, **925**, 144–9.

Butler, J., Dzielendziak, A., Lea, J.S., Ward, T.H., and Hoey, B.M. (1990). Contrasting cytotoxic mechanisms of similar antitumour diaziridinylbenzoquinones. *Free Rad. Res. Commun.*, **8**, 231–9.

Butler, W.,L., and Norris, K.H. (1963). Lifetime of the long-wavelength chlorophyll fluorescence. *Biochim. Biophys. Acta*, **66**, 72–7.

Buxton, G.V., Greenstock, C.L., Helman, W.P., and Ross, A.B. (1988). Critical review of rate constants of reactions of hydrated electrons, hydrogen atoms and hydroxyl radicals ($^{\cdot}$OH/$^{\cdot}$O$^-$) in aqueous solution. *J. Phys. Chem. Ref. Data*, **17**, 513–886.

Byrne, C. J., Marshallsay, L. V., and Ward, A. D. (1990). The composition of photofrin II *J. Photochem. Photobiol., B: Biol.*, **6**, 13–27.

Cabanes, J., Sanchez-Ferrer, A., Bru, R., and Garcia-Carmona, F. (1988). Chemical and enzymic oxidation by tyrosinase of 3,4-dihydroxymandelate. *Biochem. J.*, **256**, 681–4.

Cadenas, E. (1989). Biochemistry of oxygen toxicity. *Ann. Rev. Biochem.*, **58**, 79–110.

Cadenas, E., Brigelius, R., and Sies, H. (1983). Paraquat-induced chemiluminescence of microsomal fractions. *Biochem. Pharmacol.*, **32**, 147–50.

Cadenas, E., Boveris, A., and Chance, B. (1984). Low-level chemiluminescence of biological systems. In *Free radicals in Biology*, Vol. 6 (ed. W. A. Pryor). Academic Press, San Diego.

Cadet, J. and Vigny, P. (1990). The photochemistry of nucleic acids. In *Bioorganic photochemistry, Vol. 1: Photochemistry and the nucleic acids* (ed. H. Morrison), pp. 1–272. Wiley Interscience, New York.

Cadet, J., Berger, M., Decarroz, C., Wagner, J. R., Van Lier, J. E., Ginot, Y. M., and Vigny, P. (1986). Photosensitized reactions of nucleic acids. *Biochimie*, **68**, 813–34.

Cahan, D, Bults, G., Carty, H., and Malkin, S. (1980). Photoacoustics in life sciences. *J. Biochem. Biophys. Methods*, **3**, 293–310.

Candeias, L. P. and Steenken, S. (1992). Ionization of purine nucleosides and nucleotides by 193-nm laser photolysis in aqueous solution: model studies for oxidative damage of DNA. *J. Am. Chem. Soc.*, **114**, 699–704.

Cantor, C. R. and Schimmel, P. R. (1980). *Biophysical chemistry*, W. H. Freeman, San Francisco.

Capellos, C. and Bielski, B. H. J. (1972). *Kinetic systems*, Wiley-Interscience, New York.

Cario, G. and Franck, J. (1923). Uber sensibilisierte fluoreszenz von gasen. *Z. Physik.*, **17**, 202–12.

Carmichael, A. J., Mossoba, M. M., and Riesz, P. (1983). Photogeneration of superoxide by adriamycin and daunomycin: an electron spin resonance and spin trapping study. *FEBS Lett.*, **164**, 401–5.

Carmichael, I. and Hug, G. L. (1986). Triplet–triplet absorption spectra of organic molecules in condensed phases. *J. Phys. Chem. Ref. Data*, **15**, 1–250.

Carmichael, I., Helman, W. P., and Hug, G. L. (1987). Extinction coefficients of triplet–triplet absorption spectra of organic molecules in condensed phases: a least-squares analysis. *J. Phys. Chem. Ref. Data*, **16**, 239–60.

Castellani, A. and Quercia, I. F. (eds). (1979). *Synchrotron radiation applied to biophysical and biochemical research*. Plenum Press, New York.

Catalani, L. H. and Wilson, T. (1989). Electron transfer and chemiluminescence. Two inefficient systems: 1, 4 dimethoxy-9, 10-diphenylanthracene peroxide and diphenoyl peroxide. *J. Am. Chem. Soc.*, **111**, 2633–9.

Cavalieri, E. L. and Rogan, E. G. (1990). Radical cations in aromatic hydrocarbon carcinogenesis. *Free Rad. Res. Commun.*, **11**, 77–87.

Cerutti, P. A. (1985). Prooxidant states and tumour promotion. *Science*, **227**, 375–81.

Chacon, J. N., Chedekel, M. R., Land E. J., and Truscott, T. G. (1987). Chemically induced Parkinson's disease: intermediates in the oxidation of 1-methyl-4-phenyl-1, 2, 3, 6-tetrahydropyridine to the 1-methyl-4-phenyl-pyridinium ion. *Biochem. Biophys. Res. Commun.*, **144**, 957–64.

Chacon, J. N., Chedekel, M. R., Land, E. J., and Truscott, T. G. (1988*a*). Pulsed irradiation studies of some reactions of melanin precursors. In '*Photosensitisation: molecular, cellular and medical aspects*' (ed. G. Moreno, R. H. Pottier, and T. G. Truscott), pp. 231–4. Springer-Verlag, Heidelberg, Germany.

Chacon, J. N., McLearie, J., and Sinclair, R. S. (1988*b*). Singlet oxygen yields and radical contributions in the dye-sensitised photo-oxidation in methanol of esters of poly-unsaturated fatty acids (oleic, linoleic, linolenic and arachidonic). *Photochem. Photobiol.*, **47**, 647–56.

Chacon, J.N., Chedekel, M.R., Land E.J., and Truscott, T.G. (1989). Chemically induced Parkinson's disease. II. Intermediates in the oxidation and reduction reactions of the 1- methyl-4-phenyl-2,3-dihydropyridinium ion and its deprotonated form. *Biochem. Biophys. Res. Commun.*, **158**, 63–71.

Chahidi, C., Aubailly, M., Momzikoff, A., Bazin, M., and Santus, R. (1981). Photophysical and photosensitizing properties of 2- amino-4-pteridinone: a natural pigment. *Photochem. Photobiol.*, **33**, 641–9.

Chanet, R., Waters, R., and Moustacchi, E. (1975). The induction of pyrimidine dimers in nuclear DNA after U.V.-irradiation during the synchronous cycle of *Saccharomyces cerevisiae*. *Int. J. Radiat. Biol.*, **27**, 481–5.

Chang, C.T. and Dougherty, T.J. (1978). Photoradiation therapy: kinetics and thermodynamics of porphyrin uptake and loss in normal and malignant cells in culture. *Radiat. Res.*, **74**, 498–9.

Chantrell, S.J., McAuliffe, C.A., Munn, R.W., Pratt, A.C., and Land, E.J. (1977). Excited states of protoporphyrin IX dimethyl ester: Reaction of the triplet with carotenoids. *J. Chem. Soc., Faraday Trans. I*, **73**, 858–65.

Charlesworth, P. and Truscott, T. G. (1993). The use of 5-aminolevulinic acid (ALA) in photodynamic therapy (PDT). *J. Photochem. Photobiol. B. Biol.*, in press.

Charlier, M., Hélène, C., and Carrier, W.L. (1972). Photochemical reactions of aromatic ketones with nucleic acids and their components—III. Chain breakage and thymine dimerization in benzophenone photosensitized DNA. *Photochem. Photobiol.*, **15**, 527–36.

Chatterjee, A. (1987). Interaction of ionizing radiation with matter. In *Radiation chemistry* (ed. Farhataziz and M.A.J. Rodgers), pp. 1-28. VCH Publishers, New York.

Chattopadhyay, S.K., Kumar, C.V., and Das, P.K. (1985). Triplet excitation transfer involving β-ionone. A kinetic study by laser flash photolysis. *Photochem. Photobiol.*, **42**, 17–24.

Chedekel, M.R., Land, E.J., Thompson, A., and Truscott, T.G. (1984). Early steps in the free radical polymerisation of 3,4-dihydroxyphenylalanine (Dopa) into melanin. *J. Chem. Soc., Chem. Commun.*, 1170–2.

Chen, J.M., Ho, T.I., and Mou, C.Y. (1990). Experimental investigation of excited-state electron transfer reaction: effects of free energy and solvents on rates. *J. Phys. Chem.*, **94**, 2889–96.

Cherniak, E.A., Collinson, E., and Dainton, F.S. (1964). Beta radiolysis of liquid benzene. *Trans. Faraday Soc.*, **60**, 1408–23.

Cherry, R.J., Chapman, D., and Langelaar, J. (1968). Fluorescence and phosphorescence of β-carotene. *Trans. Faraday Soc.*, **64**, 2304–7.

Chessin, M., Livingston, R., and Truscott, T.G. (1966). Direct evidence for the sensitized formation of a metastable state of β-carotene. *Trans. Faraday Soc.*, **62**, 1519–24.

Chinsky, L., Turpin, P.Y., Duquesne, M., and Brahms, J. (1978). Nucleic acid derivatives studied by preresonance and resonance Raman spectroscopy in the ultraviolet region. *Biopolymers*, **17**, 1347–59.

Choi, H.T., Askew, D., and Lipsky, S. (1982). A note on the G value for the production of the lowest excited singlet state of cyclohexane. *Radiat. Phys. Chem.*, **19**, 373–5.

Christensen, R.L. and Kohler, B.E. (1974). Excitation spectroscopy of retinal and related polyenes. *Photochem. Photobiol.*, **19**, 401–11.

Christensen, R.L. and Kohler, B.E. (1976). High-resolution optical spectroscopy of polyenes related to the visual chromophore. *J. Phys. Chem.*, **80**, 2197–200.

Christman, C.L., Carmichael, A.J., Mossoba, M.M. and Riesz, P. (1987). Evidence for free radicals produced in aqueous solutions by diagnostic ultrasound. *Ultrasonics*, **25**, 31–4.

Chrysochoos, J. (1968). Pulse radiolysis of phenylalanine and tyrosine. *Radiat. Res.*, **33**, 465–79.

Chuaqui, C.A. and Petkau, A. (1987). Chemical reactivity and biological effects of superoxide radicals. *Radiat. Phys. Chem.*, **30**, 365–73.

Cilento, G. (1982). Electronic excitation in dark biological processes. In *Chemical and biological generation of excited states* (ed. W. Adam and G. Cilento), pp. 277–307. Academic Press, New York.

Cilento, G. and Adam, W. (1988). Photochemistry and photobiology without light, *Photochem. Photobiol.*, **48**, 361–8.

Claesson, S. (1976). Techniques for flash photolysis. In *Spectroscopy of the excited state* (ed. B. DI Bartolo), NATO ASI B12, pp. 95–116. Plenum, New York.

Clar, E. and John, Fr. (1930). Über eine neue klasse tiefgefärbter radiakalischer kohlenwasserstoffe und über das vermeintliche pentacen von E. Philippi. *Ber. Dtsch. Chem. Ges.*, **63**, 2967–77.

Cleaver, J.E., Cortès, F., Karentz, D., Lutze, L.H., Morgan, W.F., Player, A.N., Vuksanovic, L., and Mitchell, D.L. (1988). The relative biological importance of cyclobutane and (6–4) pyrimidine–pyrimidone dimer photoproducts in human cells: evidence from a xeroderma pigmentosum revertant. *Photochem. Photobiol.*, **48**, 41–9.

Closs, G.L. and Miller, J.R. (1988) Intramolecular long distance electron transfer in organic molecules. *Science*, **420**, 440–7.

Cogdell, R.J. and Frank, H.A. (1987). How carotenoids function in photosynthetic bacteria. *Biochim. Biophys. Acta*, **895**, 63–79.

Cogdell, R.J., Land, E.J., and Truscott, T.G. (1983). The triplet extinction coefficients of some bacterial carotenoids. *Photochem. Photobiol.*, **38**, 723–5.

Cohen, G. (1985). The Fenton reaction. In *Handbook of methods for oxygen radical research* (ed. R.A. Greenwald), pp. 55–64. CRC Press, Boca Raton, Florida.

Cole, C.A., Forbes, P.D., and Davies, R.E. (1986). An action spectrum for UV photocarcinogenesis. *Photochem. Photobiol.*, **43**, 275–84.

Cole, R.S. and Sinden, R.S. (1975). Psoralen cross-links in DNA: Biological consequences and cellular repair. In *Radiation research: biomedical, chemical and physical perspectives* (ed. O.F. Nygaard, H.I. Adler, and W.K. Sinclair), pp. 582–9. Academic Press, New York.

Commoner, B., Townsend, J., and Pake, G.E. (1954). Free radicals in biological materials. *Nature*, **174**, 689–91.

Compton, A.H. (1923). A quantum theory of the scattering of X-rays by light elements. *Phys. Rev.*, **21**, 207 and 483–502.

Condon, E.U. and Morse, P.M. (1929). *Quantum mechanics*, McGraw-Hill, New York.

Conn, P.F., Schalch, W., and Truscott, T.G. (1991). The singlet oxygen and carotenoid interaction. *J. Photochem. Photobiol., B: Biol.*, **11**, 41–7.

Conn, P.F., Lambert, C., Land, E.J., Schalch, W., and Truscott, T.G. (1992). Carotene–oxygen radical interactions. *Free. Rad. Res. Commun.*, **16**, 401–8.

Connolly, J.S. (1969). The photochemical addition of alcohols to purine. In *Ph.D. Chemistry, Brandeis University*, University Microfilms Inc., Ann Arbor, Michigan, 69-20720.

Connolly, J.S. and Bolton, J.R. (1988). Intramolecular electron transfer: history and some implications for artificial photosynthesis. In *Photoinduced electron transfer. Part D.* (ed. M.A. Fox and M. Chanon), pp. 303–93. Elsevier, Amsterdam.

Connolly, J. and Linschitz, H. (1968). Photoaddition of alcohols to purine. *Photochem. Photobiol.*, **7**, 791–806.

Connolly, J.S., Janzen, A.F., and Samuel, E.B. (1982). Fluorescence lifetimes of chlorophyll *a*: Solvent, concentration and oxygen dependence. *Photochem. Photobiol.*, **36**, 559–63.

Coohill, T.P. (1984). Action spectra for mammalian cells *in vitro*. In *Topics in photomedicine* (ed. K.C. Smith), pp. 1–37. Plenum, New York.

Cooksey, C.J., Jimbow, K., Land, E.J., and Riley, P.A. (1992). Reactivity of *ortho*-quinones involved in tyrosinase-dependent cytotoxicity: differences between alkythio- and alkoxy-substituents. *Melanoma Res.*, **2**, 283–93.

Cooksey, C.J., Land, E.J., Riley, P.A., Sarna, T., and Truscott, T.G. (1987). On the interaction of anisyl-3,4-quinone with oxygen. *Free Rad. Res. Commun.*, **4**, 131–8.

Corey, E.J. and Taylor, W.C. (1964). A study of the peroxidation of organic compounds by externally generated singlet oxygen molecules *J. Am. Chem. Soc.*, **86**, 3881–2.

Corin, A.F., Blatt, E., and Jovin, T.J. (1987). Triplet-state detection of labeled proteins using fluorescence recovery spectroscopy. *Biochemistry*, **26**, 2207–17.

Corradini, M.G., Napolitano, A., and Prota, G. (1986). A biosynthetic approach to the structure of eumelanins. The isolation of oligomers from 5,6-dihydroxy-1-methyindole. *Tetrahedron*, **42**, 2083–8.

Corwin, A.H. Chivvis, A.B., Poor, R.W., Whitten, D.G., and Baker, E.W. (1968). The interpretation of porphyrin and metalloporphyrin spectra. *J. Am. Chem. Soc.*, **90**, 6577–83.

Cosgrove, S.A., Guite, M.A., Burnell, T.B., and Christensen, R.L. (1990). Electronic relaxation in long polyenes. *J. Phys. Chem.*, **94**, 8118–24.

Costalat, R., Blais, J., Ballini, J.-P., Moysan, A., Cadet, J., Chalvet, O., and Vigny, P. (1990). Formation of cyclobutane thymine dimers photosensitized by pyridopsoralens: A triplet–triplet energy transfer mechanism. *Photochem. Photobiol.*, **51**, 255–62.

Craig, B.B., Gorman, A.A., Hamblett, I., and Kerr, C.W. (1984). The relaxed triplet state of the 1,2,4,5-tetracyanobenzene–benzene complex: Triplet energy transfer equilibration with benzophenone. *Radiat. Phys. Chem.*, **23**, 111–15.

Craw, M. and Lambert, C. (1983). The characterisation of the triplet state of crocetin, a water soluble carotenoid, by nanosecond laser flash photolysis. *Photochem. Photobiol.*, **38**, 241–3.

Craw, M., Bensasson, R.V., Ronfard-Haret, J.C., Sa e Melo, M.T., and Truscott, T.G. (1983). Some photophysical properties of 3-carbethoxypsoralen, 8-methoxypsoralen and 5-methoxypsoralen triplet states. *Photochem. Photobiol.*, **37**, 611–5.

Craw, M., Truscott, T.G., Dall'Acqua, F., Guiotto, A., Vedaldi, D., and Land, E.J. (1984*a*). Methylangelicins: possible correlation between photophysical properties of the triplet excited state and photobiological activity. *Photobiochem. Photobiophys.*, **7**, 359–65.

Craw, M., Chedekel, M.R., Truscott, T.G., and Land, E.J. (1984*b*). The photochemical interaction between the triplet state of 8-methoxypsoralen and the melanin precursor L-3,4-dihydroxyphenylalanine. *Photochem. Photobiol.*, **39**, 155–9.

Craw, M., Redmond, R., and Truscott, T.G. (1984*c*). Laser flash photolysis of haematoporphyrins in some homogeneous and heterogeneous environments. *J. Chem. Soc., Faraday Trans. I*, **80**, 2293–8.

Creed, D. (1984*a*). The photophysics and photochemistry of the near-UV absorbing amino acids—1. Tryptophan and its simple derivatives. *Photochem. Photobiol.*, **39**, 537–62.

Creed, D. (1984*b*). The photophysics and photochemistry of the near-UV absorbing amino acids. II Tyrosine and its simple derivatives. *Photochem. Photobiol.*, **39**, 363–75.

Creed, D. (1984*c*). The photophysics and photochemistry of the near-UV absorbing amino acids—III. Cystine and its simple derivatives. *Photochem. Photobiol.*, **39**, 577–83.

Cremer, R. J., Perryman, P. W., and Richards, D. H. (1958). Influence of light on the hyperbilirubinaemia of infants. *Lancet*, **1**, 1094–7.

Croke, D. T., Blau, W., OhUigin, C., Kelly, J. M., and McConnell, D. J. (1988). Photolysis of phosphodiester bonds in plasmid DNA by high intensity UV laser irradiation. *Photochem. Photobiol.*, **47**, 527–36.

Cubeddu, R., Ramponi, R, Liu, W.-Q., and Docchio, F., (1989). Time-gated fluorescence spectroscopy of the tumour localizing fraction of HPD in the presence of cationic surfactant. *Photochem. Photobiol.*, **50**, 157–63.

Cudina, I. and Josimovic, L. J. (1987). The effect of oxygen on the radiolysis of tyrosine in aqueous solution. *Radiat. Res.*, **109**, 206–15.

Cummings, J., Kerr, D. J., and Kaye, S. B. (1987*a*). Occurrence of circulating 7-deoxyaglycone metabolites of 4′-deoxydoxorubicin in man. *Cancer Chemother. Pharmacol.*, **20**, 263–4.

Cummings, J., Milroy, R., Banham, S. W., and Kaye, S. B. (1987*b*). Method for the determination of 4-demethoxydaunorubicin, its quinone and hydroquinone metabolites in human plasma and urine by high-performance liquid chromatography. *Cancer Chemerother. Pharmacol.*, **19**, 296–300.

Cundall, R. B. and Dale, R. E. (eds) (1983). *Time-resolved fluorescence spectroscopy in biochemistry and biology*. Nato ASI Series, **69**, Plenum, New York.

Cunningham, M. L., Johnson, J. S., Giovanazzi, S. M., and Peak, M. J. (1985*a*). Photosensitized production of superoxide anion by monochromatic (290–405 nm) ultraviolet irradiation of NADH and NADPH coenzymes. *Photochem. Photobiol.*, **42**, 125–8.

Cunningham, M. L., Krinsky N. I., Giovanazzi, S. M., and Peak M. J. (1985*b*). Superoxide anion is generated from cellular metabolites by solar radiation and its components. *J. Free Rad. Biol. Med.*, **1**, 381–5.

Curie, P. and Debierne, A. (1901). Sur la radio-activité induite provoquée par les sels de radium. *C. R. Acad. Sci. Paris*, **132**, 548–51.

Curley, Jr, R. W. and Fowble, J. W. (1988). Photoisomerization of retinoic acid and its photoprotection in physiologic-like solutions. *Photochem. Photobiol.*, **47**, 831–5.

Cutler, R. G. (1984). Antioxidants, aging, and longevity. In *Free radicals in biology*, vol. VI (ed. W. A. Pryor), pp. 371–428. Academic Press, New York.

Czapski, G. and Bielski B. H. J. (1993). Absorption spectra of the OH and O^- radicals in aqueous solution. *Radiat. Phys. Chem.*, **41**, 503–5.

Czapski, G. and Ilan, Y. A. (1978). On the generation of the hydroxylation agent from superoxide radical. Can the Haber–Weiss reaction be the source of OH radicals. *Photochem. Photobiol.*, **28**, 651–3.

Czapski, G., Goldstein, S., and Meyerstein, D. (1988). What is unique about superoxide toxicity as compared to other biological reductants?—A hypothesis. *Free Rad. Res. Commun.*, **4**, 231–6.

Czochrabska, B., Kawczynski, W., Bartosz, G., and Shugar, D. (1984). Oxidation of excited-state NADH and NAD dimer in aqueous medium: involvement of O_2^{-} as a mediator in the presence of oxygen. *Biochim. Biophys. Acta*, **801**, 403–9.

Dallinger, R.F., Guanci, J.J., Woodruff, W.H., and Rodgers, M.A.J. (1979). Vibrational spectroscopy of the electronically excited state: Pulse radiolysis/time resolved resonance Raman study of triplet β-carotene. *J. Am. Chem. Soc.*, **101**, 1355–7.

Daniels, M. (1976). Excited states of the nucleic acids: bases, mononucleosides, and mononucleotides. In *Photochemistry and photobiology of nucleic acids*, Vol. I *Chemistry* (ed.S.Y. Wang), pp. 23–108. Academic Press, New York.

Daniels, M. (1983). Recent developments in the excited states of nucleic acids. *Photochem. Photobiol.*, **37**, 691–3.

Daniels, M. and Hauswirth, W. (1971). Fluorescence of the purine and pyrimidine bases of the nucleic acids in neutral aqueous solution at 300K. *Science*, **171**, 675–7.

Danziger, R.M., Bar Eli, K.H., and Weiss, K. (1967). The laser photolysis of methylene blue. *J. Phys. Chem.*, **71**, 2633–40.

Darwent, J.R., Douglas, P., Harriman, A., Porter, G., and Richoux, M.-C. (1982). Metal phthalocyanines and porphyrins as photosensitisers for reduction of water to hydrogen. *Coord. Chem.*, **44**, 83–126.

Das, P.K. and Becker, R.S. (1979). Triplet state photophysical properties and intersystem crossing quantum efficiencies of homologues of retinals in various solvents. *J. Am. Chem. Soc.*, **101**, 6348–53.

Davies, A.K., Navaratnam, S., and Phillips, G.O. (1976). Photochemistry of chlorpromazine [2-Chloro-*N*-(3-dimethyl-aminopropyl) phenothiazine] in propan-2-ol solution. *J. Chem. Soc., Perkin Trans. II*, 25–9.

Davies, A.K., Land, E.J., Navaratnam, S., Parsons, B.J., and Phillips, G.O. (1979). Pulse radiolysis study of chlorpromazine and promazine free radicals in aqueous solution. *J. Chem. Soc., Faraday Trans. I*, **75**, 22–35.

Davies, M.J. (1988). Detection of peroxyl and alkoxyl radicals by reaction of hydroperoxides with heme-proteins by electron spin resonance spectroscopy. *Biochim. Biophys. Acta*, **964**, 28–35.

Dawe, E.A. and Land, E.J. (1975). Radical ions derived from photosynthetic polyenes. *J. Chem. Soc., Faraday Trans. I*, **71**, 2162–9.

Dawson, W. and Abrahamson, E.W. (1962). Population and decay of the lowest triplet state in polyenes with conjugated heteroatoms: retinene. *J. Phys. Chem.*, **66**, 2542–7.

Dayhaw-Barker, P. and Truscott, T.G. (1988). Direct detection of singlet oxygen sensitized by nalidixic acid: The effect of pH and melanin. *Photochem. Photobiol.*, **47**, 765–7.

DeCosta, D.P. and Pincock, J.A. (1989). Control of product distribution by Marcus type electron-transfer rates for the radical pair generated in benzylic ester photochemistry. *J. Am. Chem. Soc.*, **111**, 8948–50.

DeCoster, B., Christensen, R.L., Gebhand, R., Lugtenburg, J., Farhoosh, R., and Franck, H.A. (1992). Low-lying electronic states of carotenoids. *Biochim. Biophys. Acta*, **1102**, 107–14.

Decuyper-Debergh, D. (1988). Actions mutagène et cytotoxique de l'oxygène singulet. PhD. Thesis, University of Liège, Belgium.

DeFelippis, M.R., Murphy, C.P., Faraggi, M., and Klapper, M.H. (1989). Pulse radiolytic measurement of redox potentials: The tyrosine and tryptophan radicals. *Biochemistry*, **28**, 4847–53.

DeFelippis, M.R., Faraggi, M., and Klapper, M.H. (1990). Evidence for through-bond long-range electron transfer in peptides. *J. Am. Chem. Soc.*, **112**, 5640–2.

DeFelippis, M.R., Murthy, C.P., Broitman, F., Weinraub, D., and Faraggi, M.

(1991). Electrochemical properties of tyrosine phenoxy and tryptophan indole radicals in peptides and amino acid analogues. *J. Phys. Chem.*, **95**, 3416–19.

De Haas, M.P. and Warman, J.M. (1982). Photon-induced molecular charge separation studied by nanosecond time-resolved microwave conductivity. *Chem. Phys.*, **73**, 35–53.

Demas, J.N. (1983). *Excited state lifetime measurements.* Academic Press, New York.

Denny, R.W. and Nickon, A. (1973). Sensitized photo-oxygenation of olefins. *Org. React.*, **20**, 133–336.

Descartes, R. (1637). *Discours de la méthode pour bien conduire sa raison et chercher la vérité dans les sciences.* App. I: La Dioptrique. App. III: Les Météores. Paris (trans. P. J. Olscamp (1965): *Discourse on method, optics, geometry and meterology.* Indiana University Press, Indianapolis).

Dessaignes, J.P. (1810). Mémoire sur les phosphorescences. *J. Physique*, 70, 109–28.

Devasagayam, T.P.A., Di Mascio, P., Kaiser, H., and Sies, H. (1991). Singlet oxygen induced single-strand breaks in plasmid pBR322 DNA: the enhancing effect of thiols. *Biochim. Biophys. Acta*, **1088**, 409–12.

Dexter, D.L. (1953). A theory of sensitized luminescence in solids. *J. Chem. Phys.*, **21**, 836–50.

Diesenhofer, J. and Michel, H. (1988). The crystal structure of the photosynthetic reaction center from *Rhodopseudomonas viridis.* In *The photosynthetic bacterial reaction center*, Nato ASI Ser. A, **149**, pp. 1–3.

Diesenhofer, J. and Michel, H. (1989*a*). The photosynthetic reaction center from the purple bacterium *Rhodopseudomonas viris. Science*, **245**, 1463–73.

Diesenhofer, J. and Michel, H. (1989*b*). The photosynthetic reaction centre from the purple bacterium *Rhodopseudomonas sphaeroides* and *Rhodopseudomonas rubrum. Chem. Scr.*, **29**, 205–20.

Diesenhofer, J., Epp, O., Miki, K., Huber, R., and Michel, H. (1984). X-Ray structure analysis of a membrane protein complex. Electron density map at 3 angstrom resolution and a model of the chromophores of the photosynthetic reaction center from *Rhodopseudomonas viridis. J. Mol. Biol.*, **180**, 385–98.

Dillon, J. and Atherton, S.J. (1990). Time resolved spectroscopic studies on the intact human lens. *Photochem. Photobiol.*, **51**, 465–8.

Di Mascio, P., Wefers, H., Do-Thi, H.-P., Lafleur, M.V.M., and Sies, H. (1989). Singlet molecular oxygen causes loss of biological activity in plasmid and bacteriophage DNA and induces single-strand breaks. *Biochim. Biophys. Acta*, **1007**, 151–7.

Dobrowolski, D.C., Ogilby, P.R., and Foote, C.S. (1983). Chemistry of singlet oxygen. 39. 9,10-Dicyanoanthracene sensitized formation of singlet oxygen. *J. Phys. Chem.*, **87**, 2261–3.

Dodd, N.J.F. and Mukherjee, T. (1984). Free radical formation from anthracycline antitumour agents and model systems 1. Model naphthoquinones and anthraquinones. *Biochem. Pharmacol.*, **33**, 379–85.

Donahue, J.M. and Waddell, W.H. (1984). The *trans* → *cis* photoisomerization of protonated retinyl schiff bases. *Photochem. Photobiol.*, **40**, 399–401.

Dorfman, L.M. and Jou, F.Y. (1973). Optical absorption spectrum of the solvated electron in ethers and in binary liquid systems. In *Electron in fluids. The nature of metal-ammonia solutions* (ed. J. Jortner and N.R. Kestner), pp. 447–59. Springer, Berlin.

Dougherty, T.J., Kaufman, J.E., Goldfarb, A., Weishaupt, K.R., Boyle, D., and Mittleman, A. (1978). Photoradiation therapy for the treatment of malignant tumours. *Cancer Res.*, **38**, 2628–35.

Dougherty, T.J., Weishaupt, K.R., and Boyle, D.J. (1981). Photoradiation therapy of malignant tumours. In *Principles and practice of oncology* (ed. V. DeVita, S. Hellman, and S. Rosenberg, Jr.). J.B. Lippincott, Philadelphia.

Dougherty, T.J., Potter, W.R., and Bellnier, D. (1990). Photodynamic therapy for the treatment of cancer: current status and advances. In *Photodynamic therapy of neoplastic disease*, Vol. 1 (ed. D. Kessel), pp. 1–19. CRC Press, Boca Raton, Florida.

Draper, J.W. (1841). On some analogies between phenomena of chemical rays and those of radiant heat. *Philos. Mag.*, **19**, 195–210.

Dubbelman, T.M.A.R., De Goeij A.F.P.M., and Van Steveninck, J. (1978). Photodynamic effects of protoporphyrin on human erythrocytes: Nature of the cross-linking of membrane proteins. *Biochim. Biophys. Acta*, **511**, 141–51.

Dubbelman, T.M.A.R., Smeets, M., and Boegheim, J.P.J. (1988). Cell models. In *Photosensitization, molecular, cellular, and medical aspects* (ed. G. Moreno, R. Pottier, and T.G. Truscott), NATO ASI series, pp. 157–70. Springer, Berlin.

Duran, N., Farias Furtado, S.T., Faljoni-Alario, A., Campa, A., Brunet, J.E., and Freer, J. (1984). Singlet oxygen generation from the peroxidase-catalysed aerobic oxidation of an activated -CH_2-substrate. *J. Photochem.*, **25**, 285–94.

Dzielendziak, A., Butler, J., Hoey, B.M., Lea, J.S., and Ward, T.H. (1990). Comparison of the structural and cytotoxic activity of novel 2,5-bis(carboethoxyamino)-3, 6-diaziridinyl-1,4-benzoquinone analogues. *Cancer Res.*, **50**, 2003–8.

Eaton, W. and Szabo, A. (eds) (1991). Protein dynamics. *Chem. Phys.*, **158**, 191–531.

Eberson, L. (1987). Electron transfer reactions in organic chemistry. In *Reactivity and structure concepts in organic chemistry*, Vol. 25, pp. 1–234. Springer-Verlag, Berlin.

Ebert, G. (1980). *Biopolymere*. pp. 14–15. Darmstadt, Steinkopf.

Edelson, R., Berger, C. Gasparro, F., Jegasothy, B., Heald, P., Wintroub, B., *et al.* (1987). Treatment of cutaneous T-cell lymphoma by extracorpeal photochemotherapy. *N. Eng. J. Med.*, **316**, 297–303.

Eimhjellen, K.E., Aasmundrud, O., and Jensen, A. (1963). A new bacterial chlorophyll. *Biochem. Biophys. Res. Commun.*, **10**, 232–7.

Einhorn, C., Einhorn, J., and Luche, J.-L. (1989). Sonochemistry—The use of ultrasonic waves in synthetic organic chemistry. *Synthesis—J. Synth. Org. Chem.*, 787–813.

Einstein, A. (1905). Uber einen die erzeugung und verwandlung des lichtes betreffenden heuristisches gesichtspunkt. *Ann. Phys.*, **17**, 132–48.

Einstein, A. (1906). Zur theorie die lichterzeugung und licht absorption. *Ann. Phys.*, **20**, 199–206.

Einstein, A. (1912). Theormodynamic foundation of the law of photochemical equivalents. *Ann. Physik.*, **37**, 832–8.

Einstein, A. (1917). Zur quantentheorie der strahlung. *Phys. Zeit.*, **18**, 121–8.

Elad, D. (1976). Photoproducts of purines. In *Photochemistry and photobiology of nucleic acids*, Vol. I *Chemistry* (ed. S.Y. Wang), pp. 357–80. Academic Press, New York.

Elliot, A.J., Simsons, A.S., and Sopchyshyn, F.C. (1984). Radiolysis of solutions containing organo-disulphides, *Radiat. Phys. Chem.*, **23**, 377–84.

El Sayed, M.A. (1963). Spin-orbit coupling and the radiationless processes in nitrogen heterocyclics. *J. Chem. Phys.*, **38**, 2834–8.

El Sayed, M. (1982). Time-resolved chromophore resonance Raman and protein fluorescence of intermediates in some photobiological changes. In *Trends in photo-*

biology (ed. C. Hélène, M. Charlier, Th. Montenay-Garestier, and G. Laustriat), pp. 1–10. Plenum Press, New York.

Emmerson, P.T. and Willson, R.L. (1968). Pulse radiolysis of aqueous solutions of thymine and triacetoneamine *N*-oxyl. *J. Phys. Chem.*, **72**, 3669–71.

Emmerson, P.T., Fielden, E.M., and Johansen, I. (1971). A possible steric factor in the radiosensitization of anoxic bacteria to X-rays by *N*-oxyl radicals. *Int. J. Radiat. Biol.*, **19**, 229–36.

Encinas, M.V. and Lissi, E.A. (1986). Quenching of pyrene fluorescence by tryptophan in micellar solutions. *Photochem. Photobiol.*, **44**, 579–85.

Ennever, J.F. (1988). Phototherapy for neonatal jaundice. *Photochem. Photobiol.*, **47**, 871–6.

Epe, B., Mützel, P., and Adam, W. (1988). DNA damage by oxygen radicals and excited state species: a comparative study using enzymatic probes *in vitro*. *Chem.-Biol. Interact.*, **67**, 149–65.

Epe, B., Hegler, J., and Wild, D. (1989). Singlet oxygen as an ultimately reactive species in *Salmonella typhimirium* DNA damage induced by methylene blue/visible light. *Carcinogenesis*, **10**, 2019–24.

Epstein, J.H. and Grekin, D.A. (1981). Inhibition of ultraviolet-induced carcinogenesis by all-*trans* retinoic acid. *J. Invest. Dermat.*, **76**, 178–80.

Errera, M. (1952). Etude photochimique de DNA, II Etude des produits de la photolyse. *Biochim. Biophys. Acta*, **8**, 115–24.

Evans, D.F. (1960). Magnetic perturbation of singlet–triplet transitions. Part IV. Unsaturated compounds. *J. Chem. Soc.*, 1735–45.

Evans, D.F. (1961). Magnetic perturbation of singlet–triplet transitions. Part VI Octa-1,3,5,7-tetraene. *J. Chem. Soc.*, 2566–9.

Evans, D.F. (1980). Singlet–triplet transitions in square planar nickel complexes which quench singlet oxygen. *J. Chem. Soc., Chem. Commun.*, 1134.

Everaert, J. and Rentzepis, P.M. (1982). Picosecond fluorescence kinetics of *n*-all-*trans*-retinylidene-*n*-butylamine. *Photochem. Photobiol.*, **36**, 543–50.

Exner, O. (1975). *Dipole moments in organic chemistry*, p. 108. Thieme, Stuttgart.

Fang, H.L.B., Thrash, R.J., and Leroi, G.E. (1977). Observation of 'hidden' electronic states by two-photon excitation spectroscopy: confirmation of the low-energy ^1A_g state of diphenyloctatetraene. *J. Chem. Phys.*, **67**, 3389–91.

Fang, H.L.B., Thrash, R.J., and Leroi, G.E. (1978). Observation of the low-energy ^1A_g state of diphenylhexatriene by two-photon excitation spectroscopy. *Chem. Phys. Lett.*, **57**, 59–63.

Fano, U. (1954). Principles of radiological physics. In *Radiation biology* (ed. A. Hollaender), pp. 1–144. McGraw-Hill, New York.

Faraggi, M. and Klapper, M.H. (1988). Intramolecular long-range electron transfer in the α-hemoglobin subunit. *J. Am. Chem. Soc.*, **110**, 5753–6.

Faraggi, M. and Klapper, M.H. (1991). Intramolecular electron transfer reactions in peptides and proteins. In *Excess electrons in dielectric media* (ed. C. Ferradini, and J.-P. Jay-Gerin), pp. 397–423. CRC Press, Boca Raton.

Faraggi, M. and Tal, Y. (1975). The reaction of the hydrated electron with amino acids, peptides and proteins in aqueous solution. II. Formation of radicals and electron transfer reactions. *Radiat. Res.*, **62**, 347–56.

Faraggi, M., Hemmerich, P., and Pecht, I. (1975). O_2- affinity of flavin radical species as studied by pulse radiolysis. *FEBS Lett.*, **51**, 47–51.

Faraggi, M., DeFilippis, M.R., and Klapper, M.H. (1989). Long-range electron transfer between tyrosine and tryptophan in peptides. *J. Am. Chem. Soc.*, **111**, 5141–5.

Farmilo, A. and Wilkinson, F. (1973). On the mechanism of quenching of singlet oxygen in solution. *Photochem. Photobiol.*, **18**, 447–50.

Farrel, M.K., Balistreri, W.F., and Suchy, F.J. (1982). Serum sulfated lithocholate as an indicator of cholestasis during parenteral nutrition in infants and children. *J.P.E.N.*, **6**, 30–3.

Farrington, J.A., Ebert, M., Land, E.J., and Fletcher, K. (1973). Bipyridylium quaternary salts and related compounds. V. Pulse radiolysis studies of the reaction of paraquat radical with oxygen. Implications for the mode of action of bipyridyl herbicides. *Biochim. Biophys. Acta*, **314**, 372–81.

Farrington, J.A., Ebert, M., and Land, E.J. (1978). Bipyridylium quaternary salts and related compounds. Part 6. Pulse radiolysis studies of the reaction of paraquat radical analogues with oxygen. *J. Chem. Soc., Faraday Trans. I*, **74**, 665–75.

Fasman, G.D. (ed.) (1976). *Handbook of biochemistry and molecular biology*, p. 188. CRC Press, Cleveland, Ohio.

Faulds, D., Balfour, J.A., Chrisp, P., and Langtry, H.D. (1991). Mitoxantrone. A review of its pharmacodynamic and pharmokinetic properties, and therapeutic potential in the chemotherapy of cancer. *Drugs*, **41**, 400–49.

Favaudon, V., Tourbez, H., Houée-Levin, C., and Lhoste, J.-M. (1990). $CO_2^{\cdot-}$ radical induced cleavage of disulphide bonds in proteins. A γ-ray and pulse radiolysis mechanistic investigation. *Biochem.*, **29**, 10 978–89.

Favre, A. (1990). 4-Thio-uridine as intrinsic photoaffinity probe of nucleic acid structure and interactions. In *Bioorganic photochemistry*, Vol. 1 Photochemistry and the nucleic acids (ed. H. Morrison), pp. 379–425. Wiley Interscience, New York.

Feigl, E.O. (1988). EDRF—a protective factor? *Nature*, **331**, 490–1.

Fendler, J.H. (1982). *Membrane mimetic chemistry*. Wiley, New York.

Fenton, H.J.H. (1894). Oxidation of tartaric acid in the presence of iron. *J. Chem. Soc.*, **65**, 899–910.

Fenton, H.J.H. (1899). Oxidation of certain organic acids in the presence of ferrous salts. *Proc. Chem. Soc.*, **15**, 224.

Ferguson, J. and Johnson, B.E. (1986). Photosensitivity due to retinoids—clinical and laboratory studies. *Br. J. Dermatology*, **115**, 275–83.

Ferguson, J. and Johnson, B.E. (1989). Retinoid associated phototoxicity and photosensitivity. *Pharmacol. Ther.*, **40**, 123–35.

Ferradini, C., Foos, J., Houée, C., and Pucheault, J. (1978). The reaction between superoxide anion and hydrogen peroxide. *Photochem. Photobiol.*, **28**, 697–700.

Fielden, E.M. and Hart, E.J. (1967). Primary radical yields in pulse-irradiated alkaline aqueous solution. *Radiat. Res.*, **32**, 564–80.

Finkelstein, E., Rosen, G.M., and Rauckman, E.J. (1980). Spin trapping. Kinetics of the reaction of superoxide and hydroxyl radicals with nitrones. *J. Am. Chem. Soc.*, **102**, 4994–9.

Finkelstein, E., Rosen, G.M., and Rauckman, E.J. (1981). Production of hydroxyl radical by decomposition of superoxide spin-trapped adducts. *Mol. Pharmacol.*, **21**, 262–5.

Firey, P.A. and Rodgers, M.A.J. (1987). Photo-properties of a silicon naphthalocyanine: A potential photosensitiser for photodynamic therapy. *Photochem. Photobiol.*, **45**, 535–8.

Firey, P.A. and Rodgers, M.A.J. (1988). Photochemical properties of erythrocyte ghosts containing porphyrin. *Photochem. Photobiol.*, **47**, 615–9.

Firey, D.A., Ford, W.E., Sounick, J.R., Kenney, M.E., and Rodgers, M.A.J. (1988*a*). Silicon naphthalocyanine triplet state and oxygen: A reversible energy-transfer reaction. *J. Am. Chem. Soc.*, **110**, 7626–30.

Firey, P.A., Jones, T.W., Jori, G., and Rodgers, M.A.J. (1988*b*). Photoexcitation of zinc phthalocyanine in mouse myeloma cells: The observation of triplet states but not of singlet oxygen. *Photochem. Photobiol.*, **48**, 357–60.

Fisher, G.J. and Johns, H.E. (1976). Pyrimidine photodimers. In *Photochemistry and photobiology of nucleic acids*, Vol. I *Chemistry* (ed. S.Y. Wang), pp. 225–94. Academic Press, New York.

Fisher, G.J. and Land, E.J. (1983). Photosensitization of pyrimidines by 2-methylnaphthoquinone in water: A laser flash photolysis study. *Photochem. Photobiol.* **37**, 27–32.

Fisher, G.J. Varghese, A.J., and Johns, H.E. (1974). Ultraviolet-induced reactions of thymine and uracil in the presence of cysteine. *Photochem. Photobiol.*, **20**, 109–20.

Fisher, M.M. and Weiss, K. (1974). Laser photolysis of retinal and its protonated and unprotonated n-butylamine schiff base. *Photochem. Photobiol.*, **20**, 423–32.

Fitzpatrick, T.B., Szabó, G., Seiji, M., and Quevedo, W.C. (1979). Biology of the melanin pigmentary system. In *Dermatology in general medicine* (2nd edn) (ed. T.B. Fitzpatrick, A.Z. Eisen, K. Wolff, I.M. Freedberg, and K.F. Austen), pp. 131–63. McGraw-Hill, New York.

Fleming, G.R. and Siegman, A.E. (eds) (1986). *Ultrafast phenomena V*. Springer Verlag, New York.

Flohé, L., Beckman, R., Giertz, H., and Loschen, G. (1985). Oxygen-centered free radicals as mediators of inflammation. In *Oxidative stress* (ed. H. Sies), pp. 403–35. Academic Press, London.

Foner, S.N. and Hudson, R.L. (1956). Metastable oxygen molecules produced by electrical discharges. *J. Chem. Phys.*, **25**, 601–2.

Fong, K.L., McCay, P.B., Poyer, J.L., Misra H.P., and Keele, B.B. (1976). Evidence for superoxide-dependent reduction of Fe^{3+} and its role in enzyme-generated hydroxyl radical formation. *Chem. Biol. Interact.*, **15**, 77–89.

Foote, C.S. (1976). Photosensitized oxidation and singlet oxygen: consequences in biological systems. In *Free radicals in biology*, Vol. II (ed. W.A. Pryor), pp. 85–133. Academic Press, New York.

Foote, C.S. (1979). Detection of singlet oxygen in complex systems: A critique. In *Oxygen: Biochemical and clinical aspects* (ed. W.S. Caughey), pp. 603–26. Academic Press, New York.

Foote, C.S. (1991). Definition of Type I and Type II photosensitised oxidation. *Photochem. Photobiol.*, **54**, 659.

Foote, C.S. and Peters, J.W. (1971). Chemistry of singlet oxygen. XIV. A reactive intermediate in sulfide photooxidation. *J. Am. Chem. Soc.*, **93**, 3795–6.

Forbes, P.D., Davies, R.E., and Urbach, F. (1977). Experimental ultraviolet carcinogenesis. In *Research in photobiology* (ed A. Castellani), pp. 469–78. Plenum Press, New York.

Ford, W.E., Rihter, B.D., Kenney, M.E., and Rodgers, M.A.J. (1989). Photoproperties of alkoxy-substituted phthalocyanines with deep-red optical absorbance. *Photochem. Photobiol.*, **50**, 277–82.

Forni, L.G. and Willson, R.L. (1986). Thiyl free radicals and the oxidation of ferrocytochrome *c*. *Biochem. J.*, **240**, 905–7.

Förster, T. (1948). Zwischenmolekulare energiewanderung und fluoreszenz. *Ann. Physik.*, **2**, 55–75.

Förster, T. (1959). Transfer mechanisms of electronic excitation. *Disc. Faraday Soc.*, **27**, 7–17.

Förster, T. (1975). Excimers and exciplexes. In *The exciplex* (ed. M. Gordon and W.R. Ware), pp. 1–21. Academic Press, New York.

Fox, M. A. (1990). Photoinduced electron transfer. *Photochem. Photobiol.*, **52**, 617–27.

Foyt, D. C. (1982). Data acquisition and analysis in pulse radiolysis. In *The study of fast processes and transient species by electron pulse radiolysis* (ed. J. H. Baxendale and F. Busi), pp. 199–240. D. Reidel, Dordrecht.

Frank, J. (1925). Elementary processes of photochemical reactions. *Trans. Faraday Soc.*, **21**, 536–42.

Franck, J. and Rabinowitsch, E. (1934). Some remarks about free radicals and the photochemistry of solutions. *Trans. Faraday Soc.*, **30**, 120–31.

Frank, H. A. and Violette, C. A. (1989). Monomeric bacteriochlorophyll is required for triplet energy transfer between the primary donor and the carotenoid in photosynthetic bacterial reaction centers. *Biochim. Biophys. Acta*, **976**, 222–32.

Frankenberg, D. (1984). Interpretation of the dose and LET dependence of RBE values for lethal lesions in yeast cells. *Radiat. Res.*, **97**, 329–40.

Frankenberg-Schwager, M., Frankenberg, D., and Harbich, R. (1984) Repair of DNA double-strand breaks as a determinant of RBE of alpha particles. *Br. J. Cancer.*, **49**, 169–73.

Frauenfelder, H. Sligar, S. G., and Wolynes, P. G., (1991). The energy landscapes and motions of proteins. *Science*, **254**, 1598–603.

Fridovich, I. (1978). Superoxide radicals, superoxide dismutases and the aerobic lifestyle. *Photochem. Photobiol.*, **28**, 733–41.

Fridovich, I. (1983). Superoxide radical: an endogenous toxicant. In *Ann. Rev. Pharmacol. Toxicol.*, Vol. 23 (ed. R. George, R. Okun, and A. K. Cho), pp. 239–57. Annual Reviews Inc., Palo Alto, California.

Frimer, A. A. (1983). Organic reactions involving the superoxide anion. In *The chemistry of functional groups, peroxides* (ed. S. Patai), pp. 429–61. Wiley, New York.

Frimer, A. A. (ed.) (1985). *Singlet* O_2, CRC Press, Boca Raton, Florida.

Fujita, S. and Steenken, S. (1981). Pattern of OH radical addition to uracil and methyl- and carboxyl-substituted uracils. Electron transfer of OH adducts with *N,N,N′,N′*-tetramethyl-*p*-phenylene diamine and tetranitromethane. *J. Am. Chem. Soc.*, **103**, 2540–5.

Fusi, F., Agati, G. and Pratesi, R. (1985). The role of ZZ–EZ isomerisation in the photochemistry of bilirubin bound to human serum albumin *in vitro*. *Lettere Al Nuovo Cimento*, **44**, 477–81.

Gan, E. V., Haberman, H. F., and Menon, I. A. (1974). Oxidation of NADH by melanin and melanoproteins. *Biochim. Biophys. Acta*, **370**, 62–9.

Garfield, E. (1979). The hazards of sun bathing, *Current Contents* (Feb 19), 5–12.

Garnier-Suillerot, A., Tosi, L., and Paniago, E. (1984). Kinetic and mechanism of vesicle lipoperoxide decomposition by Fe (II). *Biochim. Biophys. Acta*, **794**, 307–12.

Gates, F. L. (1930). A study of the bactericidal action of ultraviolet light. III. The absorption of ultraviolet light by bacteria. *J. Gen. Physiol.*, **14**, 31–42.

Gaudiano, G. and Koch, T. H. (1991). Redox chemistry of anthracycline antitumour drugs and use of captodative radicals as tools for its elucidation and control. *Chem. Res. Toxicol.*, **4**, 2–16.

Gauduel, Y., Martin, J. L., Migus, A., Yamada, N., and Antonetti, A. A. (1986). Femtosecond study of electron localization and solvation in pure water. In *Ultrafast phenomena V*, Springer series in Chemical Physics, Vol. 46 (ed. G. R. Fleming and A. E. Siegman), pp. 308–11. Springer-Verlag, Berlin.

Gay-Lussac, H. L. (1815). Du cyanogène ou radical d'acide prussique. *Ann. Chim*, **95**, 172–99.

Geacintov, N. E. and Brenner, H. C. (1989). The triplet state as a probe of dynamics and structure in biological macromolecules. *Photochem. Photobiol.*, **50**, 841–58.

Geacintov, N. E., Prusik, T., and Khosrofian, J. M. (1976). Properties of benzopyrene–DNA complexes investigated by fluorescence and triplet flash photolysis techniques. *J. Am. Chem. Soc.*, **98**, 6444–52.

Gebhard, R., Van der Hoef, K., Violette, C. A., de Groot, H. J. M., Franck, H. A., and Lugtenburg, J. (1991). C MAS NMR evidence for a 15,15′-*Z* configuration of the spheroidene chromophore in the *rhodo-bacter spheroides* reaction center: synthesis of ^{13}C- and ^{2}H-labelled spheroidenes. *Pure Appl. Chem.*, **63**, 115–22.

Gebicki, J. M. and Bielski, B. H. J. (1981). Comparison of the capacities of the perhydroxyl and the superoxide radicals to initiate chain oxidation of linoleic acid. *J. Am. Chem. Soc.*, **103**, 7020–2.

Georghiou, S., Nordlund, T. M., and Saim, A. M. (1985). Picosecond fluorescence decay time measurements of nucleic acids at room temperature in aqueous solution. *Photochem. Photobiol.*, **41**, 209–12.

Gerschman, R., Gilbert, D. L., Nye, S. W., Dwyer, P., and Fenn, W. O. (1954). Oxygen poisoning and X-irradiation: A mechanism in common. *Science*, **119**, 623–6.

Gey, K. F. and Puska, P. (1989). Plasma vitamin E and A inversely correlated to mortality from ischemic heart disease in cross-cultural epidemiology. *Ann. N. Y. Acad. Sci.*, **570**, 268–82.

Gey, K. F., Brubacher, G. B., and Stähelin, H. B. (1987). Plasma levels of antioxidant vitamins in relation to ischemic heart disease and cancer. *Am. J. Clin. Nutr.*, **45**, 1368–77.

Ghiggino, K. P., Bennett, L. E., and Henderson, R. W. (1988). Photochemical properties of porphyrin C: An agent for use in tumour phototherapy. *Photochem. Photobiol.*, **47**, 65–72.

Ghiron, C., Bazin, M., and Santus, R. (1988). Characterization of the indole triplet excited state in proteins using laser flash photolysis. *J. Biochem. Biophys. Methods*, **15**, 337–48.

Gibson, S. L., Cohen, H. J., and Hilf, R. (1984). Evidence against the production of superoxide by photoirradiation of hematoporphyrin derivative. *Photochem. Photobiol.*, **40**, 441–8.

Gillbro, T. and Cogdell, R. J. (1989). Carotenoid fluorescence. *Chem. Phys. Lett.*, **158**, 312–16.

Girotti, A. W. (1990). Photosensitised lipid peroxidation in biological membranes. In *Photodynamic therapy of neoplastic disease*, Vol. 1 (ed. D. Kessel), pp. 229–45. CRC Press, Boca Raton, Florida.

Goldschmidt, C. R. and Ottolenghi, M. (1970). Absorption spectrum of the pyrene excimer. *J. Phys. Chem.*, **74**, 2041–2.

Goldstein, S. and Czapski, G. (1987). Oxidation–reduction reactions of iron bleomycin in the absence and presence of DNA. *Free Rad. Res. Commun.*, **2**, 259–70.

Goldstein, S., Michel, C., Bors, W., Saran, M., and Czapski, G. (1988). A critical reevaluation of some assay methods for superoxide dismutase activity. *Free Rad. Biol. Med.*, **4**, 295–303.

Gollnick, K. (1968). Type II photooxygenation reactions in solution. *Adv. Photochem.*, **6**, 1–122.

Gomberg, M. (1900). An instance of trivalent carbon: triphenylmethyl. *J. Am. Chem. Soc.*, **22**, 757–71.

Gomer, G. J. (1984). Preclinical examination of ocular photoradiation therapy. In *Porphyrins in tumour phototherapy* (ed. A. Andreoni and R. Cubeddu), pp. 447–54. Plenum, New York.

Gorman, A. A. and Hamblett, I. (1983). Lifetime, triplet–triplet absorption spectrum and relaxation energy of an acyclic conjugated triene triplet: A pulse radiolysis study. *Chem. Phys. Lett.*, **97**, 422–6.

Gorman, A.A. and Rodgers, M.A.J. (1992). Current perspectives of singlet oxygen detection in biological environments. *J. Photochem. Photobiol. B: Biol.*, **14**, 159–76.

Gorman, A.A., Lovering, G. and Rodgers, M.A.J. (1978). A pulse radiolysis study of the triplet sensitized production of singlet oxygen: determination of energy transfer efficiences *J. Am. Chem. Soc.*, **100**, 4527–32.

Gorman, A.A., Lovering, G., and Rodgers, M.A.J. (1979). The entropy-controlled reactivity of singlet oxygen ($^1\Delta_g$) towards furans and indoles in toluene. Variable temperature study by pulse radiolysis. *J. Am. Chem. Soc.*, **101**, 3050–5.

Gorman, A.A., Gould, I.R., and Hamblett, I. (1981). Lifetimes and energies of the relaxed triplet states of cyclopentadiene and tetramethylbutadiene. *J. Am. Chem. Soc.*, **103**, 4553–8.

Gorman, A.A., Hamblett, I., and Harrison, R.J. (1984*a*). A pulse radiolysis study of the azulene triplet state. *J. Am. Chem. Soc.*, **106**, 6952–5.

Gorman, A.A., Hamblett, I., and Jensen N.-H. (1984*b*). Reversible triplet energy transfer between neo-alloocinene and anthracene. *Chem. Phys. Lett.*, **111**, 293–6.

Gorman, A.A., Hamblett, I., Lambert, C., Spencer, B., and Standen, M.C. (1988). Identification of both pre-equilibrium and diffusion limits for reaction of singlet oxygen, O_2 ($^1\Delta_g$), with both physical and chemical quenchers: variable temperature, time-resolved infra red luminescence studies. *J. Am. Chem. Soc.*, **110**, 8053–9.

Gorman, A.A., Hamblett, I., and Land, E.J. (1989). A pulse radiolysis based singlet oxygen luminescence facility. *J. Am. Chem. Soc.*, **111**, 1876–7.

Görner, H. (1990*a*). Phosphorescence of nucleic acids and DNA components at 77 K. *J. Photochem. Photobiol., B: Biol.*, **5**, 359–77.

Görner, H. (1990*b*). Transients of uracil and thymine derivatives and the quantum yields of electron ejection and intersystem crossing upon 20 ns photolysis at 248 nm. *Photochem. Photobiol.*, **52**, 935–48.

Görner, H., Wala, M., and Schulte-Frohlinde, D. (1992). Strand breakage in poly(C), poly(A), single-strand and double-stranded DNA induced by nanosecond laser excitation at 193 nm. *Photochem. Photobiol.*, **55**, 173–84.

Gould, I.R., Ege, D., Mattes, S.L., and Farid, S. (1987). Return electron transfer within geminate radical ion pairs. Observation of the Marcus inverted region. *J. Am. Chem. Soc.*, **109**, 3794–6.

Gould, I.R., Moody, R., and Farid, S. (1988). Electron transfer reactions in the Marcus inverted region: difference in solvation and electronic coupling between excited charge-transfer complexes and geminate radical ion pairs. *J. Am. Chem. Soc.*, **110**, 7242–4.

Gould, I.R., Moser, J.E., Armitage, B., Farid, S., Goodman, J.L., and Herman, M.S. (1989). Electron-transfer reactions in the Marcus inverted region. Charge recombination versus charge shift reactions. *J. Am. Chem. Soc.*, **111**, 1917–9.

Gouterman, M. (1978). Optical spectra and electronic structure of porphyrins and related rings. In *The porphyrins*, Vol. 11 (ed. D. Dolphin), pp. 1–165. Academic Press, New York.

Graham, D.G., Tiffany, S.M., Bell, W.R., and Gutknecht, W.F. (1978). Autoxidation versus covalent binding of quinones as the mechanism of toxicity of dopamine, 6- hydroxydopamine and related compounds towards C1300 neuroblastoma cells *in vitro*. *Mol. Pharmacol.*, **14**, 644–53.

Grant, J.L., Kramer, V.J., Ding, R., and Kispert, L.D. (1988). Carotenoid cation radical: electrochemical, optical and EPR study. *J. Am. Chem. Soc.*, **110**, 2151–7.

Gray, H. G. (1986). Long range electron-transfer in blue copper proteins. *Chem. Soc. Rev.*, **15**, 17–30.
Gray, L. H. (1929). The absorption of penetrating radiation. *Proc. R. Soc. London*, **A122**, 647–68.
Gray, P. J., Phillips, D. R., and Wedd, A. G. (1982). Photosensitized degradation of DNA by daunomycin *Photochem. Photobiol.*, **36**, 49–57.
Green, B. I., Lamola, A. A., and Shank, V. V. (1981). Picosecond primary processes of bilirubin bound to human serum albumin. *Proc. Natl. Acad. Sci. USA*, **78**, 2008–12.
Greenaway, F. T., Redmond, R. W., and Ledbetter, J. W. (1991). Hydrated electron formation on laser excitation of *P*-pyridoxyl amino acids and proteins. *Photochem. Photobiol.*, **54**, 667–72.
Greenberg, J. M., Malhotra, V., and Ennever, J. F. (1987). Wavelength dependence of the quantum yield for the structural isomerisation of bilirubin. *Photochem. Photobiol.*, **46**, 453–6.
Greenstock, C. L. and Dunlop, I. (1973). Pulse radiolysis studies of nitrofurans: Chemical radiosensitization. *Radiat. Res.*, **56**, 428–40.
Greenstock, C. L. and Dunlop, I. (1975). Reactions of nucleic acid radicals with radiation modifiers. In *Fast processes in radiation chemistry and biology* (ed. G. E. Adams, E. M. Fielden, and B. D. Michael), pp. 247–58. Wiley, New York.
Greenstock, C. and Wiebe, R. H. (1981).Kinetic studies of peroxides and peroxy radicals, and their reactions with biological molecules. In *Oxygen and oxy-radicals in chemistry and biology* (ed. M. A. J. Rodgers and E. L. Powers), p. 119–31. Academic Press, New York.
Greenstock, C. L., Adams, G. E., and Willson, R. L. (1970). Electron transfer studies of nucleic acid derivatives in solutions containing radiosensitizers. In *Radiation protection and sensitization* (ed. H. L. Moroson and M. Quintiliani), pp. 65–71. Taylor & Francis, London.
Greenstock, C. L., Ruddock, G. W., and Neta, P. (1976). Pulse radiolysis and ESR studies of the electron-affinic properties of nitroheterocyclic radiosensitizers. *Radiat. Res.*, **66**, 472–84.
Grigoryants, V. M., Lozovoy, V. V., Chernosov, Yu. D., Shebolaev, I. V., Aryutinov, V. A., Anisimov, O. A., and Molin, Yu. N. (1989). Pulse radiolytic system with picosecond time resolution referred to Cherenkov radiation. *Radiat. Phys. Chem.*, **34**, 349–52.
Grootveld, M. and Halliwell, B. (1986). An aromatic hydroxylation assay for hydroxyl radicals utilizing high performance liquid chromatography (HPLC). Use to investigate the effect of EDTA on the Fenton reaction. *Free Rad. Res. Commun.*, **1**, 243–50.
Grossweiner, L. I., Swenson, G. W., and Zwicker, E. F. (1963). Photochemical production of the hydrated electron. *Science*, **141**, 805–6.
Grotthus, T. (1815). Uber einen neuen lightsauger nebst einigen allegeminen betrachtungen über die phosphoreszens und die farben. *J. Chem. Physik.*, **14**, 133–92.
Groves, J. T. (1980). Mechanisms of metal-catalysed oxygen insertion. In *Metal-ion activation of dioxygen* (ed. T. G. Spiro), pp. 125–62. Wiley Interscience, New York.
Grubbé, E. H. (1933). Priority in the therapeutic use of X-rays. *Radiology*, **21**, 156–62.
Gudgin, E., Lopez-Delgado, R., and Ware, W. R. (1983). Photophysics of tryptophan in H_2O, D_2O and in nonaqueous solvents. *J. Phys. Chem.*, **87**, 1559–65.
Guéron, M., Eisinger, J., and Lamola, A. A. (1974). Excited states of nucleic acids. In *Basic principles in nucleic acid chemistry* (ed. P. O. P. T'so), pp. 311–98. Academic Press, New York.
Guiotto, A., Rodighiero, P., Pastorini, G., Bordini, F., Baccichetti, F., Carlassare, F., Vedaldi, D., Dall'Acqua, F., and Land, E. J. (1981). A new water soluble derivative

of 4,5′-dimethylangelicin as a potential agent for photo-chemotherapy. *Il Farmaco*, **36**, 536–50.

Gust, D. and Moore, T.A. (1989). Mimicking photosynthesis. *Science*, **244**, 35–41.

Gust, D. and Moore, T.A. (1991). Mimicking photosynthetic electron and energy transfer. In *Advances in photochemistry* (ed. D. Volman, G. Hammond, and D. Neckers), pp. 1–63. Wiley, New York.

Gust, D., Moore, T.A., Bensasson, R.V., Mathis, P., Land, E.J., Chachaty, C., Moore, A.L., Liddell, P.A., and Nemeth, G.A. (1985). Stereodynamics of intramolecular triplet energy transfer in carotenoporphyrins. *J. Am. Chem. Soc.*, **107**, 3631–40.

Haber, F. and Weiss, J.J. (1934). The catalytic decomposition of hydrogen peroxide by iron salts. *Proc. R. Soc. London*, **A147**, 332–51.

Haber, F. and Willstäter, R. (1931). Unpaarigkeit und radikalketten in reaktions mechanismus organischer und enzymatischer vorgange. *Ber. Deutsch. Chem. Ges.*, **64**, 2844–56.

Hagerman, P.J. (1988). Flexibility of DNA. *Ann. Rev. Biophys. Chem.*, **17**, 265–86.

Haley, J.L., Fitch A.N., Goyal, R., Lambert, C., Truscott, T.G., Chacon, J.N., Stirling, D., and Schalch W. (1992). The S_1 and T_1 energy levels of all-*trans*-β-carotene. *J. Chem. Soc., Chem. Commun.*, 1175–6.

Haley, L.V. and Koningstein, J.A. (1983). Space and time-resolved resonance-enhanced vibrational Raman spectroscopy from femtosecond—lived excited state of β-carotene. *Chem. Phys.*, **77**, 1–9.

Hall, R.D., Buettner, G.R., and Chignell, C.F. (1991). The biphotonic photoionization of chlorpromazine during conventional flash photolysis: Spin trapping results with 5,5-dimethyl-1-pyrroline-*N*-oxide. *Photochem. Photobiol.*, **54**, 167–73.

Halliwell, B. (1976). An attempt to demonstrate a reaction between superoxide and hydrogen peroxide. *FEBS Lett.*, **72**, 8–10.

Halliwell, B. (1978). Superoxide-dependent formation of hydroxyl radical in the presence of iron salts. Its role in degradation of hyalorunic acid by a superoxide generating system. *FEBS Lett.*, **96**, 238–42.

Halliwell, B. and Gutteridge, J.M.C. (1981). Formation of a thiobarbituric-acid-reactive substance from deoxyribose in the presence of iron salts. *FEBS Lett.*, **128**, 347–52.

Halliwell, B. and Gutteridge, J.M.C. (1984). Oxygen toxicity, oxygen radicals, transition metals and disease. *Biochem. J.*, **219**, 1–14.

Halliwell, B. and Gutteridge, J.M.C. (1985). The importance of free radicals and catalytic metal ions in human diseases. *Molec. Aspects Med.*, **8**, 89–193.

Halliwell, B. and Gutteridge, J.M.C. (1989). *Free radicals in biology and medicine* (2nd edn). Oxford University Press, Oxford.

Halliwell, B. and Gutteridge, J.M.C. (1990). Role of free radicals and catalytic metal ions in human disease: an overview. In *Methods in Enzymology*., Vol. 186, Oxygen radicals in biological systems, Part B Oxygen radicals and antioxidants (ed. L. Packer and A.N. Glazer), pp. 1–85. Academic Press, San Diego.

Hamaguchi, H., Okamoto, H., Tasumi, M., Mukai, Y., and Koyama, Y. (1984). Transient raman spectra of the all-*trans* and 7-, 9-, 11- and 13-mono-*cis* isomers of retinal and the mechanism of the *cis–trans* isomerization in the lowest excited triplet state. *Chem. Phys. Lett.*, **107**, 355–9.

Hammond, G.S. and Saltiel, J. (1963). Mechanisms of photoreactions in solution. XVIII. Energy transfer with non vertical transitions. *J. Am. Chem. Soc.*, **85**, 2516–17.

Hammond, G.S., Saltiel, J., Lamola, A.A., Turro, N.J., Bradshaw, J.S., Cowan, D.O., Counsell, R.C., Vogt, V., and Dalton, C. (1964). Mechanisms of photochemical reactions in solution XXII. Photochemical *cis–trans* isomerization. *J. Am. Chem. Soc.*, **86**, 3197–217.

Hanzlik, C.A., Knox, R.S., Gibson, S.L., and Hilf, R. (1989). Picosecond fluorescence of R3230AC mammary carcinoma mitochondria after treatment with hematoporphyrin derivative and photofrin II *in vivo*. *Photochem. Photobiol.*, **50**, 45–53.

Harbour, J.R., and Issler, S.L. (1982). Involvement of the azide radical in the quenching of singlet oxygen by azide anion in water. *J. Am. Chem. Soc.*, **104**, 903–5.

Hariharan, P.V., Achey, P.M., and Cerutti, P.A. (1977). Biological effect of thymine ring saturation in coliphage ϕX-174-DNA. *Radiat. Res.*, **69**, 375–8.

Harrigan, E.T., Chakrabarti, A., and Hirota, N. (1976). Single crystal EPR, zero field ODMR and phosphorescence studies of the T_1 state of coumarin. *J. Am. Chem. Soc.*, **98**, 3460–5.

Harriman, A. and Richoux, M.-C. (1980). Attempted photoproduction of hydrogen using sulphonated phthalocyanines as chromophores for three component systems. *J. Chem. Soc., Faraday Trans. II*, **76**, 1618–26.

Harriman, A., Maiya, B.G., Mavia, T., Hemmie, G., Sessler, J.L., and Mallorck, T.E. (1989). Metallotexaphyrins: A new family of photosensitizers for efficient generation of singlet oxygen. *J. Chem. Soc., Chem. Commun.*, 314–16.

Hart, E.J. and Boag, J.W. (1963). Absorption spectrum of the hydrated electron. *Nature*, **197**, 45–7.

Hasegawa, K. and Patterson, L.K. (1978). Pulse radiolysis studies in model lipid systems: formation and behavior of peroxy radicals in fatty acids. *Photochem. Photobiol.*, **28**, 817–23.

Hashimoto, H. and Koyama, Y. (1988). Time resolved resonance Raman spectroscopy of triplet β-carotene produced from all-*trans*, 7-*cis*, 9-*cis* 13-*cis*, and 15-*cis* isomers and high-pressure liquid chromatography analyses of photoisomerization via the triplet state. *J. Phys. Chem.*, **92**, 2101–8.

Hashimoto, H., Koyama, Y., Hirata, Y., and Mataga, N. (1991). S_1 and T_1 species of β-carotene generated by direct photoexcitation from all-*trans*, 9-*cis*, 13-*cis*, and 15-*cis* isomers as revealed by picosecond transient absorption and transient Raman spectroscopies. *J. Phys. Chem.*, **95**, 3072–6.

Hashimoto, S., Seki, H., Masuda, T., Imamura, M., and Kondo, M. (1981). Dimer formation in irradiated aqueous solutions of lysozyme studied by light-scattering intensity measurements. *Int. J. Radiat, Biol.*, **40**, 31–46.

Hashimoto, S. Kira, A., Imamura, M., and Masuda, T. (1982). Lysozyme dimer formation on lysozyme oxidation with Br_2^- as studied by fluorescence evolution. *Int. J. Radiat. Biol.*, **41**, 303–14.

Hasty, N., Merkel, P.B., Radlik, P. and Kearns, D.R. (1972). Role of azide in singlet oxygen reactions: reaction of azide with singlet oxygen. *Tetrahedron Lett.*, 49–52.

Hausmann, W. (1911). The sensitizing action of haematoporphyrin. *Biochemische Zeitschrift*, **30**, 276–316.

Hauswirth, W. and Daniels, M. (1976). Excited states of the nucleic acids:polymeric forms. In *Photochemistry and photobiology of nucleic acids*, Vol. I *Chemistry* (ed. S.Y. Wang), pp. 109–67. Academic Press, New York.

Hayon, E. (1969). Spectroscopic observation of the triplet excited state of uracil in solution. *J. Am. Chem. Soc.*, **91**, 5397–8.

Hayon, E. and Simic, M. (1971). Pulse radiolysis study of cyclic peptides in aqueous solution. Absorption spectrum of the peptide radical -NHCHO-. *J. Am. Chem. Soc.*, **93**, 6781–96.

Hayon, E. and Simic, M. (1974). Acid–base properties of free radicals in solution. *Acc. Chem. Res.*, **7**, 114–21.

Heelis, P.F. (1982). The photophysical and photochemical properties of flavins (isoalloxazines). *Chem. Soc. Rev.*, **11**, 15–39.

Heelis, P.F. and Sancar, A. (1986). Photochemical properties of *Escherichia coli* DNA photolyase: a flash photolysis study. *Biochemistry*, **25**, 8163–6.

Heelis, P.F., Parsons, B.J., Phillips, G.O., and McKellar, J.F. (1978). A laser flash photolysis study of the nature of flavin mononucleotide triplet states and the reactions of the neutral form with amino acids. *Photochem. Photobiol.*, **28**, 169–73.

Hefetz, Y., Dunn, D.A., Deutsch, T.F., Buckley, L., Hillenkamp, F., and Kochevar, I.E. (1992). Laser photochemistry of DNA: two-photon absorption and optical breakdown using high-intensity, 532 nm radiation. *J. Am. Chem. Soc.*, **112**, 8528–32.

Heihoff, K., Redmond, R.W., Braslavsky, S.E., Rougée, M., Salet, C., Favre, A., and Bensasson, R.V. (1990). Quantum yield of triplet and $O_2(^1\Delta_g)$ formation of 4-thiouridine in water and acetonitrile. *Photochem. Photobiol.*, **51**, 635–41.

Held, K.D., Harrop, H.A., and Michael, B.D. (1984). Effect of oxygen and sulphydryl-containing compounds on irradiated transforming DNA. III. Reaction rates. *Int. J. Radiat. Biol.*, **45**, 627–36.

Held, K.D., Harrop, H.A., and Michael, B.D. (1985). Pulse radiolysis studies of the interactions of the sulfhydryl compound dithiothreitol and sugars. *Radiat. Res.*, **103**, 171–85.

Hélène, C. (1987). Specific gene regulation by oligodeoxynucleotides covalently linked to intercalating agents. In *DNA-ligand interactions* (ed. W. Guschlbauer and W. Saenger), pp. 127–40. Plenum, New York.

Hélène, C. and Toulmé, J. (1989). Control of gene expression by oligodeoxynucleotides covalently linked to intercalating agents and nucleic acids cleaving reagents. *Topics Mol. Struct. Biol.*, **12**, 137–72.

Hélène, C., Santus, R., and Douzou, P. (1966). Photoionisation and biphotonic processes of nucleic acid derivatives in frozen solutions. *Photochem. Photobiol.*, **5**, 127–33.

Henderson, B.W., and Dougherty, T.J. (1992). How does photodynamic therapy work? *Photochem. Photobiol.*, **55**, 145–57.

Henderson, B.W. and Miller, A.C. (1986). Effects of reactive oxygen and radical species on cell survival following photodynamic treatment *in vitro*: comparison to ionising radiation. *Radiat. Res.*, **108**, 196–205.

Henderson, B.W., Dougherty, T.J., and Malone, P.B. (1984). Studies on the mechanism of tumour destruction by photoradiation therapy. In *Porphyrin localization and treatment of tumours* (ed. D.R. Doiron and C.J. Gomer), pp. 601–12. Alan R. Liss, New York.

Henderson, B.W., Waldow, S.M., Mang, T.S., Potter, W.R., and Dougherty, T.J. (1985). Tumour destruction and kinetics of tumour cell death in two experimental mouse tumours following photodynamic therapy. *Cancer Res.*, **45**, 572–6.

Henglein, A. (1976). Pulse radiolysis and polarography. *Electroanal. Chem.*, **9**, 163–244.

Henglein, A. and Korman. C. (1985). Scavenging of OH radicals produced in the sonolysis of water. *Int. J. Radiat. Biol.*, **48**, 251–8.

Henglein, A., Guttiérrez, M., and Ulrich, R. (1988). Oxidation of iodide by the intense acoustic bursts of an extracorporeal lithotripter. *Int. J. Radiat. Biol.*, **54**, 123–6.

Henri, V. (1914). Etude de l'action métabiotique des rayons ultraviolets. Productions de formes de mutation de la bactériedie charbonneuse. *Compt. Rend.*, **158**, 1032–5.

Herkstroeter, W.G. (1975*a*). Determination of triplet-energy levels in azomethine dyes, energy-transfer measurements. *J. Am. Chem. Soc.*, **97**, 3090–6.

Herkstroeter, W.G. (1975*b*). The triplet energies of azulene, β-carotene and ferrocene. *J. Am. Chem. Soc.*, **97**, 4161–7.

Herkstroeter, W.G. and Hammond, G.S. (1966). Mechanisms of photochemical reactions in solution. XXXIX. Study of energy transfer by kinetic spectrophotometry. *J. Am. Chem. Soc.*, **88**, 4769–77.

Hertz, H.R. (1887). Über einen einfluss des ultravioletten lichten auf die electrische entladung. *Ann. Physik.*, **31**, 983–1000.

Herzberg, G. (1950). Molecular spectra and molecular structure. I. Spectra of diatomic molecules (2nd edn). Van Nostrand, New York.

Hetherington, A.M. and Johnson, B.E. (1984). Photohemolysis. *Photodermatology*, **1**, 255–60.

Hicks, M. and Gebicki, J.M. (1978). A quantitative relationship between permeability and the degree of peroxidation in ufasome membranes. *Biochem. Biophys. Res. Commun.*, **80**, 704–8.

Hildenbrand, K. and Schulte-Frohlinde, D. (1989). E.s.r. studies on the mechanism of hydroxyl radical-induced strand breakage of polyuridylic acid. *Int. J. Radiat. Biol.*, **55**, 725–38.

Hilderson, H.J. (ed.) (1988). *Subcellular biochemistry*, Vol. 13 *Fluorescence studies on biological membranes*. Plenum Press, New York.

Hiller K.-O. and Asmus K.-D. (1983). Formation and reduction reactions of α-amino radicals derived from methionine and its derivatives in aqueous solution. *J. Phys. Chem.*, **87**, 3682–8.

Hiller, K.-O., Hodd, P.L., and Willson, R.L. (1983). Antiinflammatory drugs: Protection of a bacterial virus as an *in vitro* biological measure of free radical activity, *Chem. Biol. Interactions*, **47**, 293–305.

Hirayama, K. (1955). Absorption spectra and chemical structures. I. Conjugated polyenes and *p*-polyphenyls. *J. Am. Chem. Soc.*, **77**, 373–9.

Hochstrasser, R.M. and Johnson, C.K. (1988). Biological processes studied by ultrafast laser techniques. In *Ultrashort laser pulses and applications* (ed. W. Kaiser), Topics in applied physics, Vol. 60, pp. 357–417. Springer-Verlag, Berlin.

Hoey, B.M. and Butler, J. (1984). The repair of oxidised amino acids by antioxidants. *Biochim. Biophys. Acta*, **791**, 212–8.

Hoey, B.M., Butler, J., Lea, J.S., and Sarna, T. (1988*a*). A comparison of the free radical properties of several anthracycline anti-tumour drugs and some of their analogues. *Free Rad. Res. Commun.*, **5**, 169–76.

Hoey, B.M., Butler, J., and Swallow, A.J. (1988*b*). Reductive activation of mitomycin C. *Biochemistry*, **27**, 2608–14.

Hoff, A.J. (1984). Electron spin polarization of photosynthetic reactants. *Q. Rev. Biophys.*, **17**, 152–282.

Hoff, A.J. (ed.) (1989). *Advanced EPR. Applications in biology and biochemistry*, Elsevier, Amsterdam.

Hoffman, B.M., Natan, M.J., Nocek, J.M., and Wallin, S.A. (1991). Long-range electron transfer within metal-substituted protein complexes. In *Long-range electron transfer in biology* (ed. G. Palmer), pp. 85–108. Springer-Verlag, Berlin.

Hoffman, M.Z. and Hayon, E. (1972). One-electron reduction of the disulfide linkage in aqueous solution. Formation, protonation, and decay kinetics of the $RSSR^-$ radical. *J. Am. Chem. Soc.*, **94**, 7950–7.

Hoffman, M.Z. and Hayon, E. (1973). Pulse radiolysis study of sulfhydryl compounds in aqueous solution. *J. Phys. Chem.*, **77**, 990–6.

Holden, J.L., Dewey, D.L., and Galpine, A.R. (1984). Cytotoxicity produced by the oxidation of 4-hydroxyanisole by tyrosinase. In *Hydroxyanisole: Recent advances in anti-melanoma therapy* (ed P.A. Riley), pp. 101–10. IRL Press, Oxford.

Holick, M.F., MacLaughlin, J.A., Parrish, J.A., and Anderson, R.R. (1982). The photochemistry and photobiology of vitamin D3. In *The science of photomedicine* (ed. J.A. Regan and J.A. Parrish), pp. 195–218. Plenum Press, New York.

Holmes, D.E., Ingalls, R.B., and Myers, L.S. (1967). ESR spectra of free radicals induced in nucleic acid purine and pyrimidine bases and selected analogues by exposure to hydrogen atoms. *Int. J. Radiat. Biol.*, **13**, 225–34.

Holroyd, R.A. (1966). Radical formation in the radiolysis of branched hydrocarbons. *J. Am. Chem. Soc.*, **88**, 5381–6.

Holroyd, R.A. (1987). The electron: its properties and reactions. In *Radiation Chemistry* (ed. Farhataziz and M.A.J. Rodgers), pp. 201–35. VCH Publishers, New York.

Holroyd, R.A. and Klein, G.W. (1965). Radical yields in the radiolysis of unsaturated hydrocarbons. *J. Phys. Chem.*, **69**, 194–202.

Holtom, G.R. and McLain, W.M. (1976). Two-photon excitation spectra of the low energy excited state of diphenylhexatriene and diphenyloctatetraene. *Chem. Phys. Lett.*, **44**, 436–9.

Holten, D., Gouterman, M., Parson, W.W., Windsor, M.W., and Rockley, M.G. (1976). Electron transfer from photoexcited singlet and triplet bacteriopheophytin. *Photochem. Photobiol.*, **23**, 415–23.

Honig, B. and Ebrey, T.G. (1974). Structure and spectra of the chromophore of the visual pigments. *Ann. Rev. Biophys. Bioeng.*, **3**, 151–77.

Hore, P. (1985). Flash photolysis NMR. In *Primary photo-processes in biology and medicine*, NATO, ASI series A, Vol. 85 (ed. R.V. Bensasson, G. Jori, E.J. Land, and T.G. Truscott), pp. 111–22. Plenum Press, New York.

Hore, P.J., Joslin, C.G., and MacLauchlan, K.A. (1979*a*). *Electron Spin Resonance*, Specialist Periodical Reports, Vol. 5 (ed. P.B. Ayscough), pp. 1–45. The Chemical Society, London.

Hore, P.J., Joslin, C.G., and MacLauchlan, K.A., (1979*b*). The role of chemically-induced dynamic electron polarization (CIDEP) in chemistry. *Chem. Soc. Rev.*, **8**, 29–61.

Horie, T. and Vanderkooi, J.M. (1982). Phosphorescence of tryptophan from parvalbumin and actin in liquid solution. *FEBS Lett.*, **147**, 69–73.

Horrocks, A.R. and Wilkinson, F. (1968). Triplet state formation efficiencies of aromatic hydrocarbons in solution. *Proc. R. Soc.*, **A306**, 257–73.

Houée-Levin, C., Gardes-Albert, M., Ferradini, C., Faraggi, M., and Klapper, M. (1985). Pulse radiolysis study of daunorubicin redox cycles. Reduction by e_{aq}^- and $COO^{\cdot -}$ free radicals. *FEBS Lett.*, **179**, 46–50.

Houée-Levin, C., Gardes-Albert, M., and Ferradini, C. (1986). Pulse radiolysis study of daunorubicin redox reactions; redox cycles or glycosidic cleavage? *J. Free Rad. Biol. Med.*, **2**, 89–97.

Howard, J.A. and Ingold, K.U. (1967). Absolute rate constants for hydrocarbon autoxidation V. The hydroperoxy radical in chain propagation and termination. *Can. J. Chem.*, **45**, 785–92.

Howard, J.A. and Ingold, K.U. (1968). The self-reaction of *sec*-butylperoxy radicals. Confirmation of the Russell mechanism. *J. Am. Chem. Soc.*, **90**, 1056–8.

Hudson, B.S. and Kohler, B.E. (1972). A low-lying weak transition in the polyene α,ω-diphenyloctatetraene. *Chem. Phys. Lett.*, **14**, 299–304.

Hudson, B.S. and Kohler, B.E. (1974). Linear polyene electronic structure and spectroscopy *Ann. Rev. Phys. Chem.*, **25**, 437–60.

Hudson, B.S., Kohler, B.E., and Shulta, K. (1982). Linear polyene electronic structure and potential surfaces in excited states, Vol. 6 (ed. E.C. Lin). Academic Press, New York.

Hug, G.L. (1981). Optical spectra of nonmetallic inorganic transient species in aqueous solution. *Nat. Stand Ref. Data Ser. Nat. Bur. Stand. (USA)*, **61**, 1–159.

Huie, R.E., Brault, D., and Neta, P. (1987). Rate constants for one-electron oxidation by the $CF_3O_2^{\cdot}$, $CCl_3O_2^{\cdot}$ and $CBr_3O_2^{\cdot}$ radicals in aqueous solutions. *Chem. Biol. Interactions*, **62**, 227–35.

Hünerbein, M., Stern, J., Freidrich, E.A., Sinn, H., Grashew, G., Maier-Borst, W., and Schlag, P. (1991). The preferential accumulation of macromolecular bound photosensitizers in tumors. *Antibody, Immunoconjugates and Radiopharmaceuticals*, **4**, 483–91.

Huppert, D., Rentzepis, P.M., and Tollin, G. (1976). Picosecond kinetics of chlorophyll and chlorophyll/quinone solutions in ethanol. *Biochim. Biophys. Acta*, **440**, 356–64.

Hurley, J.K., Sinai, N., and Linschitz, H. (1983). Actinometry in monochromatic flash photolysis: The extinction coefficient of triplet benzophenone and quantum yield of triplet zinc tetraphenyl porphyrin. *Photochem. Photobiol.*, **38**, 9–14.

Hurst, J.R. and Schuster, G.B. (1983). Non radiative relaxation of singlet oxygen in solution. *J. Am. Chem. Soc.*, **105**, 5756–60.

Hush, N.S. (1982). Parameters of electron-transfer kinetics. In *Mechanistic aspects of inorganic reactions*, ACS Symposium series 198 (ed. D.B. Rorabacker and J.F. Endicott), pp. 301–29. American Chemical Society, Washington, DC.

Hush, N.S. and Cheung, A.S. (1975). Ionization potentials and donor properties of nucleic acid bases and related compounds. *Chem. Phys. Lett.* **34**, 11–13.

Hutchinson, F. (1985). Chemical changes induced in DNA by ionizing radiation. *Prog. Nucleic Acid Res. Mol. Biol.*, **32**, 115–54.

Ilan, Y.A., Meisel, D. and Czapski, G. (1974). The redox potential of the O_2–$O_2^{\cdot -}$. species system in aqueous media. *Isr. J. Chem.*, **12**, 891–5.

Ilan, Y.A., Czapski, G., and Meisel, D. (1976). The one-electron transfer redox potentials of free radicals. I. The oxygen/superoxide system. *Biochim. Biophys. Acta*, **430**, 209–24.

Infelta, P.P., de Haas, M.P., and Warman, J.M. (1977). The study of the transient conductivity of pulse irradiated dielectric liquids on a nanosecond timescale using microwaves. *Radiat. Phys. Chem.*, **10**, 353–65.

Ingold, K.U. (1969). Peroxy radicals. *Acc. Chem. Res.*, **2**, 1–9.

Ippen, H. (1969). Topical agents for protection against ultraviolet radiation. In *The biologic effects of ultraviolet radiation* (ed. F. Urbach), pp. 681–7. Pergamon Press, Oxford.

Isaacs, S.T., Shen, C.-K.J., Hearst, J.E., and Rapoport, H. (1977). Synthesis and characterization of new psoralen derivatives with superior photoreactivity with DNA and RNA. *Biochemistry*, **16**, 1058–64.

d'Ischia, M. and Prota, G. (1987). Photooxidation of 5,6-dihydroxy-1-methyl-indole. *Tetrahedron*, **43**, 431–4.

Isied, S.S., Ogawa, M.Y., and Wishart, J.F. (1992). Peptide-mediated intramolecular electron transfer: long range distance dependence. *Chem. Rev.*, **92**, 381–94.

Isler, O. (ed.). Gutmann, H., and Salmis, U. (co-ed.) (1971). *Carotenoids*. Birkhauser Verlag, Basel and Stuttgart.

Ito, S. (1986). Re-examination of the structure of eumelanin. *Biochim. Biophys. Acta*, **883**, 155–61.

Ito, S. and Fujita, K. (1984). Oxygen-dependent conjugation of dopa with cysteine catalysed by iron–EDTA complex. *Biochem. Pharmacol.*, **33**, 2193–7.

Iu, K.-K. and Ogilby, P.R. (1987). A time-resolved study of singlet molecular oxygen ($^1\Delta_g O_2$) formation in a solution-phase photosensitised reaction: a new experimental technique to examine the dynamics of quenching by oxygen. *J. Phys. Chem.*, **91**, 1611–17.

Iwaoka, T. and Kondo, M. (1974). Mechanistic studies on the photooxidation of chlorpromazine in water and ethanol. *Bull. Chem. Soc. Japan*, **47**, 980–6.

Iwaoka, T. and Kondo, M. (1977). Photoionization of chloropromazine hydrochloride in binary mixed solvent systems. *Bull. Chem. Soc. Japan*, **50**, 1–5.

Jabben, M., Garcia, N.A., Braslavsky, S.E., and Schaffner, K. (1986). Photophysical parameters of chlorophylls *a* and *b*: Fluorescence and laser-induced optoacoustic measurements. *Photochem. Photobiol.*, **43**, 127–31.

Jablonski, A. (1933). Efficiency of anti-Stokes fluorescence. *Nature*, **131**, 839–40.

Jablonski, A. (1935). Uber den mechanismus der photolumineszenz von farbstoff-phosphoren. *Z. Physik.*, **94**, 38–46.

Jagger, J. (1976). Ultraviolet inactivation of biological systems. In *Photochemistry and photobiology of nucleic acids*, Vol. II *Chemistry* (ed. S.Y. Wang), pp. 147–86. Academic Press, New York.

Jagger, J. (1983). Physiological effects of near ultraviolet radiation on bacteria. In *Photochemical and photobiological reviews*, Vol. 7 (ed. K.C. Smith), pp. 1–75. Plenum, New York.

Janata, E. and Schuler, R.H. (1980). Direct observation of the protonation of acetone ketyl radical by conductometric pulse radiolysis. *J. Phys. Chem.*, **84**, 3351–3.

Janzen, E.G. (1980). A critical review of spin trapping in biological systems. In *Free radicals in biology*, Vol. IV (ed. W.A. Pryor), pp. 115–54. Academic Press, New York.

Janzen, E.G. and Bolton, J.R. (eds) (1990). Spin-trapping and aminoxyl radical chemistry. *Free Rad. Res. Commun.*, **9**, 159–408 and **10**, 1–136.

Jay-Gerin, J.-P. and Ferradini, C. (1991). Electron solvation in polar liquids. In *Excess electrons in dielectric media* (ed. C. Ferradini and J.-P. Jay-Gerin), pp. 259–85. CRC Press, Boca Raton.

Jensen, N.-H., Wilbrandt, R., Houée-Levin, C., and Bensasson, R.V. (1982). Isomerization and structure of retinals in the triplet state. In *Primary photo-processes in biology and medicine* (ed. R.V. Bensasson, G. Jori, E.J. Land, and T.G. Truscott), NATO ASI Series A: Life Sciences **85**, pp. 105–9. Plenum Press, New York.

Jensen, N.-H., Wilbrandt, R., and Bensasson, R.V. (1989). Sensitized photoisomerization of all-*trans*- and 11-*cis*-retinal. *J. Am. Chem. Soc.*, **111**, 7877–88.

Jerumanis, S. and Martel, A. (1970). Réactions photochimiques de la caféine avec les éthers. *Can. J. Chem.*, **48**, 1716–21.

Johnson, B.E. (1988). Skin and related photoreactions. In *Photosensitisation: Molecular, cellular and medical aspects*. (ed. G. Moreno, R.H. Pottier, and T.G. Truscott), pp. 253–64. Springer-Verlag, Berlin.

Johnson, B.E. and Daniels, Jr, F. (1969). Lysosomes and the reactions of skin to ultraviolet radiation. *J. Invert. Dermatol.* **53**, 85–94.

Johnson, D.W. and Salmon, G.A. (1977). Pulse radiolysis of methanolic solutions of tetranitromethane. *J. Chem. Soc., Faraday Trans. I*, **73**, 256–62.

Johnston, B.H. and Hearst, J.E. (1981). Low-level, psoralen–deoxyribonucleic acid cross-links induced by single laser pulses. *Biochemistry*, **20**, 739–45.

Johnston, B.H., Johnson, M.A., Moore, C.B., and Hearst, J.E. (1977). Psoralen–DNA photoreaction: controlled production of mono- and diadducts with nanosecond ultraviolet laser pulses. *Science*, **197**, 906–8.

Johnston, B.H., Kung, A.H., Moore, C.B., and Hearst, J.E. (1981). Kinetics of formation of deoxyribonucleic acid cross-links by 4′-(aminomethyl)-4,5′8-trimethylpsoralen. *Biochemistry*, **20**, 735–8.

Jonah, C.D. (1975). A wide-time range pulse radiolysis system of picosecond time resolution. *Rev. Sci. Instrum.*, **46**, 62–6.

Jones, R.A., Parsons, B.J., Navaratnam, S., and Phillips, G.O. (1985). The photodegradation of azapropazone: Its relevance to photodermatology. In *Primary photo-processes in biology and medicine* (ed. R.V. Bensasson, G. Jori, E.J. Land,

and T. G. Truscott), pp. 61–4. Plenum Press, New York.

Jones, R. A., Navaratnam, S., Parsons, B. J., and Phillips, G. O. (1988). One-electron oxidation and reduction of azapropazone and phenylbutazone derivatives in aqueous solutions: A pulse radiolysis study. *Photochem. Photobiol.*, **48**, 401–8.

Jonsson, P.-E., Agrup, G., Arnbjornsson, E., Hafstrom, Lo., and Rorsman, H. (1980). Treatment of malignant melanoma with Dacarbazin (DTIC-DOME) with special reference to urinary excretion of 5-*S*-cysteinyldopa. *Cancer*, **45**, 245–8.

Jori, G. (1992). Far-red absorbing photosensitizers: their use in the photodynamic therapy of tumours. *J. Photochem. Photobiol., A: Chem.*, **62**, 371–8.

Jori, G. and Reddi, E. (1990). Strategies for tumour targeting by photodynamic sensitizers. In *Photodynamic therapy of neoplastic disease*, Vol. II (ed. D. Kessel), pp. 117–30. CRC Press, Boca Raton, Florida.

Jori, G. and Spikes, J. D. (1984). Photobiochemistry of porphyrins. In *Topics in photomedicine* (ed. K. C. Smith), pp. 183–318. Plenum Press, New York.

Jori, G., Galaiazzo, G., and Scoffone, E. (1969). Photodynamic action of porphyrins on amino acids and proteins. I. Selective photooxidation of methionine in aqueous solution. *Biochemistry*, **8**, 2868–75.

Jorns, M. S., Baldwin, E. T., Sancar, G. B., and Sancar, A. (1987). Action mechanism of *Escherichia coli* DNA photolyase. II Role of the chromophores in catalysis. *J. Biol. Chem.*, **262**, 486–91.

Jovanovic, S. V., Jankovic, I., and Josimovic, L. (1992). Electron-transfer reactions of alkyl peroxy radicals. *J. Am. Chem. Soc.*, **114**, 9018–21.

Jovanovic, S. V. and Simic, M. G. (1985*a*). Repair of tryptophan radicals by antioxidants. *J. Free Rad. Biol. Med.*, **1**, 125–9.

Jovanovic, S. V. and Simic, M. G. (1985*b*). Tryptophan metabolites as antioxidants. *Life Chemistry Reports*, **3**, 124–30.

Jovanovic, S. V. and Simic, M. G. (1986). One-electron redox potentials of purines and pyrimidines. *J. Phys. Chem.*, **90**, 974–8.

Jovanovic, S. V. and Simic, M. G. (1988). Redox properties of oxy and antioxidant radicals. In *Oxygen radicals in biology and medicine* (ed. M. G. Simic, K. A. Taylor, J. F. Ward, and C. von Sonntag), pp. 115–23. Plenum Press, New York.

Jovanovic, S. V. and Simic, M. G. (1989). The DNA guanyl radical: kinetic and mechanisms of generation and repair. *Biochim. Biophys. Acta*, **1008**, 39–44.

Kai, Y. and Imakubo, K. (1979). Temperature dependence on the phosphorescence lifetimes of heterogeneous tryptophan residues in globular proteins between 293 and 77 K. *Photochem. Photobiol.*, **29**, 261–5.

Kakitani, T. and Mataga, N. (1987). Comprehensive study on the role of coordinated solvent mode played in electron transfer reactions in polar solutions. *J. Phys. Chem.*, **91**, 6277–85.

Kaklika, K. A., Kissel, T., Thayer, A. L., Burns P. A., and Schaap, A. P. (1979). Mechanisms of 1,2 dioxetane decomposition: The role of electron transfer. *Photochem. Photobiol.*, **30**, 35–44.

Kalyanasundaram, K. (1987). *Photochemistry of microheterogeneous systems.* Academic Press, New York.

Kanne, D., Straub, K., Rapoport, H., and Hearst, J. E. (1982). Psoralen-deoxyribonucleic acid photoreaction. Characterization of the monoaddition products from 8-methoxypsoralen and 4,5′,8-trimethylpsoralen. *Biochemistry*, **21**, 861–71.

Kanner, J., Harel, S., and Granit, R. (1991). Nitric oxide as an antioxidant. *Arch. Biochem. Biophys.*, **289**, 130–6.

Kanofsky, J. R. (1984). Singlet oxygen production by lactoperoxydase; halide dependence and quantitation of yield. *J. Photochem.*, **25**, 105–13.

Kanofsky, J. R. (1989). Singlet oxygen production by biological systems. *Chem. Biol. Interactions*, **70**, 1–28.

Kanofsky, J.R., Hoogland, H., Wever, R. and Weiss, S.J. (1988). Singlet oxygen production by human eosinophils. *J. Biol. Chem.*, **263**, 9692–6.

Kappus, H. (1985). Lipid peroxidation: Mechanism, analysis, enzymology and biological relevance. In *Oxidative stress* (ed. H. Sies), pp. 284–90. Academic Press, London.

Karam, L.R., Bergtold, D.S. and Simic, M.G. (1991). Biomarkers of OH radical damage *in vivo*. *Free Rad. Res. Comm.*, **12–13**, 11–16.

Karam, L.R., Dizdaroglu, M., and Simic, M.G. (1984). OH radical-induced products of tyrosine peptides. *Int. J. Radiat. Biol.*, **46**, 715–24.

Karas, J.L., Leiber, C.M., and Gray, H.B. (1988). Free energy dependence of the rate of long-range electron transfer in proteins. Reorganizational energy in ruthenium-modified myoglobin. *J. Am. Chem. Soc.*, **110**, 599–600.

Kasha, M. (1950). Characterization of electronic transitions in complex molecules. *Disc. Faraday Soc.*, **9**, 14–19.

Katahira, M., Sakaguchi-Katahira, R., Hayashi, F., Uesugi, S., and Kyogoku, Y. (1991). Photochemically induced dynamic nuclear polarization studies of oligonucleotide duplexes. *J. Am. Chem. Soc.*, **113**, 8647–51.

Kavarnos, G.J. and Turro, N.J. (1986). Photosensitisation by reversible electron transfer: theories, experimental evidence and examples. *Chem. Rev.*, **86**, 401–49.

Kawanishi, S., Inoue, S., Sano, S., and Aiba, H. (1986). Photodynamic guanine modification by hematoporphyrin is specific for single-stranded DNA with singlet oxygen as mediator. *J. Biol. Chem.*, **261**, 6090–5.

Kearns, D.R. (1971). Physical and chemical properties of singlet molecular oxygen. *Chem. Rev.*, **71**, 395–427.

Keene, J.P. (1960). Kinetics of radiation-induced chemical reactions. *Nature*, **188**, 843–4.

Keene, J.P. (1963). Optical absorptions in irradiated water. *Nature*, **197**, 47–8.

Keene, J.P., Kessel, D., Land, E.J., Redmond, R.W., and Truscott, T.G. (1986). Direct detection of singlet oxygen sensitized by haematoporphyrin and related compounds. *Photochem. Photobiol.*, **43**, 117–20.

Keir, W.F., Land, E.J., MacLennan, A.H., McGarvey, D.J., and Truscott, T.G. (1987). Pulsed radiation studies of photodynamic sensitisers: The nature of DHE. *Photochem. Photobiol.*, **46**, 587–9.

Kellogg, R.E. (1969). Mechanism for chemiluminescence from peroxy radicals *J. Am. Chem. Soc.*, **91**, 5433–6.

Kemp, T.J. and Martins, L.J.A. (1981). Triplet state of 5-nitro-2-furoic acid by laser flash photolysis: Spectrum, lifetime and reactivity. *J. Chem. Soc., Faraday Trans. I*, **77**, 1425–35.

Kemp, T.J., Parker, A.W., and Wardman, P. (1985). Electron-transfer quenching of triplet state thymine and uracil. *J. Chem. Soc., Chem. Commun.*, 1377–8.

Kemp, T.J., Parker, A.W., and Wardman, P. (1987*a*). Interaction of triplet-state nucleic acid bases with electroaffinic molecules in solution by laser flash photolysis. *J. Chem. Soc., Perkin Trans. II*, 397–403.

Kemp, T.J., Parker, A.W., and Wardman, P. (1987*b*). Radiosensitising nitroimidazole drugs: triplet state energy and reactivity. *Photochem. Photobiol.*, **45**, 663–6.

Kennedy, J.C., Pottier, R.H., and Pross, D.C. (1990). Photodynamic therapy with endogenous protoporphyrin IX; Basic principles and present clinical experience. *J. Photochem. Photobiol., B: Biol.*, **6**, 143–8.

Kennedy, K.A., McGurl, J.D., Leondaridis, L., and Alabaster, O. (1985), pH Dependence of mitomycin C-induced cross-linking activity in EMT6 tumor cells. *Cancer Res.*, **45**, 3541–7.

Kessel, D. (1989). Determinants of photosensitization by purpurins. *Photochem. Photobiol.*, **50**, 169–74.

Kessel, D., Thompson, P., Saatio, K., and Nantwi, K. (1987). Tumour localization and photosensitization by sulphonated derivatives of tetraphenylporphine. *Photochem. Photobiol.*, **45**, 787–90.

Kessler, R.W. and Wilkinson, F. (1981). Diffuse reflectance triplet–triplet absorption spectroscopy of aromatic hydrocarbons chemisorbed on γ-alumina. *J. Chem. Soc., Faraday Trans. I*, **77**, 309–20.

Khan, A.U. (1984). Discovery of enzyme generation of $^1\Delta_g$ molecular oxygen: Spectra of (0, 0) $^1\Delta_g \rightarrow {}^3\Sigma_g$-IR emission. *J. Photochem.*, **25**, 327–34.

Khan, A.U. (1985). Singlet molecular oxygen spectroscopy: chemical and photosensitized. In *Singlet O_2* (ed. A.A. Frimer), pp. 39–79. CRC Press, Boca Raton, Florida.

Khoroshilova, E.V., Repeyev, Y.A., and Nikogosyan, D.N. (1990). UV photolysis of aromatic amino acids and related dipeptides and tripeptides. *J. Photochem. Photobiol. B: Biol.*, **7**, 159–72.

Kiefer, J. (1985). Cellular and subcellular effects of very heavy ions. *Int. J. Radiat. Biol.*, **38**, 267–75.

Kim, S.-T., Hartman, R.F., and Rose, S.D. (1990). Solvent dependence of pyrimidine dimer splitting in a covalently linked dimer-indole system. *Photochem. Photobiol.*, **52**, 789–94.

Kimel, S., Tromberg, B.J., Roberts, W.G., and Berns, M.W. (1989). Singlet oxygen generation of porphyrins, chlorins and phthalocyanines. *Photochem. Photobiol.*, **50**, 175–83.

King, P.A., Anderson, V.E., Edwards, J.O., Gustafson, G., Plumb, R.C., and Suggs, J.W. (1992). A stable solid that generates hydroxyl radical upon dissolution in aqueous solutions: reaction with proteins and nucleic acid. *J. Am. Chem. Soc.*, **114**, 5430–2.

Kinoshita, S., Seki, T., Fu Liu, T., and Kushida, T. (1988). Fluorescence of haematoporphyrin in living cells and in solution. *J. Photochem. Photobiol., B: Biol.*, **2**, 195–208.

Kira, A. and Thomas, J.K. (1974). Equilibria between triplet states of aromatic hydrocarbons. *J. Phys. Chem.*, **78**, 196–9.

Kittler, L. and Löber, G. (1977). Photochemistry of the nucleic acids. In *Photochemical and photobiological reviews*, Vol. 2 (ed. K.C. Smith), pp. 39–131. Plenum Press, New York.

Kittler, L., Löber, G., Gollnick, F.A., and Berg, H. (1980). Redox processes during photodynamic damage III. Redox mechanism of photosensitization and radical reaction. *Bioelectrochem. Bioenerg.*, **7**, 503–11.

Klapper, M.H. and Faraggi, M. (1979). Applications of pulse radiolysis in protein chemistry. *Q. Rev-Biophy.*, **12**, 465–519.

Klein, R., Tatischeff, I., Bazin, M., and Santus, R. (1981). Photophysics of indole. Comparative study of quenching, solvent, and temperature effects by laser flash photolysis and fluorescence. *J. Phys. Chem.*, **85**, 670–7.

Kleyer, D.L. and Koch, T.H. (1983). Spectroscopic observation of the tautomer of 7-deoxydaunomycinone from elimination of daunosamine from daunomycin hydroquinone. *J. Am. Chem. Soc.*, **105**, 2504–5.

Kliger, D.S. (1989). New approaches in laser photolysis measurements. *Acc. Chem. Res.*, **22**, 178–83.

Kliger, D.S. and Albrecht, A.C. (1970). Polarized spectroscopy of excited states of substituted anthracenes on a nanosecond time scale. *J. Chem. Phys.*, **53**, 4059–65.

Kligman, A.M. and Kaidbey, K.H. (1982). Phototoxicity to benoxaprofen. *Eur. J. Rheumat. Inflamm.*, **5**, 124–37.

Klug-Roth, D. and Rabani, J. (1976). Pulse radiolysis studies on reactions of aqueous superoxide radicals with copper (II) complexes. *J. Phys. Chem.*, **80**, 588–91.

Knox, C.N., Land, E.J., and Truscott, T.G. (1986). Singlet oxygen generation by furocoumarin triplet states. I. Linear furocoumarins (psoralens). *Photochem. Photobiol.*, **43**, 359–63.

Knox, C.N., Land, E.J., and Truscott, T.G. (1988). Triplet state properties and triplet-state-oxygen interactions of some linear and angular furocoumarins. *J. Photochem. Photobiol. B: Biol.*, **1**, 315–21.

Kobayashi, H., Ueda, T., Kobayashi, T., Tagawa, S., and Tabata, Y. (1981). Performance and improvements of an electron accelerator producing a picosecond single electron pulse. *Nucl. Inst. Meth.*, **179**, 223–8.

Kobayashi, T. (1982). Picosecond laser spectroscopy. *Meth. Exp. Phys.* **20**, 163–95.

Koch, W.H. and Chedekel, M.R. (1986). Photoinitiated DNA damage by melanogenic intermediates *in vitro*. *Photochem. Photobiol.*, **44**, 703–10.

Kochevar, I.E., Morison, W.L., Lamm, J.L., McAuliffe, D.J., Western, A., and Hood, A.F. (1986). Possible mechanism of piroxicam-induced photosensitivity. *Arch. Dermatol.*, **122**, 1283–7.

Kollias, N., Sayre, R.M., Zeise, L., and Chedekel, M.R. (1991). Photoprotection by melanin. *J. Photochem. Photobiol., B: Biol.*, **9**, 135–60.

König, K., Schneckenburger, H., Rück, A. and Steiner, R. (1993). *In vivo* photoproduct formation during PDT with ALA-induced endogenous porphyrins. *J. Photochem. Photobiol. B: Biol.*, in press.

Koo, J.Y. and Schuster, G.B. (1978). Chemiluminescence of diphenoyl peroxide. Chemically initiated electron exchange luminescence. A new general mechanism for chemical production of electronically excited states. *J. Am. Chem. Soc.*, **100**, 4496–503.

Koo, J.Y., Schmidt, S.P., and Schuster, G.B. (1978). Bioluminescence of the firefly: key steps in the formation of the electronically excited state for model systems. *Proc. Natl. Acad. Sci. USA*, **75**, 30–3.

Koppenol, W.H. (1978). Free energies of some interconversion reactions. *Photochem. Photobiol.*, **28**, 431–3.

Koppenol, W.H. (1989). Generation and thermodynamic properties of oxyradicals. In *CRC Critical Reviews in Membrane Lipid Oxidation*, Vol. 1, pp. 1–13. CRC Press, Boca Raton.

Koppenol, W.H. (1990*a*). What is in a name? Rules for radicals. *Free Rad. Biol. Med.*, **9**, 225–7.

Koppenol, W.H. (1990*b*). Oxyradical reactions: from bond-dissociation energies to reduction potentials. *FEBS Lett.*, **264**, 165–7.

Koppenol, W.H. and Butler, J. (1977). Mechanism of reactions involving singlet oxygen and the superoxide anion. *FEBS Lett.*, **83**, 1–6.

Koppenol, W.H. and Liebman, J.F. (1984). The oxidising nature of the hydroxyl radical. A comparison with the ferryl ion (FeO^{2+}). *J. Phys. Chem.*, **88**, 99–101.

Koppenol, W.H., Butler, J., and van Leeuwen, J.W. (1978). The Haber–Weiss cycle. *Photochem. Photobiol.*, **28**, 655–60.

Kornhauser, A. (1976). UV-induced DNA protein crosslinks *in vitro* and *in vivo*. *Photochem. Photobiol.*, **23**, 457–60.

Kosonocky, W.F., Harrison, S.E., and Stander, R. (1965). Observations of the triplet state in phthalocyanines. *J. Chem. Phys.*, **43**, 831–3.

Koyama, Y. (1991). Structure and functions of carotenoids in photosynthetic systems. *J. Photochem. Photobiol., B: Biol.*, **9**, 265–80.

Koyama, Y., Takii, T., Saiki, K., and Tsukida, K. (1983). Configuration of the carotenoid in the reaction centers of photosynthetic bacteria—2. Comparison of the resonance Raman lines of the reaction centers with those of 14 different *cis–trans* isomers of β-carotene. *Photobiochem. Photobiophys.*, **5**, 139–50.

Koyama, Y., Takatsuka, I., Kanaji, M., Tomimoto, K., Kito, M., Shimamura, T.,

Yamashita, J., Saiki, K., and Tsukida, K. (1990). Configurations of carotenoids in the reaction center and the light-harvesting complex of *Rhodospirillum rubrum*. Natural selection of carotenoid configurations by pigment protein complexes. *Photochem. Photobiol.*, **51**, 119–28.

Kraljic, I. and El Moshni, S. (1978). A new method for the detection of singlet oxygen in aqueous solutions. *Photochem. Photobiol.*, **28**, 577–81.

Krasnovskii, A.A., Neverov, Jr, K.V., Egorov, S.Yu., and Recter, B. (1988). Photophysical parameters of pheophorbide *a*: Phosphorescence and generation of singlet oxygen. *Opt. Spectros. (USSR)*, **64**, 471–4.

Krasnovsky, A.A., Sukhorukov, V.L., and Potapenko, A.Ya. (1983). Photogeneration of singlet oxygen by psoralens. *Bull. Exp. Biol. Med. (USSR)*, **96**, 59–61.

Krasnovsky, A.A., Sukhorukov, V.L., Egorov, S.Yu., and Potapenko, A.Ya. (1986). Generation and quenching of singlet molecular oxygen by furocoumarins. Direct measurements. *Studia Biophysica*, **114**, 149–58.

Krinsky, N.I. and Deneke, S.M. (1982). Interaction of oxygen and oxy-radicals with carotenoids. *J. Natl. Cancer Inst.*, **69**, 205–9.

Kristiansen, M., Scurlock, R.D., Iu, K.-K., and Ogilby, P.R. (1991). Charge transfer state and singlet oxygen ($^1\Delta_g O_2$). *J. Phys. Chem.*, **95**, 5190–7.

Kuki, A. (1991). Electron tunneling paths in proteins. In *Long-range electron transfer in biology* (ed. G. Palmer), pp. 49–83. Springer-Verlag, Berlin.

Kuppusamy, P. and Zweier, J.L. (1989). Characterization of free radical generation by xanthine oxidase. Evidence for hydroxyl radical generation. *J. Biol. Chem.*, **264**, 9880–4.

Kwan, M., Niinikoski, J., and Hunt, T.K. (1971). *In vivo* measurements of oxygen tension in the cornea, aqueous humor, and anterior lens of the open eye. *Invest. Ophtal.*, **11**, 108–14.

Lachish, U., Schafferman, A., and Stein, G. (1976). Intensity dependence in laser flash photolysis experiments: hydrated electron formation from ferrocyanide, tyrosine, and tryptophan. *J. Chem. Phys.*, **64**, 4205–11.

Ladenberg, E. (1908). On the initial velocity and number of photoelectric electrons produced by light of different wavelengths. *Phys. Z.*, **8**, 590–4.

Laderoute, K., Wardman, P., and Rauth, A.M. (1988). Molecular mechanisms for the hypoxia-dependent activation of 3-amino-1,2,4-benzotriazine-1,4-dioxide (SR 4233). *Biochem. Pharmacol.*, **37**, 1487–95.

Lafferty, J., Roach, A.C., Sinclair, R.S., Truscott, T.G., and Land, E.J. (1977). Absorption spectra of radical ions of polyenes of biological interest. *J. Chem. Soc., Faraday Trans. I*, **73**, 416–29.

Lafferty, J., Truscott, T.G., and Land, E.J. (1978). Electron transfer reactions involving chlorophylls *a* and *b* and carotenoids. *J. Chem. Soc., Faraday Trans. I*, **74**, 2760–2.

Lafleur, M.V.M., Nieuwint, A.W.R., Aubry, J.M., Kortbeek, H., Arwert, F., and Joenje, H. (1987). DNA damage by chemically generated singlet oxygen. *Free Rad. Res. Commun.*, **2**, 343–50.

Lai, T.-I., Liu, B.T., and Lim, E.C. (1982). Photophysical properties of biologically important molecules related to proximity effects: psoralens. *J. Am. Chem. Soc.*, **104**, 7631–5.

Lakowicz, J.R. (1981). Fluorescence spectroscopic investigations of the dynamic properties of biological membranes. In *Spectroscopy in biochemistry*, Vol. 1 (ed. J. Ellis Bell), pp. 195–245. CRC Press, Boca Raton.

Lakowicz, J.R. (1983). *Principles of fluorescence spectroscopy*. Plenum, New York.

Lakowicz, J.R., Laczko, G., Gryczynski, I., Szmacinski, H., and Wiczk, W. (1988). Gigahertz frequency-domain fluorometry: resolution of complex decays, picosecond processes and future developments. *J. Photochem. Photobiol. B: Biol.*, **2**, 295–311.

Lambert, C., Rodgers, M.A.J., Jori, G., Reddi, E., and Spikes, J.D. (1985). The photophysics of some liposome-bound porphyrins. In *Primary photoprocesses in biology and medicine* (ed. R. Bensasson, G. Jori, E.J. Land, and T.G. Truscott), pp. 325–8. Plenum, New York.

Lambert, C., Chacon, J.N., Chedekel, M.R., Land, E.J., Riley, P.A., Thompson, A., and Truscott, T.G. (1989). A pulse radiolysis investigation of the oxidation of indolic melanin precursors: Evidence for indolequinones and subsequent intermediates. *Biochim. Biophys. Acta*, **993**, 12–20.

Lambert, C., Sarna, T., and Truscott, T.G. (1990*a*). Rose Bengal radicals and their reactivity. *J. Chem. Soc., Faraday Trans.*, **86**, 3879–82.

Lambert, C., Land, E.J., Riley, P.A., and Truscott, T.G. (1990*b*). A pulse radiolysis investigation of the oxidation of methoxylated metabolites of indolic melanin precursors. *Biochim. Biophys. Acta*, **1035**, 319–24.

Lambert, C., Truscott, T.G., Land, E.J., and Riley, P.A. (1991). The role of azide concentration in pulse radiolysis studies of oxidation: 3,4-dihydroxyphenylalanine. *J. Chem. Soc., Faraday, Trans.*, **87**, 2939–42.

Lambert, J. (1760). *Photometria siva de mensura et gradibus luminis, colorum et umbrae*. Augusberg.

Lamola, A.A. and Eisinger, J. (1971). Excited states of nucleotides in water at room temperature. *Biochim. Biophys. Acta*, **240**, 313–25.

Lamola, A.A., Yamane, T., and Trozzolo, A.M. (1973). Cholesterol hydroperoxide formation in red cell membranes and photohemolysis in Erythropoietic Protoporphyria. *Science*, **179**, 1131–3.

Lamola, A.A., Flores, J., and Doleiden, F.H. (1982). Quantum yield and equilibrium position of the configurational photoisomerization bilirubin of bound to human serum albumin. *Photochem. Photobiol.*, **35**, 649–54.

Land, E.J. (1968). Extinction coefficients of triplet–triplet transitions. *Proc. R. Soc.*, **A305**, 457–71.

Land, E.J. (1969). Absorption spectrum of the triplet state of duroquinone. *Trans. Faraday Soc.*, **65**, 2815–22.

Land, E.J. (1988). Pulsed irradiation studies of some reactive intermediates of melanogenesis. *Rev. Chem. Intermediates*, **10**, 219–40.

Land, E.J. (1991). Time-resolved luminescence-pulse radiolysis determination of the fractions of porphyrin photosensitizer triplets giving rise to singlet oxygen on quenching by ground state triplet oxygen. *J. Photochem. Photobiol., A: Chem.*, **61**, 165–70.

Land, E.J. and Ebert, M. (1967). Pulse radiolysis studies of aqueous phenol. Water elimination from dihydroxycyclohexadienyl radicals to form phenoxyl. *Trans. Faraday Soc.*, **63**, 1181–90.

Land, E.J. and Porter, G. (1961). A stable anilino radical. *J. Chem. Soc.*, 3540–2.

Land, E.J. and Prütz, W.A. (1979). Reaction of azide radicals with amino acids and proteins. *Int. J. Radiat. Biol.*, **36**, 75–83.

Land, E.J. and Swallow, A.J. (1971). One-electron reactions in biochemical systems as studied by pulse radiolysis. IV Oxidation of dihydronicotinamide-adenine-dinucleotide. *Biochim. Biophys. Acta*, **234**, 34–42.

Land, E.J. and Truscott, T.G. (1979). Triplet excited state of coumarin and 4′5′-dihydropsoralen: Reaction with nucleic acid bases and amino acids. *Photochem. Photobiol.*, **29**, 861–6.

Land, E.J., Porter, G., and Strachan, E. (1961). Primary photochemical processes in aromatic molecules. Part 6—The absorption spectra and acidity constants of phenoxyl radicals. *Trans. Faraday Soc.*, **57**, 1885–93.

Land , E.J., Sykes, A., and Truscott, T.G. (1970). The metastable state of β-carotene excited by pulse radiolysis. *J. Chem. Soc., Chem. Commun.*, 332.

Land, E.J., Sykes, A., and Truscott, T.G. (1971). The '*in vitro*' photochemistry of biological molecules-II. The triplet state of β-carotene and lycopene excited by pulse radiolysis. *Photochem. Photobiol.*, **13**, 311–20.

Land, E.J., Rushton, F.A.P., Beddoes, R.L., Bruce, J.M., Cernik, A.J., Dawson, S.C., and Mills, O.S. (1982). A[2 + 2] photoadduct of 8-methoxypsoralen and thymine: X-ray crystal structure; a model for the reaction of psoralens with DNA in the phototherapy of psoriasis. *J. Chem. Soc., Chem. Commun.*, 22–3.

Land, E.J. Mukherjee, T., Swallow, A.J., and Bruce, J.M. (1983*a*). Reduction of the naphthazarin molecule as studied by pulse radiolysis. 1. Addition of a single electron. *J. Chem. Soc., Faraday Trans. I*, **79**, 391–404.

Land, E.J., Mukherjee, T., Swallow, A.J., and Bruce, J.M. (1983*b*). Reduction of the naphthazarin molecule as studied by pulse radiolysis. 2. Second one-electron step. *J. Chem. Soc., Faraday Trans. I*, **79**, 405–15.

Land, E.J., Mukherjee, T., Swallow, A.J., and Bruce, J.M. (1983*c*). One-electron reduction of adriamycin: properties of the semiquinone. *Arch. Biochem. Biophys.*, **225**, 116–21.

Land, E.J., Mukherjee, T., Swallow, A.J., and Bruce, J.M. (1985). Possible intermediates in the action of adriamycin—A pulse radiolysis study. *Br. J. Cancer*, **51**, 515–23.

Land, E.J., Thompson, A., Truscott, T.G., Subbarao, K.V., and Chedekel, M.R. (1986). Photochemistry of melanin precursors: Dopa, 5-*S*-cysteinyldopa and 2,5-*S*, *S*'-dicysteinyl dopa. *Photochem. Photobiol.*, **44**, 697–702.

Land, E.J., McDonagh, A.F., McGarvey, D.J., and Truscott, T.G. (1988). Photophysical studies of tin(IV)-protoporphyrin: Potential phototoxicity of a chemotherapeutic agent proposed for the prevention of neonatal jaundice. *Proc. Natl. Acad. Sci., USA*, **85**, 5249–53.

Land, E.J., Cooksey, C.J., and Riley, P.A. (1990). Reaction kinetics of 4-methoxy *ortho* benzoquinone in relation to its mechanism of cytotoxicity: A pulse radiolysis study. *Biochem. Pharmacol.*, **39**, 1133–5.

Langkilde, F.W., Jensen, N.-H., Wilbrandt, R., Brouwer, A.M., and Jacobs, H.J.C. (1987*a*). UV absorption and Raman spectra of the ground states and time-resolved resonance Raman spectra of the lowest excited triplet states of the *E* and *Z* isomers of 2,5,-dimethyl-1,3,5-hexatriene. Indication of the non-equilibrium of excited rotamers in the lowest triplet state. *J. Phys. Chem.*, **91**, 1029–40.

Langkilde, F.W., Jensen, N.-H., and Wilbrandt, R. (1987*b*). Time-resolved resonance Raman spectroscopy of the lowest excited states of the *E* and *Z* isomers of 1,3,5-hexatriene. *J. Phys. Chem.*, **91**, 1040–7.

Langlois, R., Ali, H., Brasseur, N., Wagner, J.R., and Van Lier, J.E. (1986). Biological activities of phthalocyanines IV. Type II sensitised oxidation of L-tryptophan and cholesterol by sulphonated metallo phthalocyanines. *Photochem. Photobiol.*, **44**, 117–23.

Larson, S. (1981). Electron transfer in chemical and biological systems, orbital rules for non adiabatic transfer. *J. Am. Chem. Soc.*, **103**, 4034–40.

Lavalette, D. Bensasson, R., Amand, B., and Land, E.J. (1971). On the experimental determination of triplet–triplet extinction coefficients and oscillator strengths. *Chem. Phys. Lett.*, **10**, 331–3.

Lea, J.S., Rushton, F.A.P., Land, E.J., and Swallow, A.J. (1989). The reductive deglycosylation of adriamycin in aqueous medium: a pulse radiolysis study. *Free Rad. Res. Commun.*, **8**, 241–9.

Leach, S. and Migirdicyan, E. (1966). Fluorescence à longue durée de vie. In *Actions chimiques et biologiques des radiations*, Vol. 9 (ed. M. Haïssinsky), pp. 119–86, Masson, Paris.

Lebedeva, L. I., Ostrovskaya, R. M., and Tsimmerman, V. G. (1978). Possible role of protein damage in the origin of UV-induced iso-chromatid breaks. *Genetika*, **14**, 256–66.

Le Doan, T., Perrouault, L., Asseline, U., Thuong, N. T., Rivalle, C., Bisagni, E., and Hélène C. (1991). Recognition and photoinduced cleavage and crosslinking of nucleic acids by oligonucleotides covalently linked to ellipticine. *Antisense Research and Development*, **1**, 43–54 (Mary AnnLiebert, Inc. Publishers).

Lee, P. C. C. and Rodgers, M. A. J, (1987). Laser flash photolysis studies of rose bengal sensitized photodynamic interactions of nucleotides and DNA. *Photochem. Photobiol.*, **45**, 79–86.

Legrand-Poels, S., Vaira, D., Pincemail, J., Van de Vorst, A., and Piette, J. (1990). Activation of human immunodeficiency virus type I by oxidative stress. *Aids Res.*, **12**, 1389–97.

Legros, B., Vandereecken, P., and Soumillon, J.Ph. (1991). Electron transfer photoinduced from naphtholate anions: anion oxidation potentials and use of Marcus free energy relationships. *J. Phys. Chem.*, **95**, 4752–61.

Leismann, H., Scharf, H.-D., Strassburger, W., and Wollmer, A. (1983). Determination of subnanosecond fluorescence decays of chlorobenzene, tryptophan and the benzene–triethylamine exciplex using a nanosecond flash lamp. *J. Photochem.*, **21**, 275–80.

Lenard, P. (1902). Light electric effect. *Ann. Phys.*, **8**, 149–98.

Le Nouen, D., Marko, J., Vermeersch, G., Febvay-Garot, N., Lablache Combier, A., Perrée-Fauvet M., and Gaudemer, A. (1990). Porphyrin–tyrosine interactions: photo-CIDNP and NMR studies. *J. Phys. Org. Chem.*, **3**, 69–76.

Leone, S. R. (1989). Time-resolved FTIR emission studies of molecular photofragmentation. *Acc. Chem. Res.*, **22**, 139–44.

Leonhardt, H. and Weller, A. (1963). Elektronenubertragugreaktionen des ageregIen perylens. *Ber. Bunsenges. Phys. Chem.*, **67**, 791–5.

Le Pecq, J. B (1975). Cationic fluorescent probes of polynucleotides. in *Biochemical fluorescence*: *Concepts*, Vol. 2 (ed. R. F. Chen and H. Edelhoch), pp. 711–36. Marcel Dekker, New York.

Levich, V. G. (1965). Present state of the theory of oxidation–reduction in solution *Adv. Electrochem. Eng.*, **4**, 249–371.

Levine, J. I. (1990). *Medications that increase sensitivity to light. A 1990 listing*. US Department of Health and Human Services, Center for Devices and Radiological Health, HHS Publication FDA, 91–8280.

Lewis, G. N. (1926). The conservation of photons. *Nature*, **118**, 874–5.

Lewis, G. N. and Kasha, M. (1944). Phosphorescence and the triplet state. *J. Am. Chem. Soc.*, **66**, 2100–16.

Lewis, G. N. and Kasha, M. (1945). Phosphorescence in fluid media and the reverse process of singlet triplet absorption. *J. Am. Chem. Soc.*, **67**, 994–1003.

Li, Z. and Galley, W. C. (1989). Evidence for ligand-induced conformational changes in proteins from phosphorescence spectroscopy. *Biophysical J.*, **56**, 353–60.

Li, Z., Lee, W. E., and Galley, W. C. (1989). The distance dependence of the trytophan-disulfide interactions at the triplet level from pulsed phosphorescence studies on a model system. *Biophysical J.*, **56**, 361–7.

Lim, E. C. (1977). Vibronic interactions and luminescence in aromatic molecules with non-binding electrons. In *Excited States*, Vol. 3 (ed E. C. Lim), pp. 305–37. Academic Press, New York.

Lind, S. C. (1921). *The chemical effects of alpha-particles and electrons*. The Chemical Catalogue Company, New York.

Lindenau, D., Hagen, U., and Schnabel, W. (1976). On the dynamics of DNA in aqueous solution. Pulse radiolysis studies using the light scattering detection method. *Radiat. Environ. Biophys.*, **13**, 287–94.

Lindig, B. A. and Rodgers, M. A. J. (1981). Rate parameters for the quenching of singlet oxygen by water-soluble and lipid-soluble substrates in aqueous and micellar systems. *Photochem. Photobiol.*, **33**, 627–34.

Lindqvist, L. (1966). Utilisation d'un laser à emission ultraviolette pulse' L'en photolyse-éclairs: Etude de l'état triplet de l'acridine. *C.R. Acad. Sci. Paris*, **263**, 852–4.

Lindqvist, L., Czochralska, B., and Grigorov, I. (1985). Determination of the mechanism of photoionization of NADH in aqueous solution on laser excitation at 355 nm. *Chem. Phys. Lett.*, **119**, 494–8.

Lindstrom, T. P. and Anders, M. W. (1978). Effects of agents known to alter carbon tetrachloride hepatotoxicity and cytochrome P-450 levels on carbon tetrachloride-stimulated lipid peroxidation and ethene expiration in the intact rat. *Biochem. Pharmacol.*, **27**, 563–7.

Linschitz, H. (1961). Chemiluminescence in porphyrin-catalysed decomposition of peroxides. In *Light and life* (ed. W. D. McElroy and B. Glass), pp. 173–82. Johns Hopkins Press, Baltimore.

Lipke, H., Sugumaran, M., and Henzel, W. (1983). Mechanisms of sclerotization in dipterans. *Adv. Insect Physiol.*, **17**, 1–84.

Lippert, E. (1957). Spektroskopische bestimmung des dipolmomentes aromatischer verbindungen im ersten angeregten singulettzustand. *Z. Elektrochem.*, **61**, 962–75.

Lippert, E. (1961). Umwandlung von elektonenanregungsenergie *Angew. Chem.*, **73**, 695–706.

Lipsky, S. (1981). Ionization and excitation in non-polar organic liquids. *J. Chem. Ed.*, **58**, 93–101.

Lipson, R. L. and Baldes, E. J. (1960). The photodynamic properties of a particular haematoporphyrin derivative. *Arch. Dermatol.*, **82**, 508–16.

Liptay, W. (1974). Dipole moments and polarizabilities of molecules in excited electronic states. In Excited states, Vol. I (ed. E. C. Lim). Academic Press, New York, London.

Little, J. B. (1991). Rôle de la distribution d'énergie dans l'ADN sur la mutagénèse des radiations ionisantes. *J. Chim. Phys.*, **88**, 1077–94.

Little, J. B., Benjamin, M. S., Li, C.-Y., and Yandell, D. W. (1991). Molecular mechanisms of radiation induced mutagenesis in human cells. In *Fundamental and applied radiobiology* (ed. C. B. Seymour, and C. Mothersill), pp. 126–31. Taylor and Francis, London.

Liu, L. F. (1989). DNA topoisomerase poisons as antitumor drugs. *Ann. Rev. Biochem.*, **58**, 351–75.

Lloyd, D. C. (1985). Comment on the paper by Roberts and Holt. *Int. J. Radiat, Biol.*, **48**, 940–2.

Lo, K. K. N., Land, E. J., and Truscott, T. G. (1982). Primary intermediates in the pulsed irradiation of retinoids. *Photochem. Photobiol.*, **36**, 139–45.

Loeb, L. L. (1989). Endogenous carcinogenesis: molecular oncology into the twenty-first century. *Cancer Research*, **49**, 5489–96.

Long, D. A. (1989). Linear and nonlinear Raman effects: the principles. *Chemistry in Britain*, **25**, 592–6.

Long, F. H., Lu, H., and Eisenthal, K. B. (1990). Femtosecond studies of the presolvated electron: An excited state of the solvated electron? *Phys. Rev. Lett.*, **64**, 1469–72.

Longworth, J. W. (1971). Luminescence of polypeptides and proteins. In *Excited states of proteins and nucleic acids* (ed. R. F. Steiner and I. Weinryb), pp. 319–484. Plenum Press, New York.

Longworth, J. W. (1982). Light, colors and origins of spectroscopy. In *The Science of photomedicine* (ed. J. D. Reagan and J. A. Parrish), pp. 21–67. Plenum Press, New York.

Longworth, J. W. (1983). Intrinsic fluorescence of proteins. In *Time-resolved fluorescence spectroscopy in biochemistry and biology* (ed. R. B. Cundall and R. E. Dale) NATO ASI Ser. A: 69, pp. 651–725, Plenum, New York.

Lorberbaum, Z., Sperling, J., and Elad, D. (1976). The selective photo-initiated alkylation of purines in DNA with 2-propylamine. *Photochem. Photobiol.*, **24**, 389–91.

Lown, J. W. (1985). Molecular mechanisms of action of anticancer agents involving free radical intermediates. *Adv. Free. Rad. Biol. Med.*, **1**, 225–64.

Lutz, M. (1977). Antenna chlorophyll in photosynthetic membranes. A study by resonance Raman spectroscopy. *Biochim. Biophys. Acta*, **460**, 408–30.

Lutz, M., Chinsky, L., and Turpin, P.-Y. (1982). Triplet states of carotenoids bound to reaction centers of photosynthetic bacteria: time-resolved resonance Raman spectroscopy. *Photochem. Photobiol.*, **36**, 503–15.

Machtalere, G., Houée-Levin, C., Gardes-Albert, M. Ferradini, C., and Hickel, B. (1988). Pulse radiolysis study of the reduction mechanism of an anti-tumour antibiotic, mitomycin C. *C.R. Acad. Sci. Paris*, **307**, 17–22.

MacKie, R. M. (ed.) (1983). *Malignant melanoma, advances of a decade* (Pigment Cell. Vol. 6). Karger, Basel.

Macpherson, A. N., Kessel, D., Morgan, A. R., Munro, I., and Truscott, T. G. (1990). A photophysical study of some purpurins. *J. Chem. Soc., Faraday Trans.*, **86**, 3081–5.

Maillard, B., Ingold, K. U., and Scaiano, J. C. (1983). Rate constants for the reactions of free radicals with oxygen in solution. *J. Am. Chem. Soc.*, **105**, 5095–9.

Makino, K., Mossoba, M. M., and Riesz, P. (1983). Formation of $OH^{\cdot}$ and $H^{\cdot}$ in aqueous solutions by ultrasound using clinical equipment. *Radiat. Res.*, **96**, 416–21.

Malawista, S. E., Trock, D. H., and Edelson, R. L. (1991). Treatment of rheumatoid arthritis by extracorporeal photochemotherapy. A pilot study. *Arthritis Rheum.*, **34**, 646–54.

Mallet, L. (1927). Phénomènes de luminescence au cours de réactions oxydantes en solutions aqueuses. *C.R. Acad. Sci. (Paris) Série C*, **185**, 352–4.

Manring, L. E. and C. S. Foote (1982). Chemistry of singlet oxygen XXXVII One electron-oxidation of tetramethylphenylenediamine by singlet oxygen. *J. Phys. Chem.*, **86**, 1257–9.

Mantione, M. J. and Pullman, B. (1964). Sur le mécanisme de la photodimérisation de la thymine. *Biochim. Biophys. Acta*, **91**, 387–98.

Mantulin, W. W. and Song, P.-S. (1973). Excited states of skin-sensitizing coumarins and psoralens. Spectroscopic studies. *J. Am. Chem. Soc.*, **95**, 5122–9.

Marcus, R. A. (1964). Chemical and electrochemical electron-transfer theory. *Ann. Rev. Phys. Chem.*, **15**, 155–96.

Marcus, R. A. (1965). On the theory of electron-transfer reactions. VI. *J. Chem. Phys.*, **43**, 677–704.

Marcus, R. A. (1982). Electron, proton and related transfer. *Faraday Disc. Chem. Soc.*, **74**, 7–15.

Marcus, R. A., and Sutin, N. (1985). Electron transfers in chemistry and biology. *Biochim. Biophys. Acta*, **811**, 265–322.

Markert, M., Andrews, P. C., and Babior, B. M. (1984). Measurement of $O_2^{-\cdot}$ production by human neutrophils. The preparation and assay of NADPH oxidase-containing particles from human neutrophils. In *Oxygen radicals in biological*

systems. Methods in Enzymology, Vol. 105, (ed. L. Packer), pp. 358–65. Academic Press, New York.
Marletta, M.A., Yoon, P.S., Iyengar, R., Leaf, C.D., and Wishnock, J.S. (1988). Macrophage oxidation of L-arginine to nitrite and nitrate: nitric oxide is an intermediate. *Biochemistry*, **27**, 8706–11.
Martin, J.L., Migus, A., Poyart, C., Lecarpentier, Y., Astier, R., and Antonetti, A. (1983). Femtosecond photolysis of Co-ligated protoheme and hemoproteins: appearance of deoxy species with a 350 fsec time constant. *Proc. Natl. Acad. Sci., USA*, **80**, 173–7.
Mason, H.S. (1948). The chemistry of melanin. Mechanism of the oxidation of dihydroxyphenylalanine by tyrosinase. *J. Biol. Chem.*, **172**, 83–92.
Mason, R.P. (1982). Free-radical intermediates in the metabolism of toxic chemicals. In *Free radicals in biology*, Vol. V (ed. W.A. Pryor), pp. 161–222. Academic Press, New York.
Matheson, I.B.C. and Lee, J. (1979). Chemical reaction rates of amino acids with singlet oxygen. *Photochem. Photobiol.*, **29**, 879–81.
Matheson, I.B.C. and Rodgers, M.A.J. (1982). Crocetin, a water soluble carotenoid monitor for singlet molecular oxygen. *Photochem. Photobiol.*, **36**, 1–4.
Matheson, I.B.C., Etheridge, R.D., Kratowich, N.R., and Lee, J. (1975). The quenching of singlet oxygen by amino acids and proteins. *Photochem. Photobiol.*, **21**, 165–71.
Matheson, M.S. and Dorfman, L.M. (1960). Detection of short-lived transients in radiation chemistry. *J. Chem. Phys.*, **32**, 1870–1.
Mathews, J.I., Braslavsky, S.E., and Camilleri, P. (1980). The photophysics of bonellin: A chlorin found in marine animals. *Photochem. Photobiol.*, **32**, 733–8.
Mathews-Roth, M.M. (1986). Carotenoids quench evolution of excited species in epidermis exposed to UV-B (290–320 nm) light. *Photochem. Photobiol.*, **43**, 91–3.
Mathews-Roth, M.M. and Krinsky, N.I. (1984). Effect of dietary fat level on UV-B induced skin tumors, and anti-tumor action of β-carotene. *Photochem. Photobiol.*, **40**, 671–3.
Mathews-Roth, M.M. and Krinsky, N.I. (1987). Carotenoids affect development of UV-B induced skin cancer. *Photochem. Photobiol.*, **46**, 507–9.
Mathies, R. (1979). Biological applications of resonance Raman spectroscopy in the visible and ultraviolet: visual pigments, purple membrane and nucleic acids. In *Chemical and biochemical applications of lasers*, Vol. IV (ed. C.B. Moore). Academic Press, New York.
Mathies, R., Freedman, T.B., and Stryer, L. (1977). Resonance Raman studies of the confirmation of·retinal in rhodopsin and isorhodopsin. *J. Mol. Biol.*, **109**, 367–72.
Mathis, P. (1969). Etude, par spectroscopie du transfert d'énergie chlorophyll-carotenoide. *Photochem. Photobiol.*, **9**, 55–63.
Mathis, P. and Rutherford, A.W. (1984). Effect of phenolic herbicides on the oxygen-evolving side of photosystem II. Formation of the carotenoid cation. *Biochim. Biophys. Acta*, **767**, 217–22.
Mathis, P., Satoh, K., and Hansson, Ö. (1989). Kinetic evidence for the function of Z in isolated photosystem II reaction centers. *FEBS Lett.*, **251**, 241–4.
Mattay, J. (ed.) (1991). Photoinduced electron transfer Vol. I, II, and III. In *Top. Curr. Chem.*, **156**, **158**, **159**. Springer-Verlag, Berlin.
Mattay, J. and Vondenhof, M. (1991). Contact and solvent-separated radical ion pairs in organic photochemistry. In *Photoinduced electron transfer*, Vol. III (ed. J. Mattay), pp. 219–55. *Top. Curr. Chemi.*, **159**. Springer-Verlag, Berlin.
Mauk, A.G. (1991). Electron transfer in genetically engineered proteins. The cytochrome

c paradigm. In *Long-range electron transfer in biology* (ed. G. Palmer), pp. 131–57. Springer-Verlag, Berlin.

Maurolico, F. (1553). *Photismi de lumine et umbrae ad perspectivam et radiorum incidentiam facientes* Venice (trans. H. Crew) (1940): Photoismi de lumine of Maurolico: A chapter in late medieval optics. American Book Company, New York.

Maxwell, J.C. (1861*a*). The theory of molecular vortices applied to magnetic phenomena. *Philos. Mag.*, **21**, 161–75.

Maxwell, J.C. (1861*b*). The theory of molecular vortices applied to electric currents. *Philos. Mag.*, **21**, 281–91 and 338–48.

Mayer, E., Gardiner, D.J., and Hester, R.E. (1973). Resonance Raman spectra of vitamin B12 and dicyanocobalamin. *Biochim. Biophys. Acta*, **297**, 568–70.

Mayo, K.H. and Chen, M.-J. (1989). Human platelet factor 4 monomer–dimer–tetramer equilibria investigated by 1H NMR spectroscopy. *Biochemistry*, **28**, 9469–78.

Mayo, S.L., Ellis, Jr, W.R., Crutchley, R.J., and Gray, H.B. (1986). Long-range electron transfer in heme proteins. *Science*, **233**, 948–52.

Mazière, J.C., Morliere, P., and Santus, R. (1991). The role of the low density lipoprotein receptor pathway in the delivery of lipophilic photosensitizers in the photodynamic therapy of tumours. *J. Photochem. Photobiol., B: Biol.*, **8**, 351–60.

McBrien, D.C.H. and Slater, T.F. (eds) (1982). *Free radicals, lipid peroxidation and cancer*, NFCR Cancer Symposium 1, Uxbridge, UK.

McCapra, F. (1990). Chemiluminescence and bioluminescence. *J. Photochem. Photobiol.*, A: Chem., **51**, 21–8.

McCarthy, R.L and MacLachlan, A. (1960). Transient benzyl radical reactions produced by high energy radiation. *Trans. Faraday. Soc.*, **56**, 1187–200.

McCord, E.F., Morden, K.M., Pardi, A., Tinoco, I., and Boxer, S.G. (1984*a*). Chemically induced dynamic nuclear polarization studies of guanosine in nucleotides, dinucleotides, and oligonucleotides. *Biochemistry*, **23**, 1926–34.

McCord, E.F., Morden, K.M., Tinoco, I., and Boxer, S.G. (1984*b*). Chemically induced dynamic nuclear polarization studies of yeast $tRNA^{Phe}$. *Biochemistry*, **23**, 1935–9.

McCord, J. (1979). Superoxide, superoxide dismutase and oxygen toxicity. *Rev. Biochem. Toxicol.*, 109–24.

McCord, J.M. and Fridovich, I. (1968). The reduction of cytochrome *c* by milk xanthine oxidase. *J. Biol. Chem.*, **243**, 5753–60.

McCord, J.M. and Fridovich, I. (1969). Superoxide dismutase, an enzymic function for erythrocuprein, *J. Biol. Chem.*, **244**, 6049–55.

McCormick, J., Fisher, J.R., Pachlatko, J.P., and Eisenstark, A. (1976). Characterisation of a cell-lethal product from the photoxidation of tryptophan: hydrogen peroxide. *Science*, **191**, 468–9.

McCubbin, I. and Phillips, D. (1986). The photophysics and photostability of zinc(II) and aluminium(III) sulphonated napthalocyanines. *J. Photochem.*, **34**, 187–95.

McGarvey, D.J., Szekeres, P.G., and Wilkinson, F. (1992). The efficiency of singlet oxygen generation by substituted naphthalenes in benzene. Evidence for the participation of charge–transfer interactions. *Chem. Phys. Lett.*, **199**, 314–19.

McGlynn, S.P., Azumi, T., and Kinoshita, M. (1969). In *Molecular spectroscopy of the triplet state*, p. 159. Prentice Hall, Englewood Cliffs, New Jersey.

McLaren, A.D. and Hidalgo-Salvatierra, O. (1964). Quantum yields for enzyme inactivation and the amino acid composition of proteins. *Photochem. Photobiol.*, **3**, 349–52.

McLauchlan, K.A. (1985). Flash photolysis electron spin resonance. *Chemistry in Britain* (September), 825–32.

McLean, A. J. and Rodgers, M. A. J. (1992). A variable-temperature study of the quenching of benzophenone triplet with oxygen: involvement of exciplexes. *J. Am. Chem. Soc.*, **114**, 3145–7.

McLean, A. J. and Truscott, T. G. (1990). Efficiency of triplet-photosensitised singlet oxygen generation in benzene. *J. Chem. Soc., Faraday Trans.*, **86**, 2671–2.

McLean, A. J., McGarvey, D. J., Truscott, T. G., Lambert, C. R., and Land, E. J. (1990). Effect of oxygen-enhanced intersystem crossing on the observed efficiency of formation of singlet oxygen. *J. Chem. Soc., Faraday Trans.*, **86**, 3075–80.

McLendon, G. (1988). Long distance electron transfer in proteins and model systems, *Acc. Chem. Res.*, **21**, 160–7.

McLendon, G. (1991). Control of biological electron transport via molecular recognition and binding: The 'Velcro' model. In *Long-range electron transfer in biology* (ed. G. Palmer), pp. 159–74. Springer-Verlag, Berlin.

McLendon, G., Miller, J. R., Simolo, K., Taylor, K., Mauk, A. G., and English, A. M. (1986). Thermal and photoinduced long distance electron transfer in proteins and in model systems. In *Excited states and reactive intermediates, photochemistry, photophysics and electrochemistry* (ed. A. B. P. Lever), ACS Ser., Vol. 307, pp. 150–65. Washington.

McLendon, G., Pardue, K., and Bak, P. (1987). Electron transfer in the cytochrome *c*/cytochrome b_2 complex: Evidence for 'conformational gating'. *J. Am. Chem. Soc.*, **109**, 7540–1.

McRae, E. G. (1957). Theory of solvent effects on molecular electronic spectra. Frequency shifts. *J. Phys. Chem.*, **61**, 562–72.

McVie, J., Sinclair, R. S., and Truscott, T. G. (1978). Triplet states of copper and metal-free phthalocyanines. *J. Chem. Soc., Faraday Trans. II*, **74**, 1870–9.

McVie, J., Sinclair, R. S., Tait, D., Truscott, T. G., and Land, E. J. (1979). Electron transfer reactions involving porphyrins and carotenoids. *J. Chem. Soc., Faraday Trans. I*, **75**, 2869–72.

McWilliams, R. S., Cross, W. G., Kaplan, J. G., and Birnboim, H. C. (1983). Rapid rejoining of DNA strand breaks in resting human lymphocytes after irradiation by low doses of ^{60}Coγ rays or 14.6 MeV neutrons. *Radiat. Res.*, **94**, 499–507.

Medinger, T. and Wilkinson, F. (1965). Mechanism of fluorescence quenching in solution. Part 1. Quenching by bromobenzene *Trans. Faraday Soc.*, **61**, 620–30.

Meffert, H., Diezel, W., and Sonnichsen, N. (1976). Stable lipid peroxidation products in human skin: Detection, ultraviolet light-induced increase, pathogenic importance. *Experientia*, **32**, 1397–8.

Meisel, D. and Czapski, G. (1975). One-electron transfer equilibria and redox potentials of radicals studied by pulse radiolysis. *J. Phys. Chem.*, **79**, 1503–9.

Meisel, D. and Neta, P. (1975). One-electron redox potentials of nitro compounds and radiosensitizers. Correlation with spin densities of their radical anions. *J. Am. Chem. Soc.*, **97**, 5198–203.

Merenyi, G., Lind, J., and Shen, X. L., (1988). Electron transfer from indoles, phenol, and sulfite (SO_3^{2-}) to chlorine dioxide *J. Phys. Chem.*, **92**, 134–7.

Merkel, P. B. and Kearns, D. R. (1971). Direct measurement of the lifetime of singlet oxygen in solution. *Chem. Phys. Lett.*, **12**, 120–2.

Meyer-Betz, F. (1913). Investigations on the biological (photodynamic) action of haematoporphyrin and other derivatives of blood and bile pigments. *Deutsch. Arch. Klin. Med.*, **112**, 476–503.

Mialocq, J. C. (1988). La formation de l'électron solvaté en photochimie. *J. Chim. Phys.*, **85**, 31–45.

Mialocq, J. C. Sutton, J., and Goujon, P. (1980). Picosecond study of electron ejection in aqueous phenol and phenolate solutions. *J. Chem. Phys.*, **72**, 6338–45.

Michael, B. D., Adams, G. E., Hewitt, H. B., Jones, W. B. G., and Watts, M. E. (1973). A posteffect of oxygen in irradiated bacteria: A submillisecond fast mixing study. *Radiat. Res.*, **54**, 239–51.

Michael, B. D., Harrop, H. A., Maughan, R. L., and Patel, K. B. (1978). A fast kinetics study of the modes of action of some different radiosensitizers in bacteria. *Br. J. Cancer*, **37**, Suppl. III, 29–33.

Michael, B. D. Harrop, H. A., and Maughan, R. L. (1979). Fast response methods in the radiation chemistry of lethal damage in intact cells. In *Radiation research* (ed. S. Okada, M. Imamura, T. Terasima, and H. Yamaguchi), pp. 288–96, Japanese Association for Radiation Research, Tokyo.

Michael, B. D., Davies, S., and Held, K. D. (1986). Ultrafast chemical repair of DNA single and double strand precursors in irradiated V79 cells. In *Mechanisms of DNA damage and repair: Implications for carcinogenesis and risk assessment* (ed. M. G. Simic, L. Grossman, and A. C. Upton), pp. 89–100. Plenum Press, New York.

Michel, H. (1982). Three-dimensional crystals of a membrane protein complex. The photosynthetic reaction centre from *Rhodopseudomonas viridis*. *J. Mol. Biol.* **158**, 567–72.

Michl, J. (1972*a*). Photochemical reactions of large molecules. I. A single physical model of photochemical reactivity. *Molec. Photochem.*, **4**, 243–55.

Michl, J. (1972*b*). Photochemical reactions of large molecules. II. Application of the model to organic photochemistry. *Molec. Photochem.*, **4**, 257–86.

Michl, J. (1972*c*). Photochemical reactions of large molecules. III. Use of correlation diagrams for the prediction of energy barriers. *Molec. Photochem.*, **4**, 287–314.

Midden, W. R. and Wang, S. Y. (1983). Singlet oxygen generation for solution kinetics: clean and simple. *J. Am. Chem. Soc.*, **105**, 4129–35.

Migus, A., Gauduel, Y., Martin J. L., and Antonetti, A. (1987). Excess electrons in liquid water: first evidence of a prehydrated state with femtosecond lifetime. *Phys. Rev. Lett.*, **58**, 1559–62.

Milder, S. J. and Kliger, D. S. (1985). Spectroscopy and photochemistry of thiouracils: implications for the mechanism of photocrosslinking in *t*RNA. *J. Am. Chem. Soc.*, **107**, 7365–73.

Milgrom, L. (1984). Vampires, plants, and crazy kings. *New Scientist*, No 1407, 24 April, 9–13.

Mill, T., Hendry, D. G., and Richardson, H. (1980). Free radical oxidants in natural waters. *Science*, **207**, 886–7.

Miller, J. R., Calcaterra, L. Y., and Closs, G. L. (1984). Intramolecular long-distance electron transfer in radical anions. The effects of free energy and solvent on the reaction rates. *J. Am. Chem. Soc.*, **106**, 3047–9.

Millikan, R. A. (1916). A direct photoelectric determination of Planck's 'h'. *Phys. Rev.*, **7**, 355–93.

Mills, D. M., Lewis, A., Harootunian, A., Huang, J., and Smith, B. (1984). Time-resolved X-ray absorption spectroscopy of carbon monoxide–myoglobin recombination after laser photolysis. *Science*, **223**, 811–13.

Miranda, M., Botti, D., and Dicola, M. (1984). Possible genotoxicity of melanin synthesis intermediates: Tyrosinase reaction products interact with DNA *in vitro*. *Mol. Gen. Genet.*, **193**, 395–9.

Mitchell, D. L. (1988). The relative phototoxicity of (6–4) photoproducts and cyclobutane dimers in mammalian cells. *Photochem. Photobiol.*, **48**, 51–7.

Mittal, J. P. and Hayon, E. (1974). Interaction of hydrated electrons with phenylalanine and related compounds. *J. Phys. Chem.*, **78**, 1790–4.

Moan, J. (1984). Fluorescence of porphyrins in cells. In *Porphyrins in tumour phototherapy* (ed. A. Andreoni and R. Cubeddu), pp. 109–24. Plenum, New York.

Moan, J. and Kessel, D. (1988). Photoproducts formed from photofrin in cells. *J. Photochem. Photobiol., B: Biol.*, **1**, 429–36.

Moan, J., Pettersen, E.O., and Christensen, T. (1979). The mechanism of photodynamic inactivation of human cells *in vitro* in the presence of haematoporphyrin. *Br. J. Cancer*, **39**, 398–407.

Moan, J., Steen, N.B., Feren, K., and Christensen, T. (1981). Uptake of haematoporphyrin derivative and photosensitization of C3H cells with different oncogenic potential. *Cancer Lett.*, **14**, 291–6.

Moan, J., Berg, K., and Kvam, E. (1990). Effects of photodynamic treatment on DNA and DNA-related cell functions. In *Photodynamic therapy of neoplastic disease*, Vol. I (ed. D. Kessel), pp. 197–209. CRC Press, Boca Raton, Florida.

Moncada, S., Palmer, R.M.J., and Higgs, E.A., (1989). Biosynthesis of nitric oxide from L-arginine, A pathway for the regulation of cell function and communication *Biochem. Pharmacol.*, **38**, 1709–15.

Moore, A.L., Dirks, G., Gust, D., and Moore, T.A. (1980). Energy transfer from carotenoid polyenes to porphyrins: A light-harvesting antenna. *Photochem. Photobiol.*, **32**, 691–5.

Moore, A.L., Joy, A., Tom, R., Gust, D., Moore, T.A., Bensasson, R.V., and Land E.J. (1982). Photoprotection by carotenoids during photosynthesis: motional dependence of intramolecular energy transfer. *Science*, **216**, 982–4.

Moore, D.E. and Chappuis, P.P. (1988). A comparative study of the photochemistry of the non-steroidal anti-inflammatory drugs, naproxen, benoxaprofen and indomethacin. *Photochem. Photobiol.*, **47**, 173–80.

Moore, J.N., Hansen, P.A., and Hochstrasser, R.M. (1988). Iron–carbonyl bond geometries of carboxymyoglobin and carboxyhemoglobin in solution determined by picosecond time-resolved infrared spectroscopy. *Proc. Natl. Acad. Sci. USA*, **85**, 5062–6.

Moore, M.R. and Brodie, M.J. (1985). The porphyrias. *Medicine International*, **14**, 604–8.

Moore, T.A. (1983). Photoacoustic spectroscopy and related techniques applied to biological materials. *Photochem. Photobiol. Rev.*, **7**, 187–221.

Mopper, K. and Zhou, X. (1990). Hydroxyl radical photoproduction in the sea and its potential impact on marine processes. *Science*, **250**, 661–4.

Morgan, A.R., Nonis, S., Keck, R.W., and Selman, S.H. (1988). New sensitizers for PDT. In *Photosensitization: molecular, cellular and medical aspects* (ed. G. Moreno, R. Pottier, and T.G. Truscott), pp. 487–90. Springer Verlag, Berlin.

Morlière, P., Bazin, M., Dubertret, L., Santus, R., Sa e Melo, T., Huppe, G., Haigle, J., Forlot, P., and Bernard, A. (1991). Photoreactivity of 5-geranoxypsoralen and lack of photoreaction with DNA. *Photochem. Photobiol.*, **53**, 13–9.

Moureau, C., Dufraisse, C., and Dean, P.M. (1926). Sur un hydrocarbure coloré: le rubrène. *C.R. Hebd. Séances Acad. Sci.*, **182**, 1440–3.

Mozumder, A.J. (1969). Neutralization of isolated ion pair in polar media. *J. Chem. Phys.*, **50**, 3153–61, 3162–70.

Mozumder, A.J. and Magee, J.L. (1966). Theory of radiation chemistry. VII Structure and reactions in low LET tracks. *J. Chem. Phys.*, **45**, 3332–41.

Mukherjee, T., Cercek, B., Dodd, N.J.F., and Swallow, A.J. (1987). Determination of the ionic charge of semiquinones by pulsed conductivity and ESR techniques. *Radiat. Phys. Chem.*, **30**, 271–7.

Mukherjee, T., Land, E.J., Swallow, A.J., Bruce, J.M., Beaumont, P.C., and Parsons, B.J. (1988). Aspects of the oxidation of naphthazarin as studied by pulse radiolysis. *J. Chem. Soc., Faraday Trans. I*, **84**, 3423–34.

Mukherjee, T., Land, E.J., Swallow, A.J., and Bruce, J.M. (1989). One-electron reduction of adriamycin and daunomycin: Short-term stability of the semiquinones. *Arch. Biochem. Biophys.*, **272**, 450–8.

Mulazzani, Q.G., Ciano, M., D'Angelantonio, M., Venturi, M. and Rodgers, M.A.J. (1988). Singlet molecular oxygen: not a major product of the reaction between tris (2,2'-bipyridine) ruthenium (3^+) and superoxide radical anions. *J. Am. Chem. Soc.*, **110**, 2451–7.

Muller, E., Boiteux, S., Cunningham, R.P., and Epe, B. (1990). Enzymatic recognition of DNA modifications induced by singlet oxygen and photosensitizers. *Nucleic Acids Research*, **18**, 5969–73.

Muller, H.J. (1927). Artificial transmutation of the gene. *Science*, **66**, 84–7.

Murphy, M.E. and Sies, H. (1990). Visible-range low-level chemiluminescence in biological systems. In *Methods in Enzymology*, Vol. 186, Oxygen radicals in biological systems, Part B Oxygen radicals and antioxidants (ed. L. Packer and A.N. Glazer), pp. 595–610. Academic Press, San Diego.

Murphy, M.E., Piper, H.M., Watanabe, H., and Sies, H. (1991). Nitric oxide production by cultured aortic endothelial cells in response to thiol depletion and replenishment. *J. Biol. Chem.*, **266**, 19 378–83.

Murray, R.W. (1979). Chemical sources of singlet oxygen. In *Singlet oxygen* (ed. H.H. Wasserman and R.W. Murray), Chapter 3. Academic Press, New York.

Musajo, L., Bordin, F., Caporale, G., Marciani, S., and Rigatti, G. (1967*a*). Photoreactions at 3655 Å between pyrimidine bases and skin-photosensitizing furocoumarins. *Photochem. Photobiol.*, **6**, 711–9.

Musajo, L., Bordin, F., and Bevilacqua, R. (1967*b*). Photoreactions of 3655 Å linking the 3, 4-double bond of furocoumarins with pyrimidine bases. *Photochem. Photobiol.*, **6**, 927–31.

Muus, L.T., Atkins, P.W., McLauchlan, K.A., and Pedersen, J.B. (eds) (1977). *Chemically induced magnetic polarization*. Reidel, Dordrecht.

Nadezhdin, A., and Dunford, H.B. (1979). Oxidation of nicotinamide adenine dinucleotide by hydroperoxyl radical. A flash photolysis study, *J. Phys. Chem.*, **83**, 1957–61.

Nadler, W. and Marcus, R.A. (1987). Dynamical effects in electron transfer reactions. II. Numerical solution *J. Chem. Phys.*, **86**, 3906–24.

Naish, S. unpublished results quoted in Riley, P.A. (1985). Radicals and melanoma. *Phil. Trans. R. Soc. London* **B 311**, 679–89.

Nakano, M., Takayama, K., Shimizu, Y., Tsuji, Y., Inaba, H , and Migita, T. (1976). Spectroscopic evidence for the generation of singlet oxygen in self reaction of *sec*-peroxy radicals. *J. Am. Chem. Soc.*, **98**, 1974–5.

Naqui, A., Chance, B., and Cadenas, E. (1986). Reactive oxygen intermediates in biochemistry. *Ann. Rev, Biochem.*, **55**, 137–66.

Navaratnam, S., Parsons, B.J., Phillips, G.O., and Davies, A.K. (1978). Laser flash photolysis study of the photoionization of chlorpromazine and promazine in solution. *J. Chem. Soc., Faraday Trans. I*, **74**, 1811–19.

Navaratnam, S., Hughes, J.L., Parsons, B.J., and Phillips, G.O. (1985). Laser flash and steady-state photolysis of benoxaprofen in aqueous solution. *Photochem. Photobiol.*, **41**, 375–80.

Neidle, S. and Waring, MJ. (eds) (1983). *Molecular aspects of anti-cancer drug action* (Topics in molecular and structural biology, 3). Macmillan, London.

Neta, P., and Ross, A.B. (1988). Representative rate constants for radical reactions of biologically important molecules. In *Chemical kinetics of small organic radicals*, Vol. IV. *Reactions in special systems* (ed. Z.B. Alfassi), pp. 187–221. CRC Press, Boca Raton, Florida.

Neta, P., Hoffman, M.Z., and Simic, M. (1972). Electron spin resonance and pulse radiolysis studies of the reactions of OH and O^- radicals with aromatic and olefinic compounds. *J. Phys. Chem.*, **76**, 847–53.

Neta, P., Holdren, G.R., and Schuler, R.H. (1971). On the rate constants for reaction of hydrogen atoms in aqueous solution. *J. Phys. Chem.*, **75**, 449–54.

Neta, P., Huie, R.E., and Ross, A.B. (1990). Rate constants for reactions of peroxyl radicals in fluid solutions. *J. Phys. Chem. Ref. Data*, **19**, 413–513.

Neville, A.G., Brown, C.E., Rayner, D.M., Lusztyk, J., and Ingold, K.U. (1989). Direct measurement of the rate constant for β-scission of the cumyloxyl radical by laser flash photolysis with time-resolved IR detection. *J. Am. Chem. Soc* **111**, 9269–70.

Neville, A.G., Brown, C.E., Rayner, D.M., Lusztyk, J., and Ingold, K.U. (1991). First direct detection of transient free radicals in solution by time-resolved infrared spectroscopy. Kinetic studies on some acyl radicals. *J. Am. Chem. Soc.*, **113**, 1869–70.

Newton, I. (1672). A new theory about light and colours. *Phil. Trans. R. Soc. London.*, **6**, 3075–87.

Nichols, E.L. and Merritt, E. (1910). Distribution of energy in fluorescence spectra. *Phys. Rev.*, **30**, 328–46.

Nielsen, S.O., Michael, B.D., and Hart, E.J. (1976). Ultraviolet absorption spectra of e^-_{aq}, H, OH, D, OD from pulse radiolysis of aqueous solutions. *J. Phys. Chem.*, **80**, 2482–8.

Nieuwint, A.W.R., Aubry, J.M., Arwert, F., Kortbeek, H., and Herzberg, S. (1985). Inability of chemically generated singlet oxygen to break the DNA backbone. *Free Rad. Res. Commun.*, **1**, 1–9.

Nikogosyan, D.N. (1990). Two quantum UV photochemistry of nucleic acids: comparison with conventional low-intensity UV photochemistry and radiation chemistry. *Int. J. Radiat. Biol.*, **57**, 233–99.

Nikogosyan, D.N. and Angelov, D.A. (1981). Formation of free radicals in water under high-power laser UV irradiation. *Chem. Phys Lett.*, **77**, 208–10.

Nikogosyan, D.N. and Görner, H. (1992). Photolysis of aromatic amino acids in aqueous solution by nanosecond 248 and 193 nm laser light. *J. Photochem. Photobiol., B: Biol.*, **13**, 219–34.

Nikogosyan, D.N. and Letokhov, V.S. (1983). Non linear laser photophysics, photochemistry and photobiology of nucleic acids. *Riv. Nuovo Cimento*, **6**, 1–74.

Nilges, M.J. and Swartz, H.M. (1984). Quinone and semiquinone intermediates in the reaction of 4-hydroxyanisole and tyrosinase. In *Hydroxyanisole. Recent advances in anti-melanoma therapy* (ed. P.A. Riley), pp. 25–34. IRL Press, Oxford.

Nilges, M.J., Swartz, H.M., and Riley, P.A. (1984). Identification by electron spin resonance of free radicals formed during the oxidation of 4-hydroxyanisole catalysed by tyrosinase. *J. Biol. Chem.*, **259**, 2446–51.

Niu, Q. and Mendenhall, G.D. (1990). Structural effects on the yields of singlet molecular oxygen ($^1\Delta_g$) from alkylperoxyl radical recombination. *J. Am. Chem. Soc.*, **112**, 1656–7.

Nohl, H. (1990). Is redox-cycling ubiquinone involved in mitochondrial oxygen activation? *Free Rad. Res. Commun.*, **8**, 307–15.

Norrish, R.G.W. and Porter, G. (1949). Chemical reactions produced by very high light intensities. *Nature*, **164**, 658.

Novak, J.R. and Windsor, M.W. (1967). Laser photolysis and spectroscopy in the nanosecond time range: Excited singlet state absorption in coronene. *J. Chem. Phys.*, **47**, 3075–6.

Nucifora, G., Smaller, B., Remko, R., and Avery, E.C. (1972) Transient radicals of DNA bases by pulse radiolysis. Effects of cysteine and cysteamine as radioprotectors. *Radiat. Res.*, **49**, 96–111.

O'Connor, D.V. and Phillips, D. (1984). *Time-correlated single photon counting*. Academic Press, London.

Ogilby, P.R. and Foote, C.S. (1982). Chemistry of singlet oxygen, 36. Singlet molecular oxygen ($^1\Delta_g$) luminescence in solution following pulsed laser excitation. Solvent deuterium isotope effects on the lifetime of singlet oxygen. *J. Am. Chem. Soc.*, **104**, 2069–70.

Ogilby, P.R., Kristiansen, M., and Clough, R.L. (1990). Singlet oxygen formation in a solid organic polymer upon irradiation of the oxygen-polymer charge transfer band. *Macromolecules*, **23**, 2698–704.

Okamura, T., Sancar, A., Heelis, P.F., Hirata, Y., and Mataga, N. (1989). Doublet-quartet intersystem crossing of flavin radical in DNA photolyase. *J. Am. Chem. Soc.*, **111**, 5967–9.

O'Neill, P. (1983). Pulse radiolytic study of the interaction of thiols and ascorbate with OH adducts of dGMP and dG: Implications for DNA repair processes. *Radiat. Res.*, **96**, 198–210.

O'Neill, P. and Chapman, P.W. (1985). Potential repair of free radical adducts of dGMP and dG by a series of reductants. A pulse radiolytic study. *Int. J. Radiat. Biol.*, **47**, 71–80.

O'Neill, P., Al-Kazwini, A.T., Land, E.J., and Fielden, E.M. (1989*a*). Diffuse reflectance pulse radiolysis of solid DNA: The effects of hydration. *Int. J. Radiat. Biol.*, **55**, 531–7.

O'Neill, P., Al-Kazwini, A.T., Land, E.J., and Fielden, E.M. (1989*b*). Reflectance pulse radiolysis of DNA in the presence of misonidazole. *Int. J. Radiat. Biol.*, **55**, 883.

O'Neill, P., Al-Kazwini, A.T., Land, E.J., and Fielden, E.M. (1990). Early chemical events in the development of radiation damage of DNA-novel approaches. In *The early effects of radiation on DNA*. Nato ARW, Springer-Verlag, Berlin.

Onsager, L. (1936). Electric moments of molecules in liquids. *J. Am. Chem. Soc.*, **58**, 1486–93.

Onsager, L. (1938). Initial recombination of ions. *Phys. Rev.*, **54**, 554–7.

Opitz, J. and Schulte-Frohlinde, D. (1987). Laser-induced photoionization and single-strand break formation for polynucleotides and single-stranded DNA in aqueous solution: model studies for the direct effect of high energy radiation on DNA. *J. Photochem.*, **39**, 145–63.

Oraevsky, A.A. and Nikogosyan, D.N. (1985). Picosecond two-quantum UV photochemistry of thymine in aqueous solution. *Chem. Phys.*, **100**, 429–45.

Orrenius, S. (1985). Oxidative stress studies in intact mammalian cells. *Phil. Trans. R. Soc. London.*, **B 311**, 673–7.

Ortiz, F.M., Serrano, J.T., Lopez, J.N.R., Castellanos, R.V., Teruel, J.A.L., and Garcia-Canovas, F. (1988). Oxidation of 3,4-dihydroxymandelic acid by tyrosinase. *Biochim. Biophys. Acta.*, **957**, 158–63.

Oshino, N., Jamieson, D., and Chance, B. (1975). The properties of hydrogen peroxide production under hyperoxic and hypoxic condition of perfused rat liver. *Biochem. J.*, **146**, 53–65.

Ottensmeyer, F.P. and Whitmore, G.F. (1968). Coding properties of ultraviolet photoproducts of uracil. II Phenotypic reversion of the amber mutation: Implication of the uracil hydrate. *J. Mol. Biol.*, **38**, 17–24.

Packer, J.E., Slater, T.F., and Willson, R.L. (1978). Reactions of the carbon tetrachloride-related peroxy free radical ($CCl_3O_2^{\cdot}$) with amino acids: pulse radiolysis evidence. *Life Sciences*, **23**, 2617–20.

Packer, J.E., Slater, T.F., and Willson, R.L. (1979). Direct observation of a free radical interaction between vitamin E and vitamin C. *Nature*, **278**, 737–8.

Packer, J.E., Mahood, J.S., Mora-Arellano, V.O., Slater, T.F., Willson, R.L., and

Wolfenden, B.S. (1981). Free radical and singlet oxygen scavengers: Reaction of a peroxy-radical with β-carotene, diphenyl furan and 1,4-diazobicyclo (2,2,2)-octane *Biochem. Biophys. Res. Commun.*, **98**, 901–6.

Palecek, E. (1991). Local supercoil-stabilized DNA structure. *Crit. Rev. Biochem. Mol. Biol.*, **26**, 151–226.

Palmer, G.A. (ed.) (1991). Long-range electron transfer in biology. In *Structure and bonding*, Vol. 75. Springer-Verlag, Berlin.

Palumbo, A., d'Ischia, M., Misuraca, G., Prota, G., and Schultz, T.M. (1988). Structural modifications in biosynthetic melanins induced by metal ions. *Biochim. Biophys. Acta*, **964**, 193–9.

Palumbo, M.C., Garcia, N.A. and Arguello, G.A. (1990). The interaction of singlet molecular oxygen O_2 ($^1\Delta_g$) with indolic derivatives. Distinction between physical and reactive quenching. *J. Photochem. Photobiol., B: Biol.*, **7**, 33–42.

Palumbo, P., d'Ischia, M., and Prota, G. (1987). Tyrosinase-promoted oxidation of 5,6-dihydroxyindole-2-carboxylic acid to melanin. Isolation and characterization of oligomer intermediates. *Tetrahedron*, **43**, 4203–6.

Pandey, R.K., and Dougherty, T.J. (1988). Synthesis and photosensitizing activity of a di-porphyrin ether. *Photochem. Photobiol.*, **47**, 769–77.

Pandey, R.K. and Dougherty, T.J. (1989). Synthesis and photosensitizing activity of porphyrins joined with ester linkages. *Cancer Res.*, **49**, 2042–7.

Paneth, F. and Hofeditz, W. (1929). Uber die darstellung von freiem methyl. *Ber.*, **62**, 1335–47.

Papp, S. and Vanderkooi, J.M. (1989). Tryptophan phosphorescence at room temperature as a tool to study protein structure and dynamics. *Photochem. Photobiol.*, **49**, 775–84.

Parker, A.W., Hester, R.E., Phillips, D., and Umapathy, S. (1992). Time-resolved resonance Raman spectroscopic investigations of the photochemistry of ubiquinone. *J. Chem. Soc., Faraday Trans.*, **88**, 2649–53.

Parker, C.A. (1964). Phosphorescence and delayed fluorescence from solutions. *Adv. Photochem.*, **2**, 305–83.

Parker, J. (1987). Optical monitoring of singlet oxygen generation during photodynamic treatment of tumours. *IEEE (Inst. Electr. Electron. Eng.) Circ. Dev.*, 10–21.

Parker, J.G. and Stanbro, W.D. (1982). Optical determination of the collisional lifetime of singlet molecular oxygen [$O_2(^1\Delta_g)$] in acetone and deuterated acetone. *J. Am. Chem. Soc.*, **104**, 2067–9.

Parrish, J.A. (1982). The scope of photomedicine. In *The science of photomedicine* (ed. J.D. Regan and J.A. Parrish), pp. 3–17. Plenum Press, New York.

Pasteur, L. (1861). Animalcules infusoires vivant sans gaz oxygène libre et déterminant des fermentations. *C. R. Acad. Sci.*, **52**, 344–7.

Pasteur, L. (1863). Recherches sur la putréfaction. *C. R. Acad. Sci.*, **56**, 1189–94.

Patel, K.B. and Willson, R.L. (1973). Semiquinone free radicals and oxygen. Pulse radiolysis of one electron transfer equilibria. *J. Chem. Soc., Faraday Trans. I*, **69**, 814–25.

Patrick, M.H. and Rahn, R.O. (1976). Photochemistry of DNA and polynucleotides: photoproducts. In *Photochemistry and photobiology of nucleic acids*, Vol. II *Biology*, (ed. S.Y. Wang), pp. 35–95. Academic Press, New York.

Patterson, L.K. (1981). Studies of radiation induced peroxidation in fatty acid micelles. In *Oxygen and oxyradicals in chemistry and biology* (ed. M.A.J. Rodgers and E.L. Powers), pp. 89–95, Academic Press, New York.

Patterson, L.K. (1987). Instrumentation for measurement of transient behavior in radiation chemistry. In *Radiation chemistry, principles and applications* (ed. Farhataziz and M.A.J. Rodgers), pp. 65–96. VCH Publishers, New York.

Patterson, L.K., Porter, G., and Topp, M.R. (1970) Oxygen quenching of singlet and triplet states. *Chem. Phys. Lett*, **7**, 612–14.

Patterson, M.S., Madsen, S.J., and Wilson, B.C. (1990). Experimental tests of the feasibility of singlet oxygen luminescence monitoring *in vivo* during photodynamic therapy. *J. Photochem. Photobiol., B: Biol.*, **5**, 69–84.

Paul, H., Small, Jr, R.D., and Scaiano, J.C. (1978). Hydrogen abstraction by *tert*-butoxy radicals. A laser photolysis and electron spin resonance study. *J. Am. Chem. Soc.*, **100**, 4520–7.

Pauling, L. (1929). Photoionization in liquids and crystals and the dependence of the frequency of X-ray absorption edges on chemical constitution. *Phys. Rev.*, **34**, 954–63.

Pauling, L. and Pauling, P. (1975). *Chemistry*. W.H. Freeman, San Francisco.

Peak, J.G., Peak, M.J., and Tuveson, R.W. (1983). Ultraviolet action spectra for aerobic and anaerobic inactivation of *Escherichia Coli* strains specifically sensitive and resistant to near ultraviolet radiations. *Photochem. Photobiol.*, **38**, 541–3.

Peak, J.G., Peak, M.J., and MacCoss, M. (1984). DNA breakage caused by 334-nm ultraviolet light is enhanced by naturally occurring nucleic acid components and nucleotide coenzymes. *Photochem. Photobiol.*, **39**, 713–6.

Peak, J.G., Peak, M.J., Sikorski, R.S., and Jones, C.A. (1985). Induction of DNA–protein crosslinks in human cells by ultraviolet and visible radiations: action spectrum. *Photochem. Photobiol.*, **41**, 295–302.

Peak, M.J. and Peak, J.G. (1982). Single-strand breaks induced in *Bacillus subtilis* DNA by ultraviolet light: action spectrum and properties. *Photochem. Photobiol.*, **35**, 675–80.

Peak, M.J., Peak, J.G., and Jones, C.A. (1985). Different (direct and indirect) mechanisms for the induction of DNA-protein crosslinks in human cells by far- and near-ultraviolet radiations (290 and 405 nm). *Photochem. Photobiol.*, **42**, 141–6.

Peak, M.J., Peak, J.G., and Carnes, B.A. (1987). Induction of direct and indirect single-strand breaks in human cell DNA by far- and near-ultraviolet radiations: action spectrum and mechanism. *Photochem. Photobiol.*, **45**, 381–7.

Peckler, S., Graves, B., Kanne, D., Rapoport, H., Hearst, J.E., and Kim, S.-H. (1982). Structure of a psoralen–thymine mono-adduct formed in photoreaction with DNA. *J. Mol. Biol.*, **162**, 157–72.

Pelizzetti, E., Meisel, D., Mulac, W.A., and Neta, P. (1979). On the electron transfer from ascorbic acid to various phenothiazine radicals. *J. Am. Chem. Soc.*, **101**, 6954–9.

Peng, Q. Farrants, G., Madslien, K., Bommer, J.C., Moan, J., Danielsen, H.E., and Nesland, J.M. (1991). Subcellular localization, redistribution and photobleaching of derivatives of sulphonated aluminium phthalocyanines in a human melanoma cell line. *Int. J. Cancer*, **49**, 290–5.

Perone, S.P. and Drew, H.D. (1971). Electrochemical and other analytical methods for flash-photolysis studies. In *Analytical photochemistry and photochemical analysis* (ed. J.M. Fitzgerald), pp. 213–36. Marcel Dekker, New York.

Perrin, J. (1929). La fluorescence des solutions. *Ann. Phys. (Paris)*, **12**, 169–75.

Perrouault, L., Asseline, U., Rivalle, C., Thuong, N.T., Bisagni, E., Giovannangeli, C., Le Doan, T., and Hélène, C. (1990). Sequence-specific artificial photo-induced endonucleases based on triple helix-forming oligonucleotides. *Nature*, **344**, 358–60.

Petasne, R.G. and Zika, R.G. (1987). Fate of superoxide in coastal sea water. *Nature*, **325**, 516–18.

Peters, G. and Rodgers, M. A. J. (1981) Single electron transfer from NADH analogues to singlet oxygen. *Biochim. Biophys. Acta*, **637**, 43–52.

Peters, K. S. and Snyder, G. J. (1988). Time-resolved photoacoustic calorimetry: Probing the energetics and dynamics of fast chemical and biochemical reactions *Science*, **241**, 1053–7.

Peto, R., Doll, R., Buckley, J. D., and Sporn, M. B. (1981). Can dietary beta-carotene materially reduce human cancer rates? *Nature*, **290**, 201–8.

Petrich, J. W., Chang, M. C., McDonald, D. B., and Fleming, G. R. (1983). On the origin of nonexponential fluorescence decay in tryptophan and its derivatives. *J. Am. Chem. Soc.*, **105**, 3824–32.

Phillips, D., Anissimov, V., Karpukhin, O., and Shiliapintokh, V. (1967). Energy transfer from chemiluminescent species in polymers. *Nature*, **215**, 1163–5.

Pietronigro, D. D., Jones, W. B. G., Kalty, K., and Demopoulos, H. B. (1977). Interaction of DNA and liposomes as a model for membrane-mediated DNA damage. *Nature*, **267**, 78–9.

Pietronigro, D. D., McGinness, J. E., Koren, M. J., Crippa, R., Saligman, M. L., and Demopoulos, H. B. (1979). Spontaneous generation of adriamycin semiquinone radicals at physiologic pH. *Physiol. Chem. Phys.*, **11**, 405–14.

Piette, J. (1991). Biological consequences associated with DNA oxidation mediated by singlet oxygen. *J. Photochem. Photobiol., B: Biol.*, **11**, 241–60.

Piette, J., Decuyper, J., Merville-Louis, M.-P., and Van de Vorst, A. (1986). Biomolecular photoalterations mediated by phenothiazine derivatives. *Biochemie*, **68**, 835–42.

Pileni, M.-P., Santus, R., and Land, E. J. (1978). On the photosensitizing properties of *N*-formylkynurenine and related compounds. *Photochem. Photobiol.*, **28**, 525–9.

Pileni, M. P., Giraud, M., and Santus, R. (1979). Kynurenic acid—II. Photosensitizing properties. *Photochem. Photobiol.*, **30**, 257–61.

Planck, M. (1900). Zur theorie des gesetzes der energieverteilung im normalspectrum. *Verk. Deut. Phys. Ges.*, **2**, 237–45.

Planck, M. (1901). Ueber das gesetz der energieverteilung im normalspectrum. *Ann. Phys.*, **309**(4), 553–63.

Poppe, W. and Grossweiner, L. I. (1975). Photodynamic sensitization by 8-methoxypsoralen via the singlet oxygen mechanism. *Photochem. Photobiol.*, **22**, 217–19.

Porter, G. (1950). Flash photolysis and spectroscopy. A new method for the study of free radical reactions. *Proc. R. Soc.*, **A200**, 284–300.

Porter, G. (1967). Flash photolysis and primary processes in the excited state. In *Fast reactions and primary processes in chemical kinetics* (ed. S. Claesson), pp. 141–64. Almqvist and Wiksell, Stockholm.

Porter, G. and Topp, M. R. (1968). Nanosecond flash photolysis and the absorption spectra of excited singlet states. *Nature*, **220**, 1228–9.

Porter, G. and Topp, M. R. (1970). Nanosecond flash photolysis. *Proc. R. Soc.*, **A315**, 163–84.

Porter, G. and Wilkinson, F. (1961*a*). Energy transfer from the triplet state. *Proc. R. Soc.*, **A264**, 1–18.

Porter, G. and Wilkinson, F. (1961*b*) Primary photochemical processes in aromatic molecules. Part 5—Flash photolysis of benzophenone in solution. *Trans. Faraday Soc.*, **57**, 1686–91.

Posener, M. L. Adams, G. E., Wardman, P., and Cundall, R. B. (1976). Mechanism of tryptophan oxidation by some inorganic radical anions: A pulse radiolysis study. *J. Chem. Soc., Faraday Trans. II*, **72**, 2231–9.

Poston, P.E. and Harris, J.M. (1990). Excited-state calorimetry studies of triplet benzophenone using time-resolved photothermal beam deflection spectroscopy. *J. Am. Chem. Soc.*, **112**, 644–50.

Potashnik, R., Goldschmidt, C.R., and Ottolenghi, M. (1971). Triplet state formation in the quenching of fluorescence by molecular oxygen. *Chem. Phys. Lett.*, **9**, 424–5.

Potter, E.D., Herek, J.L., Pederson, S., Liu, Q., and Zewail, A.H. (1992). Femtosecond laser control of a chemical reaction. *Nature*, **355**, 66–8.

Pottier, R. and Truscott, T.G. (1986). The photochemistry of haematoporphyrin and related systems. *Int. J. Radiat. Biol.*, **50**, 421–52.

Pouliquen, J. and Wintgens, V. (1988). Transferts d'électrons photoinduits. Complexes donneur-accepteur d'électrons. *J. Chim. Phys.*, **85**, 455–78.

Powis, G. (1987). Anthracycline metabolism and free radical formation. In *Metabolism and action of anti-cancer drugs* (ed. G. Powis and R.A. Prough) pp. 211–60. Taylor and Francis, London.

Powis, G. (1989). Free radical formation by antitumour quinones. *Free Rad. Biol. Med.*, **6**, 63–101.

Pratesi, R. (1987). Semiconductor (diode) lasers: basic principles and potential applications in the biomedical field. *Photobiochem. Photobiophys.* **Suppl**, 57–75.

Previtali, C.M. (1984). Laser flash photolysis of indole in the presence of amino acids. *Photochem. Photobiol.*, **40**, 689–92.

Priestley, J. (1775). Experiments and observations on different kind of air. Vol. II, Sections III–V, pp. 29–103. In *The discovery of Oxygen. Part I.* The Alembic Club, Edinburgh 1923. In *Foundations of Anesthesiology*, Vol. 1 A. Faulconer and T.C., Keys), pp. 39–70. Charles C. Thomas, Springfield, Illinois 1965.

Profio, A.E. (1984). Laser excited fluorescence of HPD for diagnosis of cancer. *IEEE Journal of Quantum Electronics*, **20**, 1502–6.

Prota, G. (1988). Progress in the chemistry of melanins and related metabolites. *Med. Res. Rev.*, **8**, 525–56.

Prota, G., Crescenzi, S., Misuraca, G., and Nicolaus, R.A. (1970). New intermediates in phaeomelanogenesis *in vitro*. *Experientia*, **26**, 1058–9.

Prütz, W.A. (1987). Radical transformations in peptides and proteins. In *Radiation research* (ed. E.M. Fielden, J.F. Fowler, J.H. Hendry, and D. Scott), pp. 134–9. Taylor and Francis, London.

Prütz, W.A., Sommermeyer, K., and Land, E.J. (1966). Light emission after pulse radiolysis of aqueous solutions of dyes, *Nature*, **212**, 1043–4.

Prütz, W.A., Land, E.J., and Sloper, R.W. (1981). Charge transfer in peptides. Effects of temperature, peptide length and solvent conditions upon intramolecular one-electron reactions involving tryptophan and tyrosine. *J. Chem. Soc., Faraday Trans. I*, **77**, 281–92.

Prütz, W.A., Siebert, F., Butler, J., Land, E.J., Menez, A., and Montenay-Garestier, T. (1982), Charge transfer in peptides. Intramolecular radical transformations involving methionine, tryptophan and tyrosine. *Biochim. Biophys. Acta*, **705**, 139–49.

Prütz, W.A., Butler, J., and Land, E.J. (1983). Phenol coupling initiated by one-electron oxidation of tyrosine units in peptides and histone. *Int. J. Radiat. Biol.*, **44**, 183–96.

Prütz, W.A., Monig, H., Butler, J., and Land, E.J. (1985*c*). Reactions of nitrogen dioxide in aqueous model systems: oxidation of tyrosine units in peptides and proteins. *Arch. Biochem. Biophys.*, **243**, 125–34.

Prütz, W.A., Butler, J., and Land, E.J. (1985*b*). Methionyl → tyrosyl radical transitions initiated by Br_2^- in peptide model systems and ribonuclease A. *Int. J. Radiat. Biol.*, **47**, 149–56.

Prütz, W.A., Butler, J., Land, E.J., and Swallow, A.J. (1986). Unpaired electron migration between aromatic and sulphur peptide units. *Free Rad. Res. Commun.*, **2**, 69–75.

Prütz, W.A., Butler, J., Land, E.J., and Swallow, A.J. (1989). The role of sulphur peptide functions in free radical transfer: a pulse radiolysis study. *Int. J. Radiat. Biol.*, **55**, 539–56.

Pryor, W.A. (1976). The role of free radical reactions in biological systems. In *Free radicals in biology*, Vol. 1, pp. 1–49. Academic Press, New York.

Pryor, W.A. and Tang, R.H. (1978). Ethylene formation from methional. *Biochem. Biophys. Res. Commun.*, **81**, 498–503.

Pullman, B. and Jortner, R. (eds) (1990) *Molecular basis of specificity in nucleic acid-drug interactions* (Jerusalem symposia on quantum chemistry and biochemistry, Vol. 23). Kluwer Academic Publishers, Dordrecht.

Raghavan, N.V., Das, P.K., and Bobrowski, K. (1981). Transient phenomena in the pulse radiolysis of retinyl polyenes. 1. Radical anions. *J. Am. Chem. Soc.*, **103**, 4569–73.

Rahhal, S. and Richter, H.W. (1988). Reduction of hydrogen peroxide by the ferrous ion chelate of diethylenetriamine-*N*, *N*, *N′*, *N″*, *N″*-pentaacetate. *J. Am. Chem. Soc.*, **110**, 3126–33.

Rahn, R.O. and Patrick, M.H. (1976). Photochemistry of DNA; secondary structure, photosensitization, base substitution, and exogenous molecules. In *Photochemistry and photobiology of nucleic acids*, Vol. II *Biology*, (ed. S.Y. Wang), pp. 97–145. Academic Press, New York.

Ramamurthy, V. and Liu, R.S.H. (1976). Photochemistry of polyenes. IX. Excitation, relaxation, and deactivation of dienes, trienes and higher polyenes in the vitamin A series in the sensitized isomerization reaction. *J. Am. Chem. Soc.*, **98**, 2935–42.

Rao, P.S., Simic, M., and Hayon, E. (1975). Pulse radiolysis study of imidazole and histidine in water. *J. Phys. Chem.*, **79**, 1260–3.

Raper, H.S. (1928). The aerobic oxidases. *Physiol. Rev.*, **8**, 245–82.

Rassmussen-Taxdal, D.S., Ward, G.E., and Figge, F.H.J. (1955) Fluorescence of human lymphatic and cancer tissues following high doses of intravenous haematoporphyrin. *Cancer*, **8**, 78–81.

Rau, H., Frank, R., and Greiner, G. (1986). Rate dependence of electron transfer on Donor–Acceptor separation and on free enthalpy change. The $Ru(bpy)_3^{2+}$/viologen^{2+} system. *J. Phys. Chem*, **90**, 2476–81.

Raubach, R.A. and Guzzo, A.V. (1971) Singlet–triplet absorption spectrum of all-*trans*-retinal. *J. Phys. Chem.*, **75**, 983–4.

Raubach, R.A. and Guzzo, A.V. (1973). Photoisomerization pathways in the visually important polyenes. 1. The retinals. *J. Phys. Chem.*, **77**, 889–92.

Rayner, D.M., Szabo, A.G., Loufty, R.O., and Yip, R.W. (1980). Singlet energy transfer between nucleic acid bases and dyes in intercalation complexes. *J. Phys. Chem.*, **84**, 289–93.

Razi Naqvi, K. (1983). Remarks concerning the neglect of photoselection effects in laser kinetic spectroscopy: implications for laser actinometry and the comparative method for the determination of triplet quantum yields. *An. Quim.*, **79**, 76–9.

Reddi, E. and Jori, G. (1988). Steady-state and time-resolved spectroscopic studies of photodynamic sensitizers: Porphyrins and phthalocyanines. *Rev. Chem Intermed.*, **10**, 241–68.

Reddi, E., Jori, G., Rodgers, M.A.J., and Spikes, J.D. (1983). Flash photolysis studies of haemato- and copro-porphyrins in homogeneous and microheterogeneous aqueous dispersions. *Photochem. Photobiol.*, **38**, 639–45.

Redmond, R.W. and Braslavsky, S.E. (1988). Time-resolved thermal lensing and phosphorescence studies on photosensitized singlet molecular oxygen formation. Influence of the electronic configuration of the sensitizer on sensitization efficiency. *Chem. Phys. Lett.*, **148**, 523–9.

Redmond, R.W., Land, E.J., and Truscott, T.G. (1984). Aggregation effects in the photophysical properties of porphyrins in relation to mechanisms involved in photodynamic therapy. In *Methods in porphyrin photosensitisation* (ed. D. Kessel), pp. 293–302, Plenum, New York.

Redmond, R.W., Heihoff, K., Braslavsky, S.E., and Truscott, T.G. (1987). Thermal-lensing and phosphorescence emission studies of the quantum yield and lifetime of singlet oxygen ($^1\Delta_g$) sensitised by hematoporphyrin and related porphyrins in deuterated and non deuterated ethanols. *Photochem. Photobiol.*, **45**, 209–13.

Redpath, J.L. (1973). Pulse radiolysis of dithiothreitol. *Radiat Res.*, **54**, 364–74.

Rehm, D. and Weller, A. (1970). Kinetics of fluorescence quenching by electron and hydrogen-atom transfer. *Isr. J. Chem.*, **8**, 259–71.

Reichardt, C. (1988). *Solvents and solvent effects in organic chemistry*. VCH, Weinheim.

Reichardt, C. (1990) *Solvents and solvent effects in organic chemistry* (2nd edn). VCH, Weinheim.

Reszka, K., Hartley, J.A., and Lown, J.W. (1990). Photosensitization by selected anticancer agents. *Biophys. Chem.*, **35**, 313–23.

Ribeiro, D.T., Bourre, F., Sarasin, A., Di Mascio, P., and Menck, C.F.M. (1992). DNA synthesis blocking lesions induced by singlet oxygen are targeted to deoxyguanosines. *Nucleic Acids Res*, **20**, 2465–9.

Richards, JT. and Thomas, J.K. (1970). Laser and flash photolytic studies on the effects of various solvents and solutes on the excited singlet and triplet states of *N*, *N*, *N'*, *N'*,-tetramethyl paraphenylene diamine (TMPD). *Trans. Faraday Soc.*, **66**, 621–32.

Richardson, O.W. and Compton, K.T. (1912). The photoelectric effect. *Phys. Rev.*, **34**, 393–6.

Richter, A.M., Kelly, B., Chow, J., Liu, D.J., Towers, G.H.N., Dolphin, D., and Levy, J.G. (1987). Preliminary studies on a more effective phototoxic agent than hematoporphyrin. *J. Natl. Cancer Inst.* **79**, 1327–32.

Richter, A. M., Waterfield, A., Jain, A.K., Sternberg, E., Dolphin, D., and Levy, J.G. (1990). *In vitro* evaluation of phototoxic properties of four structurally related benzoporphyrin derivatives. *Photochem. Photobiol.*, **52**, 495–500.

Riesz, P., Berdahl, D., and Christman, C. L. (1985). Free radical generation by ultrasound in aqueous and nonaqueous solutions. *Environmental Health Perspectives*, **64**, 233–52.

Rigler, R. and Claesens, F. (1986). Picosecond time domain spectroscopy of structure and dynamics in nucleic acids. In *Structure and dynamics of RNA. Series A: Life Sciences*, Vol. 110 (ed. P.H. von Knippenberg and C.W. Hilbers), pp. 45–54. Plenum Press, New York.

Riley, P.A. (ed.) (1984) *Hydroxyanisole. Recent advances in anti-melanoma therapy*. IRL Press, Oxford.

Riley, P.A. (1985). Radicals and melanoma. *Phil Trans. R. Soc. London*, **B311**, 679–89.

Roberts, J.E. (1984). The photodynamic effects of chlorpromazine, promazine and hematoporphyrin on lens protein. *Invest. Ophthalmol. Vis. Sci.*, **25**, 746–50.

Roberts, Jr, J.L. and Sawyer, D.T. (1981). Facile degradation by superoxide anion of carbon tetrachloride, chloroform, methylene chloride and *p*, *p'*-DDT in aprotic media. *J. Am. Chem. Soc.*, **103**, 712–14.

Roberts, P.B. and Fielden, E.M. (1971). Pulse radiolysis studies of the radiosensitizer nor-pseudopelletierine-*N*-oxyl (NPPN). II. Reactions involving biological radicals. *Int. J. Radiat. Biol.*, **20**, 363–71.

Roberts, W.G., Smith, K.M., McCullough, J.J., and Berns, M.W. (1989). Skin photosensitivity and photodestruction of several potential photodynamic sensitizers. *Photochem. Photobiol.*, **49**, 431–8.

Rockley, M.G. and Devlin, J.P. (1977). Observation of a nonlinear photoacoustic signal with potential application to nanosecond time resolution. *Appl. Phys. Lett.*, **31**, 24–5.

Rodgers, M.A.J. (1983*a*). Molecular excited states in liquid systems. In *The study of fast processes and transient species by electron pulse radiolysis* (ed. J.H. Baxendale and F. Busi), pp. 535–50. Reidel, Dordrecht.

Rodgers, M.A.J.(1983*b*). Solvent-induced deactivation of singlet oxygen: additivity relationships in nonaromatic solvents *J. Am. Chem. Soc.*, **105**, 6201–5.

Rodgers, M.A.J. (1985). The photoproperties of porphyrins in model biological environments. In *Photodynamic therapy of tumours and other diseases* (ed. G. Jori and C. Perria), pp. 21–35. Edizioni Libreria Progetto, Padova.

Rodgers, M.A.J. (1988). On the problems involved in detecting luminescence from singlet oxygen in biological specimens. *J. Photochem. Photobiol., B: Biol*, **1**, 371–3.

Rodgers, M.A.J. and Bates, A.L. (1980) Kinetic and spectroscopy features of some carotenoid triplet states: sensitization by singlet oxygen. *Photochem. Photobiol.*, **31**, 533–7.

Rodgers, M.A.J. and Snowden, P.T. (1982). The lifetime of O_2 ($^1\Delta_g$) in liquid water as determined by time resolved infrared luminescence. *J. Am. Chem. Soc.*, **104**, 5541–3.

Roeder, B. and Wabnitz, H. (1987). Time-resolved fluorescence spectroscopy of haematoporphyrin, mesoporphyrin, pheophorbide *a* and chlorin e_6 in ethanol and aqueous solutions. *J. Photochem. Photobiol., B: Biol.* **1**, 103–13.

Rogan, E.G., Cavalieri, E.L., Tibbels, S.R., Cremonesi, P., Warner, C.D., Nagel, D.L., Tomer, K.B., Cerny, R.L., and Gross, M.L. (1988). Synthesis and identification of benzo[*a*]pyrene–guanine nucleoside adducts formed by electrochemical oxidation and by horseradish peroxidase catalysed reaction of benzo[*a*]pyrene with DNA. *J. Am. Chem. Soc.*, **110**, 4023–9.

Ronfard-Haret, J.C., Averbeck, D., Bensasson, R.V., Bisagni, E., and Land, E.J. (1982). Some properties of the triplet excited state of the photosensitizing furocoumarin: 3-carbethoxypsoralen. *Photochem. Photobiol.*, **35**, 479–89.

Ronfard-Haret, J.C., Averbeck, D., Bensasson, R.V., Bisagni, E., Land, E.J., and Moron, J. (1987). Correlation between the triplet photophysical properties and the photobiological actions in yeast of two monofunctional pyridopsoralens. *Photochem. Photobiol.*, **44**, 235–9.

Rorsman, H., Agrup, G., Hansson, C., and Rosengren, E. (1993). Biochemical recorders of malignant melanoma. In *Malignant melanoma, advances of a decade* (Pigment Cell Vol. 6) (ed. R.M. MacKie), pp. 93–115. Karger, Basel.

Rosen, G.M. and Halpern, H.J. (1990). Spin trapping biologically generated free radicals: correlating formation with cellular injury. In *Methods in enzymology*, Vol. 186, Oxygen radicals in biological systems, Part B Oxygen radicals and antioxidants (ed. L. Packer and A.N. Glazer), pp. 611–21. Academic Press, San Diego.

Rosencwaig, A. (1980). *Photoacoustics and photoacoustic spectroscopy*. J. Wiley, New York.

Rosenfeld, T., Alchalel, A., and Ottolenghi, M. (1973). Primary photoprocesses in retinol. *Chem. Phys. Lett.*, **20**, 291–7.

Rosenfeld, T., Alchalel, A., and Ottolenghi, M. (1976) Intersystem crossing, ionic dissociation and *cis–trans* isomerization mechanisms in the photolysis of retinol and related molecules. In *Excited states of biological molecules* (ed. J.B. Birks), pp. 540–54. Wiley, New York.

Rosenfeld, T., Honig, B., Ottolenghi, M., Hurley, J., and Ebrey, T.G. (1977). *Cis–trans* isomerization in the photochemistry of vision. *Pure Appl. Chem.*, **49**, 341–51.

Rosenthal, I. and Frimer, A.A. (1976). The quenching effect of iodide ion on singlet oxygen. *Photochem. Photobiol.*, **23**, 209–11.

Rougée, M. and Bensasson, R.V. (1986). Détermination des constantes de vitesse de désactivation de l'oxygène singulet ($^1\Delta_g$) en présence de biomolécules. *C.R. Acad. Sci. Paris*, **302**, 1223–6.

Rougée, M., Ebbesen, T., Ghetti, F., and Bensasson, R.V. (1982). Kinetics and mechanism of porphyrin-sensitized reduction of methyl viologen. *J. Phys. Chem.*, **86** 4404–12.

Rougée, M., Bensasson, R.V., Land, E.J., and Pariente, R. (1988). Deactivation of singlet molecular oxygen by thiols and related compounds, possible protectors against skin photosensitivity. *Photochem. Photobiol.*, **47**, 485–9.

Rousseau, J., Langlois, R., Ali, H., and Van Lier, J.E. (1990) Biological activities of phthalocyanines XII: synthesis, tumor uptake and biodistribution of ^{14}C-labeled disulfonated and trisulfonated gallium phthalocyanine in C3H mice. *J. Photochem. Photobiol., B: Biol.*, **6**, 121–32.

Roza, L., Baan, R.A., Van der Leun, J.C., and Kligman, L. (1989). UVA hazards in skin associated with the use of tanning equipment. *J. Photochem. Photobiol., B: Biol.*, **3**, 281–8.

Rush, J.D., Maskos, Z., and Koppenol, W.H. (1990). Distinction between hydroxyl radical and ferryl species. In *Methods in enzymology*, Vol. 186, Oxygen radicals in biological systems, Part B Oxygen radicals and antioxidants (ed. L. Packer and A.N. Glazer), pp. 148–56. Academic Press, San Diego.

Russell, G.A. (1957). Deuterium-isotope effects in the autoxidation of aralkyl hydrocarbons. Mechanism of the interaction of peroxy radicals. *J. Am. Chem. Soc.*, **79**, 3871–7.

Rutherford, E. (1889). Uranium radiation and the electrical conduction produced by it. *Philos. Mag.*, **47** (5th Series), 109–63.

Rutherford, E. (1911). The scattering of α and β particles by matter and the structure of the atom. *Philos Mag.*, **21** (6th Series), 669–88.

Sa e Melo, M.T., Averbeck, D., Bensasson, R.V., Land, E.J., and Salet, C. (1979). Some furocoumarins and analogs: Comparison of triplet properties in solution with photobiological activities in yeast. *Photochem. Photobiol.*, **30**, 645–51.

Sa e Melo, T., Macanita, A., Prieto, M., Bazin, M., Ronfard-Haret, J.C., and Santus, R. (1988). Photophysical behaviour of 5-methoxypsoralen in dioxane–water mixtures. *Photochem. Photobiol.*, **48**, 429–37.

Saito, I., Matsuura, T., and Inoue, K. (1983). Formation of superoxide ion via one-electron transfer from electron donors to singlet oxygen. *J. Am. Chem. Soc.*, **105**, 3200–6.

Salet, C. and Bensasson, R. (1975). Studies on the thymine and uracil triplet excited state in acetonitrile and water. *Photochem. Photobiol.*, **22**, 231–5.

Salet, C. and Moreno, G. (1990). Photosensitisation of mitochondria. Molecular and cellular aspects. *J. Photochem. Photobiol., B: Biol*, **5**, 133–50.

Salet, C., Bensasson, R., and Becker, R.S. (1979). Triplet excited states of pyrimidine nucleosides and nucleotides. *Photochem. Photobiol.*, **30**, 325–9.

Salet, C., Sa e Melo, M.T., Bensasson, R., and Land, E.J. (1980). Photophysical properties of aminomethylpsoralen in presence and absence of DNA. *Biochim. Biophys. Acta*, **607**, 379–83.

Salikhov, K.M., Molin, Yu. N., Sagdeev, R.Z., and Buchachenko, A.L. (1984). *Spin polarization and magnetic effects in radical reactions*. Elsevier, New York.

Saltiel, J. and Atwater, B.W. (1988). Spin-statistical factors in diffusion-controlled reactions. *Adv. Photochem.*, **14**, 1–90.

Saltiel, J. and Hammond, G.S. (1963). Mechanisms of photoreactions in solution. XVIII. Energy transfer with non vertical transitions. *J. Am. Chem. Soc.*, **85**, 2516–7.

Saltiel, J. Marchand, G.R., Kirkor-Kaminska, E., Smothers, W.K., Mueller, W.B., and Charlton, J.L. (1984). Nonvertical triplet excitation transfer to *cis*- and *trans*-stilbene. *J. Am. Chem. Soc.*, **106**, 3144–51.

Samuni, A. and Neta, P. (1873). Electron spin resonance study of the reaction of hydroxyl radicals with pyrrole, imidazole and related compounds. *J. Phys. Chem.*, **77**, 1629–35.

Sancar, A. (1987). DNA repair *in vitro*, Photobiochem. and Photobiophys., Suppl. In *Photophysics to photobiology* (ed. A. Favre, R. Tyrrell, and J. Cadet), pp. 301–15. Elsevier, Amsterdam.

Sanche, L., Bades, G., and Caron, L. (1982). Transmission of 0–15 eV monoenergetic electrons through aliphatic and alicyclic hydrocarbon films. *J. Chem. Phys.*, **76**, 4016–27.

Santus, R., Hélène, C., Ovadia, J., and Grossweiner, L.I. (1972). Splitting of thymine dimer by hydrated electrons. *Photochem. Photobiol.*, **16**, 65–7.

Santus, R., Bazin, M., and Aubailly, M. (1980). Nature, identification and properties of intermediates produced by UV excitation of indole derivatives at low and room temperatures. Some applications to tryptophan-containing proteins. *Rev. Chem. Intermediates*, **3**, 231–83.

Saran, M., Tait, D., Bors, W., and Michel, C. (1983). Formation and reactivities of alkoxy radicals. In *Oxy radicals and their scavenger systems*, Vol. I *Molecular aspects* (ed. G. Cohen and R.A. Greenwald), pp. 20–5. Elsevier, New York.

Saran, M., Michel, C., and Bors, W. (1990). Reaction of NO with $O_2^{\cdot-}$. Implications for the action of endothelium-derived relaxing factor (EDRF). *Free Rad. Res. Commun.*, **10**, 221–6.

Sarna, T., Hyde, J.S., and Swartz, H.M. (1976). Ion-exchange in melanin: an electron spin resonance study with lanthanide probes. *Science*, **192**, 1132–4.

Sarna, T., Pilas, B., Land, E.J., and Truscott, T.G. (1986). Interaction of radicals from water radiolysis with melanin. *Biochim. Biophys. Acta*, **883**, 162–7.

Sato, S., Iwaizumi, M., Handa, K., and Tamura, Y. (1977). Electron spin resonance study on the mode of generation of free radicals of daunomycin, adriamycin and carboquone in NAD(P)H-microsome system. *Gann*, **68**, 603–8.

Sauer, M.C., Romero, C., and Schmidt, K.H., (1987). Dynamics of geminate ions in the pulse radiolysis of cyclohexane and *n*-hexane containing aromatic solutes studied by streak camera measurements of fluorescence and absorption. *Radiat. Phys. Chem.*, **29**, 361–73.

Sawyer, D.T. (1990). The chemistry of dioxygen species (O_2, $O_2^{\cdot-}$, $HOO^{\cdot}$ and HOOH) and their activation by transition metals. *Int. Rev. Exp. Pathol.*, **31**, 109–30.

Sawyer, D.T., Nanni, E.J., and Roberts, J.L. (1982*a*). The reaction chemistry of superoxide ion in aprotic media. *Adv. Chem. Ser.*, **201**, 585–600.

Sawyer, D.T., Chiericato, Jr, G., Angelis, C.T., Nanni, Jr, E. J., and Tsuchiya, T. (1982*b*). Effects of media and electrode materials on the electrochemical reduction of dioxygen. *Anal. Chem.*, **54**, 1720–4.

Sawyer, D.T., Yamaguchi, K., and Calderwood, T.S. (1985). Electrochemical generation of superoxide ion and other oxy radicals. In *Handbook of methods for oxygen radical research* (ed. R.A. Greenwald), pp. 65–9. CRC Press, Boca Raton, Florida.

Scaiano, J.C. (1987). Kinetic studies of alkoxyl radicals. In *Oxygen radicals in biology and medicine* (ed. M.G. Simic, KA. Taylor, J.F. Ward, and C. von Sonntag), pp. 59–66. Plenum Press, New York.

Scandola, F., Balzani, V., and Schuster, G.B. (1981). Free-energy relationships for reversible and irreversible electron-transfer processes. *J. Am. Chem. Soc.*, **103**, 2519–23.

Schenck, C.C., Diner, B., Mathis, P., and Satoh, K. (1982). Flash-induced carotenoid radical cation formation in photosystem II. *Biochim. Biophys. Acta*, **680**, 216–27.

Schenck, G.O. (1984). Zur theorie der photosensibilisierten reaktion mit molekularem sauerstoff. *Naturwissenshaften*, **35**, 28–9.

Schenck, GO. (1949). Über die photosyntheses des askaridols und verwander peroxyde. *Angew. Chem.*, **61**, 434.

Schenck, G.O. and Koch, E. (1960). Zwischenreaktionen bei photosensibilisierten prozessen in lösungen. *Z. Elektrochem.*, **64**, 170–7.

Schenck, G.O. and Ziegler, K. (1944). Die synthese des ascaridols. *Naturwissenschaften*, **32**, 157.

Schmidt, R and Afshari, E. (1990). Effect of solvent on the phosphorescence rate constant of singlet molecular oxygen ($^1\Delta_g$). *J. Phys. Chem.*, **94**, 4377–8.

Schmidt, R. and Brauer, H.-D. (1987). Radiationless deactivation of singlet oxygen ($^1\Delta_g$) by solvent molecules. *J. Am. Chem. Soc.*, **109**, 6976–81.

Schneider, J.E., Price, S., Maidt, L., Gutteridge, J.M.C., and Floyd, R.A. (1990) Methylene blue plus light mediates 8-hydroxy-2′-deoxyguanosine formation in DNA preferentially over strand breakage. *Nucl. Acids Res.*, **18**, 631–5.

Schoenlein, R.W., Peteanu, L.A., Mathies, R.A., and Shank, C.V. (1991). The first step in vision: femtosecond isomerization of rhodopsin. *Science*, **54**, 412–15.

Scholes, G. and Simic, M. (1968). Radiolysis of aqueous solutions of DNA and related substances: reaction of hydrogen atoms. *Biochim. Biophys. Acta*, **166**, 255–8.

Schöneich, C., Asmus, K.-D, Dillinger, U., and Bruchhausen, F.v. (1989*a*). Thiyl radical attack on polyunsaturated fatty acids: A possible route to lipid peroxidation. *Biochem. Biophys. Res. Commun.*, **161**, 113–20.

Schöneich, C., Bonifacic, M., and Asmus, K.-D. (1989*b*). Reversible H-atom abstraction from alcohols by thiyl radicals: determination of absolute rate constants by pulse radiolysis. *Free Rad. Res. Commun.*, **6**, 393–405.

Schöneich, Ch., Bonifacic, M., Dillinger, U., and Asmus, K.-D. (1991). Hydrogen abstraction by thiyl radicals from activated C—H bonds of alcohols, ethers and polyunsaturated fatty acids. In *Sulfur-centered reactive intermediates in chemistry and biology*. (ed. C. Chatgilialoglu and K.-D. Asmus) NATO-ASI Series A, Life Sciences. Vol. 197, pp. 367–76. Plenum Press, New York.

Schroder, E.W. and Black, P.H. (1980). Retinoids: tumor preventers or tumor enhancers. *J. Natl Canc. Inst.*, **65**, 671–4.

Schrödinger, E. (1926). Quantisierung als eigenwertproblem. *Ann. Phys.*, **79**, 361–76.

Schuchmann, M.N., Schuchmann, H.-P., Hess, M., and von Sonntag, C. (1991). $O_2^{-\cdot}$

addition to ketomalonate leads to decarboxylation: a chain reaction in oxygenated aqueous solutions. *J. Am. Chem. Soc.*, **113**, 6934–7.

Schuler, R.H. and Kuntz, R.R. (1963). Methyl radical production in the radiolysis of hydrocarbons. J. Phys. Chem., **67**, 1004–11.

Schuler, R.H., Patterson, L.K., and Janata, E. (1980). Yield for the scavenging of OH radicals in the radiolysis of N_2O-saturated aqueous solutions. *J. Phys. Chem.*, **84**, 2088–9.

Schulte-Frohlinde, D., Simic, M.G., and Görner, H. (1990). Laser-induced strand break formation in DNA and polynucleotides. *Photochem. Photobiol.*, **52**, 1137–51.

Schulte-Herbrüggen, T. and Cadenas, E. (1985). Electronically excited states generation during the lipoxygenase-catalyzed aerobic oxidation of arachidonate. The effect of reduced glutathione. *Photobiochem. Photobiophys.*, **10**, 35–51.

Schwarz, H.A. (1969). Applications of the spur diffusion model to the radiation chemistry of aqueous solutions. *J. Phys. Chem.*, **75**, 1928–37.

Scurlock, R.D. and Ogilby, P.R. (1987). Effect of the solvent on the rate constant for the radiative deactivation of singlet molecular oxygen ($^1\Delta_gO_2$). *J. Phys. Chem.*, **91**, 4599–602.

Scurlock, R., Rougée, M., and Bensasson, R.V. (1990). Redox properties of phenols, their relationships to singlet oxygen and to their inhibitory effects on benzo(a)pyrene-induced neoplasia. *Free Rad. Res. Commun.*, **8**, 251–8.

Sealy, R.C., Felix, C.C., Hyde, J.S., and Swartz, H.M. (1980). Structure and reactivity of melanins: influence of free radicals and metal ions. In *Free radicals in biology* IV (ed. W.A. Pryor), pp. 209–59. Academic Press, New York.

Sealy, R.C., Puzyna, W., Kalyanaraman, B., and Felix, C.C. (1984). Identification by electron spin resonance spectroscopy of free radicals produced during autoxidative melanogenesis. *Biochim. Biophys. Acta*, **800**, 269–76.

Sekaki, A., Gardes-Albert, M., Houee-Levin, C., Ferradini, C., Rivalle, C., Bisagni, E., and Hickel, B., (1989). Influence of oxygen and superoxide free radicals on the oxidative activation of several antitumour drugs related to ellipticine. A pulse radiolysis study. In *Medical, biochemical and chemical aspects of free radicals* (ed. O. Hayashi, E. Niki, M. Kondo, and T. Yoshikawa), pp. 97–100. Elsevier, Amsterdam.

Seki, H. and Schnabel, W. (1982). On the free radical-induced aggregation of ribonuclease—A pulse radiolysis study using the light scattering detection method. *Z. Naturforsch. (C)*, **37**, 63–9.

Seliger, H.H. and McElroy, W.D. (1960). Pathways of energy transfer in bioluminescence. *Radiat. Res.*, Suppl. **2**, 528–38.

Setlow, J.K. (1966). The molecular basis of biological effects of ultraviolet radiation and photoreactivation. In *Current topics in radiation research*, Vol. II (ed. M. Ebert and A. Howard), pp. 195–248. North Holland, Amsterdam.

Sevilla, M.D., Becker, D., Yan, M., and Summerfield, S.R. (1991). Relative abundance of primary ion radicals in γ-irradiated DNA: cytosine vs thymine anions and guanine vs adenine cations. *J. Phys. Chem.*, **95**, 3409–15.

Shank, C.V. (1983). Measurement of ultrafast phenomena in the femtosecond time domain. *Science*, **219**, 1027–31.

Shen, X., Lind, J., and Merenyi, G. (1987). One-election oxidation of indoles and acid-base properties of the indolyl radicals *J. Phys. Chem.*, **91**, 4403–6.

Shetlar, M.D., Christensen, J., and Hom, K. (1985). Photochemical addition of amino acids and peptides to DNA. *Photochem. Photobiol.*, **39**, 125–33.

Shimomura, O. (1982). Mechanism of bioluminescence. In *Chemical and biological generation of electronically excited states* (ed. W. Adam and G. Cilento), pp. 249–76. Academic Press, New York.

Shreve, A. P., Trautman, J. K., Owens, T. G., and Albrecht, A. C. (1990). Two-photon excitation spectroscopy of thylakoid membranes from *Phaeodactylum tricornutum*: evidence for an *in vivo* two-photon-allowed carotenoid state. *Chem. Phys. Lett.*, **170**, 51–6.

Shreve, A. P., Trautman, J. K., Owens, T. G., and Albrecht, A. C. (1991*a*). A femtosecond study of the electronic state dynamics of fucoxanthin and implications for photosynthetic carotenoid-to-chlorophyll energy transfer mechanisms. *Chem. Phys.*, **154**, 171–8.

Shreve, A. P., Trautman, J. K., Owens, T. G., and Albrecht, A. C. (1991*b*). Determination of the S_2 lifetime of β-carotene. *Chem. Phys. Lett.*, **178**, 89–96.

Siddiqi, M. A. and Bothe, E. (1987). Single- and double-strand break formation in DNA irradiated in aqueous solution: dependence on dose and OH radical scavenger concentration. *Radiat. Res.*, **112**, 449–63.

Siders, F. and Marcus, R. A. (1981). Quantum effects in electron transfer reactions. *J. Am. Chem. Soc.*, **103**, 741–7.

Sik, R. H., Paschall, C. S., and Chignell, C. F. (1983). The phototoxic effects of benoxaprofen and its analogs on human erythrocytes and rat peritoneal mast cells. *Photochem. Photobiol.*, **38**, 411–15.

Silva, E., Rückert, V., Lissi, E., and Abuin, E. (1991). Effects of pH and ionic micelles on the riboflavin—sensitized photoprocesses of tryptophan in aqueous solution. *J. Photochem. Photobiol., B: Biol.*, **11**, 57–68.

Simic, M. (1990). Pulse radiolysis in the study of oxygen radicals. In *Methods in enzymology*, Vol. 186, Oxygen radicals in biological systems, Part B Oxygen radicals and antioxidants (ed. L. Packer and A. N. Glazer), pp. 89–100. Academic Press, San Diego.

Simic, M. G. (1991). DNA damage, environmental toxicants, and rate of aging. *Environ. Carcino. Ecotox. Rev.*, **C9** (1), 113–51.

Simic, M. G. (1992). Urinary biomarkers and the rate of DNA damage in carcinogenesis and anticarcinogenesis. *Mutation Res.* **267**, 277–90.

Simic, M. G. and Bergtold, D. S. (1991). Dietary modulation of DNA damage in humans. *Mutation Research*, **250**, 17–24.

Simic, M. G. and Dizdaroglu, M. (1985). Formation of radiation-induced cross-links between thymine and tyrosine: Possible model for cross-linking of DNA and proteins by ionizing radiation. *Biochemistry*, **24**, 233–6.

Simic, M. and Hayon, E. (1971). Reductive deamination of oligo peptides by solvated electrons in aqueous solution. *Radiat. Res.*, **48**, 244–55.

Simic, M. and Powers, E. L. (1974). Correlation of the efficiencies of some radiation sensitizers and their redox potentials. *Int. J. Radiat. Biol.*, **26**, 87–90.

Simic, M. G., Jovanovic, S. V. and Niki, E. (1992). Mechanisms of lipid oxidative processes and their inhibition. *A.C.S. Symp. Ser.*, **500**, 14–32.

Simpson, M. S. C., Beeby, A., Bishop, S. M., MacRobert, A. J., Parker, A. W., and Phillips, D. (1992). Time resolved spectroscopic studies of sulphonated aluminium phthalocyanine triplet states. In *SPIE—The International Society for Optical Engineering—Symposia—Time resolved laser spectroscopy in biochemistry, III*, pp. 19–25, January 1992. Los Angeles, California, USA.

Singh, A., Koroll, G. W., and Cundall, R. B. (1982). Pulse radiolysis of aqueous solutions of sodium azide: Reactions of azide radical with tryptophan and tyrosine. *Radiat. Phys. Chem.*, **19**, 137–46.

Slater, T. F. (1987). Free radicals and tissue injury: fact and fiction. *Br. J. Cancer*, **55**, Suppl. VIII, 5–10.

Slifkin, M. A. (1971). *Charge transfer interactions of biomolecules*. Academic Press, London.

Sloper, R.W. and Truscott, T.G. (1980). Excited states of bilirubin. *Photochem. Photobiol.*, **31**, 445–50.

Sloper, R.W. and Truscott, T.G. (1982). The quantum yield for bilirubin photoisomerisation. *Photochem. Photobiol.*, **35**, 743–5.

Sloper, R.W., Truscott, T.G., and Land, E.J. (1979). The triplet state of 8-methoxypsoralen. *Photochem. Photobiol.*, **29**, 1025–9.

Smith, G.J. (1992). Singlet oxygen produced by UV irradiation of wool keratin. *J. Photochem. Photobiol., B: Biol.*, **12**, 173–8.

Smith, K.C. (1966). Physical and chemical changes induced in nucleic acids by ultraviolet light. *Radiat. Res.*, Suppl. 6, 54–79.

Smith, K.C. (1976*a*). The radiation-induced addition of proteins and other molecules to nucleic acids. In *Photochemistry and photobiology of nucleic acids*, Vol. 2 (ed. S.Y. Wang), pp. 187–218. Academic Press, New York.

Smith, K.C. (ed.) (1976*b*). *Aging, carcinogenesis and radiation biology: The role of nucleic acid addition reactions*. Plenum, New York.

Smith, K.M. (1975). *Porphyrins and metalloporphyrins*. Elsevier, Amsterdam.

Smithen, C.E., Clarke, E.D., Dale, J.A., Jacobs, R.S., Wardman, P., Watts, M.E., and Woodcock, M. (1980). Novel (nitro-1-imidazolyl) alkanolamines as potential radiosensitizers with improved therapeutic properties. In *Radiation sensitizers: Their use in the clinical management of cancer* (ed. L.W. Brady), pp. 22–32. Masson, New York.

Snel, M.M.E., Kaptein, R., and de Kruijff, B. (1991). Interaction of apocytochrome *c* and derived polypeptide fragments with sodium dodecyl sulfate micelles monitored by photochemically induced dynamic nuclear polarization 1H NMR and fluorescence spectroscopy. *Biochemistry*, **30**, 3387–95.

Snyder, S.H. and Bredt, D.S. (1992). Biological roles of nitric oxide. *Sci. American*, May, 28–35.

Sohal, R.S., Svensson, I., and Brunk, U.T. (1990). Hydrogen peroxide production by liver mitochondria in different species. *Mech. Ageing Dev.*, **53**, 209–15.

Solar, S., Getoff, N., Surdhar, P.S., Armstrong, D.A., and Singh, A. (1991). Oxidation of tryptophan and *N*-methyl indole by $N_3^{\cdot}$, $Br_2^{\cdot -}$, and $(SCN)_2^{\cdot -}$ radicals in light- and heavy-water solutions: a pulse radiolysis study. *J. Phys. Chem.*, **95**, 3639–43.

Sommer, A., Paltauf, F., and Hermetter, A. (1990). Dipolar solvent relaxation on a nanosecond time scale in ether phospholipid membranes as determined by multi frequency phase and modulation fluorometry *Biochemistry*, **29**, 11 134–40.

Song, P.-S. and Tapley, K.J. (1979). Photochemistry and photobiology of psoralens. *Photochem. Photobiol.*, **29**, 1177–97.

Song, P.-S., Harter, M.L., Moore, T.A., and Herndon, W.C. (1971). Luminescence spectra and photocycloaddition of the excited coumarins to DNA bases. *Photochem. Photobiol.*, **14**, 521–30.

von Sonntag, C. (1987*a*). *The chemical basis of radiation biology*. Taylor and Francis, London.

von Sonntag, C. (1987*b*). Pulse radiolysis of nucleic acids and their base constituents: an updating review. *Radiat. Phys. Chem.*, **30**, 313–30.

von Sonntag, C. (1990). Free-radical reactions involving thiols and disulphides. In *Sulfur-centered reactive intermediates in chemistry and biology* (ed. C. Chatgilialoglu and K-D Asmus), pp. 359–66. Plenum, New York.

von Sonntag, C. and Schuchman, H.-P. (1991). The elucidation of peroxyl radical reactions in aqueous solution with the help of radiation-chemical methods. *Angew. Chem., Int. Ed. Engl.*, **30**, 1229–53.

Sonoda, M., Murali, Krishna, C., and Riesz, P. (1987). The role of singlet oxygen in the photohemolysis of red blood cells sensitised by phthalocyanine sulfonates. *Photochem. Photobiol.*, **46**, 625–31.

Specht, K.G., Kittler, L., and Midden, W.R. (1988). A new biological target of furocoumarins: Photochemical formation of covalent adducts with unsaturated fatty acids. *Photochem. Photobiol.*, **47**, 537–41.

Speiser, S.R., Van der Werf, R., and Kommandeur, J. (1973). Photoquenching: The dependence of the primary quantum yield of a monophotonic laser-induced photochemical process on the intensity and duration of the exciting pulse. *Chem. Phys.*, **1**, 297–305.

Spikes, J.D. (1977). Photosensitisation. In *The science of photobiology* (ed. K.C. Smith), pp. 87–112. Plenum Press, New York.

Spikes, J.D. and Jori, G. (1987). Photodynamic therapy of tumours and other diseases using porphyrins. *Lasers in Medical Science*, **2**, 3–15.

Stadler, L.J. and Uber, F.M. (1942). Genetic effects of ultraviolet radiation on maize. IV. Comparison of monochromatic radiations. *Genetics*, **27**, 84–118.

Star, W.M., Marijnissen, H.P.A., van den Berg-Blok, A.E., Versteeg, J.A., Franken, K.A., and Reinhold, H.S. (1986). Destruction of rat mammary tumour and normal tissue micro-circulation by haematoporphyrin derivative photosensitization observed *in vivo* in sandwich observation chambers. *Cancer Res.*, **46**, 2532–40.

Stark, J. (1908). Further remarks upon thermal and chemical absorption in the band spectrum. *Z. Physik*, **9**, 889–94 (see also (1912) *Ann. Phys.*, **38**, 407–30).

Steenken, S. (1989). Purine bases, nucleosides and nucleotides: Aqueous solution redox chemistry and transformation reactions of their radical cations and e^- and OH adducts. *Chem. Rev.*, **89**, 503–20.

Steenken, S. and Neta, P. (1982). One-electron redox potentials of phenols. Hydroxy- and aminophenols and related compounds of biological interest. *J. Phys. Chem.*, **86**, 3661–7.

Steenken, S., Telo, J.P., and Candeias, L.P. (1992). One-electron-reduction potentials of pyrimidine bases, nucleosides and nucleotides in aqueous solution. Consequences for DNA redox chemistry. *J. Am. Chem. Soc.*, **114**, 4701–9.

Steiner, R.F. and Kubota, Y. (1983). Fuorescent dye–nucleic acid complexes. In *Excited states of biopolymers* (ed. R.F. Steiner and Y. Kubota), pp. 203–54. Plenum Press, New York.

Stephenson, L.M. (1980). The mechanism of the singlet oxygen ene reaction. *Tetrahedron Lett.*, **21**, 1005–8.

Stern, L. (1984). Bilirubin metabolism in the newborn: Its mechanisms and relationship to kernicterus. In *Neonatal jaundice* (ed. F.F. Rubaltelli and G. Jori), pp. 1–12. Plenum Press, New York.

Stern, R.S., Thibodeau, L.A., Kleinerman, R.A., Parrish, J.A., and Fitzpatrick, T.B. (1979). Risk of cutaneous carcinoma in patients treated with oral methoxsalen photochemotherapy for psoriasis. *New Engl. J. Med.*, **300**, 809–13.

Sternberg, E., Dolphin, D., Scurlock, R., Rougée, M., and Bensasson, R.V. (1993). Photophysical properties of benzoporphyrin derivatives, candidate photosensitizers for photochemotherapy. *Photochem. Photobiol.*, submitted.

Stevens, B. and Algar, B.E. (1968). The photoperoxidation of unsaturated organic molecules. II. The autoperoxidation of aromatic hydrocarbons. *J. Phys. Chem.*, **72**, 3468–74.

Stob, S. and Kaptein, R. (1989). Photo-CIDNP of the amino-acids. *Photochem. Photobiol.*, **49**, 565–77.

Stob, S., Scheek, R.M., and Kaptein, R. (1989). Photo-CIDNP in guanine derivatives, a mechanistic study. *Photochem. Photobiol.*, **49**, 717–23.

Strambini, G.B. and Gonnelli, M. (1985). The indole nucleus triplet-state lifetime and its dependence on solvent microviscosity. *Chem. Phys. Lett.*, **115**, 196–200.

Strickler, S.J. and Berg, R.A. (1962). Relationship between absorption intensity and fluorescence lifetime of molecules. *J. Chem. Phys.*, **37**, 814–22.

Stuglik, Z. and Zagorski, Z.P. (1981). Pulse radiolysis of neutral iron(II) solutions: oxidation of ferrous ions by $OH^{\cdot}$ radicals. *Radiat. Phys. Chem.*, **17**, 229–33.

Sturgeon, R.J. and Schulman, S.G. (1977). Electronic absorption spectra and protolytic equilibria of doxorubicin: direct spectrophotometric determination of microconstants. *J. Pharm. Sci.*, **66**, 958–61.

Sugimoto, H., Sawyer, D.T., and Kanofsky, J.R. (1988). Transient formation of hydrogen tetraoxide [HOOOOH] from the dimerisation of perhydroxyl ($HOO^{\cdot}$): concerted homolytic dissociation to singlet oxygen ($^1O_2$) and hydrogen peroxide. *J. Am. Chem. Soc.*, **110**, 8707–8.

Sugumaran, M. (1986). Tyrosinase catalyses an unusual oxidative decarboxylation of 3,4-dihydroxymandelate. *Biochemistry*, **25**, 4489–92.

Sugumaran, M. and Lipke, H. (1983*a*). Quinone methide formation—A new reaction catalysed by cuticular polyphenol oxidase(s) from *Sarcophaga bullata. Fed. Proc. Fed. Am. Soc. Exp. Biol.*, **42**, 1828.

Sugumaran, M. and Lipke, H. (1983*b*). Quinone methide formation from 4-alkylcatechols—A novel reaction catalysed by cuticular polyphenol oxidase. *FEBS Lett.*, **155**, 65–8.

Summerfield, F.W. and Tappel, A.L. (1984). Effects of dietary polyunsaturated fats and vitamin E on aging and peroxidative damage to DNA. *Arch. Biochem. Biophys.*, **233**, 408–16.

Suppan, P. (1990). Solvatochromic shifts: the influence of the medium on the energy of electronic states. *J. Photochem. Photobiol., A: Chem.*, **50**, 293–330.

Surdhar, P.S. and Armstrong, D.A. (1986). Redox potential of some sulfur-containing radicals. *J. Phys. Chem.*, **90**, 5915–17.

Surdhar, P.S., Mezyk, S.P., and Armstrong, D.A. (1989). Reduction potential of the $CO_2^{\cdot -}$ radical anion in aqueous solutions. *J. Phys. Chem.*, **93**, 3360–3.

Sutherland, B.M. (1981). Photoreactivation. *Bioscience*, **31**, 439–44.

Sutin, N. (1973). Oxidation–reduction in coordination compounds. In *Inorganic biochemistry*, Vol. 2 (ed. G.L. Eichhorn), pp. 611–53. Elsevier, New York.

Sutin, N. and Brunschwig, B.S. (1990). Some aspects of electron transfer in biological systems. In *Electron transfer in biology and the solid state. Inorganic compounds with unusual properties*. (ed. M.K. Johnson, R.B. King, D.M. Kurtz, C. Kutal, M.L. Norton, and R.A. Scott), Advances in Chemistry Series, Vol. 226, pp. 65–88. American Chemical Society, Washington, DC.

Svingen, B.A. and Powis, G. (1981). Pulse radiolysis studies of antitumour quinones: radical lifetimes, reactivity with oxygen and one-electron reduction potentials. *Arch. Biochem. Biophys.*, **209**, 119–26.

Swallow, A.J. (1969). Hydrated electrons in sea water. *Nature*, **222**, 369–70.

Swallow, A.J. (1991). Effects of irradiation on food proteins. In Developments in food proteins—7 (ed. Hudson, B.J.F.), pp. 195–229. *Elsevier Applied Science*, London.

Sykes, A.G. (1991). Plastocyanin and the blue copper proteins. In *Long-range electron transfer in biology* (ed. G. Palmer), pp. 175–224. Springer-Verlag, Berlin.

Sykes, A. and Truscott, T.G. (1969). A metastable state of retinol (vitamin A). *J. Chem. Soc., Chem. Commun.*, 929–30.

Sykes, A. and Truscott, T.G. (1970). The *in vitro* photochemistry of biological molecules. Part 1. Energy transfer reactions involving retinol (Vitamin A). *Trans. Faraday. Soc.*, **66**, 679–86.

Szabo, A.G. (1989). The fluorescence properties of aromatic amino acids: Their role in the understanding of enzyme structure and dynamics. In *The enzyme catalysis process. Energetics, mechanism and dynamics* (ed. A. Cooper, J.L. Houben, and L.C. Chien), pp. 123–39. Plenum, New York.

Szabo, A.G., Riddell, W.D., and Yip, R.W. (1970). Detection and chemistry of the triplet state of thymine. *Can J. Chem.*, **48**, 694–6.

Szabo, A.G., Wierzchowski, K.L., and Berens, K. (1975). Triplet–triplet absorption studies of the intersystem crossing mechanism of 2-aminopurines. *J. Luminescence*, **10**, 331–43.

Szent-Györgi, A., Isenberg, I., and Baird, S.L. (1960). On the electron donating properties of carcinogens. *Proc. Natl. Acad. Sci. USA*, **46**, 1444–9.

Tabata, Y., Kobayashi, H., Washio, M., Tagawa, S., and Yoshida, Y. (1985). Pulse radiolysis with picosecond time resolution. *Radiat Phys. Chem.*, **26**, 473–9.

Takamura, K., Sakamoto, M., and Hayakawa, Y. (1979). Polarographic study: The reaction of K vitamins with thiols and the consecutive determination of the K vitamins. *Anal. Chim. Acta*, **106**, 261–9.

Takasugi, M., Guendouz, A., Chassignol, M., Decout, J.L., Lhomme, J., Thuong, N.T., and Hélène, C. (1991). Sequence-specific photoinduced cross-linking of the two strands of double-helical DNA by a psoralen covalently linked to a triple helix-forming oligonucleotide. *Proc. Natl. Acad. Sci. USA*, **88**, 5602–6.

Takemura, T., Das, P.K., Hug, G., and Becker, R.S. (1976). Emission spectroscopy and state ordering of retinals. *J. Am. Chem. Soc.*, **98**, 7099–101.

Takemura, T., Hug, G., Das, P.K., and Becker, R.S. (1978). Visual pigments 9. Fluorescence of dimers of retinals. *J. Am. Chem. Soc.*, **100**, 2631–4.

Takemura, T., Ohta, N., Nakajima, S., and Sakata, I. (1989). Critical importance of the triplet lifetime of photosensitizer in photodynamic therapy of tumor. *Photochem. Photobiol.*, **50**, 339–44.

Tal, Y. and Faraggi, M. (1975). The reaction of the hydrated electron with amino acids, peptides and proteins in aqueous solution. II. Formation of radicals and electron transfer reactions. *Radiat. Res.*, **62**, 347–56.

Tamba, M. and Badiello, R. (1985). Reactions of selenium dioxide free radicals with amino acids and enzymes. *Int. J. Radiat. Biol.*, **48**, 259–70.

Tamba, M. and Quintiliani, M. (1984). Kinetic studies of reactions involved in hydrogen transfer from glutathione to carbohydrate radicals. *Radiat. Phys. Chem.*, **23**, 259–63.

Taube, H. (1970). *Electron transfer reactions of complex ions in solution*. Academic Press, New York.

Taube, H. (1984). Electron transfer between metal complexes—a retrospective view (Nobel lecture). *Angew. Chem.*, **23**, 329–39.

Taube, H. and Gould, E.S. (1969). Organic molecules as bridging groups in electron-transfer reactions. *Acc. Chem. Res.*, **2**, 321–9.

Tauber, A.I. and Babior, B.M. (1978). $O_2^{-\cdot}$ and host defence: the production and fate of $O_2^{-\cdot}$ in neutrophils. *Photochem. Photobiol.*, **28**, 701–9.

Telfer, A., De Las Rivas, J., and Barber, J. (1991). β-Carotene within the isolated photosystem II reaction center photooxidation and irreversible bleaching of this chromophore by oxidized P680. *Biochim. Biophys. Acta*, **1060**, 106–14.

Terner, J., Spiro, T.G., Nagumo, M., Nicol, M.F., and EL-Sayed, M.A. (1980). Resonance Raman spectroscopy in the picosecond time scale: The carboxy-hemoglobin photointermediate. *J. Am. Chem. Soc.*, **102**, 3238–9.

Therien, M. J. Chang, J., Raphael, A. L., Bowler, B. E., and Gray, H. B. (1991). Long-range electron transfer in metalloproteins. In *Long-range electron transfer in biology* (ed. G. Palmer), pp. 109–29. Springer-Verlag, Berlin.

Thomas, J.K. (1984). *The chemistry of excitation at interfaces*, ACS Monograph No. 181.

Thomas, J.K. and Bensasson, R.V. (1967). Direct observation of high ion and radical concentrations in the radiolysis of water and ethanol. *J. Chem. Phys.*, **46**, 4147–8.

Thomas, J.P., Hall, R.D., and Girotti, A.W. (1987). Singlet oxygen intermediacy in the photodynamic action of membrane-bound hematoporphyrin derivative. *Cancer Lett.*, **35**, 295–302.

Thompson, A.M. (1992). The oxidizing capacity of the Earth's atmosphere: Probable past and future changes. *Science*, **256**, 1157–65.

Thompson, A., Land, E.J., Chedekel, M.R., Subbarao, K.V., and Truscott, T.G. (1985). A pulse radiolysis investigation of the oxidation of the melanin precursors 3,4-dihydroxyphenylalanine (dopa) and the cysteinyldopas. *Biochim. Biophys. Acta*, **843**, 49–57.

Thrash, R.J., Fang, H.L.B., and Leroi, G.E. (1977). The Raman excitation profile spectrum of β-carotene in the preresonance region: evidence for a low-lying singlet state. *J. Chem. Phys.*, **67**, 5930–3.

Tralau, C.J., MacRobert, A.J., Coleridge-Smith, P.D., Barr, H., and Bown, S.G. (1987). Photodynamic therapy with phthalocyanine sensitisation: Quantitative studies in a transplantable fibrosarcoma of rats. *Br. J. Cancer*, **55**, 389–95.

Tran, C.D. and Beddard, G.S. (1981). Excited state properties of bilirubin and its photo properties using picosecond fluorescence and circularly polarized luminescence spectroscopy. *Biochim. Biophys. Acta*, **678**, 497–504.

Trautman, J.K., Schreve, A.P., Violette, C.A., Franck, H.A., Owens, T.G., and Albrecht, A.C. (1990). Femtosecond dynamics of energy transfer in B800-850 light-harvesting complexes of *rhodobacter sphaeroides*. *Proc. Natl. Acad. Sci. USA*, **87**, 215–9.

Tripathi, G.N.R. and Schuler, R.H. (1983). Time resolved Raman studies of radiation produced radicals. In *Time resolved vibrational spectroscopy* (ed. G.H. Atkinson). Academic Press, New York.

Truscott, T.G. (1986). The photochemistry of haematoporphyrin and some related species. *J. Chem. Soc., Faraday Trans. II*, **82**, 2177–81.

Truscott, T.G. (1988). Photophysical and related aspects of haematoporphyrin and related systems. In *Photosensitisation: Molecular cellular and medical aspects* (ed. G. Moreno, R.H. Pottier, and T.G. Truscott), pp. 345–54. Springer Verlag, Berlin.

Truscott, T.G., Land, E.J., and Sykes, A. (1973). The *in-vitro* photochemistry of biological molecules—III. Absorption spectra, lifetimes and rates of oxygen quenching of the triplet states of β-carotene, retinal and related polyenes. *Photochem. Photobiol.*, **17**, 43–51.

Truscott, T.G., McLean, A.J., Phillips, A.M.R., and Foulds, W.S. (1988). Detection of haematoporphyrin derivative (HPD) and haematoporphyrin (HP) excited states in cell environments. *Cancer Lett.*, **41**, 31–5.

Tsamboas, D. and Orfanos, C.E. (1981). Chemotherapy of psoriasis and other skin disorders with oral retinoids. *Pharmac. Ther.*, **14**, 355–74.

Tsao, D.H.H., Casas-Finet, J.R., Maki, A.H., and Chase, J.W. (1989). Triplet state properties of tryptophan residues in complexes of mutated *Escherichia coli* single-stranded DNA binding proteins with single-stranded polynucleotides. *Biophys. J.*, **55**, 927–36.

Turro, N.J., Cox, G.S., and Paczkowski, M.A. (1985). Photochemistry in micelles. In *Photochemistry and organic photosynthesis*, Topics in Current Chemistry, Vol. 129 (ed. F.L. Boschke), pp. 57–97. Springer Verlag, Berlin.

Tyrrell, R.M. and Pidoux, M. (1987). Action spectra for human skin cells: Estimates of the relative cytotoxicity of the middle ultraviolet, near ultraviolet and violet regions of sunlight on epidermal keratinocytes. *Cancer Res.*, **47**, 1825–9.

Tyrrell, R.M. and Pidoux, M. (1988). Correlation between endogenous glutathione content and sensitivity of cultured human skin cells to radiation at defined wavelengths in the solar ultraviolet range. *Photochem. Photobiol.*, **47**, 405–12.

Tyrrell, R.M. and Pidoux, M. (1989). Singlet oxygen involvement in the inactivation of cultured human fibroblasts by UVA (334 nm, 365 nm) and near-visible (405 nm) radiations. *Photochem. Photobiol.*, **49**, 407–12.

Ueda, K., Kobayashi, S., Morita, J., and Komano, T., (1985). Site-specific DNA damage caused by lipid peroxidation products. *Biochim. Biophys. Acta*, **824**, 341–8.

Vaish, S.P. and Tollin, G. (1971). Flash photolysis of flavins. V. Oxidation and disproportionation of flavin radicals. *Bioenergetics*, **2**, 61–72.

Valduga, G., Nonell, S., Reddi, E., Jori, G., and Braslavsky, S.E. (1988). The production of singlet molecular oxygen by Zn(II) phthalocyanine in ethanol and in unilamellar vesicles. Chemical quenching and phosphorescence studies. *Photochem. Photobiol.*, **48**, 1–5.

Valentine, J.S. and Curtis, A.B. (1975). A convenient preparation of solutions of superoxide anion and the reaction of superoxide anion with a copper(II) complex. *J. Am. Chem. Soc.*, **97**, 224–6.

Valenzeno, D.P. (1987). Photomodification of biological membranes with emphasis on singlet oxygen mechanism. *Photochem. Photobiol.*, **46**, 147–60.

Valeur, B. (1993). Fluorescent probes for evaluation of local physical and structural parameters. In *Molecular luminescence spectroscopy. Methods and applications*, Part 3, (ed. S.G. Schulman), Wiley-Interscience, New York. in press.

Van Camp, J.R., Young, T., Hartman, R.F., and Rose, S.D. (1987). Photosensitization of pyrimidine dimer splitting by a covalently bound indole. *Photochem. Photobiol.*, **45**, 365–70.

Van Lith, D., Eden, J., Warman, J.M. and Hummel, A. (1986). Electron migration in hydrated DNA and collagen at low temperatures. II. The effect of additives. *J. Chem. Soc. Faraday Trans. I*, **82**, 2945–50.

Vander Donckt, E. (1970). Acid–base properties of excited states *Progr. Reaction Kinetics*, **5**, 273–99.

Vander Donckt, E. and Leitaer, D. (1972). Photochemistry of aromatic compounds. Part 5. Measurements of quantum yields of triplet state formation by flash photolysis. *J. Chem. Soc., Faraday Trans. I*, **68**, 112–20.

Vanderkooi, J.M., Calhoun, D.B., and Englander, S.W. (1987*a*). On the prevalence of room-temperature protein phosphorescence. *Science*, **236**, 568–9.

Vanderkooi, J.M., Maniara, G., Green, T.J., and Wilson, D.F. (1987*b*). An optical method for measurement of dioxygen concentration based upon quenching of phosphorescence. *J. Biol. Chem.*, **262**, 5476–82.

Van Hemmen, J.J. and Meuling, W.J.S. (1977). Inactivation of *Escherichia coli* by superoxide radicals and their dismutation product. *Arch. Biochem. Biophys.*, **182**, 743–8.

Van Lier, J.E. (1990). Phthalocyanines as sensitizers for PDT of cancer. In *Photodynamic therapy of neoplastic disease*, Vol. I (ed. D. Kessel), pp. 279–90. CRC Press, Boca Raton, Florida.

Vavilov, S. I. (1992). The dependence of the intensity of fluorescence dyes on the wavelength of the exciting light. *Philos. Mag.* **43**, 307–20.

Vedaldi, D., Dall'Acqua, F., Guiotto, A., Caffieri, S., Baccicheti, F., Knox, C. N., Truscott, T. G., and Land, E. J. (1987). Photochemical, photophysical and photobiological properties of some methylallopsoralens which are potential agents for photochemotherapy. *Biochim. Biophys. Acta*, **925**, 101–8.

Vermeersch, G., Ronfard-Haret, J. C., Bazin, M., Carillet, V., Morlière, P., and Santus, R. (1991). Type I and type II photosensitization by the antibacterial drug nalidixic acid. A laser flash photolysis study. *Photochem. Photobiol.*, **54**, 661–6.

Verweij, H. and van Steveninck, J. (1982). Model studies on photodynamic cross-linking. *Photochem. Photobiol.*, **35**, 265–7.

Vever-Bizet, C., Brault, D., and Le Doan, T., (1985). The pH dependent incorporation of dicarboxylic porphyrins into membrane models. In *Photodynamic therapy of tumours and other diseases* (ed. G. Jori and C. Perria), pp. 17–19. Edizioni Libreria Progetto, Padova.

Vever-Bizet, C., Dellinger, M., Brault, D., Rougée, M. and Bensasson, R. V. (1989). Singlet molecular oxygen quenching by saturated and unsaturated fatty-acids and by cholesterol *Photochem. Photobiol.*, **50**, 321–5.

Vigny, P. and Duquesne, M. (1976). On the fluorescence properties of nucleotides and polynucleotides at room temperature. In *Organic molecular photophysics*, Vol. 3 (ed. J. B. Birks), pp. 167–77. Wiley, New York.

Vileisis, R. A., Sorensen, K., Gonzalez-Crussi, F., and Hunt, C. E. (1982). Liver malignancy after parenteral nutrition. *J. Pediatr.*, **100**, 88–90.

Visscher, K. J., de Haas, M. P., Loman, H., Vojnovic, B., and Warman, J. M. (1987). Fast protonation of adenosine and of its radical anion formed by hydrated electron attack; a nanosecond optical and DC-conductivity pulse radiolysis study. *Int. J. Radiat. Biol.*, **52**, 745–53.

Visscher, K. J., Hom, M., Loman, H., Spoelder, H. J. W., and Verberne, J. B. (1988). Spectral and kinetic properties of intermediates induced by reaction of hydrated electrons with adenine, adenosine, adenylic and polyadenylic acid: A multicomponent analysis. *Radiat. Phys. Chem.*, **32**, 465–73.

Voet, D., Gratzer, W. B., Cox, R. A., and Doty, P. (1963). Absorption spectra of nucleotides, polynucleotides, and nucleic acids in the far ultraviolet. *Biopolymers*, **1**, 193–208.

Volkert, W. A., Kuntz, R. R., Ghiron, C. A., Evans, R. F., Santus, R., and Bazin, M. (1977). Flash photolysis of tryptophan and *N*-acetyl-L-tryptophanamide: the effect of bromide on transient yields. *Photochem. Photobiol.*, **26**, 3–9.

Wagner, J. R., Hu, C. C., and Ames, B. N. (1992). Endogeneous oxidative damage of deoxycytidine in DNA. *Proc. Natl. Acad. Sci. USA*, **89**, 3380–4.

Wagner, P. J. and Bucheck, D. J. (1970). Photodimerization of thymine and uracil in acetonitrile. *J. Am. Chem. Soc.*, **92**, 181–5.

Wala, M., Bothe, E., Görner, H., and Schulte-Frohlinde, D. (1990). Quantum yields for the generation of hydrated electrons and single-strand breaks in poly(C), poly(A) and single-stranded DNA in aqueous solution on 20 ns laser excitation at 248 nm. *J. Photochem. Photobiol., A: Chem.*, **53**, 87–108.

Walling, C. (1975). Fenton's reagent revisited. *Acc. Chem. Res.*, **8**, 125–31.

Walling, C. (1982). The nature of the primary oxidants in oxidation mediated by metal ions. In *Proc. 3rd Int. Symp. Oxidases and related redox systems* (ed. T. E. King, H. S. Mason, and M. Morrison), pp. 85–97, Pergamon, Oxford.

Walrant, P. and Santus, R. (1974*a*). *N*-formyl-kynurenine, a tryptophan photooxidation product, as a photodynamic sensitizer. *Photochem. Photobiol.*, **19**, 411–17.

Walrant, P. and Santus, R. (1974*b*). Ultraviolet and *N*-formyl-kynurenine-sensitized photoactivation of bovine carbonic anhydrase: an internal photodynamic effect. *Photochem. Photobiol.*, **20**, 455–60.

Wang, Y. L. and Taylor, D. L. (1989). *Methods in cell biology, fluorescence microscopy of living cells in culture.*, Part A, B, Vols 29, 30. Academic Press, New York.

Ward, N., Idle, M., and Boreham, J. (1980). Low serum-vitamin A and subsequent risk of cancer. *The Lancet*, 18 October, 813–5.

Wardman, P. (1982). Molecular structure and biological activity of hypoxic cell radiosensitizers and hypoxic-specific cytotoxins. In *Advanced topics in hypoxic cell radiosensitization* (ed. G. E. Adams, A. Breccia, and C. Rimondi) pp. 49–75. Plenum Press, New York.

Wardman, P. (1984). Radiation chemistry in the clinic: hypoxic cell radiosensitisers for radiotherapy. *Radiat. Phys. Chem.*, **24**, 293–305.

Wardman, P. (1987). The mechanism of radiosensitization by electron-affinic compounds. *Radiat. Phys. Chem.*, **30**, 423–32.

Wardman, P. (1989). Reduction potentials of one-electron couples involving free radicals in aqueous solution. *J. Phys. Chem. Ref. Data*, **18**, 1637–755.

Wardman, P. (1990). Bioreductive activation of quinones: redox properties and thiol reactivity. *Free Rad. Res. Commun.*, **8**, 219–29.

Wardman, P. and Clarke, E. D. (1976). One-electron reduction potentials of substituted nitroimidazoles measured by pulse radiolysis, *J. Chem. Soc., Faraday Trans. 1*, **72**, 1377–90.

Waring, M. J. (1981). DNA modification and cancer. *Ann. Rev. Biochem.*, **50**, 159–92.

Waring, M. J. (1991). Binding of antibiotics to DNA. In *Host–guest molecular interactions. From chemistry to biology.* Ciba Foundation Symp. No. 158, pp. 128–46, Wiley, Chichester.

Warman, J. M., de Haas, M. P., Paddon-Row, M. N., Cotsaris, E., Hush, N. S., Oevering, H., and Verhoeven, J. W. (1986). Light-induced giant dipoles in simple model compounds for photosynthesis. *Nature*, **320**, 615–17.

Warman, J. M., Smit, K. J., de Haas, M. P., Jonker, S. A., Paddon-Row, M. N., Oliver, A. M., Kroon, J., Oevering, H., and Verhoeven, J. W. (1991). Long-distance charge recombination within rigid molecular assemblies in nondipolar solvents. *J. Phys. Chem.*, **95**, 1979–87.

Warshel, A. and Dauber, P. (1977). Calculations of resonance Raman spectra of conjugated molecules. *J. Chem. Phys.*, **66**, 5477–88.

Wasielewski, M. R. and Kispert, L. D. (1986). Direct measurement of the lowest excited singlet state lifetime of all-*trans*-β-carotene and related carotenoids. *Chem. Phys.*, **128**, 238–43.

Wasielewski, M. R., Niemczyk, M. P., Svec, W. A., and Pewitt, E. B. (1985). Dependence of rate constants for photoinduced charge separation and dark charge recombination on the free energy of reaction in restricted-distance porphyrin–quinone molecules. *J. Am. Chem. Soc.*, **107**, 1080–2.

Wasielewski, M. R., Liddell, P. A., Barrett, D., Moore, T. A., and Gust, D. (1986). Ultrafast carotenoid to pheophorbide energy transfer in a biomimetic model for antenna function in photosynthesis. *Nature*, **322**, 570–2.

Wasserman, H. H. and Murray, R. W. (1979). *Singlet Oxygen*. Academic Press, New York.

Wasserman, H. H., Scheffer, J. R., and Cooper, J. L. (1972). Singlet oxygen reactions with 9,10-diphenylanthracene peroxide. *J. Am. Chem. Soc.*, **94**, 4991–6.

Watanabe, J., Kimoshita, S., and Kushida, T. (1987). Effects of nonzero correlation time of system-reservoir interaction on the excitation profiles of second-order optical processes in β-carotene. *J. Chem. Phys.*, **87**, 4471–7.

Watson, J.D. and Crick, F.H.C. (1953*a*). Molecular structure of nucleic acid. A structure for deoxyribose nucleic acid. *Nature*, **171**, 737–8.

Watson, J.D. and Crick, F.H.C. (1953*b*). Genetic implications of the structure of deoxyribonucleic acid nucleic acid. A structure for deoxyribose nucleic acid. *Nature*, **171**, 964–7.

Wayne, R.P. (1991). *Chemistry of atmospheres* (2nd edn). Clarendon Press, Oxford.

Webb, R.B. and Brown, M.S. (1979). Action spectra for oxygen-dependent and independent inactivation of *Escherichia coli* WP 2s from 254 to 460 nm. *Photochem. Photobiol.*, **29**, 407–9.

Webster, G.F., Kaidbey, K.H., and Kligman, A.M. (1982). Phototoxicity from benoxaprofen: *In vivo* and *in vitro* studies. *Photochem. Photobiol.*, **36**, 59–64.

Weinstein, J. and Bielski, B.H.J. (1979). Kinetics of the interaction of HO_2 and $O_2^{\cdot -}$. radicals with H_2O_2. The Haber–Weiss reaction. *J. Am. Chem. Soc.*, **101**, 58–62.

Weinstein, M., Alfassi, Z.B., DeFilippis, M.R., Klapper, M.H., and Faraggi, M. (1991). Long range electron transfer between tyrosine and tryptophan in hen egg-white lysozyme. *Biochim. Biophys. Acta*, **1076**, 173–8.

Weisehahn, G. and Hearst, J.E. (1978). DNA unwinding induced by photoaddition of psoralen derivatives and determination of dark-binding equilibrium constants by gel electrophoresis. *Proc. Natl. Acad. Sci. USA*, **75**, 2703–7.

Weisfeldt, M.L., Zweier, J., Ambrosio, G., Becker, L.C., and Flaherty, J.T. (1988). Evidence that free radicals result in reperfusion injury in heart muscle. In *Oxygen radicals in biology and medicine* (ed. M.G. Simic, K.A. Taylor, J.F. Ward, and C. von Sonntag), pp. 911–9. Plenum Press, New York.

Weishaupt, K.R., Gomer, C.J., and Dougherty, T.J. (1976). Identification of singlet oxygen as the cytotoxic agent in photoinactivation of murine tumors. *Cancer Res.*, **36**, 2326–9.

Weiss, S.J. (1986). Oxygen, ischemia and inflammation. *Acta Physiol. Scand. Suppl.*, **548**, 9–37.

Westerhof, W., Pavel, S., Kammeyer, A., Bensenberg, F.D., and Cormane, R. (1987). Melanin-related metabolites as markers of the skin pigmenting system. *J. Invest. Dermatol.*, **89**, 78–81.

Western, A., Van Camp, J.R., Bensasson, R., Land, E.J., and Kochevar, I.E. (1987). Involvement of singlet oxygen in the phototoxicity mechanism for a metabolite of piroxicam. *Photochem. Photobiol.*, **46**, 469–75.

Westhead, E.W. (1965). Photooxidation with Rose Bengal of a critical histidine residue in yeast enolase. *Biochemistry*, **4**, 2139–44.

Whillans, D.W. and Adams, G.E. (1975). Electron transfer oxidation of DNA radicals by paranitroacetophenone. *Int. J. Radiat, Biol.*, **28**, 501–10.

Whillans, D.W. and Johns, H.E. (1971). Properties of the triplet states of thymine and uracil in aqueous solution. *J. Am. Chem. Soc.*, **93**, 1358–62.

Whillans, D.W. and Johns, H.E. (1972). Triplet state studies of the nucleoside and nucleotide derivatives of uracil and thymine. *Biochim. Biophys. Acta*, **277**, 1–6.

Whillans, D.W., Adams, G.E., and Neta, P. (1975). Electron-affinic sensitization. VI. A pulse radiolysis and ESR comparison of some 2- and 5-nitroimidazoles. *Radiat. Res.*, **62**, 407–21.

Wiesenfeld, J.M. and Ippen, E.P. (1980). Dynamics of electron solvation in liquid water. *Chem. Phys. Lett.*, **73**, 47–50.

Wilbrandt, R. and Jensen, N.-H. (1981). Time-resolved resonance Raman spectroscopy: The triplet state of *all-trans* retinal. *J. Am. Chem. Soc.*, **103**, 1036–41.

Wilbrandt, R. and Jensen, N.-H. (1983). Applications of time-resolved resonance Raman spectroscopy in radiation chemistry and photobiology: structure and chemistry of carotenoids in the excited triplet state. In *Time-resolved vibrational spectroscopy*, pp. 273–85. Academic Press, New York.

Wilbrandt, R., Jensen, N.-H., and Houée-Levin, C. (1985). Resonance Raman and absorption spectra of isomeric retinals in their lowest excited triplet state, *Photochem. Photobiol.*, **41**, 175–81.

Wildes, P. D., Lichtin, N. N., Hoffman, M. Z., Andrews, L., and Linschitz, H. (1977). Anion and solvent effects on the rate of reduction of triplet excited thiazine dyes by ferrous ions. *Photochem. Photobiol.*, **25**, 21–5.

Wilkening, V. G., Lal, M., Arends, M., and Armstrong, D. A. (1968) The cobalt-60 γ-radiolysis of cysteine in deaerated solutions at pH values between 5 and 6. *J. Phys. Chem.*, **72**, 185–90.

Wilkinson, F. and Brummer, J. G. (1981). Rate constants for the decay and reactions of the lowest electronically excited singlet state of molecular oxygen in solution *J. Phys. Chem. Ref. Data*, **10**, 809–1000.

Wilkinson, F. and Ho, W. T. (1978). Electronic energy transfer from singlet molecular oxygen to carotenoids, *Spectroscopic. Lett.*, **11**, 455–63.

Wilkinson, F. and Willsher, C. J. (1984). Triplet–triplet aborption in microcrystalline benzil by diffuse reflectance laser flash photolysis. *Appl. Spectrosc.*, **38**, 897–901.

Wilkinson, F., Willsher, C. J., Warwick, P., Land, E. J., and Rushton, F. A. P. (1984). Diffuse reflectance pulse radiolysis of opaque samples. *Nature*, **311**, 40–2.

Willner, I., Rubin, S., Wonner, J., Effenberg, F., and Bäuerle, P. (1992). Photoswitchable binding of substrates to proteins: photoregulated binding of α-D-mannopyranose to concanavalin A modified by a thiophenefulgide dye. *J. Am. Chem. Soc.*, **114**, 3150–1.

Willson, R. L. (1970). The reaction of oxygen with radiation-induced free radicals in DNA and related compounds. *Int. J. Radiat. Biol.*, **17**, 349–58.

Willson, R. L. (1971). Pulse radiolysis studies on reaction of triacetoneamine-*N*-oxyl with radiation-induced free radicals. *Trans. Faraday Soc.*, **67**, 3008–19.

Willson, R. L. (1983). Free radical repair mechanisms and the interactions of glutathione and vitamins C and E. In *Radio-protectors and anticarcinogens* (ed. O. F. Nygaard and M. G. Simic), pp. 1–22. Academic Press, New York.

Willson, R. L. (1985). Organic peroxy free radicals as ultimate agents in oxygen toxicity. In *Oxidative stress* (ed. H. Sies), pp. 41–72. Academic Press, London.

Willson, R. L., Dunster, C. A., Forni, L. G., Gee, C. A., and Kittridge, K. J. (1985). Organic free radicals and proteins in biochemical injury: electron- or hydrogen-transfer reactions? *Phil. Trans. R. Soc. London*, **B311**, 545–63.

Wilson, B. C. (1989). Photodynamic therapy: Light delivery and dosage for second-generation photosensitizers. In *Photosensitizing compounds: their chemistry, biology and clinical use*, Ciba Foundation Symposium No. 146, pp. 60–73. Wiley Interscience, Chichester.

Wilson, T. (1985). Mechanisms of peroxide luminescence. In *Singlet oxygen, Vol. II, Reaction modes and products*, Part I (ed. A. A. Frimer), pp. 37–65. CRC Press, Boca Raton, Florida.

Winkler, J. R. and Gray, H. B. (1992). Electron transfer in ruthenium-modified proteins. *Chem. Rev.*, **92**, 369–79.

Winterbourn, C. C. (1987). The ability of scavengers to distinguish OH˙ production in the iron-catalysed Haber–Weiss reaction: Comparison of four assays for OH˙. *Free Rad. Biol. Med.*, **3**, 33–9.

Wintgens, W., Johnston, L. J., and Scaiano, J. C. (1988). Use of a photoreversible fulgide as an actinometer in one- and two-laser experiments. *J. Am. Chem. Soc.*, **110**, 511–17.

Wong, P.C., Griller, D., and Scaiano, J.C. (1982). A kinetic study of the reactions of *tert*-butoxyl with alkenes: hydrogen abstraction vs. addition *J. Am. Chem. Soc.*, **104**, 5106–8.

Woodward, R.B. and Hoffman, R. (1970). *The conservation of orbital symmetry.* Academic Press, New York.

Wuttke, D.S., Bjerrum, M.J., Winkler, J.R. and Gray, H.B. (1992). Electron-tunneling pathways in cytochrome-*c*. *Science*, **256**, 1007–9.

Yamashita, M., Nomera, S., Kobayashi, T., Sato, T., and Aizawa, K. (1984). Picosecond time-resolved fluorescence spectroscopy of haematoporphyrin derivative. *IEEE J. Quantum Electron*, **20**, 1363–9.

Yamazaki, I. and Piette, L.H. (1990). ESR spin-trapping studies on the reaction of Fe^{2+} ions with H_2O_2-reactive species in oxygen toxicity in biology. *J. Biol. Chem.*, **265**, 13 589–94.

Yamazaki, I. and Piette, L.H. (1991). EPR spin-trapping study on the oxidising species formed in the reaction of the ferrous ion with hydrogen peroxide. *J. Am. Chem. Soc.*, **113**, 7588–93.

Yamazaki, I., Piette, L.H., and Grover, T.A. (1990). Kinetic studies on spin trapping of superoxide and hydroxyl radicals generated in NADPH-Cytochrome P-450 reductase paraquat systems. *J. Biol. Chem.*, **265**, 652–9.

Ye, M., Nelson, J.M., and Miller, J.H. (1990). Superhelicity increases the sensitivity of plasmid DNA to strand break induction by X-rays. Abstract of papers of the 38th annual meeting of the Radiation Research Society, New Orleans, Louisiana, 7–12 April.

Yee, G.G. and Kliger, D.S. (1983). Photoacoustic spectroscopic technique for measuring excited state absorption spectra. *J. Phys. Chem.*, **87**, 1887–91.

Yee, G.G. and Kliger, D.S. (1988). Diffuse reflectance laser photolysis studies of triplet-state absorption spectra of anthracene in Shpol'skii matrices. *J. Phys. Chem.*, **92**, 6630–5.

Yoshimura, A. and Ohno, T. (1988). Lumiflavin-sensitized photooxidation of indole. *Photochem. Photobiol.*, **48**, 561–5.

Young, R.H. and Brewer, D.R. (1979). The mechanism of quenching of singlet oxygen. In *Singlet oxygen*, (ed. B. Ränby and J.F. Rabek), pp. 36–47. John Wiley, Chichester.

Young, R.H. and Martin, R.L. (1972). On the mechanism of quenching of singlet oxygen by amines. *J. Am. Chem. Soc.*, **94**, 5183–5.

Yu, S., Egorov, S.V., Zinukov, S.V., Kamalov, N.I., Krasnovskii, Jr, A.A., and Toleutaev, B.N. (1989). Singlet-oxygen photosensitized luminescence kinetics with nanosecond time resolution. *Opt. Spectrosc. (USSR)*, **65**, 530–3.

Zafiriou, C. (1974). Sources and reactions of OH and daughter radicals in seawater. *J. Geophys. Res.*, **79**, 4491–7.

Zaklika, K.A., Kissel, T., Thayer, A.L., Burns, P.A., and Schaap, A.P. (1979). Mechanism of 1,2 dioxetane decomposition: the role of electron transfer. *Photochem. Photobiol.*, **30**, 35–44.

Zamansky, G.B. and Chou, I.-N. (1988). Cytoskeletal damage induced in human cells by solar wavelengths of ultraviolet light. *Photochem. Photobiol.*, **47**, Supplement, 285.

Zavilgelsky, G.B., Gurzadyan, G.G., and Nikogosyan, D.N. (1984). Pyrimidine dimers, single-strand breaks and crosslinks induced in DNA by powerful laser UV irradiation. *Photobiochem. Photobiophys.*, **8**, 175–87.

Zavoisky, E. (1945). Paramagnetic relaxation of liquid solutions for perpendicular fields. *J. Phys. USSR*, **9**, 211–16.

Zeichmeister, L. (1962). Cis–trans *isomeric carotenes, vitamin A, and arylpolyenes.* Springer, Vienna.

Zepp, R. G., Lee Wolfe, N., Baughman, G. L., and Hollis, R. C. (1977). Singlet oxygen in natural waters. *Nature*, **267**, 421–3.

Zepp, R. G., Hoigné, J., and Bader, H. (1987). Nitrate-induced photooxidation of trace organic chemicals in water. *Environ. Sci. Technol.*, **21**, 443–50.

Zewail, A. H. (1988). Laser femtochemistry. *Science*, **242**, 1645–53.

Zewail, A. H. and Bernstein, R. B. (1988). Real-time laser femtochemistry. Viewing the transition from reagents to products *Chem. Eng. News*, 7 Nov, 24–43.

Zhang, N., Schuchmann, H.-P., and von Sonntag, C. (1991). The reaction of superoxide radical anion with dithiothreitol: a chain process. *J. Phys. Chem.*, **95**, 4718–22.

Zhou, C. (1989). Mechanisms of tumour necrosis induced by photodynamic therapy. *J. Photochem. Photobiol., B: Biol.*, **3**, 299–318.

Zhou, J. S. and Rodgers, M. A. J. (1991). Driving force dependence of rate constants of electron transfer within cytochrome *c* and uroporphyrin complex. *J. Am. Chem. Soc.*, **113**, 7728–34.

Ziegler, R. G., Subar, A. F., Craft, N. E., Ursin, G., Patterson, B. H., and Graubard, B. I. (1992). Does β-carotene explain why reduced cancer risk is associated with vegetable and fruit intake? *Cancer Res.* (Suppl.), 2060s–6s.

Zigman, S. (1977). Near UV light and cataracts. *Photochem. Photobiol.*, **26**, 437–41.

Zimerman, O. E., Cosa, J. J., and Previtali, C. M. (1990). A fluorescence study of the interaction of indolic compounds with synthetic polyelectrolytes. *Photochem. Photobiol.*, **52**, 711–16.

Zucchi, N. (1652). *Optica philosophica experimentis et ratione a fundamentia constituta.* Lugduni.

Zysset, B., Fujimoto, J. G., and Deutsch, T. F. (1990). Time-resolved measurements of picosecond optical breakdown. *Appl. Phys.*, **B48**, 139–47.

Appendix I

Units, constants, and conversions

Table A.1 The international system of units (SI)

Physical quantity	Name of unit	Symbol of unit
(a) SI base units		
length	metre	m
mass	kilogram	kg
time	second	s
electric current	ampere	A
temperature	kelvin	K
amount of substance	mole	mol
(b) SI derived units having special names and symbols		
energy	joule	$J = m^2 kg\ s^{-2}$
force	newton	$N = m\ kg\ s^{-2} = J\ m^{-1}$
power	watt	$W = m^2 kg\ s^{-3} = J\ s^{-1}$
electrical charge	coulomb	$C = s\ A$
electrical potential difference	volt	$V = m^2 kg\ s^{-3}\ A^{-1} = J\ A^{-1}\ s^{-1}$
magnetic flux density	tesla	$T = kg\ s^{-2}\ A^{-1} = V\ s\ m^{-2}$
frequency	hertz	$Hz = s^{-1}$
activity (radioactive)	becquerel	$Bq = s^{-1}$
absorbed dose (radiation)	gray	$Gy = J\ kg^{-1}$
(c) Selection of other derived SI units		
concentration		$mol\ dm^{-3}$
first-order rate constant		s^{-1}
second-order rate constant		$dm^3\ mol^{-1}\ s^{-1}$
molar absorption coefficient		$dm^3\ mol^{-1}\ cm^{-1}$

(d) SI prefixes

a	f	p	n	μ	m	c	d	k	M	G	T	P
atto	femto	pico	nano	micro	milli	centi	deci	kilo	mega	giga	tera	peta
10^{-18}	10^{-15}	10^{-12}	10^{-9}	10^{-6}	10^{-3}	10^{-2}	10^{-1}	10^3	10^6	10^9	10^{12}	10^{15}

For the use of units and symbols in physical chemistry recommended by IUPAC, see:

Mills, I., Cvitas, T., Homann, K., and Kuchitsu, K. (1988). *Quantities, units and symbols in physical chemistry*, International Union of Pure and Applied Chemistry, Blackwell Scientific Publications, Oxford.

Table A.2 Fundamental physical constants

Symbol	Constant	Value
c	Speed of light *in vacuo*	2.998×10^{8} ms^{-1}
h	Planck constant	6.63×10^{-34} J s
k	Boltzmann constant	1.38×10^{-23} JK^{-1}
N	Avogadro number	6.022×10^{23} mol^{-1}
R	Gas constant	8.31 J mol^{-1} K^{-1}
m	Mass of electron at rest	9.11×10^{-31} kg
e	Charge of electron	1.60×10^{-19} C
F	Faraday constant	9.6485×10^{4} C mol^{-1}
ε_0	Vacuum permittivity	8.8542×10^{-12} J^{-1} C^{2} m^{-1}

Table A.3 Conversion factors

To convert from:	To:	Multiply by:
calorie	erg	4.184×10^{7}
calorie	joule	4.184
cm^{-1}	joule mol^{-1}	11.962
cm^{-1}	cal mol^{-1}	2.859
cm^{-1}	electron volt	1.24×10^{-4}
electron volt	kcal mol^{-1}	23.060
electron volt	cm^{-1}	8.065×10^{3}
electron volt	erg	1.602×10^{-12}
electron volt	joule	1.602×10^{-19}
electron volt	kjoule mol^{-1}	96.483
erg	calorie	2.389×10^{-8}
erg	joule	10^{-7}
erg molecule^{-1}	joule mol^{-1}	6.022×10^{16}
erg molecule^{-1}	kcal mol^{-1}	1.440×10^{13}
joule	calorie	0.2389
joule	erg	10^{7}
joule mol^{-1}	erg molecule^{-1}	1.660×10^{-17}
joule mol^{-1}	kcal mol^{-1}	2.389×10^{-4}
kcal mol^{-1}	electron volt	0.04336
kcal mol^{-1}	erg molecule^{-1}	6.944×10^{-14}
kcal mol^{-1}	joule mol^{-1}	4.184×10^{3}
kjoule mol^{-1}	electron volt	0.01036
kjoule mol^{-1}	erg molecule^{-1}	1.660×10^{-14}
rad	Gy (joule kg^{-1})	1.0×10^{-2}
roentgen	coulomb kg^{-1}	2.5798×10^{-4}

For a more complete list of units and terms used in photochemistry and photobiology see:

Song, P.S. (1991). SI units, conversion factors and abbreviations. *Photochem. Photobiol.*, **53**, I–VI (January number).

Braslavsky, S.E. and Houk, K.N. (1988). Glossary of terms used in photochemistry. *Pure Appl. Chem.*, **60**, 1055–106.

Postface

Free radical reactions are certainly involved in many of the degenerative processes facing man, including aspects of ageing, cancer, cardiovascular diseases, atherosclerosis, neural disorders, and inflammation. Some modern therapeutic treatments such as PDT and virus photoinactivation are based on producing activated oxy-species as cytotoxic agents at specific sites, while several non-specific anti-oxidants are involved in current epidemiological trials of cancer chemoprevention. Many processes of drug metabolism involve free radical reactions and there have also been suggestions that therapeutics could be developed via light-regulated protein binding. Possibly of comparative importance, the study and mimicry of electron transfer in proteins, crucial for the understanding of the bioenergetics of respiration and photosynthesis, may lead to new electronics at the molecular level. In today's world, with increasing environmental threats such as ozone depletion, heavy metal accumulation, and the use of nuclear energy, we believe it would be a wise society that will continue to maintain and develop fundamental radiation- and photochemistry, despite the high cost of equipment such as linear accelerators and ultra-fast lasers.

Index